Handbook of Research on Digitalization Solutions for Social and Economic Needs

Richard Pettinger
University College London, UK

Brij B. Gupta
Asia University, Taiwan

Alexandru Roja
West University of Timisoara, Romania

Diana Cozmiuc
West University of Timisoara, Romania

A volume in the Advances in Human and Social Aspects of Technology (AHSAT) Book Series

Published in the United States of America by
    IGI Global
    Engineering Science Reference (an imprint of IGI Global)
    701 E. Chocolate Avenue
    Hershey PA, USA 17033
    Tel: 717-533-8845
    Fax: 717-533-8661
    E-mail: cust@igi-global.com
    Web site: http://www.igi-global.com

Library of Congress Cataloging-in-Publication Data

Names: Pettinger, Richard, editor.
Title: Handbook of research on digitalization solutions for social and
  economic needs / Richard Pettinger, Brij Gupta, Alexandru Roja, and
  Diana Cozmiuc, editors.
Description: Hershey, PA : Business Science Reference, [2023] | Includes
  bibliographical references and index. | Summary: "This book helps
  readers argue the business rationale of digitalization, introducing the
  agile operating model that has triggered digital transformation and the
  plethora of ways it has become of practical use in every country"--
  Provided by publisher.
Identifiers: LCCN 2022027246 (print) | LCCN 2022027247 (ebook) | ISBN
  9781668441022 (hardcover) | ISBN 9781668441039 (ebook)
Subjects: LCSH: Information technology--Economic aspects. | Information
  technology--Social aspects.
Classification: LCC HC79.I55 H3335 2023 (print) | LCC HC79.I55 (ebook) |
  DDC 303.48/33--dc23/eng/20220804
LC record available at https://lccn.loc.gov/2022027246
LC ebook record available at https://lccn.loc.gov/2022027247

This book is published in the IGI Global book series Advances in Human and Social Aspects of Technology (AHSAT) (ISSN: 2328-1316; eISSN: 2328-1324)

British Cataloguing in Publication Data
A Cataloguing in Publication record for this book is available from the British Library.

For electronic access to this publication, please contact: eresources@igi-global.com.

Advances in Human and Social Aspects of Technology (AHSAT) Book Series

Mehdi Khosrow-Pour, D.B.A.
Information Resources Management Association, USA

ISSN:2328-1316
EISSN:2328-1324

MISSION

In recent years, the societal impact of technology has been noted as we become increasingly more connected and are presented with more digital tools and devices. With the popularity of digital devices such as cell phones and tablets, it is crucial to consider the implications of our digital dependence and the presence of technology in our everyday lives.

The **Advances in Human and Social Aspects of Technology (AHSAT) Book Series** seeks to explore the ways in which society and human beings have been affected by technology and how the technological revolution has changed the way we conduct our lives as well as our behavior. The AHSAT book series aims to publish the most cutting-edge research on human behavior and interaction with technology and the ways in which the digital age is changing society.

COVERAGE

- Cultural Influence of ICTs
- Activism and ICTs
- Digital Identity
- Cyber Behavior
- Technology and Social Change
- Public Access to ICTs
- Human-Computer Interaction
- Technology Dependence
- ICTs and human empowerment
- End-User Computing

IGI Global is currently accepting manuscripts for publication within this series. To submit a proposal for a volume in this series, please contact our Acquisition Editors at Acquisitions@igi-global.com or visit: http://www.igi-global.com/publish/.

Titles in this Series

For a list of additional titles in this series, please visit: www.igi-global.com/book-series

Impact of Disruptive Technologies on the Socio-Economic Development of Emerging Countries
Fredrick Japhet Mtenzi (The Aga Khan University, Institute for Educational Development, Tanzania) George S. Oreku (The Open University of Tanzania, Tanzania) and Dennis M. Lupiana (Institute of Finance Management, Tanzania)
Information Science Reference • © 2023 • 335pp • H/C (ISBN: 9781668468739) • US $225.00

Exergaming Intervention for Children, Adolescents, and Elderly People
Shahnawaz Khan (Bahrain Polytechnic, Bahrain) Thirunavukkarasu Kannapiran (Karnavati University, India) Arunachalam Muthiah (Karnavati University, India) and Sharad Shetty (Karnavati University, India)
Information Science Reference • © 2023 • 300pp • H/C (ISBN: 9781668463208) • US $215.00

Adoption and Use of Technology Tools and Services by Economically Disadvantaged Communities Implications for Growth and Sustainability
Alice S. Etim (Winston-Salem State University, USA) and James S. Etim (Winston-Salem State University, USA)
Information Science Reference • © 2023 • 300pp • H/C (ISBN: 9781668453476) • US $225.00

Digital Psychology's Impact on Business and Society
Muhammad Anshari (Universiti Brunei Darussalam, Brunei) Abdur Razzaq (Universitas Islam Negeri Raden Fatah Palembang, Indonesia) Mia Fithriyah (Indonesia Open University, Indonesia) and Akmal Nasri Kamal (Universiti Brunei Darussalam, Brunei)
Information Science Reference • © 2023 • 330pp • H/C (ISBN: 9781668461082) • US $270.00

Frugal Innovation and Social Transitions in the Digital Era
Muhammad Nawaz Tunio (Mohammad Ali Jinnah University, Karachi, Pakistan) and Atia Bano Memon (University of Sindh, Badin Campus, Pakistan)
Information Science Reference • © 2023 • 208pp • H/C (ISBN: 9781668454176) • US $250.00

Handbook of Research on Implementing Digital Reality and Interactive Technologies to Achieve Society 5.0
Francesca Maria Ugliotti (Politecnico di Torino, Italy) and Anna Osello (Politecnico di Torino, Italy)
Information Science Reference • © 2022 • 731pp • H/C (ISBN: 9781668448540) • US $295.00

Exploring Ethical Problems in Today's Technological World
Tamara Phillips Fudge (Purdue University Global, USA)
Information Science Reference • © 2022 • 385pp • H/C (ISBN: 9781668458921) • US $240.00

701 East Chocolate Avenue, Hershey, PA 17033, USA
Tel: 717-533-8845 x100 • Fax: 717-533-8661
E-Mail: cust@igi-global.com • www.igi-global.com

List of Contributors

Table of Contents

Detailed Table of Contents

Chapter 1

This chapter provides recommendations for the integration of various types of patterns. Complex layers of patterns can be assembled in a holistic enterprise architecture pattern (HEAP), which is integrated by using the integration of HEAP (IoHEAP), which can support agile transformation projects. The IoHEAP is based on a layered concept of patterns and an optimized deployment process. An integration patterns model can be used as a template to integrate solution building blocks (SBB), which can be used to implement a variety of types of transformation projects. In this chapter, the focus is on the integration of various pattern types that can be used to deliver adaptable SBBs. SBBs are the backbone of Enterprise architecture (EA)-based transformation projects.

Chapter 2

When the historical development of management is examined, it is seen that every age has created a management and organizational structure suitable for its own conditions. Therefore, today, management and organizational structures are experiencing a digital transformation as a necessity of the age. From this point of view, this study was carried out as a conceptual analysis in order to conceptually examine new terms such as digital management, digital transformation, and digital leader that emerged with digitalization in management structures to explain the change in management structures by comparing them with previous management structures and to eliminate the ambiguities about the concept of digitalization. In addition, another aim of this study is to compile explanations in the literature about the metaverse, which is described as a three-dimensional virtual world, and to evaluate the innovations, opportunities, concerns, and criticisms brought by the metaverse application.

This chapter provides recommendations for classifying and using various types of patterns in a holistic enterprise architecture pattern (HEAP) that can support transformation projects. The HEAP is based on a concise, composite, and layered patterns model. A composite patterns model can be used as a template to instantiate building blocks (BB) to implement a variety of types of transformation projects. In this chapter, the focus is on various pattern standards that can be used in a holistic and adaptable BBs to support an optimal set of enterprise architectures (EA). A patterns-based EA offers BBs, like in civil engineering, that can support colossal and complex projects. For such complex projects, there is a need to create a common denominator pattern, to integrate other standard patterns, in the form of a HEAP, where the HEAP is the backbone of this research and development project (RDP). This RDP proves that the HEAP can be used for building dynamic and flexible EAs.

Chapter 4

The role of requirements engineering (RoRE) is central for the implementation of projects in general and is especially crucial for business transformation projects (or simply projects) because transformation activities incur major changes in the existing sets of archaically defined requirements. Requirements engineering (RE) is a complex part of the project because it consists of many related communication, cross-functional knowledge, and dependencies, like the need for RE to interact with executive management, business users, business architects, implementation developers, and other project actors. In this chapter, the authors will try to propose an RoRE concept (RoREC) that can support project managers (or simply managers) in transforming the enterprise and managing RE activities. RE activities are not just about assembling a huge set of business and non-functional requests and features in the form of document(s) and delivering a repository of methodology-based diagram(s).

Chapter 5

This chapter aims to create a general framework on digital maturity models and reveal the highlights of the commonly used components of digital maturity models. The emphasis will be more on detailing human-related fundamental aspects of digital maturity models rather than the technological ones. Considering the incrementally increasing options to reach collaboration technologies and swift deployment of them to different sectors, the main challenge still remains as to manage the psychological asset related parts. Although technology is generally perceived as mechanical, whose boundaries can be drawn, tasks to be assigned can be defined; the technological elements within the concept of digital maturity also intersects with the human-related dimensions of digital transformation. The study aims to advise a systematic approach by offering a theoretical framework to manage human-related factors in a digital transformation journey with an inclusive perspective positioning technology as the core medium provider and contributes to the theoretical background in the field of study.

Process mining is a paradigm shift from traditional process understanding methodologies like interviews and surveys to a data-driven understanding of the actual digital processes. It analyzes business processes by applying algorithms to the event data generated by digital systems. The chapter provides insight into various uses of process mining in different social and economic processes, with examples from past works demonstrating how practical process mining is in detecting and mitigating bottlenecks in these sectors. Then the chapter further delves into the details of process mining algorithms, key features, and metrics that can help practitioners and researchers evaluate process mining for their work. It also highlights some data quality issues in the event log that can inhibit obtaining fair results from process models. Additionally, some current limitations and concerns are described for creating awareness and building over the body of knowledge in the process and sequential mining techniques.

Information systems and/or information technology (IS/IT) outsourcing became a very common practice in developed and emerging economies. Despite IS/IT outsourcing's importance, reports on outsourcing initiatives indicate problematic situations. A large number of IS/IT outsourcing projects are being renegotiated or prematurely terminated, and many IS/IT outsourcing failures are not even publicly reported due to the fear of negative responses from the market and stakeholders. Literature suggests that the IS/IT project outsourcing is a complex maneuver. The concept of modularity has been applied in many other fields in order to manage complexity and enhance agility/flexibility; hence, four cases were analyzed using the lens of modularity in order to understand and identify the relationship between the concept of modularity and IS/IT project outsourcing. The interface aspect of modularity has emerged as the most relevant and identified in all four cases. It implies that the interface aspect should get greater attention when designing/planning a new IS/IT outsourcing project.

Chapter 8

Cozmiuc Claudia Diana, West University of Timisoara, Romania
Liviu Herman, Ioan Slavici University, Timisoara, Romania
Cristian Pitic, Ioan Slavici University, Timisoara, Romania
Andreea Bozesan, Ioan Slavici University, Timisoara, Romania
Sinel Galceava, Ioan Slavici University, Timisoara, Romania

Capability maturity and capability readiness models are designated management tools in scholarly literature. One of their applications is in shaping roadmaps, projects, and programs. Typically, individual articles tackle the two topics separately and the way the two management tools are intertwined. Scholarly literature shows the need to link the two tools. These tools tend to refer to digital transformation or include digitalization in their construction. A specific example is the TRL1-TRL9 capability maturity model used by NASA and by the European Union. This is the reference for conducting research, development, and innovation activities at the European Union. It models roadmaps and all project management techniques. A specific case study is considered in this chapter. Findings show the reference to manage research, development, and innovation activities at the European Union and NASA is a capability maturity model, a project management tool, and includes digitalization business information systems as the tools for business activity and business process management.

Chapter 9

Cozmiuc Claudia Diana, Ioan Slavici University, Timisoara, Romania
Cosmina Carmen Florica, Polytechnics University, Timisoara, Romania
Delia Albu, Ioan Slavici University, Timisoara, Romania
Avram Greti, Ioan Slavici University, Timisoara, Romania
Octavian Dondera, Ioan Slavici University, Timisoara, Romania

The past decades have seen the emergence of business ecosystems. Their historic evolution is important, as ecosystems are deemed new business models to progressively replace the old ones. Ecosystems are broadly and vaguely defined in literature review, where they represent new network business models. Empirical data about European project calls show precise definitions of ecosystem types and activities that are subject to European project funding calls. These definitions apply to new commercial entities recently created and funded by such European project calls. The chapter aims to build theory with a case study about emerging business practices with high instrumental value to other businesses, which may apply for European project calls. The methodology is a descriptive case study. Findings are the level of detail and focus on business practices exceed that of general and vague theory, thereby allowing business practice to contribute to theory and enrich it. Definitions of ecosystem entities and ecosystems management activities to be used across the European Union are this addition.

Athena Kin-kam Wong, The University of Hong Kong, Hong Kong
Dickson K. W. Chiu, The University of Hong Kong, Hong Kong

Museums offer education and enjoyment to the public through exhibitions and public programs, but what happens behind the museums can be a mystery. Thus, this study uses the visit to conservation laboratories as a behind-the-scenes tour to illustrate the conservation practices at public museums in Hong Kong and thus potential digital transformation. The value chain analysis was used to systematically exanime the environment and operations of the conservation process in depth, focusing on museum education and extension activities. Lacking human resources, safety, and access constraints often limited the capacity to offer such tours. Some suggestions on digital transformation using contemporary information and communication technologies (ICTs) were proposed for engagement improvement and expanding the audience. Scant studies research how such conservation tours facilitate learning and engagement with visitors or analyze museum operations with value-chain analysis for digital transformation, especially in East Asia.

Lemma Lessa, Addis Ababa University, Ethiopia
Mekuria Hailu, Addis Ababa University, Ethiopia

Extant literature revealed that elections conducted in traditional ways mostly result in conflicts. Intending to address such challenges, e-voting technology is being used in some countries to conduct a transparent election. However, the application of this new system encountered different challenges due to a lack of readiness to exploit its value, especially in developing countries contexts. To that end, the readiness of government, citizens, and political parties needs to be assessed before using e-voting as an electoral system. The main purpose of this study is to assess the gaps in the readiness of Ethiopia for e-voting system implementation. A qualitative research method is employed, and a thematic analysis was used to analyze the data. The finding revealed that Ethiopia is not ready in terms of information communication technology (ICT) infrastructure, human resources, and legality measures for e-voting technology. Finally, recommendations are forwarded for policymakers and practitioners for action.

Ece Özer Çizer, Yıldız Technical University, Turkey

The number of studies aiming to change consumers' health behaviors by digitizing is increasing day by day. Health behaviors have become the focus of different sciences and disciplines, both socio-cultural and technological. Especially after the COVID-19 pandemic, investments and research in this field have increased considerably. In the literature, there is a rapid increase in the number of studies on digital health behaviors. This study aims to systematically examine the studies in the literature on digitalized health behaviors. For this purpose, bibliometric analysis was applied to 357 studies in the Web of Science database using the R Studio program in order to determine the most frequently studied areas and possible gaps in the literature.

Brands are the universal symbols of the modern era, comprising a set of values with which the consumer identifies. To activate a brand is to make its promise tangible through different marketing and advertising tools. The fashion film is established in the retail sector as a form of brand communication, which can encompass fashion, photography, film, advertising, music, and art, becoming heir to different and varied artistic disciplines: a format that allows interaction with customers by breaking traditional boundaries through transmediality. The aim of this chapter is to analyse the series of three fashion films, "Zara Scenes," launched in 2019, in the context of the communication of changes in the brand's business strategy. With this purpose, the different chapters are analysed from the fashion film perspective as social media content and a brand communication tool.

Politics has long been associated with mendacity, disinformation, manipulation, and at odd with the truth. In recent times, the term post-truth has been used to further characterise politics, which implies a fresh phenomenon in the conflict between truth and politics. The chapter examines the concept of post-truth and post-truth politics. The chapter argues that the application of post-truth rhetoric in politics implies a novelty in politics and in the relationship between truth and politics, which undermines democracy. It is arguable that post-truth negatively impacts individual ability to discriminate between what is true or false, taking into consideration the volume of disinformation on the one hand and on the other hand the need to make informed decisions and choices without having to consult experts at the critical time that the stakes involved in such decisions and choices are urgent and crucial.

There are both opportunities and challenges because of technological advancements. The world as we know it has undergone a significant transformation in the digital age. And, to keep up with the changes, organizations have placed a greater emphasis on digital transformation. We are presently in the midst of a technological revolution with the intention to basically affect people, work, and the mode of interplay with one another. While the people will want to work with technology, there may additionally be a developing want of people to increase specialized skills for the way they interact with one another. To keep pace with rapid disruption, organizations should also emphasize the need to invest and reskill their people to stay competitive in the fast advancement of technology, which is altering the nature of skills and abilities required in the workplace, which also needs to make a mental shift among the employees who can make that technology useful.

    Subhanil Banerjee, PES University, Bengaluru, India
    Souren Koner, Royal School of Business (RSB), The Assam Royal Global University,
        Guwahati, India
    Arakhita Behera, School of Humanities and Social Science, K.R. Mangalam University,
        Gurgaon, India
    Suhanee Gupta, School of Humanities, K.R. Mangalam University, Gurugram, India

The chapter aims to understand the impact of mobile and broadband subscriptions on the ease of doing business and per 100 people trade volume that might be critical in facing a pandemic of recent type in the future. On the one hand, the ease of doing business has been gaining popularity over the years as an indicator of where to set business. On the other hand, the spread of mobile subscriptions and broadband and their conjugation has significantly impacted modern-day business. Moreover, trade has already proven itself as the engine of growth. With this background, the chapter intends to determine the impact of the digital duo on ease of doing business and per 100 trade volume. The study is holistic in nature. It considered 128 countries in the world regarding the variables ease of doing business ranking (dependent), per 100 capita trade volume (dependent), and per 100 broad brand penetration (independent variable) along with per 100 mobile subscriptions (independent variable) for the years 2019 and 2020, and then opted for a pooled regression analysis with robust standard errors.

Preface

BUSINESS MODEL NNOVATION

The Industrial Economy has been centered on the push business model: raw material is first extracted by process industries, transported and warehoused at several manufacturers which convert raw material into finished products via process or discrete manufacturing; finished products plus starting inventory less ending inventory become sold products, which need wholesalers and retailers to reach customers in physical locations. Economics have been of scarcity and reliance on resources. The past decades have seen the digital or Information and Communications Sector transform, typically disruptively, all sectors via new digital technologies and platform business models. The supply economy is transforming into the demand or pull economy (Van Alstyne et al, 2017). Push or pipeline business models are transformed into and disrupted by pull or as platform business models. Closed companies become open ecosystems. The logic of value creation, capture and delivery changes from one to many options. Market analysts like IDC and futurists like Frank Diana note we are today in the innovation age, where multiple innovations combine into many options meeting socio-economic trends across industries. Innovation is not only multiple and combinatorial, but also the economy driver, generating cycles of innovation and reinnovation (Madden, 2010; Powell and Snellmann, 2004). Innovation is no longer incremental to current trends, but radical and generative. In the past, almost all value was created in pipelines by tangible assets land and property, plant and equipment (Daum, 2003; Lev and Gu, 2016). The New Economy creates value in open networks, as new factors of production information, knowledge, innovation are created in networks (Ben Letaifa, 2014). Mathematical studies show intangible assets underpin value creation, capture and delivery (Damodaran, 2012; Madden, 2010), thereby shifting the paradigm of wealth creation since Adam Smith (1776). Innovation is open and network based (Chesbrough, 2012). Crowdsourcing events are organized to set requirements and organize competitions for them (Chesbrough, 2012). The advent of the Internet of Things transforms other processes, like customer relationship management and supply chain management, into network based (Kagermann et al, 2013). The issue: why do business processes happen? remains paramount. Value needs to be created for stakeholders. In the past stakeholders argued their share of value add. In times of radical change (Eccles, R.G., & Armbrester, K.; 2011, Eccles, R., & Krzus, M., 2010; Eccles, R.G., & Saltzman, D., 2011; Porter and Heppelmann, 2014, 2015) research, development and innovation efforts are intense and make possible solutions that create, capture and deliver value for all stakeholders. Highly innovative products are also customer intimate and operationally effective (Treacy and Wiersema, 2013; Porter and Heppelmann, 2014, 2015). Investing in clean supply chains and employee considerate organizations attracts consumers, who tend to reward companies that help address their health or social issues. Marketing remains centered on the customer value proposition

(Porter and Heppelmann, 2014, 2015), which means the higher the customer value the more likely the product will be purchased and remained loyal to. The most sophisticated type o open innovation is also value centered (Chesbrough, 2012). New technologies like the cloud enable the work from anywhere operating model, which is an agile model with many benefits that has helped weather the Covid crisis. Smart and connected products are ascertained to deliver enormous value (Porter and Heppelmann, 2014, 2015). Renewable energy sources operated by the Internet combine abundance, zero carbon footprint, low cost compared with fossil fuels and may shift the balance of energy supply. Autonomous vehicles ease driving and provide extra time, which may mean three hours per working day, hours for mothers, hours spent commuting. Flying vehicles may ease traffic in cities soon. Home security will be more effective at almost no cost. Smart connected products will form product-service systems and systems of systems with global value, and it maybe this value that both drives and sells solutions. In the age of digital transformation, products and services are complemented by solutions: rather than sell products, create them to fulfill multiple needs.

SOLUTIONS: ESSENTIAL TO DIGITAL TRANSFORMATION

In the IGI Encyclopedia of Digital Terminology, digital transformation has (Bhuvanij et al, 2015) several definitions, which are presented in a group manner, with each definition matching an individual article and author cited by the editor. Digital transformation goes beyond digitalization and is a much broader adoption of digital technology and cultural change; business transformation enabled by digitalization; about solving a problem or providing a new approach to customers. It is a concept that denotes the strategic potential of digitalization in practically all of the processes of organization. Digital transformation involves integrating solutions. Digital transformation is the integration of digital technology into all domains of business, essentially altering how to function and administer value to customers or consumer. It's also a cultural change that necessitates organizations to constantly challenge the status quo, experiment, and get comfortable with failure. Digital transformation is the creation of solutions by using digital elements in meeting social, organizational, sectoral, workplace, and business needs. It is the profound transformation of business and organizational activities, processes, competencies and models to fully leverage the changes and opportunities of a mix of digital technologies and their accelerating impact across society in a strategic and prioritized way, with present and future. It is transformation from classical systems to digital systems in a way that all functions of a business can be carried out on digital media. Digital transformation is the integration of digital technologies to find a solution for the needs of society and sectors through the development and transformation of existing business manners and culture. Digital transformation is the process in which individuals use new technologies in a fast and frequent way for solving problems. It refers to the process of integrating digital technologies to create new or modify existing business processes, culture, and customer experiences to meet changing business and market requirements. The change incurred by digital transformation may be the response to a changing business environment. Digital transformation is a process whereby a company makes use of smart digital technologies to redefine existing business systems, processes, procedures, culture, and customer experiences in an attempt to meet constantly changing market and business requirements. It may be viewed as a transformation of a business or an industry by the use of digital technologies to enable major business improvements, such as developing new business models, providing advanced customer experience, optimizing processes, creating and getting value in new ways, creating new digital revenue streams. Is the adoption of new digital technologies to capture change in the performance of an organization with a focus on disruptive technologies.

Digital transformation is also defined by management consultants and market analysts Bhuvanij et al, 2015; Domazet et al, 2018; Kristapsone and Bruna, 2017; Krchova and Hoesova, 2021; Mirolyubova et al, 2020; Meng and Li, 2002; Mendes, 2020; Martin and Stoica, 2010; Pavlicek et al, 2011. According to IDC, market leaders in business consulting are: KPMG, Deloitte, Accenture, BCG, CapGemeni, PWC, EY, Mc Kinsey and Company. Digital transformation is a market, according to mainstream market analysts, such as IDC (Bhuvanij et al, 2015; Domazet et al, 2018), Markets and Markets (Mirolyubova et al, 2020), Statista (Krchova and Hoesova, 2021). Digital Transformation (Bhuvanij et al, 2015; Domazet et al, 2018) means applying new technologies to radically change processes, customer experience, and value. DX allows organizations to become Digital Native Enterprise that support innovation and digital disruption rather than enhancing existing technologies and models. Digital transformation is the outcome of changes that occur with the application of digital technologies (Mirolyubova et al, 2020). The market leader in business consulting, KPMG, defines digital transformation as the journey towards the connected enterprise, with insight driven strategies and actions; innovative products and services; experience centricity by design; responsive operations and supply chain; integrated partner and supplier ecosystem; digitally enabled technology architecture; aligned and empowered workforce; seamless interactions and commerce (Meng and Li, 2002). The runner-up, Deloitte (Mendes, 2020), defines digital transformation as the journey to digitalization: shifting legacy customer, business and operating models into a new reality - where agility is the norm, human experience is the focus, technology and data are the enablers, and exponential value is the outcome. The third most important business strategy consultant, Accenture (Kristapsone and Bruna, 2017), argues that digital transformation is a pivot that combines legacy business capabilities, which are improved and transformed at the core, and new digital businesses, which are scales. BCG (Martin and Stoica, 2010) defines digital transformation as the success of six capabilities: integrated strategy with clear transformational goals; leadership commitment from CEO to middle management; deploying high caliber talent; an agile governance mindset that drives broader adoption; effective monitoring of progress toward defined outcomes; business led modular technology and data platform. PWC (Selishcheva et al, 2020) defines digital transformation as the qualitative improvement of manufacturing and business processes powered by innovation and by adapting existing business models to the modern Digital Economy. EY (Zheng et al, 2017) understands digital transformation as comprising the following capabilities: the bridge: see the future and plan with purpose; the engine room: orchestrate and accelerate like a market leader; innovation: disrupt and create like a start-up; design, test and iteration; deployment hub: plan, invest and scale up like a venture capitalist; digital factory: accelerate and industrialize transformation. Mc Kinsey (Zheng et al, 2018) defines digital transformation as the integration of advanced technologies into existing business models.

TRANFORMING FROM PRODUCTS TO SMART CONNECTED PRODUCTS AND SOLUTIONS

The issue that the digitalized economy has extended to traditional industries is now included in journal articles (Dobrolyubova et al, 2017; Jiang, 2020). Already in 2014 and 2015; Porter and Heppelmann have predicted a third wave of digital transformation, generated by intelligent connected devices or smart connected products working across the Internet of Things (Porter and Heppelmann, 2014; 2015).

A search in Web of Science by key words "smart connected products" reveals records about traditional products in various traditional industries enriched by information and communications technology with smartness and connectivity components (Abramovici et al, 2017; Azarmipour et al, 2020; Brehm and Klein, 2017; Bugeja et al, 2016; Bu et al, 2020; Chung et al, 2018; Kim et al, 2019; Li et al, 2018; Liang et al, 2021; Mani and Chouk, 2017;Park, 2016; Savarino et al, 2018; Tomiyama et al, 2019; Zheng et al, 2017, Zheng et al, 2018; Zheng et al, 2019; Zheng et al, 2020). The authors predict physical devices will be enhanced by information and communications technology and transformed to smart connected devices working across the Internet of Things. Research in top scientific journals has been conducted to find a scientific equivalent of smart connected products, and this is cyber-physical systems (Zheng et al, 2017, Zheng et al, 2018; Zheng et al, 2019; Zheng et al, 2020). Other authors find scientific definitions for the Porter and Heppelmann proposal in Harvard Business Review. In 2017; smart connected products are traditional products whose functionality is significantly enhanced by digitalization (Brehm and Klein, 2017). Smart-connected technologies, such as the Internet of things and cloud computing, are transforming how industries and enterprises do business by improving the lifecycle management of their product (Zhang et al, 2019). Smart components (i.e. computing power) are embedded in the product and are connected to a product cloud (Zhang et al, 2019). Other articles write that, with the technological development of the Internet of Things (IoT), companies are increasingly launching smart and connected products (Kim et al, 2019; Liang et al, 2021; Mani and Chouk, 2017). The advancement of information communication technologies (ICT) has begun to change both industrial and consumer products towards ones with attributes of smartness and servitization (Savarino et al, 2018). Some authors attribute smart connected products to the recent development and implementation of advanced information and communication technologies (ICT) (e.g. Internet-of-Things (IoT), Cyber-Physical System (Zheng et al, 2018). The third-wave of IT competition have embraced a promising market of low cost, high performance smart, connected products (Zheng et al, 2019). Owing to their unique capabilities, SCPs together with their generated smart services, as a solution bundle can fulfil the everchanging individual user's needs. Meanwhile, manufacturers/service providers leverage massive user generated data and product sensed data via the Internet-of-Things (IoT) for evergreen design innovation. This emerging IT-driven transdisciplinary engineering paradigm (Zheng et al, 2019) is named Smart Product-Service Systems (Smart PSS), which is an ecosystem consisting of various stakeholders as the key players for open innovation (social aspect), intelligent systems as the infrastructure to enable smartness and connectedness (technical aspect), and digital servitization as the value proposition to make higher profits (business aspect). Smart products supported by new step-changing technologies, such as Internet of Things and artificial intelligence, are now emerging in the market (Tomiyama et al, 2019). Smart products are cyber physical systems with services through Internet connection (Tomiyama et al, 2019). Smart products are software-intensive, data-driven, and service-conscious, their development clearly needs new capabilities underpinned by advanced tools, methods, and models (Tomiyama et al, 2019). Others confirm: the recent ICT innovations have begun to dramatically change traditional products towards intelligent, connected smart products (Abramovici et al, 2017). The business world is experiencing a shift away from 'physicality' due to the ubiquitous growth of the service sector and a progressive shift from selling product to offering service (Abramovici et al, 2017). The Internet of Things makes devices equipped with sensors the center of this phenomenon. Some authors (Abramovici et al, 2017) propose a typology of smart products along two dimensions: 'Product Attributes' and 'Ecosystemic attributes'. This typology includes four categories of smart products: More Efficient Products (MEPs), Augmented Products (APs), Products as a Node (PN), and Products as a Hub (PH). Fundamental changes of traditional products towards intelligent, connected

Smart Product Service Systems require radical adaption and enhancement of traditional product design approaches Savarino et al, 2018. In each category, a product acquires a certain degree of 'digital enhancement', 'embeddedness' and 'transformativeness' (Abramovici et al, 2017). Physical devices are added connectivity and intelligence and become smart connected products, such as as a smart vehicle (Tomiyama et al, 2019), a smart connected bicycle (Bu et al, 2020), smart toys (Li et al, 2018), or systems of smart connected products, such as smart cities (Srinivas et al, 2019), smart homes (Bugeja et al, 2016), smart factories (Park, 2016). Authors include additive manufacturing amongst intelligent connected devices that form systems of systems across the Internet of Things (Chung et al, 2018).

SOLUTIONS BASED ON BUSINESS PROCESS INFORMATION SYSTEMS

Scientific sources state the goal of digital transformation, the ambidextrous combination of radical innovation and incremental innovation, is not selling technology but architecting the intelligent enterprise or the digital enterprise. Intelligent enterprises apply technology and new service paradigms to the challenge of improving business performance (Quinn, 2012; Korhonen and Halen, 2017; Van den Heuvel and Tamburi, 2020; Zimmermann et al, 2018; Zimmermann et al, 2015). Intelligent enterprises rely more on intellectual assets than on physical assets. The intelligent enterprise is mentioned in scientific journals with scarcity (Bhattacharya, 2020; Lee et al, 2018; Nanda, 2019; Łobaziewicz, 2017; Szczerbicki, 2019; Oswald and Kleinemeier, 2017; Wu et al, 2010). These sources tie the intelligent enterprise to digital technologies and Industry 4.0. An alternative name for the intelligent enterprise is the digital enterprise Anagnoste, 2018; Brusakova and Shepelev, 2016; Bossert, 2016; CIMdata, 2021; Mogilko et al, 2020; Skilton, 2016; Teece and Linden, 2017; Uhl et al, 2014; Xu, 2014; Zimmermann et al, 2015; Zimmermann et al, 2016). Scientific literature tackles the intelligent enterprise and digital enterprise, mostly with analytic interventions that refer to enterprise architecture and the new technologies that make it up. Several market analyst references show the highlight of innovation to day is for the information technology in smart connected products. CIMdata is the leading market analyst for Product Lifecycle Management software, the solution for innovation. CIMdata (Xu, 2014) works with IDC and Gartner for a digital innovation collaboration platform that combines Internet of Things platforms with digital twins and digital threads to achieve smart connected products.

In conclusion, a paradigm shift from selling products manufactured via tangible asset resources to marketing solutions created mainly via abundant and network generated intangible assets now happens. Solutions may be created specifically to address society and economy needs.

ORGANIZATION OF THE BOOK

The book content shows solutions may addresses business process, business value and strategy challenges. This type of solutions is called enterprise architecture. Enterprise architecture may be called the digital enterprise or the intelligent enterprise. Oher types of solutions add to smart connected products services or other smart connected products, making them product-service systems, systems of systems or solutions. These are better alternatives to traditional products on traditional markets, which upgrade or replace existing products. These products bundle products and services into solutions and create moderately or radically new value.

Chapter 1 provides recommendations for the integration of various types of patterns. Complex layers of patterns can be assembled in a Holistic Enterprise Architecture Pattern (HEAP), which are integrated by using the Integration of HEAP (IoHEAP), that can support agile transformation projects. The IoHEAP is based on a layered concept of patterns and an optimized deployment process. An integration patterns model can be used as a template to integrate Solution Building Blocks (SBB), which can be used to implement a variety of types of transformation projects. In this article, the focus is on the integration of various pattern types that can be used to deliver adaptable SBBs. SBBs are the backbone of Enterprise Architectures (EA) based transformation projects.

Chapter 2 is a a conceptual analysis in order to conceptually examine new terms such as digital management, digital transformation and digital leader that emerged with digitalization in management structures, to explain the change in management structures by comparing them with previous management structures, and to eliminate the ambiguities about the concept of digitalization. In addition, another aim of this study is to compile explanations in the literature about the metaverse, which is described as a three-dimensional virtual world, and to evaluate the innovations, opportunities, concerns and criticisms brought by the metaverse application.

Chapter 3 provides recommendations for classifying and using various types of patterns in a Holistic Enterprise Architecture Pattern (HEAP), that can support transformation projects. The HEAP is based on a concise, composite, and layered patterns model. A composite patterns model can be used as a template to instantiate Building Blocks (BB) to implement a variety of types of transformation projects. In this article the focus is on various pattern standards that can be used in a holistic and adaptable BBs to support an optimal set of Enterprise Architectures (EA). A patterns-based EA offers BBs, like, in civil engineering, that can support colossal and complex projects. For such complex projects, there is a need to create a common denominator pattern, to integrate other standard patterns, in the form of a HEAP, where the HEAP is the backbone of this Research and Development Project (RDP). This RDP proves that the HEAP can be used for building dynamic and flexible EAs.

Chapter 4 tackles the role of Requirements engineerings, which is central for the implementation of projects in general and is especially crucial for Business Transformation Projects (or simply Projects), because transformation activities incur major changes in the existing sets of archaically defined requirements. Requirements Engineering (RE) is a complex part of the Project, because it consists of many related communication, cross-functional knowledge and dependencies, like for example, the need for RE to interact with executive management, business users, business architects, implementation developers and other Project actors. In this chapter, the authors will try to propose an RoRE Concept (RoREC) that can support Project Managers (or simply Managers) in transforming the enterprise and managing RE activities. RE activities are not just about assembling a huge set of business and non-functional requests and features, in the form of document(s) and delivering a repository of methodology-based diagram(s).

Chapter 5 aims to create a general framework on digital maturity models and reveal the highlights of the commonly used components of digital maturity models. The emphasis will be more on detailing human-related fundamental aspects of digital maturity models rather than the technological ones. Considering the incrementally increasing options to reach collaboration technologies and swift deployment of them to different sectors, the main challenge still remains as to manage the psychological asset related parts. Although, technology is generally perceived way mechanical, whose boundaries can be drawn, tasks to be assigned can be defined; the technological elements within the concept of digital maturity also intersects with the human-related dimensions of digital transformation. The study aims to advise a systematic approach by offering a theoretical framework to manage human related factors in a digital transformation journey with an inclusive perspective positioning technology as the core medium provider, and hereby, contributes to theoretical background in the field of study.

Chapter 6's topic is process mininig. Process mining is a paradigm shift from traditional process understanding methodologies like interviews and surveys to a data-driven understanding of the actual digital processes. It analyzes business processes by applying algorithms to the event data generated by digital systems. The chapter provides insight into various uses of process mining in different social and economic processes, with examples from past works demonstrating how practical process mining is in detecting and mitigating bottlenecks in these sectors. Then the chapter further delves into the details of process mining algorithms, key features, and metrics that can help practitioners and researchers evaluate process mining for their work. It also highlights some data quality issues in the event log that can inhibit obtaining fair results from process models. Additionally, some current limitations and concerns are described for creating awareness and building over the body of knowledge in the process and sequential mining techniques.

Chapter 7 covers information system of information technology outsourcing. Information systems and/or information technology (IS/IT) outsourcing became a very common practice in developed and emerging economies. Despite IS/IT outsourcing's importance, reports on outsourcing initiatives indicate problematic situations. A large number of IS/IT outsourcing projects are being re-negotiated or prematurely terminated and many IS/IT outsourcing failures are even not publicly reported due to the fear of negative responses from the market and stakeholders. Literature suggests that the IS/IT project outsourcing is a complex maneuver. The concept of modularity has been applied in many other fields in order to manage complexity and enhance agility/flexibility; hence four cases were analyzed using the lens of modularity in order to understand and identify the relationship between the concept of modularity and IS/IT project outsourcing. The interface aspect of modularity has emerged as the most relevant and identified in all four cases. It implies that the interface aspect should get greater attention when designing/planning a new IS/IT outsourcing project.

Chapter 8 covers the integration of digital technologies into capability maturity models and roadmaps. Capability Maturity and Capability Readiness Models are designated management tools in scholarly literature. One of their applications is in shaping roadmaps, projects and programs. Typically individual articles tackle the two topics separately and the way the two management tools are intertwined. Scholarly literature shows the need to link the two tools. These tools tend to refer to digital transformation or include digitalization in their construction. A specific example is the TRL1-TRL9 Capability Maturity Model used by Nasa and by the European Union. This is the reference for conducting research, development and innovation activities at the European Union. It models roadmaps and all project management techniques. A specific case study is considered in this article. Findings show the reference to manage research, development and innovation activities at the European Union and Nasa is a capability maturity model, a project management tool and includes digitalization business information systems as the tools for business activity and business process management.

Chapter 9 covers ecosystems, which have emerged in the last ten years. Their historic evolution is important, as ecosystems are deemed new business models to progressively replace the old ones. Ecosystems are broadly and vaguely defined in literature review, where they represent new network business models. Empirical data about European project calls show precise definitions of ecosystem types and activities that are subject to European project funding calls. These definitions apply to new commercial entities recently created and funded by such European project calls. The article aims to build theory with a case study about emerging business practices with high instrumental value to other businesses which may apply for European project calls. The methodology is a descriptive case study. Findings are the level of detail and focus in business practices exceeds that of general and vague theory, thereby allowing business practice to contribute to theory and enrich it. Definitions of ecosystem entities and ecosystems management activities to be used across the European Union are this addition.

Chapter 10 deals with museums, which offer education and enjoyment to the public through exhibitions and public programs, but what happens behind the museums can be a mystery. Thus, this study uses the visit to conservation laboratories as a behind-the-scene tour to illustrate the conservation practices at public museums in Hong Kong and thus potential digital transformation. The value chain analysis was used to systematically exanimate the environment and operations of the conservation process in depth, focusing on museum education and extension activities. Lacking human resources, safety, and access constraints often limited the capacity to offer such tours. Some suggestions on digital transformation using contemporary information and communication technologies (ICTs) were proposed for engagement improvement and expanding the audience. Scant studies research how such conservation tours facilitate learning and engagement with visitors or analyze museum operations with value-chain analysis for digital transformation, especially in East Asia.

Chapter 11's topic is Machine and Deep Learning (ML/DL) Algorithms, Frameworks, and Libraries. Each primary, secondary, tertiary, quaternary, and the quinary sector has huge or very huge incremental data from large-scale, small-scale industries, medium industries, or cottage industries. The data associated with each of them are very crucial from every point of view. The complex problems are increasing day by day in real-time execution which can be addressed using current trends of technology like Machine learning and Deep Learning. Machine learning is a subset of Artificial Intelligence. ML is functioning for image & speech recognition, mail filtering, Facebook tag mechanism, and many others. Deep learning is an advanced technology that is a subset of machine learning with the capacity to learn more intelligently on a large set of data. Deep learning works with multiple hidden layers to produce the predicted outcomes. Deep Learning Algorithms include Convolutional Neural networks, Recurrent Neural Networks, Long Short-Term Memory Networks, Stacked Auto-Encoders, Deep Boltzmann machines, etc.

Chapter 12 manages the readiness for Implementing E-Voting System in Ethiopia: A Gap Analysis from Supply Side. Extant literature revealed that elections conducted in traditional ways mostly result in conflicts. Intending to address such challenges, e-voting technology is being used in some countries to conduct a transparent election. However, the application of this new system encountered different challenges due to a lack of readiness to exploit its value, especially in developing countries context. To that end, the readiness of government, citizens, and political parties needs to be assessed before using e-voting as an electoral system. The main purpose of this study is to assess the gaps in the readiness of Ethiopia for e-voting system implementation. A qualitative research method is employed and a thematic analysis was used to analyze the data. The finding revealed that the Ethiopia is not ready in terms of Information Communication Technology (ICT) infrastructure, human resources, and legality measures for e-voting technology. Finally, recommendations are forwarded for policymakers and practitioners for action.

Chapter 13 referes to health behaviours, as the number of studies aiming to change consumers' health behaviors by digitizing is increasing day by day. Health behaviors have become the focus of different sciences and disciplines, both socio-cultural and technological. Especially after the Covid-19 epidemic, investments and research in this field have increased considerably. In the literature, there is a rapid increase in the number of studies on digital health behaviors. In this study, it is aimed to systematically examine the studies in the literature on digitalized health behaviors. For this purpose, bibliometric analysis was applied to 357 studies in the Web of Science database using the R Studio program in order to determine the most frequently studied areas and possible gaps in the literature.

Chapter 14 is concerned about Zara, and the way its enterprise architecture is aligned with business strategy. Brands are the universal symbols of the modern era, comprising a set of values with which the consumer identifies. To activate a brand is to make its promise tangible through different marketing and advertising tools. The fashion film is established in the retail sector as a form of brand communication, which can encompass fashion, photography, film, advertising, music and art, becoming heir to different and varied artistic disciplines; a format that allows interaction with customers by breaking traditional boundaries through transmediality. The aim of this article is to analyse the series of three fashion films, "Zara Sce-nes", launched in 2019, in the context of the communication of changes in the brand's business strategy. With this purpose, the different chapters are analysed from the fashion film perspective, as social media content and a brand communication tool.

Chapter 15 refers to democracy, where the paper examines the concept of post-truth and post-truth politics. The paper argues that the application of post-truth rhetoric in politics implies a novelty in politics and in the relationship between truth and politics which undermines democracy. It is arguable that Post-truth condition negatively impact individual's ability to discriminate between what is true or false taking into consideration the volume of disinformation on the one hand and on the other hand the need to make informed decisions and choices without having to consult experts at the critical time that the stakes involved in such decisions and choices are urgent and crucial.

Chapter 16 lists the new Skills for Sustainability in The Socially Challenging Era of Digital Transformation. The times we live in is one that is filled with both opportunities and challenges because of technological advancements. The world as we know it today has undergone a significant transformation into the digital age. And, to keep up with the changes, organizations have placed a greater emphasis on digital transformation. We are presently in the midst of a technological revolution with the intention to basically effect people, work, and the mode of interplay with one another. While the people will want to work with technology, there may be additionally a developing want people to increase specialized skills for the way they have interaction with one another. To keep pace with rapid disruption organizations should also emphasize on the need to invest and reskill their people to stay competitive in the fast advancement of technology which is altering the nature of skills and abilities required in the workplace which also needs to make a mental shift among the employees who can make that technology useful.

Richard Pettinger
University College London, UK

Brij B. Gupta
Asia University, Taiwan

Alexandru Roja
West University of Timisoara, Romania

Diana Cozmiuc
West University of Timisoara, Romania

REFERENCES

Abramovici, M., Gobel, J. C., & Savarino, P. (2017). Reconfiguration of smart products during their use phase based on virtual product twins. *CIRP Annals-Manufacturing Technology*, *66*(1), 165–168. doi:10.1016/j.cirp.2017.04.042

Acatech. (2020). *Industrie 4.0 Maturity Index. Managing the Digital Transformation of Companies – UPDATE 2020*. Retrieved on 1st December 2021; https://en.acatech.de/publication/industrie-4-0-maturity-index-update-2020/

Agarwal, N., & Brem, A. (2015). Strategic Business Transformation through Technology Convergence: Implications from General Electric's Industrial Internet Initiative. *International Journal of Technology Management*, *67*(2-4), 196–214. doi:10.1504/IJTM.2015.068224

Aheleroff, S., Xu, X., Zhong, R. Y., & Lu, Y. Q. (2021). Digital Twin as a Service (DTaaS) in Industry 4.0: An Architecture Reference Model. *Advanced Engineering Informatics*, *47*, 101225. Advance online publication. doi:10.1016/j.aei.2020.101225

Ahmed, A., Alshurideh, M., Al Kurdi, B., & Salloum, S. A. (2021). Digital Transformation and Organizational Operational Decision Making: A Systematic Review. In *Proceedings of the International Conference on Advanced Intelligent Systems and Informatics 2020*. Springer. 10.1007/978-3-030-58669-0_63

Akerberg, J., Akesson, J. F., Gade, J., Vahabi, M., Bjorkman, M., Lavassani, M., Gore, R. N., Lindh, T., & Jiang, X. L. (2021). Future Industrial Networks in Process Automation: Goals, Challenges, and Future Directions. *Applied Sciences (Basel, Switzerland)*, *11*(8), 3345. doi:10.3390/app11083345

Akimov, S. S. (2019). Business process modeling within the Digital Economy development framework. *Proceedings of the 1st International Scientific Conference Modern Management Trends and the Digital Economy: from Regional Development to Global Economic Growth*, 262-267. 10.2991/mtde-19.2019.50

Alabi, M. O., Telukdarie, A., & van Rensburg, N. J. (2019), Water 4.0: An Integrated Business Model from an Industry 4.0 Approach. *2019 IEEE International Conference on Industrial engineering and engineering Management (IEEM), International Conference on Industrial Engineering and Engineering Management IEEM*, 1364-1369

Albukhitan, S. (2020). Developing Digital Transformation Strategy for Manufacturing. *Procedia Computer Science*, *170*, 664–671. doi:10.1016/j.procs.2020.03.173

Alcacer, V., & Cruz-Machado, V. (2019). Scanning the Industry 4.0: A Literature Review on Technologies for Manufacturing Systems. *Engineering Science and Technology, an International Journal*, *22*(3). doi:10.1016/j.jestch.2019.01.006

Alcacer, V., Rodrigues, C., Carvalho, H., & Cruz-Machado, V. (2021). Tracking the maturity of industry 4.0: The perspective of a real Scenario. *International Journal of Advanced Manufacturing Technology*, *116*(7-8), 2161–2181. doi:10.100700170-021-07550-0 PMID:34248244

Anagnoste, S. (2018). Robotic Automation Process – The operating system for the digital enterprise. *Proceedings of the International Conference on Business Excellence, Sciendo*, 12(1), 54-69. 10.2478/picbe-2018-0007

Andal-Ancion, A., Cartwright, P. A., & Yip, Ge. S. (2020). The digital transformation of traditional business. *MIT Sloan Management Review, 44*(4), 34-41.

Anthony, B. Jr. (2020). Managing digital transformation of smart cities through enterprise architecture - A review and research agenda. *Enterprise Information Systems, 15*(3), 299–331. doi:10.1080/17517 575.2020.1812006

Anthony, S. D., Gilbert, C., & Johnson, M. W. (2017). Dual Transformation. How to Reposition Today's Business While Creating the Future. *Harvard Business Review*.

Attrey, A., Carblanc, A., Gierten, D., Lesher, M., Pilat, D., Wyckoff, A., & Kahin, B. (2020). Vectors of Digital Transformation. *Vestnik Mezhdunarodnykh Organizatsii – International Organizations Research Journal, 15*(3), 7-50. doi:10.17323/1996-7845-2020-03-01

Azarmipour, M., Von Trotha, C., Gries, C., Kleinert, T., & Epple, U. (2020). A Secure Gateway for the Cooperation of Information Technologies and Industrial Automation Systems. *IECON 2020: The 46th Annual Conference of the IEEE Industrial Electronics Society*, 53-58.

Babar, Z., & Yu, E. (2019). Digital Transformation - Implications for Enterprise Modeling and Analysis. *2019 IEEE 23rd International Enterprise Distributed Object Computing Workshop*, 1-8. 10.1109/EDOCW.2019.00015

Babich, O., Igolnikova, I., Levin, A., Mityuchenko, L., & Chernyshova, I. (2019). A Modern Enterprise as an Element of Digital Economy. Vision 2025: Education Excellence and Management of Innovations Through Sustainable Economic Competitive Advantage, 4987-4996.

Barba-Aragón, M. I., & Jiménez-Jiménez, D. (2020). HRM and radical innovation: A dual approach with exploration as a mediator. *European Management Journal, 38*(5), 791–803. doi:10.1016/j.emj.2020.03.007

Belyaev, S. E. (2020). Digital Economy: Distributed Project Financing Systems. *Procceding of the 2nd International Scientific and Practical Conference – Modern Management Trends and the Digital Economy: from Regional Development to Global Economic Growth (MTDE 2020)*, 1224-1230. 10.2991/aebmr.k.200502.203

Ben Letaifa, S. (2014). The uneasy transition from supply chains to ecosystems, the value-creation/value-capture dilemma. *Management Decision, 52*(2), 2014. doi:10.1108/MD-06-2013-0329

Berraies, S., & Zine El Abidine, S. (2019). Do leadership styles promote ambidextrous innovation? Case of knowledge-intensive firms. *Journal of Knowledge Management, 23*(5), 836–859. doi:10.1108/JKM-09-2018-0566

Bharadwaj, A., El Sawy, O. A., Pavlou, P. A., & Venkatraman, N. (2013). Digital business strategy: Toward a next generation of insights. *Management Information Systems Quarterly, 37*(2), 471–482. doi:10.25300/MISQ/2013/37:2.3

Bhattacharya, P. (2020). *Guarding the Intelligent Enterprise: Securing Artificial Intelligence in Making Business Decisions*. In 2020 6th International Conference on Information Management (ICIM), London, UK

Bhuvanij, K., Kasemsan, M. L. K., & Praneetpolgrang, P. (2015). An Empirical Investigation on the Required ICT Competency Readiness Towards the Digital Economy in Thailand. *Proceedings of the International Conference on Computing & Informatics*, 127-132.

Bossert, O. (2016). *A Two-Speed Architecture for the Digital Enterprise*. Emerging Trends in the Evolution of Service-Oriented and Enterprise Architectures. doi:10.1007/978-3-319-40564-3_8

Brehm, L., & Klein, B. (2017), Applying the Research on Product-Service Systems to Smart and Connected Products. *Business Information Systems Workshops, BIS 2016*, 311-319. doi:10.1007/978-3-319-52464-1_28

Brusakova & Shepelev. (2016). Innovations in the technique and economy for the digital enterprise. *2016 IEEE V Forum Strategic Partnership of Universities and Enterprises of Hi-Tech Branches*, 27-29. 10.1109/IVForum.2016.7835844

Brynjolfsson, E., & Collis, A. (2019). How Should We Measure the Digital Economy? *Harvard Business Review*.

Bu, L. G., Chen, C. H., Zhang, G., Liu, B. F., Dong, G. J., & Yuan, X. (2020). A hybrid intelligence approach for sustainable service innovation of smart and connected product: A case study. *Advanced Engineering Informatics*, *46*, 101163. Advance online publication. doi:10.1016/j.aei.2020.101163

Buck, C., Marques, C. P., & Rosemann, M. (2021). Eight Building Blocks for Managing Digital Transformation. *International Journal of Technology and Innovation Management*, *18*(5), 2150023. Advance online publication. doi:10.1142/S0219877021500231

Bugeja, J., Jacobsson, A., & Davidsson, P. (2016). On Privacy and Security Challenges in Smart Connected Homes. *2016 European Intelligence and Security Informatics Conference (EISIC)*, 172-175. DOI: 10.1109/EISIC.2016.21

Burggraf, P., Lorber, C., Pyka, A., Wagner, J., & Weisser, T. (2020). Kaizen 4.0 Towards an Integrated Framework for the Lean-Industry 4.0 Transformation. *Proceedings of the Future Technologies Conference (FTC) 2019; Advances in Intelligent Systems and Computing*, 692-709. 10.1007/978-3-030-32523-7_52

Calabrese, A., Dora, M., Ghiron, N. L., & Tiburzi, L. (2020). Industry's 4.0 transformation process: How to start, where to aim, what to be aware of. *Production Planning and Control*. Advance online publication. doi:10.1080/09537287.2020.1830315

Capaldo, A. (2007). Network structure and innovation: The leveraging of a dual network as a distinctive relational capability. *Strategic Management Journal*, *28*(6), 585–608. Advance online publication. doi:10.1002mj.621

Carcary, M., Doherty, E., & Conway, G. (2016). A Dynamic Capability Approach to Digital Transformation: A Focus on key Foundational Themes. *Proceedings on the 10th European Conference on Information Systems Management*. doi:10.339019010138 PMID:26805843

Cerezo-Narvaez, A., Otero-Mateo, M., Rodriguez-Pecci, F., & Pastor-Fernandez, A. (2018). Digital Transformation of Requirements in the Industry 4.0: Case of Naval Platforms. *Dyna*, *93*(4), 448–456. Advance online publication. doi:10.6036/8636

Chesbrough, H. (2002, Mar.). Making Sense of corporate venture capital. *Harvard Business Review*.

Chesbrough, H. (2012). Open Innovation. Where We've Been and Where We're Going. *Research Technology Management, 55*. doi:10.5437/08956308X5504085

Chung, B. D., Kim, S. I., & Lee, J. S. (2018). Dynamic Supply Chain Design and Operations Plan for Connected Smart Factories with Additive Manufacturing. *Dynamic Supply Chain Design and Operations Plan for Connected Smart Factories with Additive Manufacturing, 8*(4), 583. doi:10.3390/app8040583

Ciara, H., & Power, D. J. (2018). Challenges for digital transformation – towards a conceptual decision support guide for managers. *Journal of Decision Systems, 27*(sup1), 38–45. doi:10.1080/12460125.20 18.1468697

CIMdata. (2021). *Product Lifecycle Management (PLM)*. Definition. https://www.cimdata.com/en/resources/about-plm

Daum, J. H. (2002). *Intangible Assets oder die Kunst Mehrwert zu schaffen*. Galileo-Press.

Daum, J. H. (2003). *Intangible Assets and Value Creation*. Wiley.

Davidovski, V. (2018). Exponential Innovation through Digital Transformation. *Proceedings of the 3rd International Conference on Applications in Information Technology (ICAIT - 2018), 3rd International Conference on Applications in Information Technology (ICAIT)*, 3-5. DOI: 10.1145/3274856.327485

Dneprovskaya, N., Urintsov, A., & Afanasev, M. A. (2018). Study of the Innovative Environment of the Digital Economy. *Proceedings of the 15th International Conference on Intellectual Capital, Knowledge Management and Organziational Learning*, 67-76.

Dobrolyubova, E., Alexandrov, O., & Yefremov, A. (2017). Is Russia Ready for Digital Transformation? *Digital Transformation and Global Society (DTGS 2017), Communications in Computer and Information Science*, 431-444. doi:10.1007/978-3-319-69784-0_36

Domazet, I., Zubovic, J., & Lazic, M. (2018). Driving Factors of Serbian Competitiveness - Digital Economy and ICT. *Strategic Management, 23*(1), 20–28. doi:10.5937/StraMan1801020D

Dominic, B., Javalgi, R. G., & Cavusgil, E. (2020). International new venture performance: Role of international entrepreneurial culture, ambidextrous innovation, and dynamic marketing capabilities. *International Business Review, 29*(2), 101639. Advance online publication. doi:10.1016/j.ibusrev.2019.101639

Doucek, P., Fischer, J., & Novotny, O. (2017). Digital Economy. IDIMT – 2017 – Digitalization In Management, Society and Economy, Book Series Schriftenreihe Informatik, 33-40.

Doucek, P., & Holoska, J. (2019). Digital Economy and Industry 4.0. *Innovation and Transformation in a Digital World (IDIMT-2019), Schriftenreihe Informatik*, 33–39.

Duodu, B., & Rowlinson, S. (2021). Opening Up the Innovation Process in Construction Firms: External Knowledge Sources and Dual Innovation. *Journal of Construction Engineering and Management, 147*(8), 04021086. Advance online publication. doi:10.1061/(ASCE)CO.1943-7862.0002108

Eccles, R., & Krzus, M. (2010). *One Report: Integrated Report for aSustainable Strategy*. Wiley.

Eccles, R. G., & Armbrester, K. (2011). Two disruptive ideas combined: Integrated in cloud. *IESE Insight, 8*, 13-20.

Eccles, R. G., & Saltzman, D. (2011). Achieving sustainability throughintegrated reporting. *Stanford Social Innovation Review, 9*(3), 56–61.

Engelsberger, M., & Greiner, T. (2017). Self-organizing Service Structures for Cyber-physical Control Models with Applications in Dynamic Factory Automation A Fog/Edge-based Solution Pattern Towards Service-Oriented Process Automation. *Closer: Proceedings of the7th International Conference on Cloud Computing and Services Science*, 238-246. 10.5220/00063655502660274

Enose, N. (2014). Implementing an Integrated Security Management framework to ensure a secure Smart Grid. *IEEE 2014 International Conference on Advances in Computing, Communications and Informatics (ICACCI), 3rd International Conference on Advances in Computing, Communications and Informatics (ICACCI).*

Eruvankai, S., Muthukrishnan, M., & Mysore, A. K. (2017). Accelerating IIOT Adoption with OPC UA. *Internetworking Indonesia, 9*(1), 3–8.

Fan, I. S., & Oswin, L. (2016). Factory Automation and Information Technology Convergence in Complex Manufacturing. *Advances in Manufacturing Technology, 3*, 331-336. doi:10.3233/978-1-61499-668-2-331

Felser, M., Rentschler, M., & Kleineberg, O. (2019). Coexistence Standardization of Operation Technology and Information Technology. *Proceedings of the IEEE, 107*(6), 962–976. doi:10.1109/JPROC.2019.2901314

Fitzgerald, M., Kruschwitz, N., Bonnet, D., & Welch, M. (2013). Embracing Digital Technology: A New Strategic Imperative. *MIT Sloan Management Review.*

Fonseca, L. M. (2018). Industry 4.0 and the digital society: concepts, dimensions and envisioned benefits. *Proceedings of the International Conference on Business Excellence, 12*(1), 386-397. 10.2478/picbe-2018-0034

Foschini, L., Mignardi, V., Montanari, R., & Scotece, D. (2021). An SDN-Enabled Architecture for IT/OT Converged Networks: A Proposal and Qualitative Analysis under DDoS Attacks. *Future Internet, 13*(10), 258. doi:10.3390/fi13100258

Frank, A. G., Dalenogare, L. S., & Ayala, N. F. (2019). Industry 4.0 technologies: Implementation patterns in manufacturing companies. *International Journal of Production Economics, 210*, 15–26. doi:10.1016/j.ijpe.2019.01.004

Frank, A. G., Mendes, G. H. S., Ayala, N. F., & Ghezzi, A. (2019). Servitization and Industry 4.0 convergence in the digital transformation of product firms: A business model innovation perspective. *Technological Forecasting and Social Change, 141*, 341–351. doi:10.1016/j.techfore.2019.01.014

Frankenberger, K., Mayer, H., & Reiter, A. (2019). The Transformer's Dilemma. *Harvard Business Review.*

Garimella, P. K. (2018). IT-OT Integration Challenges in Utilities. *Proceedings on 2018 IEEE 3rd International Conference on Computing, Communication and Security (ICCCS),* 199-204. 10.1109/CCCS.2018.8586807

Gasanov, T., A., Gasanov, G., A., Feyzullayev, F., S., Bachiyev, B., A., & Eminova, E., M. (2019). Digital Economy and Breakthrough Technologies as Fundamentals of Innovative Regional Economy, Social and Cultural Transformations in the Context of Modern Globalism. *European Proceedings of Social and Behavioural Sciences, 58.* doi:10.15405/epsbs.2019.03.02.234

Gebayew, C., Hardini, I. R., & Panjaitan, G. H. A. (2018). A Systematic Literature Review on Digital Transformation. *International Conference on Information Technology Systems and Innovation (ICITSI),* 260-265. 10.1109/ICITSI.2018.8695912

Giallanza, A., Aiello, G., & Marannano, G. (2021). Industry 4.0: advanced digital solutions implemented on a close power loop test bench. *Proceedings of the 2nd International Conference on Industry 4.0 and Smart Manufacturing (ISM 2020),* 93-101. 10.1016/j.procs.2021.01.133

Gong, C., & Ribiere, V. (2021). Developing a unified definition of digital transformation. *Technovation, 102,* 102217. Advance online publication. doi:10.1016/j.technovation.2020.102217

Gurieva, L. K., Borodin, A. I., & Berkaeva, A. K. (2019). Management Model Transformation in the Digital Economy. *Proceedings of the 1st International Scientific Conference Modern Management Trends and the Digital Economy: from Regional Development to Global Economic Growth,* 383-387. 10.2991/mtde-19.2019.73

Hafsi, M., & Assar, S. (2019). Managing Strategy in Digital Transformation Context: An Exploratory Analysis of Enterprise Architecture Management Support. *IEEE 21st Conference on Business Informatics, 1,* 165-173. 10.1109/CBI.2019.00026

Hanelt, A., Bohnsack, R., Marz, D., & Marante, C. A. (2020). A Systematic Review of the Literature on Digital Transformation: Insights and Implications for Strategy and Organizational Change. *Journal of Management Studies, 58*(5), 1159–1197. doi:10.1111/joms.12639

Haskamp, T., Dremel, C., & Uebernickel, F. (2021). Towards a Critical Realist Understanding of Digital Transformation: Results of a Structured Literature Review Completed Research. *Digital Innovation and Entrepreneurship.*

Herlitschka, S., & Valtiner, D. (2017). Digital transformation: How industry and society are remodeling as the analog becomes more and more digital. *Electrotechnik und Informationstechnik, 134*(7), 340–343. doi:10.100700502-017-0518-y

Hunter, S. T., Cushenbery, L. D., & Jayne, B. (2017). Why dual leaders will drive innovation: Resolving the exploration and exploitation dilemma with a conservation of resources solution. *Journal of Organizational Behavior, 38*(8), 1183–1195. doi:10.1002/job.2195

IDC. (2020). *Explore the 3 Chapters of 3rd Platform Evolution.* Retrieved on 1st December 2021; from: https://www.idc.com/promo/thirdplatform

IDC. (2021a). *IDC MarketScape names Accenture a business consulting leader.* Retrieved on 1st December 2021; from: https://www.accenture.com/sk- en/insights/consulting/idc-research-consulting-leader

IDC. (2021b). *Digital Transformation (DX).* Retrieved on 1st December 2021; from: https://www.idc.com/itexecutive/research/dx

Industrie 4.0 Maturity Index. (2020). *Managing the Digital Transformation of Companies*. Retrieved on 1st December 2021; from: https://en.acatech.de/publication/industrie-4-0-maturity-index-update-2020/

Ivancic, L., Vuksic, V. B., & Spremic, M. (2019). Mastering the Digital Transformation Process: Business Practices and Lessons Learned. *Technology Innovation Management Review*, *9*(2), 36–50. doi:10.22215/timreview/1217

Jiang, W., Gu, Q., & Wang, G. G. (2015). To Guide or to Divide: The Dual-Side Effects of Transformational Leadership on Team Innovation. *Journal of Business and Psychology*, *30*(4), 685–699. doi:10.100710869-014-9395-0

Jiang, X. J. (2020). Digital Economy in the post-pandemic era. *Journal of Chinese Economic and Business Studies*, *18*(4), 333–339. doi:10.1080/14765284.2020.1855066

Jonkers, K., & Sachwald, F. (2018). The dual impact of 'excellent' research on science and innovation: The case of Europe. *Science & Public Policy*, *45*(2), 159–174. doi:10.1093cipolcx071

Kagermann, H. (2015). Change Through Digitization - Value Creation in the Age of Industry 4.0. Management of Permanent Change.

Kagermann, H., Wahlster, W., & Helbig, J. (2013). *Recommendations for implementing the strategic initiative Industry 4.0: Final report of the Industry 4.0 Working Group*. Retrieved on 1st December 2021; from: https://en.acatech.de/publication/recommendations-for-implementing-the-strategic-initiative-industrie-4-0-final-report-of-the-industrie-4-0-working-group/

Kalogeras, A. P., Rivano, H., Ferrarini, L., Alexakos, C., Iova, O., Rastegarpour, S., & Mbacke, A. A. (2019). Cyber Physical Systems and Internet of Things: Emerging Paradigms on Smart Cities. *MIT Sloan Management Review*, *2015*(14), 1–25.

Kantemirova, M., Alikova, Z., Dzakoev, Z., Alikova, T., & Bolatova, M. (2019). Efficiency Digital Transformation of Enterprise, Indo American. *Journal of Pharmaceutical Sciences*, *6*(5), 10654–10657.

Karpunina, E., K., Shurchkova, J., V., Konovalova, M., E., Levchenko, L., V., & Borshchevskaya, E., P. (2019). Education Excellence and Innovation. *Management through Vision 2020*, 7454-7461.

Kempegowda, S. M., & Chaczko, Z. (2018). Industry 4.0 Complemented with EA Approach: A Proposal for Digital Transformation Success. *2018 26th International Conference on Systems Engineering (ICSENG 2018)*.

Khan, A., & Wu, X. M. (2021). Bridging the Digital divide in the Digital Economy with Reference to Intellectual Property. *Journal of Law and Political Sciences*, *28*(3), 245–267.

Khitskov, E. A., Veretekhina, S. V., Medvedeva, A. V., Mnatsakanyan, O. L., Shmakova, E. G., & Kotenev, A. (2017). Digital Transformation of Society: Problems Entering in the Digital Economy. *Eurasian Journal of Analytical Chemistry*, *12*(5b), 855–873. doi:10.12973/ejac.2017.00216a

Khosrow-Pour, M. (2018). *Encyclopedia of Information Science and Technology* (4th ed.). IGI Global. doi:10.4018/978-1-5225-2255-3

Kim, S., Del Castillo, R. P., Caballero, I., Lee, J., Lee, C., Lee, D., Lee, S., & Mate, A. (2019). Extending Data Quality Management for Smart Connected Product Operations. *IEEE Access: Practical Innovations, Open Solutions*, *79*, 144663–144678. doi:10.1109/ACCESS.2019.2945124

Kleiner, G., B. (2020). Intellectual economy of the digital age. *Ekonomika i Matematiceskie Metody – Economics and Mathematical Methods, 56*(1), 18-33. doi:10.31857/S042473880008562-7

Korhonen, J. J., & Halen, M. (2017). Enterprise Architecture for Digital Transformation. *IEEE 19th Conference on Business Informatics, 1*, 349-358. 10.1109/CBI.2017.45

Kovaite, K., Sumakaris, P., & Stankeviciene, J. (2020). Digital communication Channels in Industry 4.0 Implementation: The Role of Internal Communication. *Management*, *25*(1), 171–191. doi:10.30924/mjcmi.25.1.10

Krchova, H., & Hoesova, K. S. (2021). Selected Determinants of Digital Transformation and their Influence on the Number of Women in the ICT Sector. *Entrepreneurship and Sustainability Issues*, *8*(4), 524–535. doi:10.9770/jesi.2021.8.4(31)

Kretschmer, T., & Khashabi, P. (2020). Digital Transformation and Organization Design: An Integrated Approach. *California Management Review*, *62*(4), 86–104. doi:10.1177/0008125620940296

Kristapsone, S., & Bruna, S. (2017), Indicators of the Information and Communication Technology (ICT) Sector activity in Latvia and the EU. *Digital Economy*, 277-287. doi:10.1007/978-3-319-90835-9_31

Kusakina, O. N., Vorontsova, G.,V., Momotova, O. N., Krasnikov, A. V., & Shelkoplyasova, G. S. (2019). Using Managerial Technologies in the Conditions of Digital Economy. *Perspectives on the Use of New Information and Communication Technology (ICT) in the Modern Economy*, 261-268. doi:10.1007/978-3-319-90835-9_31

Kutnjak, A., Pihir, I., & Furjan, M. T. (2019). Digital Transformation Case Studies Across Industries - Literature Review. *42nd International Convention on Information and Communication Technology, Electronics and MicroElectronics (MIPRO)*.

Laft, R. L. A. (2021). Dual-Core Model of Organizational Innovation. *Academy of Management Journal*, *21*(2).

Lee, C., Ding, T., & Lin, Z. H. (2018). Integration of ERP and Internet of Things in Intelligent Enterprise Management. *2018 1st International Cognitive Cities Conference (IC3)*, 246-247. 10.1109/IC3.2018.00-11

Lee, M. X., Lee, Y. C., & Chou, C. J. (2017). Essential Implications of the Digital Transformation in Industry 4.0. *Journal of Scientific and Industrial Research*, *76*(8), 465–467.

Lev, B., & Gu, F. (2016). *The End of and the Path Forward for Investors and Managers*. Wiley.

Levkovskyi, B., Betzwieser, B., Loffler, A., & Wittges, H. (2020). Why Do Organizations Change? A Literature Review on Drivers and Measures of Success for Digital Transformation. *AMCIS 2020 Proceedings*.

Li, C. H., & Lau, H. K. (2018). Toy Product Safety Enhancement Using Smart Product Development. *2018 IEEE Symposium on Product Compliance Engineering – Asia 2018 (IEEE ISPCE-CN 2018)*, 109-111.

Liang, Y. H., Xu, Q., & Jin, L. Y. (2021). The effect of smart and connected products on consumer brand choice concentration. *Journal of Business Research, 135*, 163–172. doi:10.1016/j.jbusres.2021.06.039

Litvinenko, I., Smirnova, I., Solovykh, N., Aliyev, V., & Li, A. (2019). The Fundamentals of Digital Economy. *Ad Alta Journal of Interdisciplinary Research, 9*(1), 30–37.

Liu D.,Y., Chen, S., & Chou, T. (2011). Resource Fit in Digital Transformation - Lessons Learned From The CBC Bank Global E-Banking Project. *Management Decision, 49*(10), 1728-1742. doi:10.1108/00251741111183852

Liu, L., Li, R., & Guo, W. (2021). Management Decision and Dual Innovation. *E3S Web Conf., International Conference on Economic Innovation and Low-carbon Development (EILCD 2021)*, 275, 10.1155/2021/3611921

Łobaziewicz, M. (2017). The Role of ICT Solutions in the Intelligent Enterprise Performance. *Conference on Information Systems Management*, 120-136. 10.1007/978-3-319-53076-5_7

Lom, M., Pribyl, O., & Svitek, M. (2016). Industry 4.0 as a Part of Smart Cities. *2016 Smart Cities Symposium Prague (SCSP)*. 10.1109/SCSP.2016.7501015

Lukoschek, C. S., Gerlach, G., Stock, R. M., & Xin, K. (2018). Leading to sustainable organizational unit performance: Antecedents and outcomes of executives' dual innovation leadership. *Journal of Business Research, 91*, 266–276. doi:10.1016/j.jbusres.2018.07.003

Madden, B. (2010a). The Life-Cycle Valuation Model as a Total System. in Wealth Creation: A Systems Mindset for Building and Investing in Businesses for the Long Term. *IEEE 2020 Annual Reliability and Maintainability Symposium (RAMS 2020)*.

Mani, Z., & Chouk, I. (2017). Drivers of consumers' resistance to smart products. *Journal of Marketing, 33*(1-2), 1–2, 76–97. doi:10.1080/0267257X.2016.1245212

Markets and Markets. (2021). *Digital Transformation Market by Technology (Cloud Computing, Big Data and Analytics, Mobility/Social Media, Cybersecurity, Artificial Intelligence), Deployment Type, Vertical (BFSI, Retail, Education), and Region - Global Forecast to 2025*. Retrieved on 1st December 2021; from: https://www.marketsandmarkets.com/Market-Reports/digital-transformation-market-43010479.html

Martin, F. M., & Stoica, E. (2010). Minimize economic crisis through digital economy, Economic World Destiny: Crisis and Globalization? Section V: Economic Information Technology in the Avant-Garde of Economic Development. *17th International Economic Conference (IECS)*, Sibiu, Romania.

Matt, C., Hess, T., & Benlian, A. (2016). Digital Transformation Strategies. *Business & Information Systems Engineering, 57*(5), 339–343. doi:10.100712599-015-0401-5

McAfee, A., & Brynjolfsson, E. (2011). Race Against the Machine. Digital Frontier Press.

McAfee, A., & Brynjolfsson, E. (2011). *The Second Machine Age*. W. W. Norton's Company.

Mendes, M. V. I. (2020). The Limitations of International Relations Regarding MNCs and the Digital Economy: Evidence from Brazil. *Review of Political Economy*, *33*(1), 67–87. doi:10.1080/09538259.2020.1730609

Meng, Q. X., & Li, M. Z. (2002). New economy and ICT development in China. *Information Economics and Policy*, *14*(2), 275–295. doi:10.1016/S0167-6245(01)00070-1

Mergel, I., Edelmann, N., & Haug, N. (2019). Defining digital transformation: Results from expert interviews. *Government Information Quarterly*, *36*(4), 101385. Advance online publication. doi:10.1016/j.giq.2019.06.002

Miao, Z. L. (2021). Digital Economy value chain: Concept, model structure, and mechanism. *Applied Economics*, *53*(37), 4342–4357. doi:10.1080/00036846.2021.1899121

Milosevic, N., Dobrota, M., & Rakocevic, S. B. (2019). Digital Economy in Europe: Evaluation of Countries' Performances. Economics of Digital Transformation, 253-272.

Mirolyubova, T., V., Karlina, T.,V., & Nikolaev, R., S. (2020). Digital Economy: Identification and Measurements Problems in Regional Economy. *Ekonomika Regiona – Economy of Region, 16*(2), 377-390. doi:10.17059/2020-2-4

Mogilko, D. Y., Ilin, I. V., Iliashenko, V. M., & Svetunkov, S. G. (2020). BI Capabilities in a Digital Enterprise Business Process Management System. In D. Arseniev, L. Overmeyer, H. Kälviäinen, & B. Katalinić (Eds.), *Cyber-Physical Systems and Control. CPS&C 2019. Lecture Notes in Networks and Systems* (Vol. 95). Springer. doi:10.1007/978-3-030-34983-7_69

Nanda, N. K. (2019). Intelligent Enterprise with Industry 4.0 for Mining Industry. *International Symposium on Mine Planning & Equipment Selection MPES 2019: Proceedings of the 28th International Symposium on Mine Planning and Equipment Selection - MPES 2019*, 213-218.

Nandico, O. F. A. (2016). Framework to support digital transformation. Emerging Trends in the Evolution of Service Oriented and Enterprise Architectures, 113-138. doi:10.1007/978-3-319-40564-3_7

Nicolescu, O., & Nicolescu, C. (2019). Relationship between Digitalized Economy and Knowledge Based Economy. *Proceedings of the 13th International Conference Management Conference: Management Strategies for High Performance (IMC 2019)*, 457-465.

Novikov, S. V., & Sazonov, A. A. (2020). Production's digital transformation analysis using Industry 4.0 technologies. *Amazon Investiga*, *9*(27), 234–243. doi:10.34069/AI/2020.27.03.25

Novikova, N. V., Strogonova, E. V., & Dianova, L. S. (2020). Regional Projection of Digital Economy. *Proceedings of the 2nd International Scientific and Practical Conference – Modern Management Trends and the Digital Economy: from Regional Development to Global Economic Growth*, 1076-1082. 10.2991/aebmr.k.200502.178

Nowakowski, E., Farwick, M., Trojer, T., Hausler, M., Kessler, J., & Breu, R. (2019). An Enterprise Architecture Planning Process for Industry 4.0 Transformations. *Proceedings of the 21st International Conference on Enterprise Information Systems (ICEIS 2019)*, 2, 572-579. 10.5220/0007680005720579

Nwaiwu, F. (2018). Review and comparison of conceptual frameworks on digital business transformation. *Journal of Competitiveness, 10*(3), 86–100. doi:10.7441/joc.2018.03.06

O'Reilly, C. A., & Tushman, M. L. (2004). The Ambidextrous Organization. *Harvard Business Review*. PMID:15077368

Osmanbegovic, E., & Piric, N. (2019). Contemporary Approaches to Measuring the Digital Economy. *6th International Scientific Conference Economy of Integration*.

Oswald, G., & Kleinemeier, M. (2017). *Shaping the Digital Enterprise. Trends and Use Cases in Digital Innovation and Transformation*. Springer.

Paes, R., Mazur, D. C., Venne, B. K., & Ostrzenski, J. (2017). A Guide to Securing Industrial Control Networks – (IT/OT) Convergence. *2017 Industry Applications Society 64th Annual Petroleum and Chemical Industry Technical Conference (PCIC)*, 2020-95.

Pan, Y. (2018). Relationship between Dual Innovation Ability and Scientific Research, Educational Sciences. *Theory into Practice, 18*(6). Advance online publication. doi:10.12738/estp.2018.6.273

Park, S. (2016). Development of Innovative Strategies for the Korean Manufacturing Industry by Use of the Connected Smart Factory (CSF). *Procedia Computer Science, 91*, 744–750. doi:10.1016/j.procs.2016.07.067

Pavlicek, A., Kacin, R., Sigmund, T., & Hubacek, J. (2011). The Position of the ICT Sector in the National Economy of the Czech Republic. *IDIMT-2011: Interdisciplinarity in Complex Systems, 36*, 147–155.

Pavlov, B., P., Garifullin, R., F., Babushkin, V., M., & Mingaleev, G., F. (2019). Digital Transformation of the Economy, Education Excellence and Innovation. *Management through Vision*, 3359-3364.

Pihir, I., Tomicic-Pupek, K., & Furjan, M. T. (2019). Digital Transformation Playground - Literature Review and Framework of Concepts. *Journal of Information and Organizational Sciences, 43*(1), 33–48. doi:10.31341/jios.43.1.3

Pokhrel, S. R., & Garg, S. (2021). Multipath Communication With Deep Q-Network for Industry 4.0 Automation and Orchestration. *IEEE Transactions on Industrial Informatics, 17*(4), 2852–2859. doi:10.1109/TII.2020.3000502

Porter, M. E., & Heppellmann, J. E. (2014). How Smart,Connected Products Are Transforming Competition. *Harvard Business Review*.

Porter, M. E., & Heppellmann, J. E. (2015). How Smart,Connected Products Are Transforming Companies. *Harvard Business Review*.

Powell, W. W., & Snellman, K. (2004). The knowledge economy. *Annual Reviews*.

Protopopova, N. I., Grigoriev, V. D., & Perevozchikov, S. Y. (2018). Information and Digital Economy as an Economic Category. *Perspectives on the Use of New Information and Communications Technology (ICT) in the Modern Economy*, 300-307. doi:10.1007/978-3-319-90835-9_35

Quinn. (2012). The intelligent enterprise, a new paradigm. *Academy of Management Perspectives, 6*(4).

Randolph, A., Boucheneb, H., Imine, A., & Quintero, A. (2015). On Synthesizing a Consistent Operational Transformation Approach. *IEEE Transactions on Computers, 64*(4). doi:10.1109/TC.2014.2308203

Reischauer, G., & Mair, J. (2018). *Platform Organizing in the New Digital Economy: Revisiting Online Communities and Strategic Responses. In Toward Permeable Boundaries of Organizations? Research in the Sociology of Organizations* (Vol. 57). Emerald Publishing Limited. doi:10.1108/S0733-558X20180000057005

Rossini, M., Cifone, F. D., Kassem, B., Costa, F., & Portioli-Staudacher, A. (2021). Being lean: How to shape digital transformation in the manufacturing sector. *Journal of Manufacturing Technology Management, 32*(9), 239–259. doi:10.1108/JMTM-12-2020-0467

Rubera, G., Chandrasekaran, D., & Ordanini, A. (2016). Open innovation, product portfolio innovativeness and firm performance: The dual role of new product development capabilities. *Journal of the Academy of Marketing Science, 44*(2), 166–184. doi:10.100711747-014-0423-4

Rueckel, D., Muehlburger, M., & Koch, S. (2020). An Updated Framework of Factors Enabling Digital Transformation, Pacific Asia. *Journal of the Association for Information Systems, 12*(4), 1–26. doi:10.17705/1pais.12401

Safiullin, A., Krasnyuk, L., & Kapelyuk, Z. (2019). Integration of Industry 4.0 technologies for "smart cities" development. *International Scientific Conference Digital Transformation on Manufacturing, Infrastructure and Service, Conference Series-Materials Science and Engineering, 497*. 10.1088/1757-899X/497/1/012089

Sarvari, P. A., Ustundag, A., Cevikcan, E., Kaya, I., & Cebi, S. (2018). Technology Roadmap for Industry 4.0, Industrial 4.0: Managing the Digital Transformation. doi:10.1007/978-3-319-57870-5_5

Savarino, P., Abramovici, M., Gobel, J. C., & Gebus, P. (2018). Design for reconfiguration as fundamental aspect of smart products. *28th CIRP Design Conference, 70*, 374-379. doi:10.1016/j.procir.2018.01.007

Savytska, O., & Salabai, V. (2021). Digital Transformations in the Conditions of Industry 4.0 Development. *Financial and Credit Activity – Problems of Theory and Practice, 3*(38), 420–426. doi:10.18371/fcaptp.v3i38.237472

Schneider, S., & Kokshagina, O. (2020). Digital transformation: What we have learned (thus far) and what is next. *Creativity and Innovation Management, 30*(2), 384-411. doi:10.1111/caim.12414

Selishcheva, T. A., Sopina, N. V., Borkova, E. A., & Ilyina, O. P. (2020). Estimation of The Influence of Digital Economy on The Competitiveness of The World, Education Excellence and Innovation Management: A 2025 Vision to Sustain Economic Development during. *Global Challenges*, 3757–3764.

Shcherbakova, T. S. (2019). Transformation of the service industry in Digital Economy. *Procceedings of the 1st International Scientific Conference Modern Management Trends and the Digital Economy: from Regional Development to Global Economic Growth*, 288-291. 10.2991/mtde-19.2019.55

Shmakov, A. V. (2019). Code of Ritual Behavior in the Context of Digital Transformation of Economy. *Terra Economicus, 17*(4), 41–61. doi:10.23683/2073-6606-2019-17-4-41-61

Sjobakk, B. (2018). The Strategic Landscape of Industry 4.0, Advances in Production Management Systems: Smart Manufacturing for Industry 4.0, APMS 2018. *IFIP Advances in Information and Communication Technology*, *536*, 122–127. doi:10.1007/978-3-319-99707-0_16

Skilton, M. (2016). *Building Digital Ecosystem Architectures, A Guide to Enterprise Architecting Digital Technologies in the Digital Enterprise*. Palgrave MacMillan.

Smirnov, V. V., Osipov, D. G., Babayeva, A. A., Grigorieva, E. V., & Perfilova, E. F. (2019), Parity of innovation and Digital Economy in the Russian management system. *Proceedings of the 1st International Scientific Conference Modern Management Trends and the Digital Economy: from Regional Development to Global Economic Growth*, 22-27. 10.2991/mtde-19.2019.5

Smith, A. (1776). *The Wealth of Nations*. Academic Press.

Song, C., Hu, L., & Yuan, H. (2018). The Study of the Impact of Technological Innovation Network on Dual Innovation. In *Applied Computational Intelligence and Mathematical Methods* (p. 670). Springer. doi:10.1007/978-3-319-67621-0_24

Srinivas, M., Benedict, S., & Sunny, B. C. (2019). IoT Cloud based Smart Bin for Connected Smart Cities – A Product Design Approach. *2019 10th International Conference on Computing, Communication and Networking Technologies (ICCCNT), 10th International Conference on Computing, Communication and Networking Technologies (ICCCNT).*

Statista. (2021). *Spending on digital transformation technologies and services worldwide from 2017 to 2025*. https://www.statista.com/statistics/870924/worldwide-digital-transformation-market-size/

Stefanic, N., Bezic, H., & Greguric, P. (2019). More than technological evolution: organizational and business impact of Industry 4.0. *Economics of Digital Transformation*, 147-160.

Stefanic, N., Bezic, H., & Greguric, P. (2019). More than technological evolution: organizational and business impact of Industry 4.0. *Economics of Digital Transformation*.

Steiber, A., Alange, S., Ghosh, S., & Goncalves, D. (2021). Digital transformation of industrial firms: An innovation diffusion perspective. *European Journal of Innovation Management*, *24*(3), 799–819. doi:10.1108/EJIM-01-2020-0018

Stjepic, A. M., Ivancic, L., & Vugec, D. S. (2020). Mastering digital transformation through business process management: Investigating alignments, goals, orchestration, and roles. *Journal of Entrepreneurship Management and Innovation*, *16*(1), 41–73. doi:10.7341/20201612

Stout, W. M. S. (2018). Toward a Multi-Agent System Architecture for Insight & Cybersecurity in Cyber-Physical Networks. *2018 52nd Annual IEEE International Carnahan Conference on Security Technology (ICCST)*, 270-274.

Sturgeon, T. J. (2021). Upgrading strategies for the Digital Economy. *Global Strategy Journal*, *11*(1), 34–57. doi:10.1002/gsj.1364

Subbotina, T. A., & Zhukova, E. F. (2019). Digital Technologies and their Role in Modern Economy of Russia, International Scientific and Practical Conference Contemporary Issues of Economic Development of Russia: Challenges and Opportunities. *European Proceedings of Social and Behavioural Sciences, 59*, 629–635. doi:10.15405/epsbs.2019.04.67

Sundaram, S., & Zeid, A. (2021). Smart Prognostics and Health Management (SPHM) in Smart Manufacturing: An Interoperable Framework. *Sensors (Basel), 21*(18), 5994. Advance online publication. doi:10.339021185994 PMID:34577203

Szczerbicki, E. (2019). Establishing intelligent enterprise through community of practice for product innovation. *Journal of Intelligent & Fuzzy Systems, 37*(6), 7169–7178. doi:10.3233/JIFS-179329

Tabrizi, B., Lam, E., Girard, K., & Irvin, V. (2019). Digital Transformation Is Not About Technology. *Harvard Business Review.*

Teece, D. J., & Linden, G. (2017). Business models, value capture, and the digital enterprise. *Journal of Organization Design, 6*(8).

Thubert, P., Palattella, M. R., & Engel, T. (2015), 6TiSCH Centralized Scheduling: when SDN Meet IoT. *2015 IEEE Conference on Standards for Communications and Networks (CSCN)*, 42-47. 10.1109/CSCN.2015.7390418

Tian, S., & Hu, Y. H. (2019). The Role of OPC UA TSN in IT and OT Convergence. *2019 Chinese Automation Congress (CAC2019)*, 2272-2276. 10.1109/CAC48633.2019.8996645

Tinkov, S., Babenko, I., & Tinkova, E. (2019). Digital Transformation of Production Systems Within the Concept "Industry 4.0". *Vision 2025: Education Excellence and Management of Innovations Towards Sustainable Economic Competitive Advantage*, 7310-7318.

Tomiyama, T., Lutters, E., Stark, R., & Abramovici, M. (2019). Development capabilities for smart products. *CIRP Annals-Manufacturing Technology, 68*(2), 727–750. doi:10.1016/j.cirp.2019.05.010

Tovma, N. (2018). Indicator of Digital Economy. *Vision 2025: Sustainable Economic Development and Application of Innovation Management*, 5449-5454.

Treacy, M., & Wiersema, F. (1993). Customer intimacy and other value disciplines. *Harvard Business Review.*

Udaltsova, N. L. (2020). Digital Transformation of Economy. *Quality - Access to Success, 21*(175), 162–165.

Uhl, A., Born, M., Koschmider, A., & Janasz, T. (2014). *Digital Capability Framework: A Toolset to Become a Digital Enterprise, Digital Enterprise Transformation.* Routledge, Taylor and Francis.

Van Alstyne, M. W., Parker, G. G., & Choudary, S. P. (2016). Pipelines, platforms, and the new rules of strategy. *Harvard Business Review.*

Van den Heuvel, W.-J., & Tamburi, D. A. (2020). Model-Driven ML-Ops for Intelligent Enterprise Applications: Vision, Approaches and Challenges. *International Symposium on Business Modeling and Software Design, BMSD 2020: Business Modeling and Software Design*, 169-181.

Van Gils, B., & Weigand, H. (2020). Towards Sustainable Digital Transformation. *IEEE 22nd Conference on Business Informatics*, 104-113. 10.1109/CBI49978.2020.00019

van Tonder, C., Schachtebeck, C., Nieuwenhuizen, C., & Bossink, B. (2020). A Framework for Digital Transformation and Business Model Innovation. *Management, 25*(2), 111–132. doi:10.30924/mjcmi.25.2.6

Vaska, S., Massaro, M., Bagarotto, E. M., & Dal Mas, F. (2021). The Digital Transformation of Business Model Innovation: A Structured Literature Review. *Frontiers in Psychology, 11*.

Vial, G. (2019). Understanding digital transformation: A review and a research agenda. *The Journal of Strategic Information Systems*, 28(2), 118–144. doi:10.1016/j.jsis.2019.01.003

Vorobieva, D., Kefeli, I., Kolbanev, M., & Shamin, A. (2018). Architecture of Digital Economy. *10th International Congress on Ultra Modern Telecommunications and Control Systems and Workshops (ICUMT)* - Emerging Technologies for Connected Society.

Weerawardena, J., Salunke, S., Haigh, N., & Mort, G. S. (2021). Business model innovation in social purpose organizations: Conceptualizing dual social-economic value creation. *Journal of Business Research*, *125*, 762–771. doi:10.1016/j.jbusres.2019.10.016

Westerman, G., Bonnet, D., & Mc Afee, A. (2011). *Digital Transformation: a Roadmap for Billion-Dollar Organizations*. Harvard Business School Press.

Westerman, G., Bonnet, D., & Mc Afee, A. (2014). The Nine Elements of Digital Business Transformation. *MIT Sloan Management Review*.

Westerman, G., Bonnet, D., & Mc Afee, A. (2015). Leading Digital: Turning Technology into Business Transformation. Harvard Business School Press.

Wu, D. J., Sun, Y., & Zhong, F. (2010). Organizational Agent Systems for Intelligent Enterprise Modelling. Electronic Markets, 10(4).

Xu, J. (2014). Digital Enterprise Strategy Planning and Implementation. In *Managing Digital Enterprise*. Atlantis Press. doi:10.2991/978-94-6239-094-2_3

Yang, Z., Zhou, X., & Zhang, P. (2015). Discipline versus passion: Collectivism, centralization, and ambidextrous innovation. *Asia Pacific Journal of Management*, *32*(3), 745–769. doi:10.100710490-014-9396-6

Yongbo, S., Jingyan, L., & Yixin, D. (2020). Analysis of the relationship between open innovation, knowledge management capability and dual innovation. *Technology Analysis and Strategic Management*, *32*(1), 15–28. doi:10.1080/09537325.2019.1632431

Zaistev, V., E. (2019). Digital Economy as a Research Object: a Literature Review. *Voprosy Gosudarstvennogo i Munitsiplanogo Upravleniya – Public Administration Issues, 3*, 107-122.

Zaychenko, I., Gorshechnikova, P., Dubgorn, A., & Levina, A. (2020). Digital Transformation of Business: Approaches and Definitions. In Education Excellence and Innovation Management: A 2025 Vision to Sustain Economic Development During Global Challenges. International Business Information Management Association (IBIMA).

Zhang, Q., Lu, X. N., Peng, Z. L., & Ren, M. L. (2019). Perspective: A review of lifecycle management research on complex products in smart-connected environments. *International Journal of Production Research, 57*(21), 6758–6779. doi:10.1080/00207543.2019.1587186

Zhao, J. (2020). Knowledge management capability and technology uncertainty: Driving factors of dual innovation. *Technology Analysis and Strategic Management, 33*(7), 783–796. doi:10.1080/09537325.2020.1841896

Zhao, J. (2021). Dual innovation: The road to sustainable development of enterprises. *International Journal of Innovation Science, 13*(4), 423–436. Advance online publication. doi:10.1108/IJIS-07-2020-0096

Zhao, Y. (2020). *A Study of the Influence of Dual Innovation on Value Co-Creation under Enterprise Innovation Ecosystem.* Contemporary Finance & Economics. http://cfejxufe.magtech.com.cn/ddcj/EN/

Zheng, P., Lin, T. J., Chen, C. H., & Xu, X. (2018). A systematic design approach for service innovation of smart product-service systems. *Journal of Cleaner Production, 201,* 657–667. doi:10.1016/j.jclepro.2018.08.101

Zheng, P., Lin, Y., Chen, C. H., & Xu, X. (2019). Smart, connected open architecture product: An IT-driven co-creation paradigm with lifecycle personalization concerns. *International Journal of Production Research, 57*(8), 2571–2584. doi:10.1080/00207543.2018.1530475

Zheng, P., Wang, Z. X., & Chen, C. H. (2019). Smart Product-Service Systems: A Novel Transdisciplinary Sociotechnical Paradigm, Transdisciplinarity Engineering for Complex socio-Technical Systems. *Advances in Transdisciplinary Engineering, 10,* 234–241. doi:10.3233/ATDE190128

Zheng, P., Xu, X., & Chen, C.-H. (2018). A Data-Driven Cyber-Physical Approach for Personalised Smart, Connected Product Co-Development in a Cloud-Based Environment. *Journal of Intelligent Manufacturing,* 1–16.

Zheng, P., Xu, X., & Chen, C. H. (2020). A data-driven cyber-physical approach for personalised smart, connected product co-development in a cloud-based environment. *Journal of Intelligent Manufacturing, 31*(1), 3–18. doi:10.100710845-018-1430-y

Zheng, P., Xu, X., Yu, S., & Liu, C. (2017). Personalized Product Configuration Framework in an Adaptable Open Architecture Product Platform. *Journal of Manufacturing Systems, 43,* 422–435. doi:10.1016/j.jmsy.2017.03.010

Zimmermann, A., Schmidt, R., Jugel, D., & Mohring, M. (2015). Adaptive Enterprise Architecture for Digital Transformation. Celesti, A.; Leitner, P. Advances in Series Oriented and Cloud Computing (ESOCC 2015). *Communications in Computer and Information Science, 567,* 308–319. doi:10.1007/978-3-319-33313-7_24

Zimmermann, A., Schmidt, R., Jugel, D., & Möhring, M. (2015). Evolving enterprise architectures for digital transformations, In Digital Enterprise Computing (DEC 2015). Bonn: Gesellschaft für Informatik e.V.

Zimmermann, A., Schmidt, R., Jugel, D., & Möhring, M. (2016). Adaptive Enterprise Architecture for Digital Transformation. In A. Celesti & P. Leitner (Eds.), *Advances in Service-Oriented and Cloud Computing. ESOCC 2015. Communications in Computer and Information Science* (Vol. 567). Springer. doi:10.1007/978-3-319-33313-7_24

Zimmermann, A., Schmidt, R., Sandkuhl, K., Jugel, D., Bogner, J., & Mohring, M. (2018). Evolution of Enterprise Architecture for Digital Transformation. *2018 IEEE 22nd International Enterprise Distributed Object Computing Workshops (EDOCW 2018), IEEE International Enterprise Distributed Object Computing Conference Workshops-EDOCW*, 87-96. 10.1109/EDOCW.2018.00023

Zimmermann, A., Schmidt, R., Sandkuhl, K., Wissotzki, M., Jugel, D., & Mohring, M. (2015). Digital Enterprise Architecture - Transformation for the Internet of Things. *Proceedings of the 2015 IEEE 19th Internatonal Enterprise Distributed Object Computing Conference Workshops and Demonstrations (EDOCW 2015), IEEE International Enterprise Distributed Object Computing Conference Workshops-EDOCW*, 130-138. 10.1109/EDOCW.2015.16

Zimmermann, A., Schmidt, R., Sandkuhl, K., Wißotzki, M., Jugel, D., & Möhring, M. (2015). Digital Enterprise Architecture - Transformation for the Internet of Things. *2015 IEEE 19th International Enterprise Distributed Object Computing Workshop*, 130-138. 10.1109/EDOCW.2015.16

Chapter 1
Integrating a Holistic Enterprise Architecture Pattern:
A Proof of Concept

Antoine Trad

https://orcid.org/0000-0002-4199-6970

IBSITM, France

ABSTRACT

This chapter provides recommendations for the integration of various types of patterns. Complex layers of patterns can be assembled in a holistic enterprise architecture pattern (HEAP), which is integrated by using the integration of HEAP (IoHEAP), which can support agile transformation projects. The IoHEAP is based on a layered concept of patterns and an optimized deployment process. An integration patterns model can be used as a template to integrate solution building blocks (SBB), which can be used to implement a variety of types of transformation projects. In this chapter, the focus is on the integration of various pattern types that can be used to deliver adaptable SBBs. SBBs are the backbone of Enterprise architecture (EA)-based transformation projects.

INTRODUCTION

This chapter provides recommendations for the integration of various types of patterns. Complex layers of patterns can be assembled in a Holistic Enterprise Architecture Pattern (HEAP), which are integrated by using the Integration of HEAP (IoHEAP), that can support agile transformation projects. The IoHEAP is based on a layered concept of patterns and optimized deployment process. An integration patterns' model can be used as a template to integrate Solution Building Blocks (SBB), which can be used to implement a variety of types of transformation projects. In this article the focus is on the integration of various pattern types that can be used to deliver adaptable SBBs. SBBs are the backbone of Enterprise Architectures (EA) based transformation projects. An EA approach offers integration models, like, in other engineering domains, that can support complex integration phases. Complex integration activities

DOI: 10.4018/978-1-6684-4102-2.ch001

need a pattern, in order to link various SBBs, where the final solution delivers a system that respects the initial requirements. These facts enable transformation projects to be automated.

RESEARCH AND DEVELOPMENT

The RDP flow acknowledged an important gap in the mentioned fields that are presented in the form of keywords, mainly due to the fact that the existing literature on transformation projects, their failure rates and on various methodologies treating business transformations, offer practically no insight into: 1) An optimal transformation pattern, like the HEAP; and 2) The profile of the Project manager as an Architect of Adaptive Business Information Systems (AofABIS), who is the principal architect and designer. Such a holistic pattern and profile should be capable to setup a business EA integration concept to facilitate the implementation phase of Projects. Previous research models confirmed the mentioned profile, that is based on the AofABIS profile, and produced a set of managerial recommendations and related factors (SAP, 2013; Trad, & Kalpić, 2016a). This RDP uses a PoC to prove the Research Question (RQ) and the related hypothesis, establish viability, show technical issues and show overall direction. To some extent it is similar to the prototyping in engineering thesis works (Camarinha-Matos, 2012), where the prototype proofs RQ's feasibility.

Research Question

This article RQ is: *Can* IoHEAP *support complex transformation projects, and more specifically in its implementation phase?* Where HEAP is the backbone and background for building dynamic and flexible EAs that is based on various patterns' structures. This chapter is strongly related to the *Business Transformation Projects based on a Holistic Enterprise Architecture Pattern (HEAP)-The Basic Construction* chapter (Trad, & Kalpić, 2022). It mainly tries to integrate various types of pattern structures.

PATTERN'S STRUCTURE

Existing Standards and the Unbundling Process

The evolution of technology standards, like dynamic linking of components, made basic patterns receptive to development and integration with design patterns and EA. Before using complex structures and standards, there is a need to unbundle and refine the legacy ICS.

Unbundling and Refinement Predispositions

Figure 1. Major pre-patterns standards (Störrle, & Knapp, 2006)

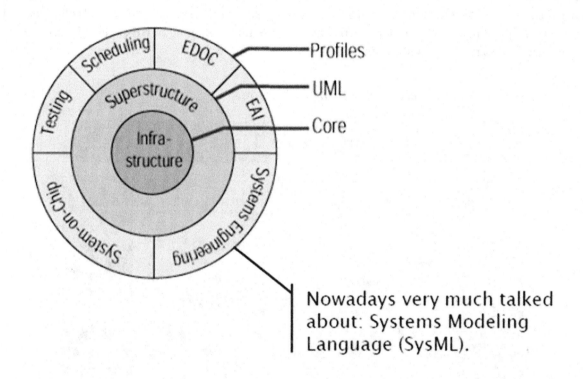

Before introducing how IoHEAP driven Projects are to be done, it is important to prepare the Entity of a concept for using the basic artefacts, like patterns, BBs, classes, composite entities, tables or services to transform the ICS by applying the Refinement Process to Extract Patterns (RP2EP). The RP2EP will support the agile and autonomic Project's components implementation phase. That needs a precise EA mapping concept that must use a standard framework (Farhoomand, 2004). Projects have limited resources and one has to establish an iterative Development and Operations (DevOps) or a Secured De-vOps (SecDevOps) model. To support HEAP, SecDevOps and RP2EP, the use of a pseudo bottom-up approach is recommended (Trad, & Kalpić, 2016a; Störrle, & Knapp, 2006); where they are all based on OO concepts and the notion of unit of work.

The Unit of work

In order to define an optimal unit of work for a basic pattern, it is needed to align all the Project's resources using a holistic 1:1 mapping concept. The authors use a concrete artefact that is an object-oriented class artefact, which is represented by a class-diagram using the extensible XML. Such a mapping concept is based on XML and on the class artefact insure the interoperability between all the Project's resources, which in turn are basic patterns (Trad, & Kalpić, 2016a; Downey, 2003).

Basic Patterns

As shown in Figure 2, this phase's main technology blocks are: 1) the basic patterns that serve as a preparation and classification phase for design patterns; and they are in focus of this article; 2) the standard design patterns; and 3) the enterprise patterns. The Project offers: 1) a real work business and technical transformation framework in the form of a set of reusable patterns, recommendations and solutions; and 2) the corresponding set of Project's patterns' architecture for the making of BBs to enable fast SecDevOps iterations (Trad, & Kalpić, 2016a).

Figure 2. This RDP's main blocks (Trad, & Kalpić, 2016a)

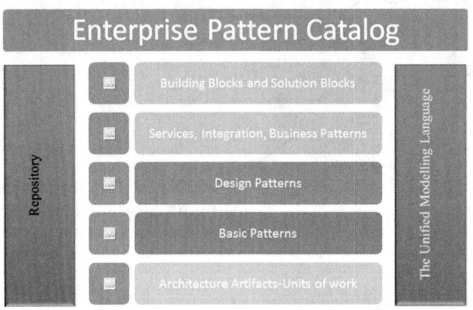

IoHEAP interacts with major patterns categories, like: 1) Low Level Patterns; 2) Generic Patterns; 3) Architecture and Design Patterns; 4) Design Patterns; 5) Enterprise Patterns; 6) Enterprise Architecture Patterns. A set of EA patterns serve to build an enterprise pattern that can serve as a Common Denominator Patterns (CDP) for EA (CDP4EA) or reference model(s). All these pattern types are used to build reference models.

Reference Models

TOGAF's technical reference model offers an interface to manage entries, like an enterprise pattern. Enterprise patterns are mainly a part of the modelling component, as shown in Figure 2. To integrate enterprise patterns, the Project must use existing standards and methodologies (Crosswell, 2014). HEAP will try to find analogous concepts and terminology, and offer a re-usable holistic pattern, that is a composite model of Related Patterns.

Related Patterns

As we have many types of patterns that tend to work together, there is a need to define the types of relations which they can have. Patterns' relationships use OO relationship types. Relationships interconnect patterns, creating a virtual composite pattern (or HEAP), indicating how the DMS solves types of Problems. Related patterns to implement the HEAP need a well-designed Enterprise Meta Model (EMM), what also ensures the Project's integrity.

The Enterprise Meta-Model

A Virtual Meta Model (VMM) is UML's expression of a formal model with a defined set of UML extensions. The EMM is a VMM variant and uses a modelling language (Dong, & Yang, 2003). The Design Pattern Modelling Language (DPML) is a notation that supports the specification of DP solutions (or ABBs) and their instantiation into UML models (or SBBs). DPML provides constructs which allow DP solutions to be modelled and integrated. DPs are described using a mixture of natural language and UML style diagrams, this causes complex scenarios in integrating DPs in the ICS. A DPML models DP to support ABBs, SBBs, aBBs, sBBs and CDPs (Mapelsden, Hosking, & Grundy, 2002). The Core of the EMM has layers representing the areas of EA to be included and rows representing the levels of abstraction, or views. This article's goal is not to present one more time the various types of patterns, but to show the optimal manner to do the preparations needed to integrate patterns in a coherent Project architecture that would use also (Trad, & Kalpić, 2016a). The Project identifies the set of main patterns and related patterns and assemble them in an Anti-Locked-in Strategy (ALS).

Figure 3. The EMM.

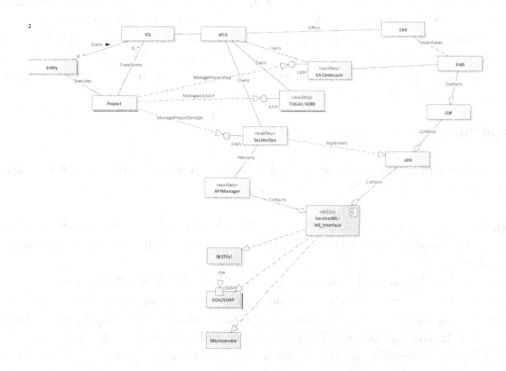

Because of this article's limitations Simple Object Access Protocol (SOAP) and BPMP will be analysed. The Project's and Entities classify and store patterns and BBs in an Architecture Continuum.

Patterns and the Architecture Continuum

Although ARCPs have not been integrated in standard EA methodologies, such as TOGAF, where in its first four main Architecture Development Method (ADM) Phases (Phases A to D), it gives a clear indication which resources should be used. There are used re-usable resources, like HEAP, which is managed by the EA Continuum, and contains the major CDP. An Entity that adopts a formal approach to apply ARCPs, must integrate them in their EA Continuum, to support EA Views and the ADM. These coordinated components and methodologies permit the usage of enterprise patterns.

ENTERPRISE PATTERNS

The Model View Controller (MVC) pattern is the most important and it offers interfaces for messaging and a related data model that serves as a messaging framework, used as an integration server. The messaging framework is essential for complex system integration (Fowler, 2003; Taleb, & Cherkaoui, 2012). Enterprise integration patterns are the base for building the EA patterns.

Enterprise Architecture Patterns

Enterprise Architecture Patterns (EAP) manage: 1) databases' concurrent access; 2) applications' user interface; and 3) legacy system' transformations. The EAP set includes the: 1) Domain Logic Patterns; 2) Data Source Architectural Patterns; 3) Object Relational Behavioral Patterns; 4) Object-Relational Structural Patterns; 5) Object-Relational Metadata Mapping Patterns; 6) Web Presentation Patterns; 7) Distribution Patterns; 8) Offline Concurrency Patterns; 9) Session State Patterns; and 10) Base Patterns. A set of EA patterns serve to build an enterprise pattern that can serve as a Common Denominator Patterns (CDP) for EA (CDP4EA) or reference model(s). EAPs are the basis of Patterns of Enterprise Application Architecture (EAA) and Enterprise Design Pattens (EDP).

Enterprise Application Archietcure Patterns

EAA is a set of pattern groups which can be used to design enterprise activities; and these groups are (Fowler, Rice, Foemmel, Hieatt, Mee, & Stafford, 2002):

- Domain Logic Patterns: Transaction Script, Domain Model, Table Module, and Service Layer.
- Data Source Architectural Patterns: Table Data Gateway, Row Data Gateway, Active Record, and Data Mapper.
- Object-Relational Behavioral Patterns: Unit of Work, Identity Map, and Lazy Load.
- Object-Relational Structural Patterns: Identity Field, Foreign Key Mapping, Association Table Mapping, Dependent Mapping, Embedded Value, Serialized LOB, Single Table Inheritance, Class Table Inheritance, Concrete Table Inheritance, and Inheritance Mappers.
- Object-Relational Metadata Mapping Patterns: Metadata Mapping, Query Object, and Repository.

- Web Presentation Patterns: Model View Controller, Page Controller, Front Controller, Template View, Transform View, and Two Step View.
- Distribution Patterns: Remote Façade, and Data Transfer Object.
- Offline Concurrency Patterns: Optimistic Offline Lock, Pessimistic Offline Lock, Coarse-Grained Lock, and Implicit Lock.
- Session State Patterns: Client Session State, Server Session State, and Database Session State.
- Base Patterns: Gateway, Mapper, Layer Supertype, Separated Interface, Registry, Value Object, Money, Special Case, Plugin, Service Stub, and Record Set.

Enterprise Design Patterns

EDP is a set of pattern groups which can be used to design enterprise activities; and these groups are (Goebl, Guenther, Klyver, & Papegaaij, 2020):

- Behavioral Patterns: Human Interest, Nurtured Trust, Powerful Questions, Listening to Understand, Hint, Tangible Presence, and Walking Your Talk.
- Practice Patterns: Evidence, Outside Inspiration, Hypotheses and Validation, Wearing Their Shoes, Dancing to Enterprise Rhythms, Corporate Politics, Focus, Shift, Refocus, Just Enough Design, and Unintended Consequences.
- Creations Patterns: Human Language, Captured Stories, Depicting Shared Understanding, Moments in time, Toolkits Sparking Change, Beauty, Tangible Futures, and Management Instruments.

The transformation of Entity's ICS(s) and its software components needs a specific pattern called the strangler pattern.

The Strangler Pattern

Instead of the very risky transform all, the best alternative is to use an agile approach based on the Io-HEAP. Such an approach recommends to observe the current component and create a component the old one; that iteratively replaces it. This approach offers an evolutive IoHEAP based plan that also controls possible problems, by reducing the risk that is associated with a brutal change. It also offers value back to the business by enabling a fast delivery of transformed features, until you the transformed component is mature and that it can replace the legacy one. Such an approach is well-known one; in fact, in 2014, it was hammered and designed about by Chris Stevenson and Andy Pols. Martin Fowler, who was inspired by The Strangler pattern, gave it this pattern the name: *The Strangler Application*. Where the basic idea is to break the legacy monolith into smaller parts, which requires precise preparations and a concise concept. The concept must support a smooth transformation process and the sustainable operation of the business, and in the same time processes incoming requirements. That needs that the Entity has the capability to transform legacy parts, by (Bocanett, 20222):

- Extracting code blocks: Frequently, this manner, which is based on copying bits of code around into the new component, is the default. This option can reintroduce bugs from the old system. It's also susceptible to the transformation effect, in which developers place high value in the former versions, which were not implemented by the same engineers, and where the reuse of code, has to be rewritten and refactored.

- Rewriting the capability: Initially, this manner can be considered as an expensive route compared to copying code, but the benefits of rewriting by capability offer an optimal Return On Investment (ROI). When rewriting, the IoHEAP delivery teams can question legacy assumptions, revisit the business problem, and improve the business process with an optimal approach. The code parts in legacy systems may not have been built on concise domain concepts or even obvious separation of concerns. The rewriting process offers a chance to correct and optimize the code parts.

Where in many Entities, a unit of code is responsible for specific operations, and that unit of code, can be a sensitive piece of the component that might be seen as a good one for extraction and eventual reuse. But in the case of rewriting, the IoHEAP team can revisit the process and consider the strategic objectives of the Entity. The IoHEAP team might replace the actual legacy part with an up to date architecture or even choose an external component; that enables new forms of activities for the Entity. If the legacy code is simple and performs a basic operation, has clear domain concepts, or had an important intellectual property value, in such a case, like in algorithms, most IoHEAP teams would opt just rewriting this part's capability. Breaking down a legacy monolith, requires a precise IoHEAP strategy, which will guide the transformation team, the implementation engineers, the team of architects, the business analysts, and other team members. This a very difficult process, because legacy monoliths have, in general no proper separation of concerns, no concise domain design, and in some cases, many technological weaknesses. By transforming the legacy monolith in terms of capability, it is possible to accomplish *business-value-oriented* implementation by prioritizing the requirements to extract resources, by using the strangler pattern, in such a way that it adds value to the business and balances important risks. For that goal there are various methods and strategies to gradually move the usage and to lift the capabilities and functions to the strangler pattern based application, by applying (Bocanett, 20222):

- Event tapping: known also as Event interception, in this strategy, whenever there are event-driven components or capabilities, there is the possibility to tap in to the stream of events; and then start to build or replace call-back functions for those events. This pattern also allows for building a parallel system for ensuring business continuity.
- Asset capture: In this case, every component manages a set of functional objects or assets, like, user accounts, transactions, historical records, or product orders. Transforming the capability of managing these assets independently into the strangler pattern based application. Such a smooth transformation is more of an art than a science, but in general, this strategy provides for a path to create services with clear domain concepts and responsibilities.
- Service bubbles: practically all applications and components are artefacts that consume a set of Application Programming Interfaces (API). Where today, the vast majority of ICSs accepts service-oriented concepts, from design to delivery. Therefore, the transformation team must explore chunking and refactoring the legacy monolith into a service-oriented concept.

The strangler pattern-based strategy examines the Entity's capabilities, to transform the legacy system, and to create small strangler-based services that encapsulate the logic of each capability in independent bubbles. Such an approach eventually leads to the creation of a mesh services, which support the entire set of capabilities of the previous legacy system. Finally, that would also lead to a IoHEAP services-based architecture, which is done by splitting software units into independent services that are organized around a business capability. By using such a strategy, the transformation team can chip away at the capabilities of the legacy system by migrating business capabilities into independent services. Knowing that the business value that is delivered is due to the prioritization and a proper mix of transformed logic and new requirements that deliver the ROI (Bocanett, 20222). The strangler pattern supports the integration of HEAP.

THE HEAP

Basics

HEAP can be used to abstract EA artefacts for Projects, where transformation engineers implement interconnected patterns. The complexity, diversity and cross-functional nature of EA and other Project's domains require that various categories of patterns should be developed and classified in various disciplines, domains, and levels of precision. The integration of various categories of patterns and their non-standardization make this topic lack of maturity (The Open Group, 2018).

Figure 4. The CDP types that are parts of a HEAP (Source: authors).

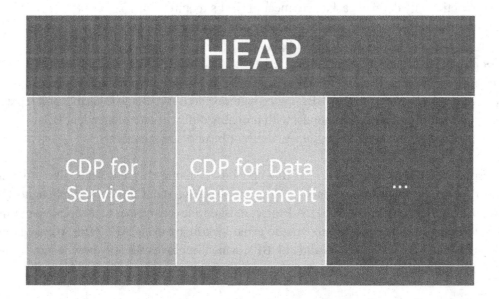

Common Denominator Patterns

The HEAP is a sum or sets of CDPs, where CDPs may have OO-like relationships; and it is in fact a pattern for integration of in-house and standard patterns. The HEAP contains:

- The CDP for Services (CDP4S).

- The CDP for Intelligence (CDP4I).
- The CDP for Knowledge (CDP4K).
- The CDP for Interfaces (CDP4I).
- The CDP for Data (CDP4D).
- …..

A Pool of Patterns

Many Entities are applying patterns to abstract their EAs at various levels ranging from software design patterns, business patterns to enterprise patterns. There is no single standard for describing HEAP, so this article can be considered as a pattern for abstracting existing major BBs and pattern categories related to IoHEAP.

IoHEAP, BBS AND METHODOLOGIES-INTEGRATION PROCESS

Generic Characteristics

Generic Building blocks (GBB) have the following generic characteristics (The Open Group, 2006a):

- It is a functionality defined package to meet Project's requirements.
- It has published interfaces to access the defined functionalities.
- It may interoperate with other Related BBs.
- The optimal BB has the following characteristics: 1) It facilitates implementation and maintenance to integrate ICS and related standards; 2) It may be assembled from other BBs, hence patterns; 3) It can be a subassembly of other BBs, hence patterns; and 4) A BB is re-usable and replaceable.
- It may have multiple implementations, with probably different inter-dependent BBs.
- It is a package of functionality (a library) used for business requirements.

One form of a GBB is the systemic BB that contains systemic characteristics, like error management, security, manageability, persistence… They are pervasive in all Project's components (Ramesh Radhakrishnan, & Radhakrishnan, 2004). A Project must define the manner to assemble patterns, functionalities, tools, and other artefacts into BBs; to avoid lock-in, by using ALS. The ALS is an Entity to define its HEAP and the way it implements its BBs, which improves the way how legacy systems are transformed, using a RP2EP.

Transformation through a Refinement Process

Entity's ICS are a collection of BBs, which are result of a RP2EP and the use of a standard set of patterns. BBs have to interoperate with other BBs, and it is important that their interfaces are stable. BBs can be defined at various levels of maturity, depending on the Project's evolution. In RP2EP early phases, a BB can be an interface to functionalities which use legacy components. BBs basics are defined in EA methodologies like TOGAF, as ABBs. As the Project advances, complex implementations replace these basic definitions of functionality, to become SBBs (The Open Group, 2006a).

It is recommended that an Entity develops its own version of the SBB, which is its Project's BB (PBB), which contains its ALS characteristics.

Architecture Building Blocks

ABBs relate to the Architecture Continuum and are managed by the ADM. Its main characteristics are to (The Open Group, 2006a): 1) Define the needed functionalities; 2) Capture Business and ICS requirements; 3) Make the Project technology aware; and 4) Manage SBBs' implement process. ABB's specifications include: 1) Fundamental functionalities and attributes; 2) Interfaces; 3) Mappings to Entity's strategy policies; and 4) Related BBs, like SBBs, with detailed information.

Solution Building Blocks

SBBs relate to the Solutions Continuum and may be either external or internal; and their main characteristics are to (The Open Group, 2006a): 1) Define which patterns will implement which set of functionalities; 2) Define pattern's implementation details; 3) Fulfil Project's business requirements; and 4) Be product- or vendor-aware, by applying the ALS. BB's specifications include: 1) Specific functionalities and attributes; 2) Interfaces; 3) Required SBBs; 3) Mapping of the used SBBs to ICS' topology and operational policies; 4) Specifications of attributes shared across the ICS; 4) Performance tuning; 5) Design drivers and constraints, including the physical architecture; and 6) Relationships between SBBs and ABBs, and the evolution of the SBB towards the PBB.

Project's Building Blocks

BBs extend the Project's EA concept, to become an PBB. A PBB supports the categorization of patterns to implement the needed transformed components. PBBs are a combination of software and platform patterns, like connectors which serve as a glue that connects various types of components. A PBB is a set of related BBs, used to put together a component and support a business service or an aBB (Ramesh Radhakrishnan, & Radhakrishnan, 2004).

Atomic Building Blocks Concept

Today's dynamic Entities have to struggle for survival, and they must be loosely interconnected in a global market. It is not a secret that a solid business environment that wants to ensure its sustainable business future must adapt itself to frequent Projects, to adapt to such a situation, aBB-based solution is proposed to support the Project's main artefacts like the ABB. Such an aBB-based strategy for frequent changes is translated in a set of solutions in the form of SBBs, supporting the continuous improvement of various business and ICS resources. Agile and loosely coupled ABBs can be used to improve the quality and success rate of the implementation and integration of the defined Project's requirements. That is achieved by simplifying and unifying of the used sets of applied DPs under HEAP's umbrella, which can be used for the Analysis, Design, Development, Tests, and Maintenance sub-phases. The optimal HEAP is based on the 1:1 mapping in which each requirement and its artefacts like aBBs/sBBs are totally independent. Standardized and simplified enterprise business architecture, enables the Project to become iterative, where its design is based on the ADM. The aBBs resources traverse through the ADM, where each phase refines the aBB's implementation's capability; such an approach uses a holistic view on the ICS that consists of: 1) A unified collection of aBBs, used to implement needed components, 2) aBBs based data and software components, and 3) Scalable technology infrastructure. The coordination of these main aICS parts is insured by the use of: 1) HEAP; 2) TOGAF/ADM and UML, and 3) Efficient tools. Managers use this concept to gain knowledge on how a Project can be managed; using aBBs, and sBBs. Such a Project has to make a choice of the optimal tooling and modelling environment based on a pseudo-Model View Control pattern. The complexity of the Project's implementation phase often causes the Project's failure and these failure rates are very high. The HEAP supports a cross-functional transformation process based on: 1) Requirements engineering; 2) Business Architecture; 3) BPMs; 4) SOA; 5) Entity's organizational structure (or organizational engineering); 6) the ICS' structure; and 7) Continuum (Trad, 2015a).

Atomic Building Blocks Structure

The aBBs comply with TOGAF's generic characteristics of BBs which have the following characteristics (The Open Group, 2011c):

- It is a package of requirements, functionalities and artefacts designed to meet Project's requirements.
- It has standardized interfaces.
- It is interoperable with other types of BBs and can be an aggregation of other aBBs.
- It defines the functionalities that will be implemented and Project's requirements.
- It ensures technology awareness and respect of standards.
- It can be used as a template to implement/instantiate sBBs.
- It is a reusable template and replaceable; that serves CDPs.
- It can have many implementations.
- It has a GUID and respects the 1:1 mapping concept.

Figure 5. BBs' management in ADM's phases (The Open Group, 2011c).

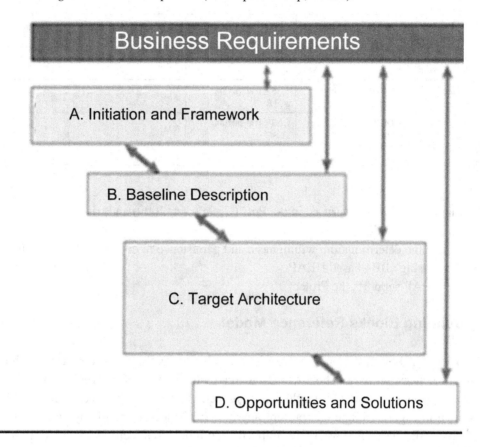

An aBBs is an architecture element, or a package of functionalities and resources designed to meet Project's transactions. The way artefacts, functionalities and development resources are combined in an aBBs might vary. The Project's team must coordinate, the design and prototype of aBBs, using the ADM's various phases, as shown in Figure 5; where these aBBs will transform the legacy components, facilitate integration and enable services' interoperability (The Open Group, 2011c). aBBs support the Project's unbundling of its monolithic environment by breaking the previous legacy components into a set of classified unique sets of services. *An aBBs is just another building brick in the Entity's wall...* The Project's team builds a PoC to define the needed sets of aBBs during the unbundling process.

Atomic Building Blocks Mapping Concept

The Project's Continuum is a bank of state-full atomic business artefacts. This is basically an alignment based on the 1:1 concept and where the aBBs are linked to various artifacts as shown in Table 1.

Table 1. The 1:1 mapping table

ID	Project artefact	Naming convention	Atomic Class	Interoperability
GUID_xxx	atomic Requirement	GUID_Transaction_aVSA	CLS_Transaction_aVSA	XMI
	atomic Contract	GUID_Transaction_aCTR	CLS _Transaction_aCTR	XMI
	atomic Use case(s)	GUID_Transaction_aUSC	CLS _Transaction_aUSC	XMI
	……..	…..	…..	…

aBB's 1:1 mapping concept is verified in the PoC (Greefhorst, 2009) and tries:

- To promote the determination, willingness, and persistence to complete the difficult implementation phase using aBBs-based HEAP.
- That the HEAP supports the Project.

Atomic Building Blocks Reference Model

The aBB structure is based on TOGAF's foundation architecture, that in turn is based on the foundation architecture of generic services. This foundation architecture is part of the Technical Reference Model (TRM), which provides a model for implementing generic platform of services. The TRM is interoperable, as shown in Figure 18, and can present any type of Project, to build any variant of architecture artifact(s), like aBBs. The TRM offers two important common architectural objectives to implement an atomic reference model(s) (The Open Group, 2011d):

- Artifact's portability, by using the Application Platform Interface, that identifies the set of aBBs, which are available for the HEAP.
- Interoperability, by using the Communications Infrastructure Interface, to identify the needed Communications Infrastructure services to support the Project.

Figure 6. Detailed Technical Reference Model (The Open Group, 2011d).

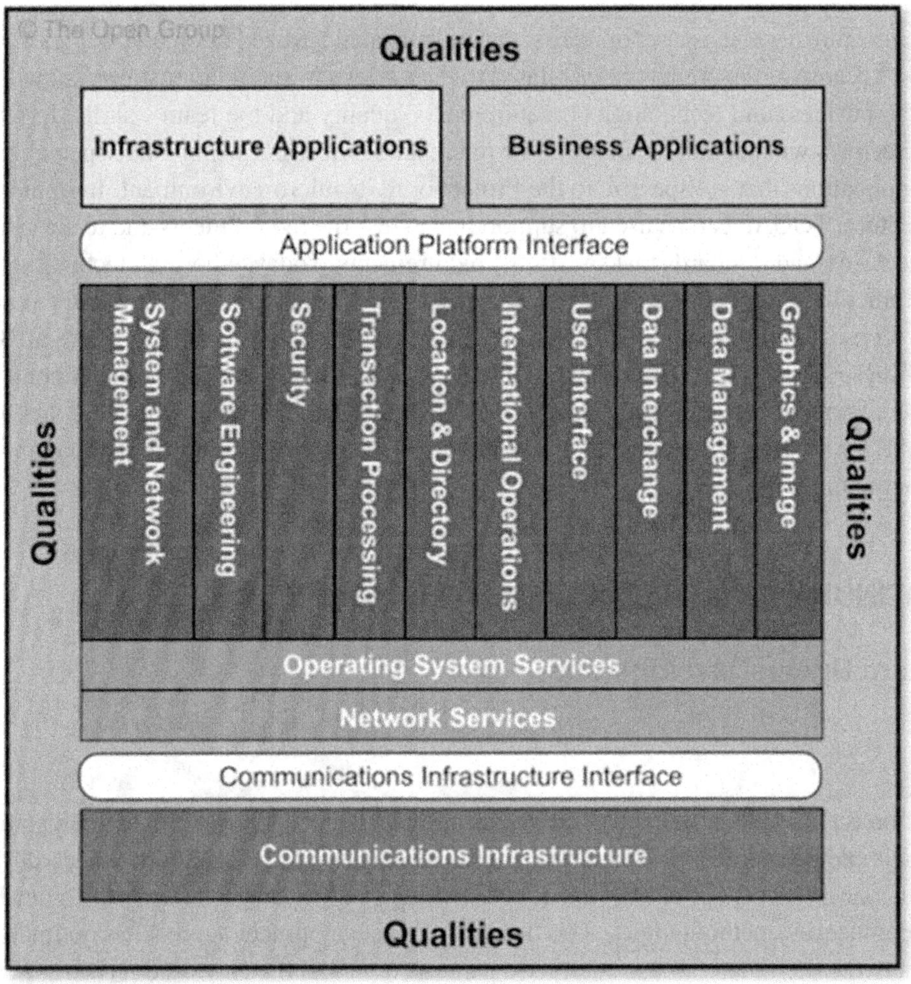

Atomic Transactions

The atomic Transactions are pseudo-sequential BPMs, where all the atomic business resources are dynamically linked. These atomic resources are generated by the related aBBs. From the Project's perspective, an atomic requirement is related to a single atomic Transaction (aTx) and its implementation is based on:

- Its relations with one or more ABB artefacts such as UML diagrams...
- Technically an aTx is composed of services, like SOA services.
- Independent applicative attributes: 1) Interaction; 2) Function; and 3) Self-description.
- Internal artefacts give its basic description.

Atomic Building Blocks and the ABB

Project's architectures, which is a set of ABBs, derive from standardized EA methodologies, where they differ greatly, because of the complexities related to their artefacts and relationships; knowing that they depend on the business and technological requirements' quality and the team's skills. In reality many Project architectures will not include many of the reference services but will include specialized services to support applications that are specific to the Project or to its micro-environment. In implementing an aBBs architecture, TOGAF can be used to support and assess the requirements and to select the needed services, interfaces, and standards that satisfy the requirements. Today, emerging technology trends are driving EA practitioners to support for Projects which are complex multi-disciplinary undertakings. The Project focuses on the creation of ABBs, which in turn are based on aBBs; where aBBs are a set of services. These services driven approach is about ensuring that the Project is under control and that it ensures the alignment between requirements, organizational (re)structure, governing and ICS; resulting in a HEAP for the actual Project. The Project's sets of ABBs, SBBs, and PBB, are managed by the Project's Continuum.

THE CONTINUUM AND THE IMPLEMENTATION LIFECYCLE

Architecture Driven Development

Tests

Transformation is about Projects that have large and complex legacy monolithic implementations where a test must be implemented after the finalization of components. Enterprise patterns need a design-driven development concept that is optimal for the transformation of silo systems that contain many blocks and use various engineering methodologies. This design driven development approach is optimal for holistic Projects, where the designers can use enterprise patterns to model the end system (Design Patterns).

The Interaction

The ADM's integration in Projects, using enterprise patterns, promotes the concept of an iterative assembling to integrate composite patterns. The usage of the enterprise patterns in the ADM's phases is done in the following phases[5]:

- In the vision phase, the concept of EA is defined.
- In the requirement's management phase, the link between the requirements items and the enterprise pattern is done.
- In the business architecture phase, the EA instance is built on business process patterns.
- In the change governance phase, the enterprise pattern is verified, and modifications are proposed.

The Design Driven Development

This Design Driven Development (DDD) that is also known as a model first design approach is based on best practices and it handles all types of requirements with the same importance (Escalona, & Koch, 2004). This design approach is optimal for holistic projects like a Project, where the designers can concentrate on building transcendent patterns (Trad, & Kalpić, 2016a).

The Tests Cycle

Types of tests (Trad, & Kalpić, 2016a):

- Test Driven Development (TDD) is a manual approach and a concept where implementers design the test first and then do the implementation. TDD is addressed as a testing approach that supports the evaluation of the design and verifies its quality. TDD privileges to design the classes to be tested, where it validates the quality of these classes in every iteration. For Projects, the implementation phase is the most important phase and the TDD offers a maintainable design concept that returns the needed statuses which help in assessing the Project's iterations' outcomes (Aniche, & Gerosa, 2010).
- Acceptance Test Driven Development (ATDD), like TDDs, ATDD design approach is also based on creating tests where tests represent the outcomes of the behaviour of various components. In the ATDD's design approach, the Project team creates acceptance level tests, then they implement the code. Afterwards, these test outcomes are verified to improve the Project's implementation quality.
- In Behaviour Driven Development (BDD) approach, acceptance tests offer the jumpstart for the Project and serve as a concept for communication between the Manager, the business environment's client, and the Project's team members. In such an agile approach, acceptance tests are implemented in prose so that non-technical team-members and clients can understand the complex implementation constructs. There is an auto-generated mapping mechanism from prose to implementation code. The prose scenarios provide enough information to automate the extraction of the needed classes and the Project's team-members suggest which code pieces can be extracted from the prose. The BDD enables a semi-automatic acceptance tests integration with the architecture iterations (Soeken, Wille, & Drechsler, 2012).

The Architecture Continuum

The Entity's or Project's Continuum of architecture artefacts includes ABBs, SBBs, and other models, important to the construction of an EA paradigm. The Continuum contains different types of existing architectures as shown in Figure 19; and the Foundation Architectures, like TOGAF and common ICS architectures, and domain (or industry)-specific architectures. The arrows, in Figure 7, represent a bi-directional relationship that exists between the different architecture artefacts and the Continuum. Organization Architectures focuses on supporting Entity's requirements, while the Foundation Architectures focuses on leveraging ABBs and other types of BBs (The Open Group, 2006a, 2006b).

Figure 7. The Architecture Continuum (The Open Group, 2006a, 2006b)

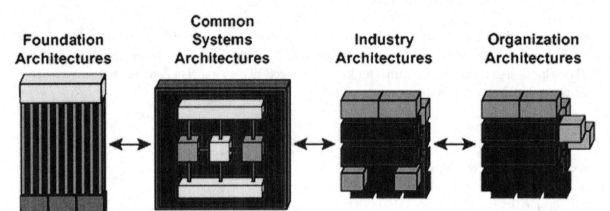

The Solution and Project Continuum

The Solutions Continuum contains the implementations of BBs at the corresponding levels of the Project. At each level, the Solutions Continuum is a population of the architecture with reference BBs (internal or external) that abstract a solution to the level's requirements. A populated Solutions Continuum can be regarded as a solutions warehouse, which can add performance value to the transformation initiative. The Solutions Continuum is illustrated in The Solutions Continuum in Figure 8 (The Open Group, 2006a, 2006b).

Figure 8. The Solutions Continuum (The Open Group, 2006a, 2006b)

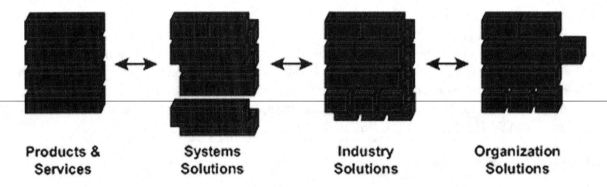

In Architecture Continuum, the arrows represent bi-directional relationships to relate different architectures. A similar concept exists in the Solutions Continuum where the rightwards arrows focus on providing solutions. The leftwards direction focuses on addressing Project's requirements. The two viewpoints are important to optimize the use of internal resources through a leverage process and the ADM.

DevOps and the ADM

BBs' design is an iterative ADM process, mainly in its Phases A, B, C, and D; where the constraints are imposed on the architecture, and the Project's feasibility one the selected BBs. Initially the needed ABBs are selected to meet the goals and objectives, then they are refined, applying an iterative process to generate set(s) of SBBs. The integration of BBs using the ADM is a transformational and iterative process. The main phases and steps are shown in Figure 9.

Figure 9. ADM Phases for BBs' implementation (The Open Group, 2006a).

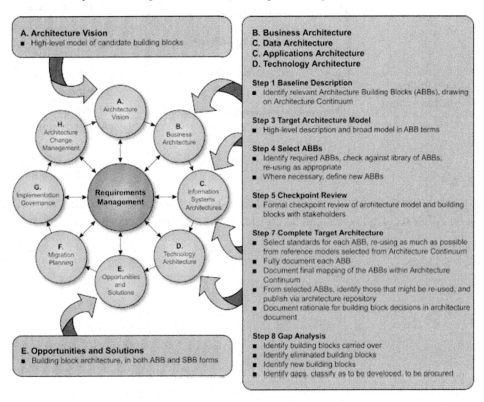

In Phase A, the first BB design starts as an abstract body within the Architecture Vision. In phases B, C, and D, BBs are implemented in the Business, Data, Applications, and Technology Architectures (The Open Group, 2006a, 2006b). ADM can be synchronized with a Secure DevOps (SecDevOps) implementation; which need strong security concepts.

1. SECURITY CONCEPTS

1.1 Atomic Security

Figure 10. API's Structure (Sheikh, 2020).

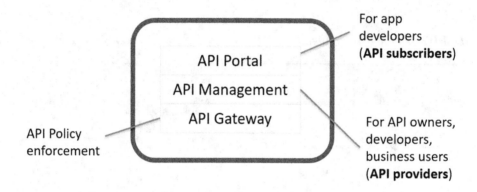

Describing the security of an Entity or a Project needs a few compendiums and therefore in this article the scope is on API's security, which can be considered as an atomic Security concept.

Figure 11. Security Management (Sheikh, 2020).

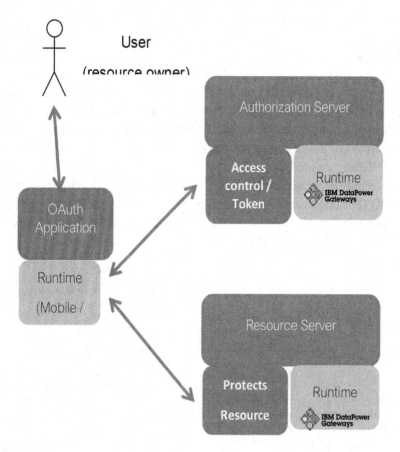

In Figure 11, the structure is fairly simple, and its security managed as shown in Figure 23 (Sheikh, 2020):

- Open Authorization (OAuth) allows the sharing of Entity's resources with external delimiters without sharing internal credentials. It uses highly secure access tokens, exchange of which does not compromise internal credentials, and uses a lightweight format compared to other secure token formats.
- API Gateway Capabilities, which comprises Authorization Server, are used to validate resource owner credentials. They integrate customizable OAuth flow with specialized scripts to meet Project's requirements.
- The use of OpenID Connect, which is a flexible user authentication for Single Sign-On (SSO) to API workloads using Entity's identities.
- JavaScript Object Notation (JSON) security provides enhanced message-level security for mobile, API, and webservices workloads; and it also provides end-to-end security between Mobile application and System of Record applications.
- A secure API Gateway exposes Entity assets to accelerate the transformation process. Where HEAP supports solutions to address common API security challenges.

API security features such as OAuth, OpenID Connect & JWT that provide authentication, authorization, and message-level security to support the Project.

Enterprise Security

Entities face a set of barriers and difficult situations, which need the management of Enterprise Security Risks (ESR), using a specialized framework to support their Projects. ESR may include CSFs related to reputation, routine operational procedures, legal and human resources management, financials, the risk of failure of internal controls systems related to the Sarbanes-Oxley Act (SOX) and global governance. The HEAP defines capabilities to protect the *Project* from attacks by 1) Localizing gaps in the infrastructures of partners; 2) Review of detection, and real-time security solutions; 3) Blocking of cumulative attacks; 4) Defining a security strategy to locate potential weaknesses; 5) Building a robust defence; 6) Integrating security in transactions; 7) Integrating an atomic Security concept; and 8) Applying qualification procedures (Clark, 2002).

Figure 12. Types of risks (Kiseleva, Karmanov, Korotkov, Kuznetsov, & Gasparian, 2018)

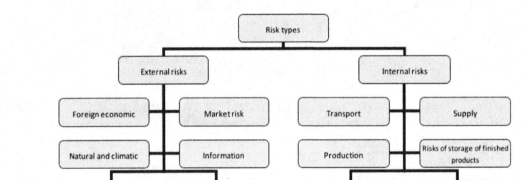

ESR' management integration is complex and needs massive use of tools and technologies to radically improve performance and ensure tangible benefits. Accounting-oriented management of ESR promotes off-shoring and ruthless growth. It can have a negative effect on *Projects* because it may promote confused and contradictory conclusions. Management of ESR is of strategic importance and if a *Project* is successful, the transformed *Entity* will excel. Transformed *Entities* with an efficient ESR management automate this management by using the *Framework,* which is in turn supported by the ADM. The *Entity* chooses a strategy to achieve its goals and tries to find ways to avoid ESR. Evaluation of ESR and the definition of the probability of hazardous events and the choice of solutions is specific to *Entity* and its eco-system. ESR are, in most cases, difficult to discover and classify, due to their diversity and complexity. There are various types of ESR that are related with each application domain. ESR's neutralization is a technical, financial, and mathematical process for the implementation of decisions for the transformation measures. The ESR's management structures ESR by using CSAs, weights them and uses delimiters to select the related CSFs. The ESR's management analyses the CSAs by applying scenarios for mitigation. ESR management system's key principles are: 1) Principle of integration using a systemic and holistic approach; 2) Principle of continuity using a set of procedures; 3) Robust atomic Security concept; and 4) Principle of validity. It provides an analysis of the ratio of costs to reduce possible ESR. Figure 2 shows an example of ESR classification that is used in economic practice (Kiseleva, Karmanov, Korotkov, Kuznetsov, & Gasparian, 2018). IoHEAP, service-oriented concepts depend on the role of various typologies.

THE ROLE OF TOPOLOGIES

Figure 13. Monolithic vs SOA vs Microservices (Kappagantula, 2018).

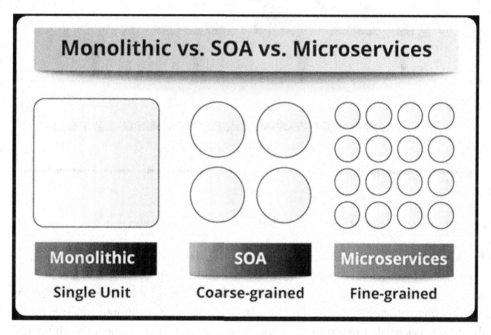

As shown in Figure 13, the scope is on layered-based architectures and the characteristics and differences between various typologies; mainly Microservices and SOA. But first there is the need to see the basic differences between the Monolithic architecture, SOA, and Microservices. In general terms, Monolithic technologies include Mainframe and other non-services technologies, which are huge containers of software modules, which are used by decentralized applications. A Service-Oriented Architecture is mainly a set of services. These services communicate with each other using various types of communication protocols. The communication can involve either simple data passing, or it can involve two or more services coordinating choreography activities. These interactions need connecting services that facilitates their communication. Microservice Architecture, is an architectural style that structures an application as a collection of micro autonomous services, modeled around a specific business domain (Kappagantula, 2018).

Microservices vs SOA

As these two categories are the major ones, when comparing Microservices vs SOA, they both rely on services as the main component, but they vary greatly in terms of service characteristics.

Service Oriented Architecture

Figure 14. Basic service types, Microservices vs SOA (Kappagantula, 2018).

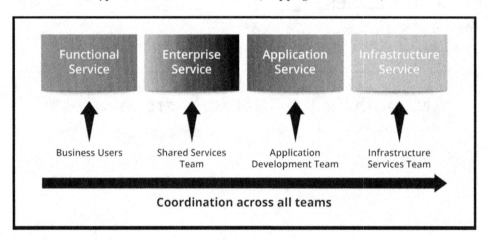

SOA defines four basic service types, as depicted below (Kappagantula, 2018):

- Business services: 1) They are coarse-grained services that define core business operations; and 2) Represented through XML, Business Process Execution Language (BPEL) and others.
- Enterprise services: 1) Implements the functionality defined by business services; and 2) They mainly rely on application services and infrastructure services to support business requests.
- Application services: 1) Fine-grained services that are confined to a specific application context; and 2) A dedicated user interface can directly invoke the services.
- Infrastructure services: 1) Implements non-functional tasks such as authentication, auditing, security, and logging; and 2) They can be invoked from either application services or enterprise services.

Microservices

As shown in Figure 15, MicroServices Architecture (MSA) is in fact, the natural evolution of SOA. At a high-level, MSA is an alternative method for architecting applications, which offers a better way to decouple ICS' components within an application boundary. If microservices were renamed as micro-components would have a better naming convention (Maguire, 2020).

Figure 15. Services of Microservices (Kappagantula, 2018).

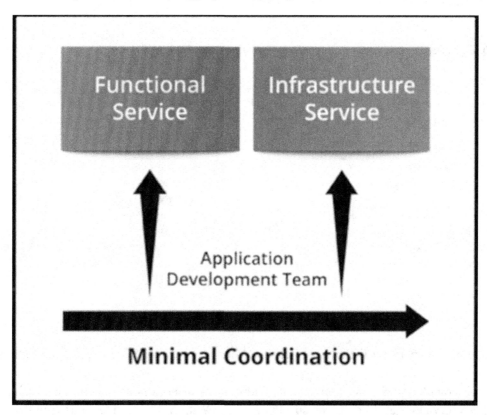

Microservices has a limited service taxonomy, and they consist of two service types as described below (Kappagantula, 2018):

- Functional services: 1) They support specific business operations; and 2) Support accessing of services, is done externally and these services are not shared with other services.
- Infrastructure services: 1) As in SOA, infrastructure services implement tasks such as auditing, security, and logging; and 2) Here, services are not unveiled to the outside world.

Major differences between SOA and MSA

Figure 16. Diffs between Microservices vs SOA (Kappagantula, 2018).

SOA	MSA
Follows "**share-as-much-as-possible**" architecture approach	Follows "**share-as-little-as-possible**" architecture approach
Importance is on **business functionality** reuse	Importance is on the concept of "**bounded context**"
They have **common governance** andstandards	They focus on **people collaboration** and freedom of other options
Uses **Enterprise Service bus (ESB)** for communication	Simple messaging system
They support **multiple message protocols**	They use **lightweight protocols** such as **HTTP/REST** etc.
Multi-threaded with more overheads to handle I/O	**Single-threaded** usually with the use of Event Loop features for non-locking I/O handling
Maximizes application service reusability	Focuses on **decoupling**
Traditional Relational Databases are more often used	**Modern Relational Databases** are more often used
A systematic change requires modifying the monolith	A systematic change is to create a new service
DevOps / Continuous Delivery is becoming popular, but not yet mainstream	Strong focus on DevOps / Continuous Delivery

As shown in Figure 16, major Differences Between Microservices and SOA, are (Kappagantula, 2018):

- Service granularity: Service components within a microservices architecture are mainly single-purpose services that do only one specific activity. With SOA, service components can vary in size anywhere from small application services to very large enterprise services. In fact, it is common to have a service component within SOA represented by a large application or even an ICS subsystem.
- Component sharing: is one of the most important characteristic of SOA. In fact, component sharing is the essence of enterprise services' concept. SOA supports the evolution of component sharing, whereas MSA minimizes on sharing through the mechanics of *bounded context*; where a *bounded context* refers to the coupling of a component and its data as a single domain specific unit with minimal dependencies' level. SOA relies on multiple services to fulfill a business requirement or request, but there is no proof that SOA based systems are slower than MSA ones.
- Middleware vs API layer: the MSA pattern, normally contains an API layer, whereas SOA has a messaging middleware component, which at the end comes to the same. The messaging middleware in SOA offers a hub of additional services' capabilities, which is not supported by MSA, which includes: 1) Mediation and routing; 2) Message enhancement; 4) Message; and 5) Protocol transformation. Whereas MSA has an API layer between services and service consumers as shown in Figure 17.
- Remote services: SOA architectures rely on messaging protocols, like: 1) AMQP/MSMQ); and 2) SOAP; as primary remote access protocols. Whereas most MSAs rely on two protocols: 1) REST; 2) Simple messaging (JMS, MSMQ); and 3) The protocol used in MSA is usually homogeneous.
- Heterogeneous interoperability: SOA promotes the use of multiple heterogeneous protocols through its messaging middleware component. MSA attempts to simplify the architecture and design patterns by reducing the hairball of choices for an optimal integration. If the Project needs to

integrate several end-systems that have different protocols, in a heterogeneous environment, then SOA is the most optimal solution. Otherwise, if the end-system's services could be exposed and accessed through a unique remote access protocol, then MSA is optimal.

Figure 17. Differences between Microservice Architecture and SOA (Kappagantula, 2018).

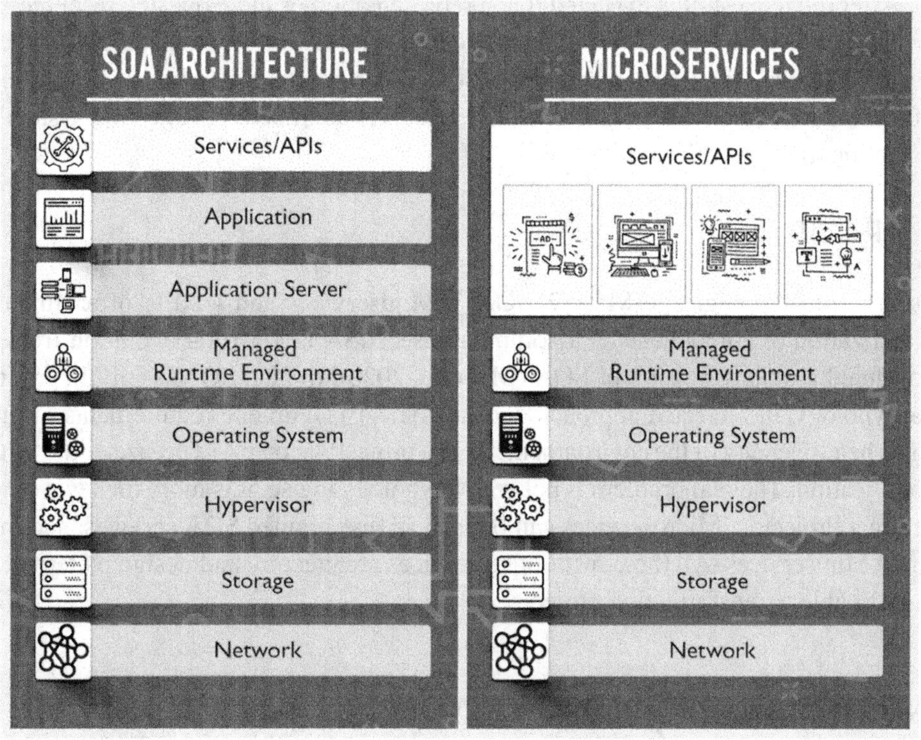

It is a complex to define which architecture and design pattern(s) is optimal for a given end-system. It mainly depends on the define architecture principles of the end-system, to be built or transformed. SOA is optimal for large and complex end-system environments that require integration with many heterogeneous applications, components, platforms… Whereas as simpler and smaller end-systems are not optimal for SOA, because basically they don't need a messaging middleware component. MSA, on the other hand, optimal for smaller and well-defined, web-based systems in which microservices enables greater control on the design, implementation and deployment processes. So, it is recommended- that as they both have different architecture and design characteristics, the Project team should make the choice depending on the purpose of the end-system to be transformed (Kappagantula, 2018).

REST

RESTful web services are services built according to REST principles. The idea is to have them designed to essentially work well on the web. But, what is REST? Let's start from the beginning by defining REST.

It is important to say that REST is an architectural style and not a toolkit. REST provides a set of design rules in order to create stateless services that are shown as resources and, in some cases, sources of specific information such as data and functionality. The identification of each resource is performed by its unique Uniform Resource Identifier (URI). REST describes simple interfaces that transmit data over a standardized interface such as HTTP and HTTPS without any additional messaging layer, such as SOAP. The consumer will access REST resources via a URI using HTTP methods (this will be explained in more detail later). After the request, it is expected that a representation of the requested resource is returned. The representation of any resource is, in general, a document that reflects the current or intended state of the requested resource. The REST architectural style describes the following six constraints (Lahoti, 2019): Uniform interface, Stateless, Cacheable, Client-server architecture, A layered system, and Code on demand (optional).

A Mixed Topology

The main architectural typologies: 1) MSA; 2) SOA; 3) Miniservices; and 4) APIs; offer similar concepts, where modularization of components is a common base. MSA is a natural evolution from SOA, and MSA is considered as a light version of SOA (Maguire, 2020). In fact, SOA is still the most dominant architectural typology, the selected approach depends on: 1) The unique requirements of the software project; 2) The heterogeneity of the environment; 3) The project's scope; 4) The size of the Entity; 5) The domain of application. The real problem is not which typology to use, it is more the level of granularity to be applied in a Project…. Microservices can be seen as fine-grained SOA services, and in turn SOA as fine grained Miniservices. All these mentioned services architecture and design patterns can be used to support and enable a transformation project.

ENABLING THE TRANSFORMATION PROJECT

This section presents the roadmap that enables the Project to avoid major problems.

Performance Issues

One of the major problems that can faced by Projects applying various typologies based on a variety of patterns, is the problem of the end-system's performance. That can be faced by applying and iterative method like the ADM combined with DevOps; and upon each major iteration the performance of the end-system is evaluated. That can improved by using Digital Integration Hubs (DIH).

The Digital Integration Hub

As shown in Figure 18, a DIH is an advanced platform architecture that aggregates multiple back-end sub-systems and datastores, in a low-latency and shared/unique data store. The datastore caches and persists IoHEAP data sets dispersed across various siloed back-end databases. The DIH makes datastores available to the end-system's applications through high-performance APIs. Applications access the DIH, by using an API service layer and enables important performance improvements by requesting data from only one DIH interface to the distributed store (Apache Ignite, 2022).

Figure 18. The DIH interaction with various datastores (Apache Ignite, 2022).

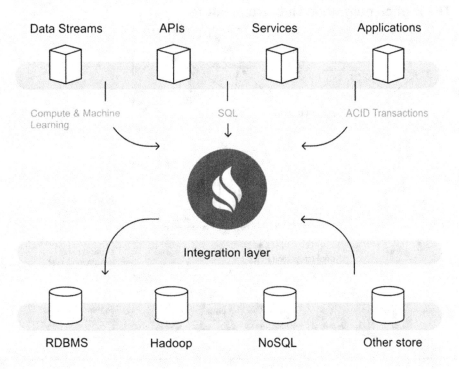

Gartner defines the DIH as an avant-garde application architecture that aggregates multiple back-end system of record datastores, int a low-latency and scale-out, high-performance datastore. A DIH typically supports access to data via an API services layer. The high-performance data store is synchronized with the back-end sources via some combination of event-based, request-based, and batch integration patterns, as shown in Figure 19 (Pezzini, 2018).

Figure 19. Gartner's view on DIH (Pezzini, 2018).

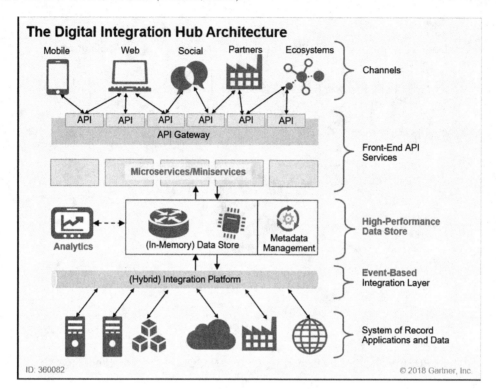

THE POC

Preparing aBBs

The proposed solution and structure are evaluated through the implementation of a real-world business transaction with the concept of aBBs, where we see the class diagram in Figure 20, proved to be applicable for Projects, and also that the granularity approach can be used to refine the 1:1 mapping.

Figure 20. The prototype's atomic business transaction

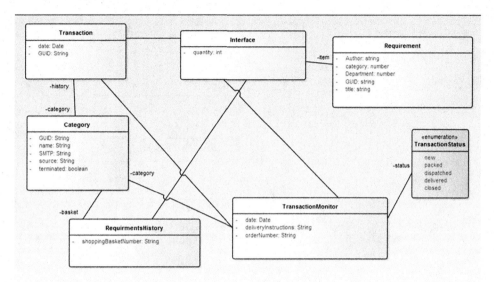

A logical view of a series of transactions based on the Service Oriented Architecture-type approach was used as shown in Figure 21, and the consumption of an atomic web service in a single transaction. From the atomic business transaction activity diagram, the resilience of events exchanged during the transaction's execution is important. All the events that are exchanged between various nodes require a strong encryption setting which is defined in the technology architecture phase. From a technological perspective, the atomic business transaction is composed of application components which are the fundamental business entities of the system, as presented in Table 1. A top-down combination of TOGAF's phases B and D resulted in the optimal construction of a transaction, based on an atomic web services approach.

Figure 21. The PoC's aTx.

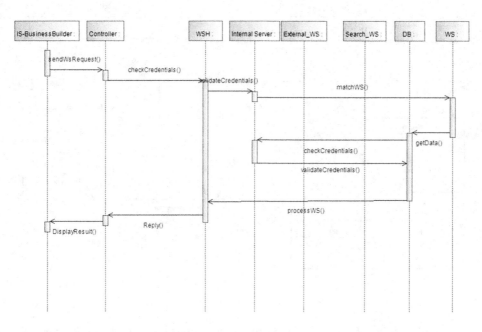

aBB's Generator

This aBB, as shown in Figure 22, generates the needed lowest levels of BBs. That corresponds to the integration of basic and design patterns.

Figure 22. The aBB generator interface.

Adding an aTx

Table 2. The aTx component.

ICS Builder	Provide business artifacts
Controller	Passes web service request
Search_WS	Web service looking for requested web services
Data Component	Contains sensitive information

This PoC assesses the steps for the aICS's integration and has the following requests: 1) to promote the optimal aICS's strategy, vision, willingness, and persistence to finalize the difficult implementation phase; and 2) to prove the business benefits of a Project. The PoC was evaluated through the implementation of an application using Microsoft's .Net and java development environments. The PoC proved that the aICS is feasible and that it is optimal for frequent Project iterations. The PoC is a business transaction that is based on an aBB's instance and is used to verify the aICS's concept. The concrete phases' artefacts, activities and tools are shown in table 2. (Trad, 2015a, 2015b).

Adding an aTx

Table 3. The PoC phases and tools relationship

Activity(s)	Architecture phase(s)	Tool(s) and standard(s)
A vision concept	phase A	UML/Archimate-Modellio
A generic GUID artefact and a 1:1 concept	phase A	WSDL
A transaction business use case	phases R/B/C	UC/BPM-Modellio
A basic business transaction in the form of BPM	phases B/C	BPM-Modellio
A set of business services	phases B/C	SoaML-Modellio
A corresponding set of modelling artefacts	all phases	UML/Archimate-Modellio
A tracing and identification pattern	phase F	Syslog+Client/SysML
A tracing platform and interface	phases F/D	Syslog/TcpMon/SoapUI-WS
A decision aggregation interface	phases F/E	syslog tool
A program plan generator	phase H	PMXML

This PoC serves to assert IoHEAP's base architecture model for implementing Projects

HEAP's Generation

The generated aBB, as shown in Figure 23, generates a business process management artefact in a standardized XML interchangeable format (XMI).

Figure 23. The business process management artefact in the interchangeable format.

```
<?xml version='1.0' encoding='UTF-8'?>

<xmi:XMI xmlns:uml='http://www.omg.org/spec/UML/20110701' xmlns:xmi='http://www.omg.org/spec/XMI/20110701' xmlns:BP
    <uml:Model xmi:type='uml:Model' xmi:id='_U999_' name='bpm4mcsrv3'>
        <packagedElement xmi:type='uml:Activity' xmi:id='_U000_' name='LaunchSrv'>

            <edge xmi:type='uml:ControlFlow' xmi:id='_U002_' visibility='public' source='_U005_' target='_U003_'>
                <weight xmi:type='uml:LiteralUnlimitedNatural' xmi:id='_U007_' value='1'/>
            </edge>

            <edge xmi:type='uml:ControlFlow' xmi:id='_U001_' visibility='public' source='_U003_' target='_U008_'>
                <weight xmi:type='uml:LiteralUnlimitedNatural' xmi:id='_U004_' value='2'/>
            </edge>

            <edge xmi:type='uml:ControlFlow' xmi:id='_U010_' visibility='public' source='_U008_' target='_U006_'>
                <weight xmi:type='uml:LiteralUnlimitedNatural' xmi:id='_U011_' value='2'/>
            </edge>

            <node xmi:type='uml:InitialNode' xmi:id='_U005_' visibility='public'>
                <outgoing xmi:idref='_U002_'/>
            </node>

            <node xmi:type='uml:OpaqueAction' xmi:id='_U003_' name='Act_001' visibility='public'>
                <incoming xmi:idref='_U002_'/>
                <outgoing xmi:idref='_U009_'/>
            </node>
            <node xmi:type='uml:OpaqueAction' xmi:id='_U008_' name='Act_002' visibility='public'>
                <incoming xmi:idref='_U009_'/>
                <outgoing xmi:idref='_U001_'/>
            </node>
            <node xmi:type='uml:ActivityFinalNode' xmi:id='_U006_' visibility='public'>
                <incoming xmi:idref='_U001_'/>
            </node>

        </packagedElement>
    </uml:Model>
</xmi:XMI>
```

The Integration Interface

Figure 24. BPM's artefact the interchangeable format

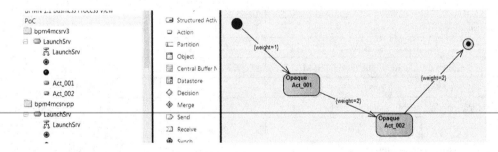

The generated XMI file, as shown in Figure 24, can be imported in an enterprise architect environment to present the end business choreography that is accessed through IoHEAP.

THE CONCLUSION AND RECOMMENDATION

The outcome of building avant-garde IoHEAP based Projects, heavily depends on the use of patterns and existing standards like the Open Group's architecture framework, where the use of generic patterns like aBBs and their solution blocks instances in transforming a traditional business environment into an agile environment simplifies and ensures the success of the implementation phase. The HEAP base Project interfaces various standards including, like: UML, SOA Markup Language (SoaML), System Markup Language (SysML), … The use of various standards, especially methodologies, depends on their maturity, like the case of GPs and UML, as they are practically integrated and automated in implementation environments. Legacy ICSs and teams are the major cause of failure for the Projects which are essential for businesses companies. This fact has motivated the authors to research various techniques to promote and recommend solutions, like the use of HEAP based architectures. DPs can be used for integrating various types of patterns, to support the IoHEAP, EA models and BPM modules (Microsoft, 2016a; Design Patterns). The RDP and the PoC propose a set of technical and managerial recommendations for the usage of IoHEAP; which can be used to model and analyse main system components of the future systems. This article presents the IoHEAP and the optimal manner to integrate it in a Project. The authors targeted the use of existing patterns, typologies and they forward the following set of recommendations:

- Define the Project's typology.
- To use the generic pattern's concept that incorporates: 1) Basic patterns; 2) Classification for patterns; and 3) Unbundling process, which dlivers the pool of services.
- A coherent IoHEAP supports a design-driven development or the design-first approach, or a mixture.
- The role of OO is essential for a IoHEAP based Project, especially in the implementation of components.
- The role of UML is central and serves as a common architecture language.
- To use GUID for Project's artefacts.
- For the data architecture, class diagrams can be mapped to table constructs where a class corresponds to a table or an OO object.
- Ensure the definition of the optimal unit of work that will align all the *Project*'s resources using the 1:1 concept.
- A 1:1 mapping concept must be applied to services, BBs, composite patterns.
- To use composite models like design patterns in HEAP.
- To apply an aBB architecture strategy and pattern that will facilitate the implementation phase of the Projects.
- The atomic approach supports a bottom-up approach that is based on the choreography of aBSs.
- aBSs are in turn built on atomic BPMs (IBM, 2014).
- DPs describe a common structure for a general design problem within a Project implementation; and supports agile Project's complexity in its implementation phase.
- Composite DPs incorporate: 1) Basic patterns; 2) Design patterns; and 3) Enterprise patterns.
- An ESB must be used as a communication highway and must impact on the composite patterns' architecture.
- Use Composite DPs to build a HEAP as a template for a Project.
- An Entity must build its own major sets of patterns to protect its business.
- Before the introduction of HEAP based architecture roadmap, the Project must prepare the environment to use basic patterns and DPs.
- To use the MVC pattern to synchronize the Project's activities.
- Apply the ADM's integration in Projects, using the HEAP, to promote the concept of an iterative assembling to integrate composite patterns.

- An aICS pattern is based on the aBBs and the 1:1 mapping approach. This fact simplifies the manipulation of the architecture artefacts, through the various phases of the ADM; where IoHEAP serves as the main interface.

REFERENCES

Aniche, M., & Gerosa, M. (2010). *How Test-Driven Development Influences Class Design: A Practitioner's Point of View. Department of Computer Science. University of Sao Paulo*. USP.

Apache Ignite. (2022). *Digital Integration Hub With Apache Ignite*. The Apache Software Foundation. https://ignite.apache.org/use-cases/digital-integration-hub.html

Arcitura (2020). *SOA Patterns*. Arcitura. https://patterns.arcitura.com/soa-patterns

Bocanett, W. (2022). *Break the monolith: Chunking strategy and the Strangler pattern-Build decoupled microservices to strangle your monolithic application*. IBM.

Buschmann, F., Meunier, R., Rohnert, H., Sommerlad, P., & Stal, M. (1996). *Pattern-Oriented Software Architecture: A System of Patterns*. Wiley.

Camarinha-Matos, L. M. (2012). *Scientific research methodologies and techniques- Unit 2: Scientific method. Unit 2: Scientific methodology. PhD program in electrical and computer engineering*. Uninova.

Capgemini. (2007). *Trends in Business transformation - Survey of European Executives*. Capgemini Consulting and The Economist Intelligence Unit.

Capgemini. (2009). *Business transformation: From crisis response to radical changes that will create tomorrow's business. A Capgemini Consulting survey*. Capgemini.

Clark, D. (2002). *Enterprise Security: The Manager's Defense Guide*. Addison-Wesley Professional.

Crosswell, A. (2014). *Bricks and the TOGAF TRM*. National Institute of Health.

Dong, J., & Yang, Sh. (2003). *Visualizing Design Patterns With A UML Profile. Department of Computer Science*. University of Texas.

Downey, K. (2003). *Architectural Design Patterns for XML Documents*. https://www.xml.com/pub/a/2003/03/26/patterns.html

Duvander, A. (2021). *API Design Patterns for API. Stoplight*. https://blog.stoplight.io/api-design-patterns-for-rest-web-services

Escalona, J., & Koch, N. (2004). *Requirements engineering for web applications – A comparative Study*. University of Seville.

Farhoomand, A. (2004). *Managing (e)business transformation*. Palgrave Macmillan. doi:10.1007/978-1-137-08380-7

Fowler, M., Rice, D., Foemmel, M., Hieatt, E., Mee, R., & Stafford, R. (2002). *Patterns of Enterprise Application Architecture*. Addison Wesley.

Goebl, W., Guenther, M., Klyver, A., & Papegaaij, B. (2020). Enterprise design patterns-35 ways to radically increase your impact on the enterprise. Intersection Group.

Greefhorst, D. (2009). *Using TOGAF as a pragmatic approach to architecture*. Informatica.

IBM. (2014). *Smart Service Oriented Architecture: Helping businesses restructure. Where are you on the Service Oriented Architecture adoption path?* IBM.

Kappagantula, S. (2018). *Microservices vs SOA — Battle Between The Top Architectures*. Edureka. https://medium.com/edureka/microservices-vs-soa-4d71c5590fc6

Kiseleva, I., Karmanov, M., Korotkov, A., Kuznetsov, V., & Gasparian, M. (2018). Risk management in business: Concept, types, evaluation criteria. *Revista ESPACIOS*.

Lahoti, S. (2019). *Defining REST and its various architectural styles*. packtpub. https://hub.packtpub.com/defining-rest-and-its-various-architectural-styles/

Maguire, J. (2020). *Microservices vs SOA vs API Comparison*. Devteam. https://www.devteam.space/blog/microservices-vs-soa-and-api-comparison/

Mapelsden, D., Hosking, J., & Grundy, J. (2002). *Design Pattern Modelling and Instantiation using DPML*. Department of Computer Science, University of Auckland.

Microsoft. (2016a). *Code Snippets for Design Patterns*. CodePlex.

Microsoft. (2016b). *Model-View-Controller*. https://msdn.microsoft.com/en-us/library/ff649643.aspx

Oracle. (2002). *Core J2EE Patterns - Data Access Object*. Oracle. https://www.oracle.com/technetwork/java/dataaccessobject-138824.html

Oracle. (2016a). *API Management in 2026*. Oracle.

Oracle. (2016b). *Evolution and Generations of API Management*. Oracle.

Perroud, T., & Inversini, R. (2013). *Enterprise Architecture Patterns*. Practical Solutions for Recurring IT-Architecture Problems. doi:10.1007/978-3-642-37561-3

Pezzini, M. (2018). *The Digital Integration Hub Turbocharges Your API Strategy*. LinkedIn. https://www.linkedin.com/pulse/digital-integration-hub-turbocharges-your-api-strategy-pezzini/

Ramesh Radhakrishnan, R., & Radhakrishnan, R. (2004). *IT Infrastructure Architecture-Building Blocks*. Sun Microsystem. https://www.opengroup.org/architecture/0404brus/papers/rakesh/abb-1.pdf

SAP. (2013). GBTM: Global Business Transformation Manager Master Certification (SAP Internal). Business Transformation Academy. SAP.

Sheikh, O. (2020). *Securing Your Digital Channels With API Gateways*. IBM.

Soeken, M., Wille, R., & Drechsler, R. (2012). Assisted Behavior Driven Development Using Natural Language Processing. Institute of Computer Science, University of Bremen Group of Computer Architecture. doi:10.1007/978-3-642-30561-0_19

Störrle, H., & Knapp, A. (2006). *UML 2.0 – Tutorial/Unified Modeling Language 2.0*. University of Innsbruck.

Taleb, M., & Cherkaoui, O. (2012, January). Pattern-Oriented Approach for Enterprise Architecture: TOGAF Framework. *Journal of Software Engineering & Applications*, 5(1), 45–50. doi:10.4236/jsea.2012.51008

The Open Group. (2006a). *Building Blocks*. The Open Group. https://pubs.opengroup.org/architecture/togaf8-doc/arch/chap 32.html

The Open Group. (2006b). *The Enterprise Continuum in Detail*. https://pubs.opengroup.org/architecture/togaf8-doc/arch/chap 18.html

The Open Group. (2011a). *Phase P: The Preliminary Phase*. The Open Group. https://pubs.opengroup.org/architecture/togaf9-doc/arch/chap 06.html

The Open Group. (2011b). *TOGAF's. The Open Group Architecture Framework*. www.open-group.com/togaf

The Open Group. (2011c). *Introduction to Building Blocks*. http://www.opengroup.org/public/arch/p4/bbs/bbs_intro.htm

The Open Group. (2011d). *Foundation Architecture: Technical Reference Model*. http://www.opengroup.org/public/arch/p3/trm/trm_dtail.htm

The Open Group. (2018). *Architecture Patterns*. The Open Group. https://pubs.opengroup.org/architecture/togaf9-doc/arch/chap 22.html

Trad, A. (2015a). *A Transformation Framework Proposal for Managers in Business Innovation and Business Transformation Projects-Intelligent aBB architecture*. Centeris.

Trad, A. (2015b). *A Transformation Framework Proposal for Managers in Business Innovation and Business Transformation Projects-An ICS's atomic architecture vision*. Centeris.

Trad, A., & Kalpić, D. (2016a). *A Transformation Framework Proposal for Managers in Business Innovation and Business Transformation Projects-Basics of a patterns based architecture*. Centeris.

Trad, A., & Kalpić, D. (2016b). *A Transformation Framework Proposal for Managers in Business Innovation and Business Transformation Projects-Design patterns based architecture*. Centeris.

Trad, A., & Kalpić, D. (2016c). *A Transformation Framework Proposal for Managers in Business Innovation and Business Transformation Projects-Enterprise patterns based architecture.* Centeris. doi:10.1016/j.procs.2016.09.158

Trad, A., & Kalpić, D. (2022). Business Transformation Projects based on a Holistic Enterprise Architecture Pattern (HEAP)-The Basic Construction. IGI.

Walker, J. (2019). *Generics & Design Patterns.* Oberlin College Computer Science. https://www.cs.oberlin.edu/~jwalker/langDesign/GDesignPat/

Chapter 2
Digitalization in Management:
The Case of the Metaverse

Fetullah Battal
Bayburt University Social Science, Turkey

Halil H. Öz
Gümüşhane University, Turkey

ABSTRACT

When the historical development of management is examined, it is seen that every age has created a management and organizational structure suitable for its own conditions. Therefore, today, management and organizational structures are experiencing a digital transformation as a necessity of the age. From this point of view, this study was carried out as a conceptual analysis in order to conceptually examine new terms such as digital management, digital transformation, and digital leader that emerged with digitalization in management structures to explain the change in management structures by comparing them with previous management structures and to eliminate the ambiguities about the concept of digitalization. In addition, another aim of this study is to compile explanations in the literature about the metaverse, which is described as a three-dimensional virtual world, and to evaluate the innovations, opportunities, concerns, and criticisms brought by the metaverse application.

DIGITALIZATION

Digitalization has become a popular concept with the rapid development of technology in recent years. The term digitalization was first used in the North American Rewiev in 1971 as "digitalization of societies". In terms of content, the potentials of computer-based humanities are examined. Digitization enables the development of a wide range of technologies accessible to all. For this reason, digitalization has led to significant changes in cultural, behavioral, demographic and lifestyles (Kupiainen, 2006, p. 287). Stolterman and Fors, who are among the first scientists working on the theoretical features of the concept of digitalization, define digitalization as a world where everything is associated through information technologies (Stolterman & Fors, 2004, p. 688). According to another definition, digitalization

DOI: 10.4018/978-1-6684-4102-2.ch002

is defined as the development of a new business model or the use of digital technologies to create a new resource and value by changing an existing business model. For example, converting a handwritten text into a digital format is described as digitization. (Bloomberg, 2018).

Digital Transformation

The concept of digital transformation is expressed as the structuring of technology, production and management process in the organization, especially with digital technologies at different degrees, including the business model (Kraus, Schiavone, Pluzhnikova, & Invernizzi, 2021). According to Abbu et al., digital transformation first includes the restructuring of production, communication and human resources management of organizations. Secondly, digitalization includes top management (Abbu, Mugge, Gudergan, & Kwiatkowski, 2020). Although there is no universally accepted definition of digital transformation in the literature, another definition surrounding the previous two definitions states that digital transformation is a company that improves customer experience information of the organization and facilitates business transactions by using new internet-related digital technologies such as artificial intelligence, cloud storage system and blockchain. and a process of transformation and change aimed at creating new business models (Warner & Wäger, 2019, p. 326). From this point of view, digital transformation refers to the innovative use of various advanced digital technologies to create value on a larger scale for internal and external stakeholders (Gong & Ribiere, 2021, p. 102).

Digitizing Management

With digitalization, the structure of organizations is faced with an organic transformation expressed as "on-site management, empowerment, flexibility in rules and procedures, horizontal communication and teamwork". With this structure explained, organizations can adapt more appropriately to learning, virtual organizational climate and digital environment in an innovative way (Tom & Stalker, 1998). With the digital transformation in organizations, personalization, improvement and visualization based on cyber-physical systems find a place in the management level (Brettel, Friederichsen, Keller, & Rosenberg, 2014). In other words, it is thought that organizational effectiveness and efficiency will increase with the opportunities that the use of the internet in organizational business and transactions can create together with digital transformation (Porter & Heppelmann, 2014). In addition, it creates a fast communication network within the organization by adapting the internet and information technologies to the organization. With digitalized management, complex systems can be standardized and thus effectively managed. Organizing and designing ways of doing business, ensuring continuity in the training and career development of employees, and using organizational resources efficiently are expressed as opportunities that digitalized management provides to the organization (Landscheidt & Kans, 2016, p. 3). One of the most important goals of digitalized management is to create organizations that improve the ability of organizations to adapt to their environment, are flexible, can use organizational resources effectively, and are in good harmony with buyers and business partners (Akgül, Akgül, & Ayer, 2018, p. 201).

Digital Leader

Digital leaders play an important role in adapting digital changes to the organization in the digital transformation process (Stana, Fischer, & Nicolajsen, 2018). Digital leaders enable businesses to gain a sustainable competitive advantage by adapting their business strategies to the digital age. A talented digital leader who has developed himself according to the needs of the digital age can help create the digital business strategy of the organization and increase organizational performance and effectiveness (Kıyak & Bozkurt, 2020). In line with the explanations made, digital leadership is defined as a leader who quickly adapts information technologies that will contribute to the development of the organization (Tanniru, Khuntia, & Weiner, 2018). In another definition, digital leadership is defined as a leadership style that is fast, beyond hierarchy, focused on innovation with an approach based on teamwork and collaboration (Abbu et al., 2020).

METAVERSE

In the last half century, a new paradigm emerges approximately every ten years. While the paradigm of the 90s was the developments in the computer and communication sector, the web networks of the 2000s, and the change of the mobile in the 2010s, the key concept of the paradigm that emerged in the 2020s is the metaverse. Although the concept of metaverse is the paradigm of the 2020s, it was first used to describe a three-dimensional visual world in the novel "Snow Crash" written by Neal Stephenson in 1992 (Lee, 2021, p. 72). Especially in 2020, with the Covid-19 virus turning into a worldwide epidemic, people's non-face-to-face relationships start a new era. It is aimed to fill this gap in social relations with the use of digital technologies. This situation increases the interest in the concept of metaverse. Because the metaverse provides access to three-dimensional visual and auditory reality, the need for people to be in the same physical environment for social interaction is no longer necessary (Choi & Kim, 2017). Metaverse creates a virtual universe that puts people at its center, creating innovations and opportunities in many areas. Considering especially in terms of accessibility, diversity and equality, the metaverse creates significant positive effects. With the increase in globalization, the communication and interaction between countries or societies is much more than in the past. However, physical distances increase the costs in this process. However, the metaverse can provide an extraordinary accessibility that can prevent these situations. Converting many events to a visual format is supported by Metaverse. In this way, uninterrupted accessibility can be ensured by preventing disruptions in education or business life due to physical distances or epidemics. Again, for many reasons such as social distances and language differences, it is not possible for different people to be together in an integrated way in the real world. However, with the metaverse, geographical distances or language differences are no longer a problem. Thus, it can be ensured that individuals from various geographies, races or languages can be effectively coordinated together. Another benefit that can be obtained with the metaverse is equality. Thanks to the metaverse, these factors that negatively affect the equality can be eliminated. For example, avatars can be created in the virtual universe, allowing users to exist equally in all areas (Duan, et al., 2021, p. 154).

REFERENCES

Abbu, H., Mugge, P., Gudergan, G., & Kwiatkowski, A. (2020). Digital leadership-Character and competency differentiates digitally mature organizations. In *2020 IEEE International Conference on Engineering, Technology and Innovation (ICE/ITMC)* (pp. 1-9). IEEE. 10.1109/ICE/ITMC49519.2020.9198576

Akgül, H., Akgül, B., & Ayer, Z. (2018). Sanayi 4.0 sürecinde gazetecilik sektöründe çalışacak person-elin mesleki yetenek ve yeterliliğine yönelik değerlendirme ve öngörüler. *Avrasya Sosyal Ve Ekonomi Araştırmaları Dergisi, 5*(8), 198–205.

Bloomberg. (2018, August 28). *Digitization, Digitalization, and Digital Transformation: Confuse Them at Your Peril.* Author.

Brettel, M., Friederichsen, N., Keller, M., & Rosenberg, M. (2014). How Virtualization, Decentralization And Network Building Change The Manufacturing Landscape: An Industry 4.0 Perspective. *International Journal of Information and Communication Engineering, 8*(1), 37–44.

Duan, H., Li, J., Fan, S., Lin, Z., Wu, X., & Cai, W. (2021). Metaverse for social good: A university campus prototype. *Proceedings of the 29th ACM International Conference on Multimedia*, 153-161. 10.1145/3474085.3479238

Gong, C., & Ribiere, V. (2021). Developing a unified definition of digital transformation. *Technovation, 102*, 102–117. doi:10.1016/j.technovation.2020.102217

Kıyak, A., & Bozkurt, G. (2020). A general overview to digital leadership concept. *Uluslararası Sosyal ve Ekonomik Çalışmalar Dergisi, 1*(1), 84–95.

Kraus, S., Schiavone, F., Pluzhnikova, A., & Invernizzi, A. C. (2021). Digital transformation in healthcare: Analyzing the current state-of-research. *Journal of Business Research, 123*, 557–567. doi:10.1016/j.jbusres.2020.10.030

Kreiss, S. B. (2014). Digitalisation and Digitization. Encyclopedia of Communication Theory and Philosophy, 3-15.

Kupiainen, J. (2006). Translocalisation over the Net: Digitalisation, information technology and local cultures in Melanesia. *E-Learning and Digital Media, 3*(3), 279–290. doi:10.2304/elea.2006.3.3.279

Landscheidt, S., & Kans, M. (2016). Automation practices in wood product industries: Lessons learned, current practices and future perspectives. In *The 7th Swedish Production Symposium SPS.* Lund University.

Lee, J. Y. (2021). A study on metaverse hype for sustainable growth. *International Journal of Advanced Smart Convergence, 10*(3), 72-80.

Porter, M. E., & Heppelmann, J. E. (2014). Wie smarte Produkte den Wettbewerb verändern. *Harvard Business Manager*, 34-60.

Porter Choi, H. S., & Kim, S. H. (2017). A content service deployment plan for metaverse museum ex-hibitions—Centering on the combination of beacons and HMDs. *International Journal of Information Management, 37*(1), 1519–1527. doi:10.1016/j.ijinfomgt.2016.04.017

Stana, R. A., Fischer, L. H., & Nicolajsen, H. W. (2018). Review for future research in digital leader-ship. *Context, 5, 6.*

Stolterman, E., & Fors, A. C. (2004). Information technology and the good life. *Information Systems Research*, 687–692.

Tanniru, M., Khuntia, J., & Weiner, J. (2018). Hospital leadership in support of digital transformation. *Pacific Asia Journal of the Association for Information Systems, 10*(1).

Tom, B., & Stalker, G. M. (1998). *The Management Of İnnovation*. Tavistock.

Warner, K., & Wäger, S. R. (2019). Building dynamic capabilities for digital transformation: An ongoing process of strategic renewal. *Long Range Planning, 52*(3), 329–349. doi:10.1016/j.lrp.2018.12.001

Chapter 3
Business Transformation Projects Based on a Holistic Enterprise Architecture Pattern (HEAP):
The Basic Construction

Antoine Trad

https://orcid.org/0000-0002-4199-6970

IBSITM, France

ABSTRACT

This chapter provides recommendations for classifying and using various types of patterns in a holistic enterprise architecture pattern (HEAP) that can support transformation projects. The HEAP is based on a concise, composite, and layered patterns model. A composite patterns model can be used as a template to instantiate building blocks (BB) to implement a variety of types of transformation projects. In this chapter, the focus is on various pattern standards that can be used in a holistic and adaptable BBs to support an optimal set of enterprise architectures (EA). A patterns-based EA offers BBs, like in civil engineering, that can support colossal and complex projects. For such complex projects, there is a need to create a common denominator pattern, to integrate other standard patterns, in the form of a HEAP, where the HEAP is the backbone of this research and development project (RDP). This RDP proves that the HEAP can be used for building dynamic and flexible EAs.

DOI: 10.4018/978-1-6684-4102-2.ch003

RESEARCH AND DEVELOPMENT

Figure 1. The mixed method flow diagram (Trad, & Kalpić, 2016a)

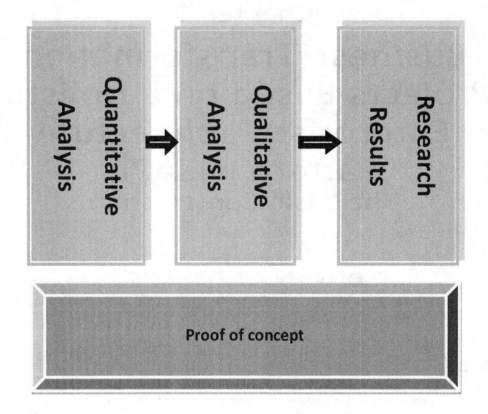

As shown in Figure 1, the RDP flow acknowledged an important gap in the mentioned fields in the form of keywords, mainly due to the fact that the existing literature on transformation projects, their failure rates and on various methodologies treating business transformations, offer practically no insight into: 1) An optimal transformation pattern; and 2) The profile of the Project manager as an Architect of Adaptive Business Information Systems (AofABIS), who is the principal patterns designer. Such a holistic pattern and profile should be capable to setup a business EA integration concept to facilitate the implementation phase of Projects. Previous research models confirmed the profile, that is based on the AofABIS profile, and produced a set of managerial recommendations and factors (SAP, 2013; Trad, & Kalpić, 2016a). This RDP uses a PoC to prove the Research Question (RQ) and the related hypothesis, establish viability, show technical issues and show overall direction. To some extent it is similar to the prototyping in engineering thesis works (Camarinha-Matos, 2012), where the prototype proofs RQ's feasibility. This chapter is strongly related to the *Business Transformation Projects based on a Holistic Enterprise Architecture Pattern (HEAP)-The implementation*, and its implementation part (Trad, & Kalpić, 2022).

Research Question

This article RQ is: *Can a HEAP support complex transformation projects, and which main categories and standards can support it?* Where HEAP is the backbone and background for building dynamic and flexible EAs.

Background

Increasingly complex, competitive, and automated business environments like mechanistic organizations or enterprises (simply Entity) are the essence for investment in dynamic architecture and transformation of flexible and efficient business environments (Capgemini, 2007, 2009). HEAP's background is based on the following assumptions (The Open Group, 2018):

- A *pattern* has already been defined as: *…an idea that has been useful in one practical context and will probably be useful in others*" (Fowler, 1996).
- In standard EA methodologies like *The Open Group's Architecture Framework* TOGAF), patterns are to be considered as a concept for using Building Blocks into Project's context; like in the case of a re-usable solution to a Project problem.
- BBs are the artefacts to use, whereas patterns direct the manner on how to use them, in a specific context, and express the trade-offs.
- The use of patterns may support EA practitioners to identify combinations of Architecture BBs (ABB) and/or Solution BBs (SBB), which have been implemented and verified to deliver successful solutions.
- Pattern techniques are acknowledged to have been established as a valuable architectural design technique by Christopher Alexander, a buildings architect, who described this approach in his book *The Timeless Way of Building*, published in 1979 (Alexander, 1975, 1979). This book introduces the ideas behind the use of patterns, and Alexander followed it with two further books, *A Pattern Language and The Oregon Experiment* (Alexander, Silverstein, Angel, Ishikawa, & Abrams, 1975) in which he expanded on his description of the features and benefits of a patterns approach to architecture.
- In EA, it is considered that Information and Communication System's (ICS) and civil engineering buildings architects, have similar issues to address and solve. Therefore, EA practitioners take interest in patterns as an architectural environment.
- There are many works, papers, projects and books that have been achieved and published on these topics, since Alexander's 1979 book; but probably the most significant and renowned is *Design Patterns: Elements of Re-usable Object-Oriented Software* (Gamma, Vlissides, Helm, & Johnson, 1994). This book describes simple and elegant solutions to specific problems in Object-Oriented (OO) software design.

A successful finalization of the implementation phase can give an important business advantage and can guarantee the transformed business company's perennity. This research phase's main blocks are:

- The basic patterns serve as a preparation and classification phase for design patterns integration (Trad, & Kalpić, 2016b).

- The design patterns that are this article's focus as shown in Figure 1.
- The enterprise patterns (Trad, & Kalpić, 2016c).
- EA methodologies like TOGAF and its Architecture Development Method (ADM), manages (GitBook, 2021): 1) Business; 2) Information System; and 3) Technology, layers. All these layers map to ArchiMate layers, as shown in Figure 2. which supports the Business Transformation Project's (simply the Project) Vision.

Figure 2. ADM's Phases mapping to ArchiMate layers (GitBook, 2021).

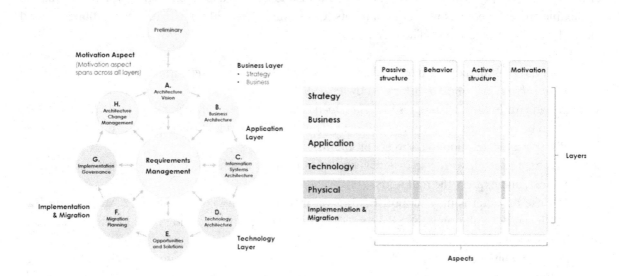

The Vision

As already mentioned, the research topic (and RQ) is about managing complexity in Projects using standardized methodologies and solutions, as shown in Figure 3, where all these methodologies have many abbreviations and terms which make the reading of such topics difficult, but unfortunately that is the nature of such fields. There is also a need that the valuable reader has extensive knowledge in EA, modelling standards, implementation of services and business engineering (Trad, 2015a).

Figure 3. The ADM's vision phase interaction with other phases (The Open Group, 2011a)

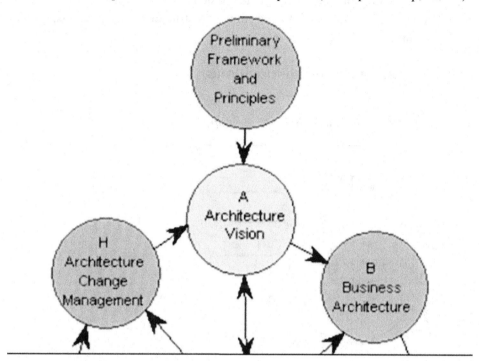

ICSs that are based on services in the form of a set of atomic services is a major transformation paradigm shift. Atomic services unbundle the ICS into independent artefacts that can interact together over the Entity; where building generic atomic Business Services (aBS) are Project's important factor (Gartner, 2013b). Such Projects are complex, and they mainly fail (Gartner, 2013a; Altman, 2014), hence the initial unbundling process to deliver the pool of aBSs and Patterns' Structure, is the main prerequisite to enable Project's success.

PATTERN'S STRUCTURE

Existing Standards

An Idiom is a basic pattern for implementation purposes, and it describes the implementation of a specific aspect of a composite model. Patterns offer the Project the possibility to identify combinations of architecture artefacts, services and SBBs that have been verified to deliver solutions for the future business environment. The Project proposes the aBBs to support such an approach. Concerning the basic pattern-based architecture, the Project uses existing technology and business engineering standards that includes, the: 1) OO Methodologies (OOM), like the Unified Modelling Language (UML); 2) object relational database standards; 3) eXtensible Markup Language (XML); 4) XML Schema Diagram (XSD); 5) OO Programming (OOP) standards; and 6) the Gang of Four (GoF) design patterns. A coherent basic pattern(s) concept can be achieved by using an optimal manner of designing the system, like the design driven development or design first approach (Escalona, & Koch, 2004). The evolution of technology standards, like dynamic linking of components, made basic patterns receptive to development and integration with design patterns and EA. Before using complex structures and standards, there is a need to unbundle and refine the legacy ICS.

Unbundling and Refinement Predispositions

Figure 4. Major pre-patterns standards (Störrle, & Knapp, 2006)

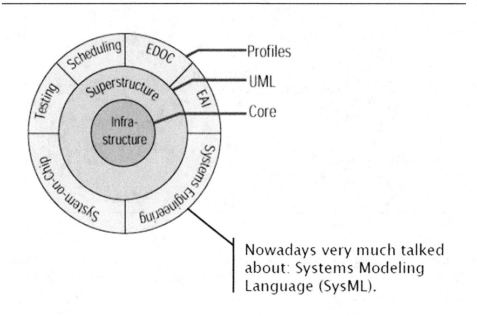

Before introducing how HEAP driven Projects are to be done, it is important to prepare the Entity of a concept for using the basic artefacts, like patterns, BBs, classes, composite entities, tables or services to transform the ICS by applying the Refinement Process to Extract Patterns (RP2EP). The RP2EP will support the agile and autonomic Project's components implementation phase. That needs a precise EA mapping concept that must use a standard framework (Farhoomand, 2004); that is the main Project principle. The RP2EP supports an agile iterative model that can map all the Project's artefacts in a linear *1:1* manner. The RP2EP tries to extract granular/atomic classes that are needed for the future services'-based patterns; so that each Project artefact can be managed independently. The RP2EP delivers a pool of atomic Project Services (aPS), which must be stored, managed, classified, and interconnected using existing interoperability standards. The RP2EP applies a structured unbundling process, by using the 1:1 mapping rules that are based on atomic BBs (aBB), or composite classes) (Fowler, 2014). As shown in Figure 4, the Project proposes the HEAP to integrate existing standards and indicates where the problem is, how to keep the Project feasible with so many types of artefacts, standards and methodologies. Projects have limited resources and one has to establish an iterative Development and Operations (DevOps) or a Secured DevOps (SecDevOps) model. To support HEAP, SecDevOps and RP2EP, the use of a pseudo bottom-up approach is recommended (Trad, & Kalpić, 2016a; Störrle, & Knapp, 2006); where they are all based on OO concepts.

The Evolution due to Object Orientation

The Project uses OO basic constructs or patterns for the implementation, in order to enable efficient and automated DevOps activities. The Project must instore technology agnostic principles, where the target implementation environment is adapted to the HEAP and its related patterns (Trad, & Kalpić, 2016a), which are mainly Unified Modelling Language (UML) profiles.

OO and the Unified Modelling Language

UML's usage must be limited to its most inter-operable artefacts (Störrle, & Knapp, 2006) like: 1) Use case diagram; 2) Class diagram, which is the most interoperable; 3) Activity diagram; 4) Communication diagram; and 5) Package diagram. OO and UML profiles support ICS's integration process.

The Integration role of the ICS and the Enterprise Service Bus

On the level of basic patterns, the UML's package diagrams are essential, where they can assist at the concept mapping and support deployment on the Enterprise Service Bus (ESB). Package diagrams hide the complexity and its heterogeneous content, where a package corresponds to an ABB or SBB. To simplify this process, there is a need for introduction of a unit of work.

The Unit of work

In order to define an optimal unit of work for a basic pattern, it is needed to align all the Project's resources using a holistic 1:1 mapping concept. The authors use a concrete artefact that is an object-oriented class artefact, which is represented by a class-diagram using the extensible XML. Such a mapping concept is based on XML and on the class artefactto insure the interoperability between all the Project's resources. These resources can conform to one of the following artefacts' standard formats (Trad, & Kalpić, 2016a; Downey, 2003):

- The UML XML format that uses the XML Interchange (XMI) format.
- Design patterns and their XML interfaces.
- The Entity Relational Modelling (ERM) using the XML Scheme Diagrams (XSD) format.
- Test Driven Development (TDD) format.
- Behaviour Driven Development (BDD) format.

All these mentioned standard formats support the data management system's integration.

The role of the Data Management Systems

UML class diagrams can map to table constructs where a class corresponds to a table or a pattern; that can be done using various persistence frameworks and by applying the Object Relational Mapping (ORM) concept. Design Patterns (DPs) can map to complex query statements that can be supported by various persistence frameworks and by applying the enterprise transformation layer. Access to data varies depending on the source of the data. Access to data storages, like a database, depends on the type of storage and the offered implementation. The Data Access Object (DAO) pattern can be used to encapsulate types of accesses and to manage the open connections with data (Oracle, 2002). All these mechanisms support the notion of a Meta Model's approach.

Fundaments

Meta, in Semite languages (Aramaic, Arabic and Hebrew) meaning transcending, after, having a *holistic* view, *beyond or death (the afterlife)*. And in engineering it refers to build blocks that contain recurrent models of implementations. There are various forms for describing patterns, and there is no global standard for that goal; but there is broad agreement on the types of Elements and characteristics that a pattern should contain. The headings, elements, and characteristics (simply *Elements)* which follow are taken from Pattern-Oriented Software Architecture presented in A System of Patterns (Buschmann, Meunier, Rohnert, Sommerlad, & Stal, 1996) and in the authors' works. The *Elements* described in the following section can be found in practically all patterns, even if different *Elements* are used to abstract them (The Open Group, 2018):

- Name: A standardized (or convened), meaningful and memorable manner to refer to the pattern, typically a single word, preferably using the *Hungarian Notation*, that is an IDentifier (ID) naming convention in software engineering, where the Name of a variable or function indicates its intention or kind, and in some dialects its type. A Name is uniquely related to a Global and Unique ID (GUID).
- GUID: Project packages or (BBs) must be identifiable using global identifiers that are automatically generated and they help in keeping the 1:1 mapping concept. This concept can be used also in the business integration between various Entities. The GUID Identifies a unique pattern, which is stored in the Project's repository,, and can be related to a type of Problem.
- Problem: A short description of the problem, using a Natural Language Programming (NLP) environment, describing the intention in applying this pattern. Also the defined goals and objectives to be achieved within the Project context and using priorities and metrics like Critical Success Factors (CSF), Critical Success Areas (CSA) and Key Performance Indicators (KPI); which can be used in a specific Project Context.
- Context: The preconditions and constraints under which the pattern can be applied; and a short description of the pattern's initial state, before the pattern is applied) to create an ABB with specific capabilities or Forces.
- Forces: A description of the relevant pattern Forces and constraints, and how they interact and how they enable reaching task's or Project's goals and objectives. The description of the Problem's details contains optimal trade-offs that must be analysed. Trade-offs make Problem solving processes difficult. They can generate the need to create a pattern. A Force is related to processing quality that EA practitioners want to reach. Of course, there are other concerns that should be reached in implementing Projects, like: 1) Security, robustness, reliability, fault-tolerance; Manageability; 2) Efficiency, performance, throughput, bandwidth requirements, space utilization; 3) Scalability (incremental scaling on-demand); 4) Extensibility, evolvability, maintainability; Modularity, independence, re-usability, openness, composability, portability; 5) Completeness and correctness; 6) Ease-of-construction; 7) Ease-of-use; and 8) Problem solving, decision making capabilities; and others.

- Decision Making Capabilities (DMC) support the patterns intelligence to treat a specific set of Problem(s). The company's or organisation's (simply referred to as an Entity's) Decision Making System (DMS) is a synchronized and classified set of DMCs that can support the Project by delivering Solutions.

- Solution is a short description and possible Heuristic Decision Tree (HDT) diagrams, on how the solutions have been achieved; and the defined goals and objectives. The description includes the Solution's: 1) Static structure and its dynamic behaviour; 2) The Project's actors; 3) Possible collaborations; 4) A guideline for implementing the solution; and 5) Resulting Context.

- Resulting Context describes the pattern's post-conditions which may require trade-offs among defined Forces. Therefore, it should contain the description of which Forces have been solved; and which other Related Patterns were used.

- Related Patterns present the relationships between the Project's used patterns, by describing the predecessor pattern(s); which after processing, results in context(s) that corresponds to the successor pattern(s), whose initial contexts correspond to the resulting context of this one; or alternative patterns, that describe a different Project Solution, for the same processed Problem. But using different Forces and constraints. It can also contain Examples or Cases of solved Problems.

- Examples or Cases: A pattern description should contain examples of its implementation; and more specifically how was the defined Problem solved, in which Project Context, and which Forces have been identified. The successful Cases can be used as Known Uses.

- Known Uses are known and confirmed applications of the pattern in the Project, which have delivered optimal Solution(s) to a Problem type. All these presented Elements, define a pattern's Rationale.

- Rationale is a detailed explanation of the pattern as a model, or a component, indicating how it works, and how it uses Forces to reach Solutions. The Solution Element describes the *external structure and behaviour* of the Solution, whereas the Rationale provides insight into its internal mechanisms.

The evolution of patterns has made it possible to create *Architecture and Design Patterns*, that are the predecessors of basic patterns.

Basic Patterns

Common Characteristics

As shown in Figure 5, this phase's main technology blocks are: 1) the basic patterns that serve as a preparation and classification phase for design patterns; and they are in focus of this article; 2) the standard design patterns; and 3) the enterprise patterns. The Project offers: 1) a real work business and technical transformation framework in the form of a set of reusable patterns, recommendations and solutions; and 2) the corresponding set of Project's patterns' architecture for the making of BBs to enable fast SecDevOps iterations (Trad, & Kalpić, 2016a).

Figure 5. This RDP's main blocks (Trad, & Kalpić, 2016a)

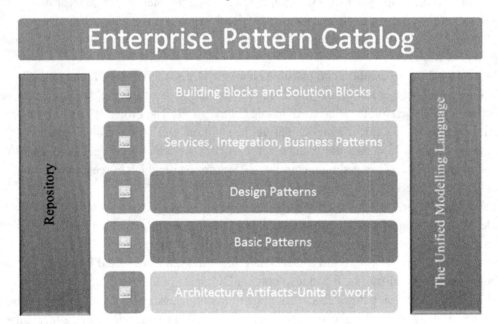

Low Level Patterns

The basic patterns are common implementation constructs, used in OO Design (OOD), Analysis (OOA) and OOP. Basic patterns for HEAP's support are the following (Trad, & Kalpić, 2016a):

- Interface Pattern Is used to design service provider classes that support a client object to use different classes in a seamless manner, without changing the client's implementation.
- Abstract Parent Class Pattern is used to design a consistent implementation of the functionalities common to a set of related classes.
- Private Methods Pattern is used to provide a design for class's behaviour so that external objects have no access to its internal behaviour parts. Behaviour logic is kept internal and encapsulated.
- Accessor Methods Pattern supports the access to the object's state using specialized methods and it prohibits the client object to directly access its attributes; that also makes the class more portable and maintainable.
- Constant Data Manager Pattern is used to design a maintainable and common repository for constant data.
- Immutable Object Pattern ensures that objects are unchangeable and enables concurrent access to the object's data parts.

Generic Patterns

Generics in ICS related domains are the common basics for implementing DPs; in some domains Generic Patterns (GP) are known as templates (also known as parametric polymorphism). Templates facilitate the phases of design and implementation of Project's components. The most known GPs are: 1) Singleton;

2) Linked List; and 3) Visitor. GP's main goals are to extend work elements and to preserve the level of abstraction (Walker, 2019). This is the minimal set to be empowered with other Project specialized basic patterns to support logging, security, data-management… The Projects denotes the basic patterns category as the enterprise's basic implementation patterns. The evolution of patterns made it possible to create *Architecture and Design Patterns* that are the predecessors of ABBs.

Architecture and Design Patterns

A Design Pattern, is mainly used in software architecture, design, or implementation Project phases. In *Pattern-Oriented Software Architecture*, A *System of Patterns*, can have the following three types of patterns (Buschmann, Meunier, Rohnert, Sommerlad, & Stal, 1996):

- An Architecture Pattern expresses a fundamental structural organization or schema for a software library, system, or component. It can be considered as an ABB when using TOGAF's terminology. A well-designed ICS should provide a set of predefined ABBs and subsystems, that specify their responsibilities, include rules for their relationships and implementation details (Perroud, & Inversini, 2013).
- A Design Pattern provides a scheme for refining, a library, subsystems or component, of a software artefact, and describes their relationships. It describes a common structure that solves a general design Problem within a specific Project Context.
- An Idiom is a low-level pattern, specific to a programming language, like C++. It describes how to implement particular aspects of components and their relationships by using specific features of the used programming language.

The previous definitions are important in the field of software design and architecture. Software design and architecture is an important layer of EA and TOGAF, but it is not the only focus and view. In this article we are concerned with various types and categories of patterns for EA based Projects. Knowing that most of these patterns are embedded in ICS components; and their usage is not visible; but their architecture and description must be understood by EA practitioners, in order to be capable of combining them. HEAP will try to find analogous concepts and terminology, and offer a re-usable holistic pattern, that is a composite model of Design Patterns.

Design Patterns

DPs represent implementation's best practices that can be used by a Project team of experienced object-oriented implementers. DPs are solutions to generic problems that implementers can use for solving standard and recurrent problems which are faced during the Project's implementation phase. These Project DPs can be obtained by trial-and-error activities. In the year 1994 a book entitled Design Patterns was published, that defines reusable OO elements for the implementation phase that initiated the concept of DPs (Gamma, Vlissides, Helm, & Johnson, 1994). DP is a composite element of an architecture that is a part of the Architecture modules (Trad, & Kalpić, 2016b; Trad, & Kalpić, 2016c).

Creational Patterns

This set of design patterns deals with common composite constructs that can be used in a Project, where their main activity is the instantiation of BBs or services. Creational patterns support a standardized mechanism to factorize the end system's BBs or services; where these patterns hide by encapsulation the implementation details on how these instances or BBs are created. This set of patterns enables the use of interfaces and promotes dynamic interaction that will make the end system agile and reduce coupling and the end system's independency on how artefacts are created. The following patterns are part of this category:

- Singleton provides a controlled BBs creation mechanism to ensure their uniqueness.
- Factory helps the client BBs to instantiate foreign objects without knowing their details.
- Abstract factory permits the creation of an instance of ABB with its related classes.
- Prototype provides a simpler way for creating an ABB by cloning an existing BB.
- Builder permits the creation of a complex BB using its type and content, while hiding the details of the BB's creation, keeping it transparent to the end BB or service.

Structural Patterns

Structural patterns manage BBs by delegating their behaviour to other BBs, what permits the creation of a layered architecture of components, using loose coupling, facilitating BB's communication and accessibility. This pattern also provides manners to structure a composite BB so that it can be instantiated in using minimum end system's resources. Here we focus on how BBs and classes are associated or can be composed to make relatively large Project structures, where inheritance mechanisms are mostly used to combine interfaces or implementations. The following patterns are part of this category (Sarcar, 2016):

- Proxy provides a placeholder for another object to control its access.
- Decorator extends the functionality of ABB to open to its clients without using inheritance.
- Adapter enables the BB's conversion of its interface to another interface which clients can use in the case of incompatible interfaces to work with it.
- Chain of responsibility avoids binding the sender and receiver BBs. The sender BB passes its request without knowing which BB will handle the request.
- Façade provides a high level interface to a subsystem of BBs or classes, making the subsystem easy to manage.
- Bridge enables the separation of ABB abstract interface from its implementation and eliminates the dependency between end-points, allowing them to be modified independently.
- Virtual proxy facilitates the mechanism for delaying the creation of ABB until it is really needed, in a way that this process is transparent to its client BBs.
- Counting proxy supports auxiliary Project activities, like logging and persistence, where it recommends the storage of these functionalities in a specialized BB.
- Aggregate enforcer enables an instantiated aggregate BB's data member to be initialized.
- Explicit object release enables the BB's resources to be released in a timely manner.
- Object cache stores the resulting data set of ABB's method call in a repository; where these data are returned to the calling BB.

- Flyweight splits BB's common data to save memory and processing resources.
- Composite permits both specific BB and composite BBs to be managed in a standard manner.

Behavioural Patterns

In this category the focus is on BB's algorithms, and it also focuses on the communication between the Project's artefacts. The following patterns are part of this category (Sarcar, 2016):

- The Observer pattern is used for in the case of a one-to-many relationship between BBs and when ABB is modified, then all its dependent BBs are automatically informed.
- Template is an abstract BB that exposes defined template(s) to execute its methods.
- Command is a data driven design pattern where ABB request is wrapped in a command and sent to the calling BB that has the appropriate mechanism to handle the command.
- Iterator is used to access the objects of a collection of BBs in sequential manner.
- State makes ABB's behaviour change, based on its state.
- Mediator simplifies the communication between multiple BBs and enables easy maintenance and loose coupling.
- Strategy permits BB behaviour to be modified at run time, where ABB represents different strategies. The BB's behaviour depends on a strategy.
- Chain of responsibility creates a receiver BB for a request, in a decoupled context.
- Visitor permits an operation to be executed on a collection of different BBs.
- Interpreter evaluates an implementation grammar or expression and generates an expression interface.
- Memento restores the state of ABB to a previous state.
- Iterator permits a client BB to access the contents of a composite BB in a repetitive manner.

HEAP will try to find analogous concepts and terminology, and offer a re-usable holistic pattern, that is a composite model of Enterprise Patterns.

Enterprise Patterns

The Model View Controller (MVC) pattern is the most important and it offers interfaces for messaging and a related data model that serves as a messaging framework, used as an integration server. The messaging framework is essential for complex system integration (Fowler, 2003; Taleb, & Cherkaoui, 2012). Enterprise integration patterns are the base for building the EA patterns.

Enterprise Architecture Patterns

Enterprise Architecture Patterns (EAP) manage: 1) databases' concurrent access; 2) applications' user interface; and 3) legacy system' transformations. The EAP set includes the: 1) Domain Logic Patterns; 2) Data Source Architectural Patterns; 3) Object Relational Behavioural Patterns; 4) Object-Relational Structural Patterns; 5) Object-Relational Metadata Mapping Patterns; 6) Web Presentation Patterns; 7) Distribution Patterns; 8) Offline Concurrency Patterns; 9) Session State Patterns; and 10) Base Patterns.

A set of EA patterns serve to build an enterprise pattern that can serve as a Common Denominator Patterns (CDP) for EA (CDP4EA) or reference model(s).

Reference Models

TOGAF's technical reference model offers an interface to manage entries, like an enterprise pattern. Enterprise patterns are mainly a part of the modelling component, as shown in Figure 2. To integrate enterprise patterns, the Project must use existing standards and methodologies (Crosswell, 2014). HEAP will try to find analogous concepts and terminology, and offer a re-usable holistic pattern, that is a composite model of Related Patterns.

Related Patterns

As we have many types of patterns that tend to work together, there is a need to define the types of relations which they can have. Patterns' relationships use OO relationship types. Relationships interconnect patterns, creating a virtual composite pattern (or HEAP), indicating how the DMS solves types of Problems. Related patterns to implement the HEAP need a well-designed Enterprise Meta Model (EMM), what also ensures the Project's integrity.

The Enterprise Meta-Model

Basics

UML based DPs are common in ICS development processes, where they support the communicating architectural solutions by implementing complex solutions, and transformation activities. A *UML profile* is a stereotyped package that contains customized model elements, which can be used for a specific domain by extending the meta-model using stereotypes, tagged values and constraints. A stereotype, denoted by <<stereotype-name>>, enables the definition of extensions to UML's vocabulary. It classifies tagged values and constraints, using meaningful names. When a stereotype is branded to a model element, the semantic meaning of the tagged values and the constraints associated with the stereotype are attached to that model element implicitly. Constraints add new semantic restrictions to a customized model element by using the *Object Constraint Language* (OCL). A Virtual Meta Model (VMM) is UML's expression of a formal model with a defined set of UML extensions. The EMM is a VMM variant and uses a modelling language (Dong, & Yang, 2003).

A Modelling Language

The Design Pattern Modelling Language (DPML) is a notation that supports the specification of DP solutions (or ABBs) and their instantiation into UML models (or SBBs). DPML provides constructs which allow DP solutions to be modelled and integrated. DPs are described using a mixture of natural language and UML style diagrams, this causes complex scenarios in integrating DPs in the ICS. As shown in Figure 6, a DPML models DP to support ABBs, SBBs, aBBs, sBBs and CDPs (Mapelsden, Hosking, & Grundy, 2002).

Figure 6. Core concepts of DPML (Mapelsden, Hosking, & Grundy, 2002)

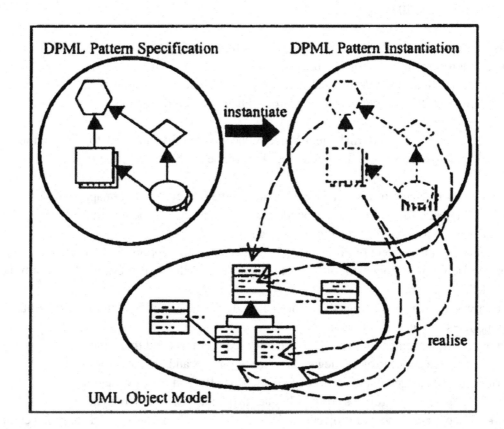

Bridging DP based BBs

Model-Driven Architecture (MDA) delivers models which are expressed in a standard notation (like UML). Such models can be DPs. MDA delivers models supported by transformation patterns. MDA models are: 1) Computation Independent Model (CIM); 2) Platform Independent Model (PIM); 3) Platform Specific Model (PSM) and Implementation Specific Model (ISM) (Kleppe, Warmer and Bast, 2003; Stojanović, 2005).

Building an EMM

The EMM is an ontology for EA concepts using EA frameworks and tools. EMM's intention is to provide extensible set of concepts and corresponding relationships, with semantics that can be mapped to the patterns (or a HEAP), ABBs, concepts, activities, and standard case tools of frameworks. The EMM can map to existing frameworks, like the Defence Architectural Framework (MODAF), the Federal Enterprise Architecture Framework (FEAF), TOGAF, and other… EMM is applied to abstract the complexity of these frameworks and a HEAP in a Project. The abstraction views are: Conceptual, Logical, and Physical. The Core of the EMM has layers representing the areas of EA to be included and rows representing the levels of abstraction, or views. As shown in Figure 7, EA's layers are (Enterprise Architecture Solutions, 2021):

- Business contains the Project's objectives, capabilities and Business Processes (BP).
- Application contains the Project's functional behaviour, supported by BPs.
- Information contains the Project's structured and unstructured data sources.
- Technology contains the Project's software and hardware components.

The EMM uses a set of external Meta-Model concepts and relationships to support the Project's management and governance, like the following (Enterprise Architecture Solutions, 2021):

- Strategy Management of Project's future state and the predefined Roadmap.
- Change Management of Project's dependencies that impact BPs, people, ICS...
- Standards Management of Project's EA standards.
- Governance Management of Project's policies, decisions, exceptions and controls.
- Service Delivery Management of Project's dependencies that exist between BPs, team members and ICS parts.
- Security Management of Project's requirements, design, and implementation of the Entity's security policies and in its EA Roadmap.
- Obligation Management of Project's regulatory requirements and their impact.
- Legal Management of Project's legal compliance for ICS and business services, …
- Performance Management of Project's CSFs/CSFs/KPIs and their evaluation.
- Cost Management of Project's costs of EA artefacts, like ICS' implementation…
- Compliance Management of Project's initial set of classes for managing and analysing compliance policies and associating these with artefacts in the core EMM.
- Skills Management of Project's skills required by Business Roles, which are provided by Actors in the business layer.
- Lifecycle Management of Project's classification schemes and lifecycle policies related to EA's artefacts.

Figure 7. Essential Meta Model Scope... (Enterprise Architecture Solutions, 2021)

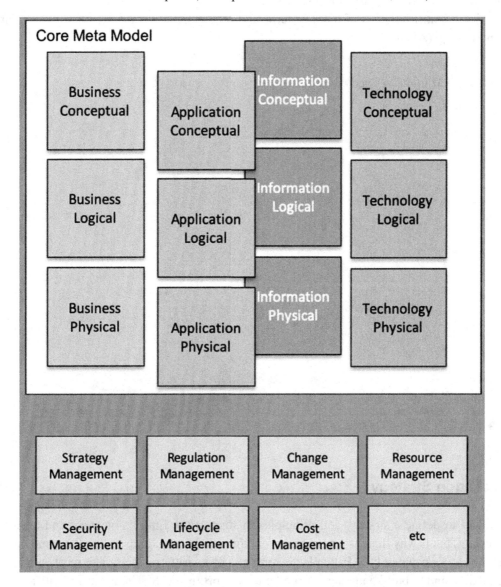

The Composite Construct

This article's goal is not to present one more time the various types of patterns, but to show the optimal manner to do the preparations needed to integrate patterns in a coherent Project architecture that would use also (Trad, & Kalpić, 2016a):

- The implementation patterns: These types of patterns understand a family of patterns that have a composite structure, like: 1) the classical design patterns and 2) the services patterns.
- The integration and enterprise patterns: These types of patterns understand patterns that have a component structure, like: 1) Integration patterns; 2) the EA patterns; and 3) the BPM patterns.

in order to deliver an EMM as shown in Figure 8.

The Project identifies the set of main patterns and related patterns and assemble them in an Anti-Locked-in Strategy (ALS).

Figure 8. The EMM in the defined context.

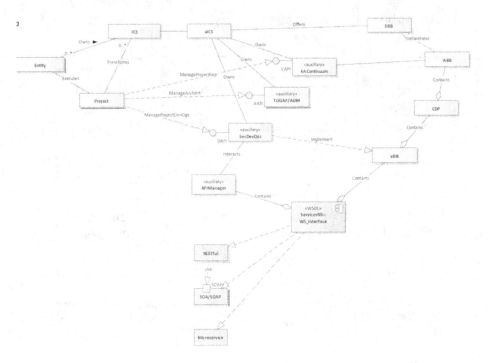

Anti-Locked-in Strategy

The ALS tries to define a strategy and principles to support the Entity's business and technological independence. That is why there is a need to respect the existing standards and in the same time it tries to use its own loose coupled HEAP, that is the umbrella of its patterns. A Project must avoid using too many standards, and to build a unique EA model to support the ALS. The Project identifies the set of main patterns and pattern types to implement the HEAP, which addresses these various topics and delivers a common concept, based on pattern types and categories.

PATTERN TYPES AND CATEGORIES

There are many of them, but the authors will take the most important ones to present this article's background. The various types of patterns addressed by the HEAP are:

- GoF Patterns (GOFP).
- Business Process Management (BPM) Patterns (BPMP).

- Application Programming Interface (API) Patterns (APIP).
- Service Oriented Architecture (SOA) Patterns (SOAP).
- ESB Patterns (ESBP).
- Enterprise Applications Integration (EAI) Patterns (EAIP).
- Cloud Computing Design (CCD) Patterns (CCDP).
- Organisational (ORG) Patterns (ORGP).
- Architecture Patterns (ARCP).
- REpresentational State Transfer (REST) Patterns (RESP).
- ….

Because of this article's limitations SOAP and BPMP will be analysed. The Project's and Entities classify and store patterns and BBs in an Architecture Continuum.

Patterns and the Architecture Continuum

Although ARCPs have not been integrated in standard EA methodologies, such as TOGAF, where in its first four main ADM Phases (Phases A to D), it gives a clear indication which resources should be used. There are used re-usable resources, like HEAP, which is managed by the EA Continuum, as shown in Figure 9 and contains the major CDP. An Entity that adopts a formal approach to apply ARCPs, must integrate them in their EA Continuum, to support EA Views and the ADM.

Figure 9. The CDP types that are parts of a HEAP (Source: authors)

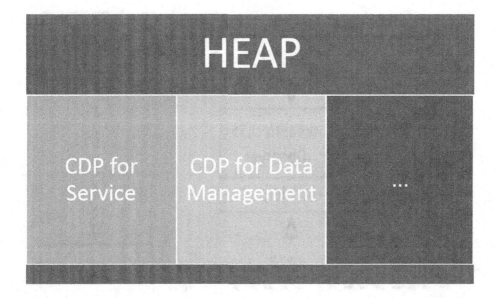

HEAP and EA Views

EA views are selected parts of models, aBBs and CDPs, representing a complete Entity and ICS architectures; where the focus is on aspects that address the: 1) Tangible concerns of one or more stakeholders; 2) Intangibles, which is mainly quality. The HEAP supports the design of such complex Entity models to support a view, using ARCP, SARP, BPMP, RESP and many others. From an EA point of view, the MVC is one of the most used and known patterns.

The Model View Control Pattern – A Central Pattern

The MVC pattern is the most complex and complete pattern that must be analysed separately and considered a category, as shown in Figure 10. The MVC decouples the: 1) Modelling of data and the domain CDPs; 2) Presentation CDP; and 3) Actions or services thar are based on Graphical User Interface (GUI) (Burbeck, S; Microsoft, 2016b). MVC's pattern main characteristics are:

- The Model manages the ABBs' and SBBs': 1) Behaviour and data CDPs; 2) Responds to Requests For Information (RFI) on its state(fullness); and 3) Responds to requests to change its state.
- The View manages user interface CDPs and special features.
- The Controller interprets various types of GUI requests, informing the Model CDP(s) and the View CDP(s).

Figure 10. The MVC pattern and interactions (Microsoft, 2016b).

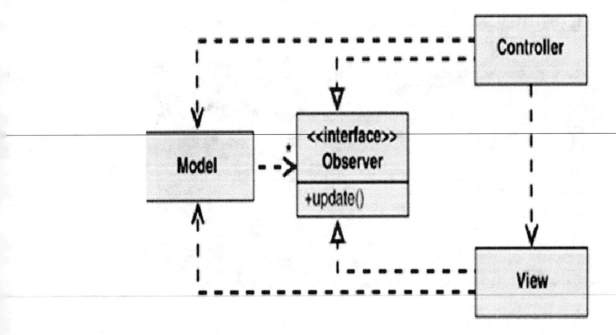

To establish a coherent HEAP it is important to model a central ARCP.

THE ENTERPRISE ARCHITECTURE PATTERN

Basics

To defined ARCP, a fundamental architecture element is the role of ICS' transformation vision; that should be crafted in an applicable framework or concept. This framework should include easy to integrate patterns. The proposed just-in-time framework can change the ICS's architecture and its implementation outcomes. The atomic ICS (aICS) can be used in HEAP, which can be applied to support aBBs of crucial importance for the implementation phase of the complex Projects; where these patterns can be adapted in a just-in-time manner, using aBBs, where an aICS is a set of aBBs (Trad, 2015a, 2015b).

Management

There is also a need to govern or control the aICS patterns for Projects. Unfortunately, adaptable aICS patterns for such projects are still in infancy age or have a hermetic approach like Microservices. An aICS pattern can be also used in the enterprise's production activities, which comes after the finalization of the implementation phase of a Project, to control and govern the resultant business system. The aICS's pattern main component is the aBB that manages the implementation of aBS. In this article the author presents a set of aICS recommendations in the form of reusable patterns to promote the optimal ICS's architecture.

The aICS Development Phases

aICS patterns present how to interact between ADM's phases and it simplifies:

- aICS' resources; as shown in Fig. 4, this article delivers the following Project artefacts: 1) The integration of aBB's concept; 2) An aICS pattern; 3) an atomic Control and Monitoring Concept; and 4) an atomic MVC (aMVC) pattern.
- The RDP integrates CSFs or independent variables in the Project for: 1) The selection; and 2) The support of the HEAP. Therefore, the HEAP supports aICS' integration and quantifies the Project's capabilities by using the following set of CSFs, like: 1) Coalition to support the vision (VIS_CSV); 2) Time for Execution (VIS_T4X); 3) Tooling adoption (VIS_TAD); 4) Proof of Concept's (PoC) capability(VIS_PCC); …
- The optimal aICS pattern for a Project should be a key solution of its implementation phase; where it builds on an existing framework like TOGAF's and where aICS' pattern application of various viewpoints is used to abstract various solution components within the Project (Taleb, & Cherkaoui, 2012). aICS' viewpoints concept specifies: 1) The requirements management phase; 2) The stakeholders risk reporting concept; 3) The sets of aBBs to be used as design models; 4) The standards to be applied; and 5) The methodologies and case-tools to be used.

Standards and Case-Tools

Project's complexity forces the team to 1) Be receptive to the implementation of stacked standards, as shown in Figure 11.; and 2) Apply evolutionary practices usage. That opens the possibility to improve the transformation integration and building of coherent and complex competitive HEAP and related CDPs, which are the basis of successful business models.

Figure 11. Project's services integration (Gartner, 2005; The Open Group, 2011b)

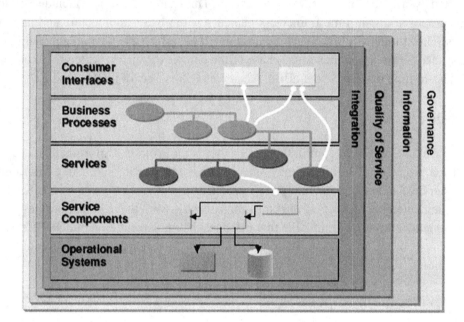

The Project interfaces various standards including the object management group's: UML, SOA Markup Language (SoaML), System Markup Language (SysML), … Whereas the use of various standards, especially methodologies, depend on their maturity, like the case of GPs and UML, as they are practically integrated and automated in implementation environments. Regardless of the applied domains, executive management understands the necessity that a Project may take many years to be finalized. Technically, this agility is built on SOA standards to be the basis of an aICS, as shown in Figure 11.

The aICS Development Phases

The ADM is a generic method for the implementation of an architecture that can be used with any type of 1) business system (re)design; 2) Project architecture; and 3) organizational restructuring and transformation. Frequent transformation iterations need a specific integration of an ADM to suit granular unbundling requirements, and that is why the *Environment* tailors the ADM to adopt the atomic resources based on the aICS pattern. The aICS pattern's ADM's integration inputs and outputs start in the Preliminary Phase.

The Preliminary Phase

This phase defines the Project's EA approach, and its outputs support the (The Open Group, 2011a): 1) Definition of the Project's key drivers; 2) Definition of requirements for an aICS based ICS; 3) Definition of EA's principles; 4) Selection of the other frameworks, and to define the relationships between these frameworks; and 5) Evaluation of the Project's EA Maturity Level.

The Vision Phase - Phase A

This Phase establishes the basis of the Project's Vision, Principles, HEAP, CDPs and other patterns, and the corresponding atomic mappings. This phase: 1) Establishes the Project's business goals and EA principles; 2) Designs an adequate strategy; 3) Sets up the aICS principles. According to existing global and EA standards, the Project's vision is mainly the jumpstart that puts forward the business benefits to be considered to the higher management. The Project's starting point (or first iteration) is to design a HEAP based Vision that has to include: 1) The business requirements' management patterns and objectives; 2) The Project's strategic drivers and stakeholders; 3) Governance's principles; and 4) The design and development of the atomic resources patterns (The Open Group, 2011a). In fact, the proposed HEAP defines sets of CDPs that map to aBBs (Trad, 2015a, 2015b) to assist the building of the Project; that is the essence of the aICS based on TOGAF. This Phase's outcome is the design of the major aBBs and their relationship to HEAP, where one aICS pattern uniquely identified by a GUID can contain many aBBs, as shown in Table 1. This is basically an alignment, based on the 1:1 mapping of all of Project's artefacts.

Table 1. The Vision Phase's main artefacts

GUID	EA Continuum	Unique Name	Managed Artefacts	Tools/Standards
GUID_xxx	aBB set, aSB set…	GUID_xxx_aICS	INPUT: None. OUTPUT: Define a basic aICS pattern.	Archimate/UML profile

The Requirements' Management Phase – Phase R

Upon a concrete business or infrastructure requirement, the Project requests a *Business Transformation Work Contract* to implement it. The Project team ensures that new requirements are implemented. Then the requirement is linked to an instance of an aBB, and its instance or an atomic Solution Block, (aSB), as shown in Table 2. The aBB is a part of an aICS pattern.

Table 2. The Requirement Phase's main artefacts

GUID	EA Continuum	Unique Name	Managed Artefacts	Tools/Standards
GUID_ xxx	aBB set, aSB set...	GUID_xxx_aICS	INPUT: Requirement contract. OUTPUT: create and add: Use Case, BPM, Class Diagram, Activity diagram, Business service.	Archimate/ XMI/UML/SysML

The Business Architecture Phase – Phase B

The business architecture phase inputs the implemented aBB and its aSB, then the Project's team implements the solution, as shown in Table 3. The outputs are: 1) A use case diagram; 2) A BPM; 3) Business rules artefacts; 4) An activity diagram; and 5) An aBS diagram. At the end, the required, aBB and aSB resources are stored in the Project's EA Continuum.

Table 3. The Business Phase's main artefacts

GUID	EA Continuum	Unique Name	Managed Artefacts	Tools/Standards
GUID_ xxx	aBB set, aSB set...	GUID_xxx_aICS	INPUT: aBB/ aSB archive. OUTPUT: modifies: Use Case, BPM, Business rules, Class Diagram, Activity diagram, Business service.	Archimate/BPM/RuleML/UML

The ICS architecture phase "C"

The ICS architecture phase inputs the aBB and its instance (aSB); then ICSs' architects (or system analysts) develop this phase's outputs, by modifying some of its artefacts, as shown in Table 4. These outputs are: 1) the update of the activity diagram; 2) the development of the atomic business services; 3) the development of the service-oriented architecture web service; 4) the deployment diagram; and 4) the development of the atomic Model-View-Control (aMVC) concept (Caplin, 2012; Nimit, 2014). When finalized, the aBB resources are stored in the aICS pattern in the Project's repository.

Table 4. The Phase's C main artefacts

GUID	EA Continuum	Unique Name	Managed Artefacts	Tools/Standards
GUID_xxx	aBB set, aSB set…	GUID_xxx_aICS	INPUT: aBB/aSB archive. OUTPUT: modifies: Use Case, BPM, Class Diagram, Activity diagram, Business service, aMVC, class diagram.	XMI/BPMN/WSDL/UML/ArchiMate/SoaML/SysML

The Technology Architecture Phase-Phase D

This Phase inputs aBBs implementation and their instances (aSBs), then the implementation process delivers outputs, as shown in Table 5. These outputs are: 1) Sequence diagram(s); 2) SOA based services' deployment concept; 3) Deployment diagram(s); and 4) An atomic MVS (aMVC) concept corresponding to an aBB. When finalized, the aBB/aSB resources are updated in the aICS pattern and then stored in the Project's repository.

Table 5. The Phase's D main artefacts

GUID	EA Continuum	Unique Name	Managed Artefacts	Tools/Standards
GUID_xxx	aBB set, aSB set…	GUID_xxx_aICS	INPUT: aBB/aSB archive. OUTPUT: modifies: Sequence diagram, web service, aMVC.	XMI/DPs/CDP/ArchiMate/SoaML/SysML

The Opportunities and Solutions Phase-Phase E

This Phase inputs aBBs and their instances (aSBs) to be modified; then the Project team implements the requested outputs, as shown in Table 6. The outputs in the form of aBB/aSB resources are stored in the Continuum. This phase compares two consecutive iterations and evaluates the Gap(s) and possible solutions in the forms of SBBs and sBBs.

Table 6. The Phase's E main artefacts.

GUID	EA Continuum	Unique Name	Managed Artefacts	Tools/Standards
GUID_ xxx	aBB set, aSB set…	GUID_xxx_aICS	INPUT: aBB/ aSB archive. OUTPUT: modifies: Sequence diagram, web service, aMVC.	XMI/UML ArchiMate

The Migration Planning Phase-Phase F

This Phase inputs aBBs and their instances (aSBs) to be modified; then the Project team implements the requested outputs, as shown in Table 7. These outputs are: 1) Use Case diagram(s); 2) BPMs; and 3) Project plan(s). When finalized, the aBB/aSB resources are stored in the aICS pattern in the Project's Continuum.

Table 7. The Phase's F main artefacts.

GUID	EA Continuum	Unique Name	Managed Artefacts	Tools/Standards
GUID_ xxx	aBB set, aSB set…	GUID_xxx_aICS	INPUT: aBB/ aSB archive. OUTPUT: modifies: Use case diagram, BPM, project plan, XP, PMXM.	XMI/PMXML/BPMN/XPDL

The Implementation Governance Phase-Phase G

This Phase inputs aBBs and their instances (aSBs) to be modified; then the Project team implements the requested outputs, as shown in Table 8.

Table 8. The Phase's G main artefacts.

GUID	EA Continuum	Unique Name	Managed Artefacts	Tools/Standards
GUID_ xxx	aBB set, aSB set…	GUID_xxx_aICS	INPUT: aBB/ aSB archive. OUTPUT: modifies: project plan, PMXML, ADM, XP.	XMI/PMXML/BPMN/XPDL

These outputs are: 1) Use Case diagram(s); 2) BPM(s); and 3) Project plan(s). When finalized, the aBB/aSB resources are stored in the aICS pattern in the Project's Continuum.

Change Governance Phase- Phase H

This Phase inputs aBBs and their instances (aSBs) to be modified; then the Project team implements the requested outputs, as shown in Table 9. These outputs are: 1) the use case diagram; 2) the business process model; and 3) Project plan(s). When finalized the aBB/aSB resources are stored in the aICS pattern in the Project's Contiunum.

Table 9. The Phase's H main artefacts

GUID	EA Continuum	Unique Name	Managed Artefacts	Tools/Standards
GUID_xxx	aBB set, aSB set…	GUID_xxx_aICS	INPUT: aBB/aSB archive. OUTPUT: modifies: Use case diagram, BPM, project plan.	XMI/PMXML/BPMN/XPDL

As SOA is fundamental for TOGAF, ADM and other disciplines, the SARP will be analysed; SARP s' are stored in the Continuum.

THE SOA PATTERN (SARP)

This design patterns catalogue (published by Arcitura Education) supports SOA standards.

Types of SARP

These patterns encompass service-oriented architecture and service technology (Arcitura, 2020):

- Foundational Inventory Patterns: Canonical Protocol, Canonical Schema, Domain Inventory, Enterprise Inventory, Logic Centralization, Service Layers, Service Normalization.
- Logical Inventory Layer Patterns, Entity Abstraction, Process Abstraction, Utility Abstraction, Micro Task Abstraction.
- Inventory Centralization Patterns: Policy Centralization, Process Centralization, Rules Centralization, Schema Centralization.
- Inventory Implementation Patterns: Canonical Resources, Cross-Domain Utility Layer, Dual Protocols, Inventory Endpoint, Service Grid, State Repository, Stateful Services, Augmented Protocols.

- Inventory Governance Patterns: Canonical Expression, Canonical Versioning, Metadata Centralization.
- Foundational Service Patterns: Agnostic Capability, Agnostic Context, Functional Decomposition, Non-Agnostic Context, Service Encapsulation.
- Service Implementation Patterns: Partial State Deferral, Partial Validation, Redundant Implementation, Service Data Replication, Service Façade, UI Mediator, Reference Data Centralization, Microservice Deployment, Containerization.
- Service Security Patterns: Exception Shielding, Message Screening, Service Perimeter Guard, Trusted Subsystem.
- Service Contract Design Patterns: Concurrent Contracts, Contract Centralization, Contract Denormalization, Decoupled Contract, Validation Abstraction, Legacy Encapsulation Patterns, File Gateway, Legacy Wrapper, Multi-Channel Endpoint.
- Service Governance Patterns: Compatible Change, Decomposed Capability, Distributed Capability, Proxy Capability, Service Decomposition, Service Refactoring, Termination Notification, Version Identification.
- Capability Composition Patterns: Capability Composition, Capability Recomposition
- Service Messaging Patterns: Asynchronous Queuing, Event-Driven Messaging, Intermediate Routing, Messaging Metadata, Reliable Messaging, Service Agent, Service Call-back, Service Instance Routing, Service Messaging, State Messaging.
- Composition Implementation Patterns: Agnostic Sub-Controller, Atomic Service Transaction, Compensating Service Transaction, Composition Autonomy.
- Service Interaction Security Patterns: Brokered Authentication, Data Confidentiality, Data Origin Authentication, Direct Authentication.
- Transformation Patterns: Data Format Transformation, Data Model Transformation, Protocol Bridging.
- REST inspired Patterns: Entity Linking, Lightweight Endpoint, Reusable Contract, Content Negotiation, Endpoint Redirection, Idempotent Capability.
- Composite Patterns Overview: Canonical Schema Bus, ESB, Federated Endpoint Layer, Official Endpoint, Orchestration, Service Broker, Three-Layer Inventory, Uniform Contract.

Services and Business Processes

Today, there are many types of services, like standard functions, business services, SOA/SOAP, REST, Microservices,… In order to simplify services integration in the 1:1 concept, the authors propose the aBB which has the following capabilities (Trad, 2015a, 2015b): 1) To facilitate the Project's implementation phase and the usage of standards; 2) To manage and assemble Project's BBs; 3) To make services agile, reusable and easily replaceable; 4) Can be implemented in many SBBs; 5) Has a GUID; 6) It asserts the Project's 1:1 mapping concept; and 6) Enables business activities' interoperability and integration. SARP embeds the aBS pattern in order to offer a simple interface which can be managed by business specialist. aBSs must be adequately classified and interconnected using interoperability standards (Fuego, 2006). SARP is one of the most important and complete set of patterns; analysing the number and content of these patterns, which shows the complexity of usage and integration of various sets of patterns, like, the BPMP.

THE BPM PATTERN

BPMPs are patterns that show how to model and connect activities together, in order to solve a Project Problem. BPs are like motorways, as we drive, we become used to similar and time proven motorways. Countries ensure that their engineers follow proven specifications therefore motorway constructions are consistent. BPMP are the specifications of motorways of BPs (Fuego, 2006):

- Basic Control Patterns: Sequence Pattern, Exclusive Choice Pattern, Simple Merge Pattern, Parallel Split and Synchronization Patterns.
- Advanced Branching and Synchronization Patterns: Multiple Choice and Synchronizing Merge Patterns, Discriminator and N-out-of-M Join patterns, and Multiple Merge Pattern.
- Structural Patterns: Arbitrary Cycles Pattern, Collaboration Pattern, Implicit Termination Pattern.
- Multiple Instance Patterns: Multiple Instances without Synchronization Pattern, Multiple Instances with Design and/or Runtime Knowledge Patterns.
- State Based Patterns: Deferred Choice Pattern and Milestone Pattern.
- Cancellation Patterns: Cancel Activity Pattern and Cancel Case Pattern.

That is why it is recommended to englobe various pattern sets, including RESP.

THE API AND REST PATTERN

The REST Pattern

The REST pattern is based on Create Read Update Delete (CRUD) based pattern, which embeds the following operations: 1) POST, which lists, paginates, filters the lists of attributes of object(s); 2) GET, which retrieves the representation of an object; 3) PATCH which updates specific attributes of an object; and 4) DELETE which deletes a specific attribute from an object.

The API Pattern

Figure 12. API Platform-Provider (Oracle, 2016a, 2016b)

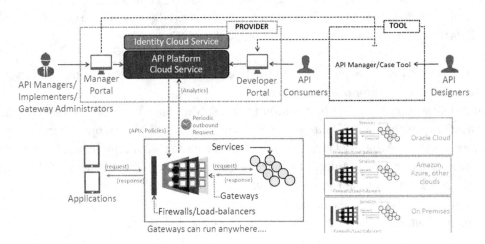

To minimize risks, Entities need to ensure their APIs are high-performance, controlled, robust, and secure. An API has a complex flow as shown in Figure 12. Its gateway is the bridge to access stored APIs, also by securing and managing traffic intra API consumers and the SBBs (or sBB) that expose the needed APIs. Such a bridge handles authentication and authorization, request routing to backends, rate limiting to avoid overloading systems and protect against security attacks, and to handle errors or exceptions (Oracle, 2016a, 2016b; Sidell, 2020).

Figure 13. API growth is exponential (Oracle, 2016a, 2016b)

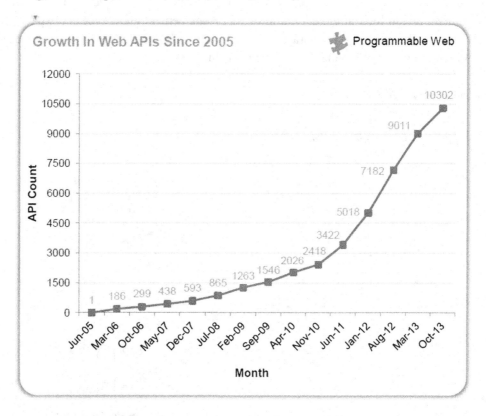

In contrast, API management refers to the process of managing APIs across their full lifecycle, including defining and publishing them, monitoring their performance, and analysing usage patterns to maximize business value (Oracle, 2016a, 2016b; Sidell, 2020). Taking in account their fulgurant growth as shown in Figure 13.

Figure 14. API-From Generation Zero to 3rd Generation API Management (Oracle, 2016a, 2016b).

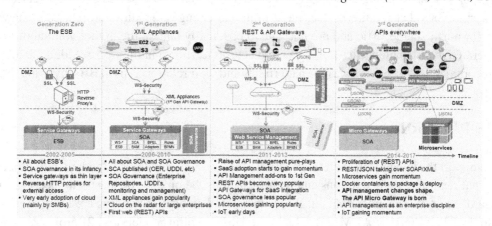

As shown in Figure 14, it is interesting to view the various phases of evolution of perspectives on API Management; and to look at the evolution of API Management and of the API space, in general. Such a view on how the space and how practices have evolved allows us to better understand why terms such as API Management have changed meaning over time.

The API Gateway Pattern

Figure 15. Gateway for Services (Sidell, 2020)

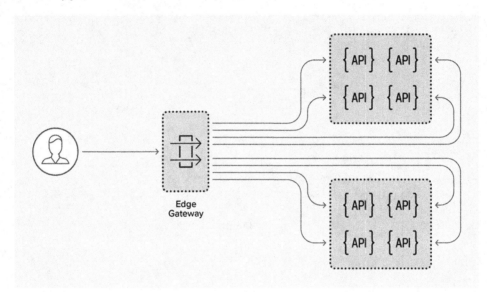

This is the most common API gateway pattern and follows a traditional Application Delivery Controller (ADC) architecture. In this pattern, the gateway handles almost everything, including (Sidell, 2020): SSL/TLS termination, Authentication, Authorization, Request routing, Rate limiting, Request/response manipulation, and Façade routing. The Gateway approach is optimal for publicly exposing aBSs from monolithic applications with centralized governance. But it is not well-suited for Microservices architectures or situations that require frequent and profound changes. Traditional interface gateways are optimized for north-south traffic and are not able to efficiently handle the higher volume of east-west traffic generated in distributed Microservices architectures as shown in Figure 15 (Sidell, 2020). That is why it is recommended to englobe various pattern sets and their relationships in BBs, by using the ESBP.

THE ESB PATTERN

The ESBs support the integration of all the mentioned technology standards. Standardized Projects must be transparent regarding their solutions and their focus must be on their business engineering choreography, regardless of the business domain. BTMs integrated enterprise patterns are using the following methodologies: 1) TOGAF's ADM that adopts the UML's spiral model; and 2) project management concepts. Once the Project's standards are established, a pre-enterprise patterns architecture blueprint must be defined. If the unbundling process is successful, the Project maps all DPs to services and BBs. These BBs or services can be called via the enterprise service bus, as shown in Figure 16.

Figure 16. The enterprise service bus integration (Netapsys, 2016)

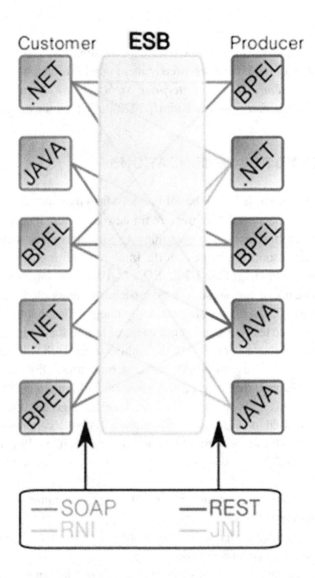

The Project identifies the set of main patterns and related patterns to implement the HEAP, which addresses these various topics and delivers a common concept.

The HEAP

The HEAP can be used to abstract EA artefacts for Projects (Trad, & Kalpić, 2022), where transformation engineers implement interconnected patterns. The complexity, diversity and cross-functional nature of EA and other Project's domains require that various categories of patterns should be developed and classified in various disciplines, domains, and levels of precision. The integration of various categories of patterns and their non-standardization make this topic lack of maturity (The Open Group, 2018).

THE POC

The proposed solution and structure are evaluated through the implementation of a real-world business transaction with the concept of the HEAP; as already mentioned this chapter is related to the *Business Transformation Projects based on a Holistic Enterprise Architecture Pattern (HEAP)-The implementation*, and its implementation chapter (Trad, & Kalpić, 2022), in which the complete PoC is presented.

THE CONCLUSION AND RECOMMENDATIONS

The outcome of building avant-garde HEAP based Projects, heavily depends on the use of patterns and existing standards like the Open Group's architecture framework, where the use of generic patterns like aBBs and their solution blocks instances in transforming a traditional business environment into an agile environment simplifies and ensures the success of the implementation phase. The HEAP base Project interfaces various standards including, like: UML, SOA Markup Language (SoaML), System Markup Language (SysML), … The use of various standards, especially methodologies, depends on their maturity, like the case of GPs and UML, as they are practically integrated and automated in implementation environments. Legacy ICSs and teams are the major cause of failure for the Projects which are essential for businesses companies. This fact has motivated the authors to research various techniques to promote and recommend solutions, like the use of HEAP based architectures. DPs can be used for integrating various types of patterns, to support the HEAP, EA models and BPM modules (Microsoft, 2016a; Design Patterns). The RDP and the PoC propose a set of technical and managerial recommendations for the usage of HEAP; which can be used to model and analyse main system components of the future systems. This chapter presents the HEAP and the optimal manner to integrate it in a Project. The authors targeted the use of existing patterns and they forward the following set of recommendations:

- To use the generic pattern's concept that incorporates: 1) Basic patterns; 2) Classification for patterns; and 3) Unbundling process.
- A coherent HEAP supports a design-driven development or the design-first approach.
- The role of OO is essential for a HEAP based Project.
- The role of UML is central and serves as a common architecture language.
- To use GUID for Project's artefacts.
- For the data architecture, class diagrams can be mapped to table constructs where a class corresponds to a table or an OO object.
- Ensure the definition of the optimal unit of work that will align all the *Project*'s resources using the 1:1 concept.
- A 1:1 mapping concept must be applied to services, BBs, composite patterns.
- To use composite models like design patterns in HEAP.

- To apply an aBB architecture strategy and pattern that will facilitate the implementation phase of the Projects.
- The atomic approach supports a bottom-up approach that is based on the choreography of aBSs.
- aBSs are in turn built on atomic BPMs (IBM, 2014).
- DPs describe a common structure for a general design problem within a Project implementation; and supports agile Project's complexity in its implementation phase.
- Composite DPs incorporate: 1) Basic patterns; 2) Design patterns; and 3) Enterprise patterns.
- An ESB must be used as a communication highway and must impact on the composite patterns' architecture.
- Apply the ADM's integration in Projects, using the HEAP, to promote the concept of an iterative assembling to integrate composite patterns.

REFERENCES

Alexander, C. (1977). *A Pattern Language*. Oxford University Press.

Alexander, C. (1979). *The Timeless Way of Building*. Oxford University Press.

Alexander, C., Silverstein, M., Angel, Sh., Ishikawa, S., & Abrams, D. (1975). *The Oregon Experiment*. Oxford University Press.

Altman, R. (2014). *Hype Cycle for Application Architecture*. Gartner.

Arcitura. (2020). *SOA Patterns*. Arcitura. https://patterns.arcitura.com/soa-patterns

Burbeck, S. (n.d.). *Application Programming in Smalltalk-80: How to use Model-View-Controller (MVC)*. University of Illinois in Urbana-Champaign (UIUC) Smalltalk Archive.

Buschmann, F., Meunier, R., Rohnert, H., Sommerlad, P., & Stal, M. (1996). *Pattern-Oriented Software Architecture: A System of Patterns*. Wiley.

Camarinha-Matos, L. M. (2012). *Scientific research methodologies and techniques- Unit 2: Scientific method. Unit 2: Scientific methodology. PhD program in electrical and computer engineering*. Uninova.

Capgemini. (2007). *Trends in Business transformation - Survey of European Executives*. Capgemini Consulting and The Economist Intelligence Unit.

Capgemini. (2009). *Business transformation: From crisis response to radical changes that will create tomorrow's business. A Capgemini Consulting survey*. Capgemini.

Caplin, M. (2012). *MVC Techniques with jQuery, JSON, Knockout, and C#*. www.codeproject.com

Crosswell, A. (2014). *Bricks and the TOGAF TRM*. National Institute of Health.

Dong, J., & Yang, Sh. (2003). *Visualizing Design Patterns With A UML Profile. Department of Computer Science*. University of Texas.

Downey, K. (2003). *Architectural Design Patterns for XML Documents*. https://www.xml.com/pub/a/2003/03/26/patterns.html

Enterprise Architecture Solutions. (2021). *Essential Meta Model*. Enterprise Architecture Solutions. https://enterprise-architecture.org/docs/introduction/essential_meta_model/

Escalona, J., & Koch, N. (2004). *Requirements engineering for web applications – A comparative Study*. University of Seville.

Farhoomand, A. (2004). *Managing (e)business transformation*. Palgrave Macmillan. doi:10.1007/978-1-137-08380-7

Fowler, M. (1996). *Analysis Patterns: Reusable Object Models*. Addison-Wesley.

Fowler, M. (2003). *Catalog of Patterns of Enterprise Application Architecture*. https://martinfowler.com/eaaCatalog

Fowler, M. (2014). *Microservices*. https://martinfowler.com/articles/microservices.html. USA.

Fuego. (2006). *BPM Process Patterns-Repeatable Designs for BPM Process Models*. Fuego.

Gamma, E., Vlissides, J., Helm, R., & Johnson, R. (1994). *Design Patterns: Elements of Re-usable Object-Oriented Software*. Addison-Wesley.

Gartner. (2005). *External Service Providers' service oriented architecture Frameworks and Offerings: Capgemini*. Gartner.

Gartner. (2013a). *Scenario Toolkit: Using EA to Support Business Transformation*. Gartner Inc.

Gartner. (2013b). *Hype Cycle for Business Process Management*. Gartner.

GitBook. (2021). *ArchiMate Guide-ArchiMate and TOGAF Layers*. GitBook. https://archimatetool.gitbook.io/project/archimate-and-togaf-layers

IBM. (2014). *Smart Service Oriented Architecture: Helping businesses restructure. Where are you on the Service Oriented Architecture adoption path?* IBM.

Kleppe, A., Warmer, J., & Bast, W. (2003). *MDA Explained: The Model Driven Architecture-Practice and Promise*. Addison-Wesley.

Mapelsden, D., Hosking, J., & Grundy, J. (2002). *Design Pattern Modelling and Instantiation using DPML*. Department of Computer Science, University of Auckland.

Microsoft. (2016a). *Code Snippets for Design Patterns*. CodePlex.

Microsoft. (2016b). *Model-View-Controller*. https://msdn.microsoft.com/en-us/library/ff649643.aspx

Netapsys. (2016). *Les ESB, exemple particulier de Mule ESB*. http://blog.netapsys.fr/les-esb-exemple-particulier-de-mule-esb

Nimit, J. (2014). *Model First Approach in ASP.Net MVC 5*. C Sharp-corner. https://www.c-sharpcorner.com/UploadFile/4b0136/model-first-approach-in-Asp-Net-mvc-5/

Oracle. (2002). *Core J2EE Patterns - Data Access Object*. Oracle. https://www.oracle.com/technetwork/java/dataaccessobject-138 824.html

Oracle. (2016a). *API Management in 2026*. Oracle.

Oracle. (2016b). *Evolution and Generations of API Management*. Oracle.

Perroud, T., & Inversini, R. (2013). *Enterprise Architecture Patterns*. Practical Solutions for Recurring IT-Architecture Problems. doi:10.1007/978-3-642-37561-3

SAP. (2013). GBTM: Global Business Transformation Manager Master Certification (SAP Internal). Business Transformation Academy. SAP.

Sarcar, V. (2016). *Java Design Patterns: A tour of 23 gang of four design patterns in Java*. Apress. doi:10.1007/978-1-4842-1802-0

Sidell, E. (2020). *Choosing the Right API Gateway Pattern for Effective API Delivery*. NGINX. https://www.nginx.com/blog/choosing-the-right-api-gateway-pa ttern/

Stojanović, Z. (2005). *A Method for Component-Based and Service-Oriented Software Systems Engineering*. Technische Universiteit Delft. doi:10.4018/978-1-59140-426-2

Störrle, H., & Knapp, A. (2006). *UML 2.0 – Tutorial/Unified Modeling Language 2.0*. University of Innsbruck.

Taleb, M., & Cherkaoui, O. (2012, January). Pattern-Oriented Approach for Enterprise Architecture: TOGAF Framework. *Journal of Software Engineering & Applications*, 5(1), 45–50. doi:10.4236/jsea.2012.51008

The Open Group. (2011a). *Phase P: The Preliminary Phase*. The Open Group. https://pubs.opengroup.org/architecture/togaf9-doc/arch/chap 06.html

The Open Group. (2011b). *TOGAF's The Open Group Architecture Framework*. www.open-group.com/togaf

The Open Group. (2018). *Architecture Patterns*. The Open Group. https://pubs.opengroup.org/architecture/togaf9-doc/arch/chap 22.html

Trad, A. (2015a). *A Transformation Framework Proposal for Managers in Business Innovation and Business Transformation Projects-Intelligent aBB architecture*. Centeris.

Trad, A. (2015b). *A Transformation Framework Proposal for Managers in Business Innovation and Business Transformation Projects-An ICS's atomic architecture vision*. Centeris.

Trad, A., & Kalpić, D. (2016a). *A Transformation Framework Proposal for Managers in Business Innovation and Business Transformation Projects-Basics of a patterns based architecture.* Centeris.

Trad, A., & Kalpić, D. (2016b). *A Transformation Framework Proposal for Managers in Business Innovation and Business Transformation Projects-Design patterns based architecture.* Centeris.

Trad, A., & Kalpić, D. (2016c). *A Transformation Framework Proposal for Managers in Business Innovation and Business Transformation Projects-Enterprise patterns based architecture.* Centeris. doi:10.1016/j.procs.2016.09.158

Trad, A., & Kalpić, D. (2022). Business Transformation Projects based on a Holistic Enterprise Architecture Pattern (HEAP)-The Implementation. IGI.

Walker, J. (2019). *Generics & Design Patterns.* Oberlin College Computer Science. https://www.cs.oberlin.edu/~jwalker/langDesign/GDesignPat/

Chapter 4
Business Transformation Projects:
The Role of Requirements Engineering (RoRE)

Antoine Trad
https://orcid.org/0000-0002-4199-6970
IBISTM, France

ABSTRACT

The role of requirements engineering (RoRE) is central for the implementation of projects in general and is especially crucial for business transformation projects (or simply projects) because transformation activities incur major changes in the existing sets of archaically defined requirements. Requirements engineering (RE) is a complex part of the project because it consists of many related communication, cross-functional knowledge, and dependencies, like the need for RE to interact with executive management, business users, business architects, implementation developers, and other project actors. In this chapter, the authors will try to propose an RoRE concept (RoREC) that can support project managers (or simply managers) in transforming the enterprise and managing RE activities. RE activities are not just about assembling a huge set of business and non-functional requests and features in the form of document(s) and delivering a repository of methodology-based diagram(s).

INTRODUCTION

This chapter's authors based their research work on a mixed method that is supported by a Heuristic Decision Tree (HDT) (Trad & Kalpić, 2020a), which is applied in this chapter's Proof of Concept (PoC). The proposed AHMM4RE based RoRE uses an empiric process that can be compared to the Behaviour-Driven Development (BDD) methods, which are optimal for complex Projects, where a requirement maps to all its related resources. In this chapter the PoC is supported by a business case, abstracted by the RoRE Pattern (RoREP) that is supported by the alignment of various existing standards, methodologies, and

DOI: 10.4018/978-1-6684-4102-2.ch004

development strategies, like the Development and Operations for RE (DevOps4RE), and other engineering procedures which are used to support requirements' development. RoRE is based on confirmed cases, academic resources, and empirical experiences for detecting and processing of requirements, business transformation, artefacts engineering and Enterprise Architecture (EA) procedures. This business requirements driven approach offers a set of possible recommendations in the form of RoREP practices. These recommendations are to be applied by Managers, business architects, analysts, and engineers to propose solutions for transformation problems. RoREP uses an EA driven approach and offers a set of patterns to support Project*s*, especially its holistic requirements management and classification, management, and technical guidance, coupled with a mapping model to link them to other Project*'s* resources (Trad, & Kalpić, 2018a, 2018b). The proposed RoRE recommendations are to be applied on how to instantiate RoREP to enable digital transformations' iterations using a robust background.

BACKGROUND

This chapter's background mixes RE, RoREP, HDT, DevOps4RE, Knowledge Management System for RE (KMS4RE), EA, ICS management, and transformation engineering fields (Ebert, Gallardo, Hernantes, & Serrano, 2016). Applying RoREP instances for the mapping of requirements to the company's or organization's (simply the Entity's) strategic Decision-Making System for RE (DMS4RE) processes, is the main strategic goal for the transformed Entity, because DMS4RE integration is done right in the beginning of the Project and not at its end where teams try to find ad-hoc solutions (Trad, 2020).

Figure 1. RoREC interactions

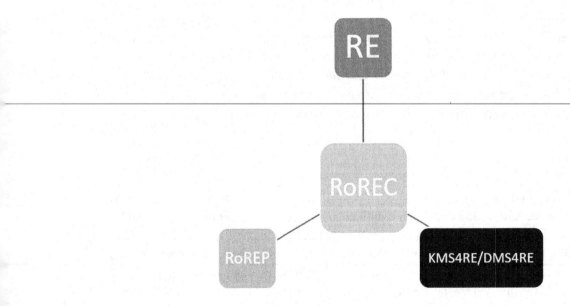

As shown in Figure 1, the RoREC is the concept that supports the interaction between RE activities, RoREP instances, and KMS4RE/DMS4RE which supports problem solving. The RoREP is a cross-domain pattern that interacts with the HDT, which manages sets of Critical Success Factors (CSF) and Areas (CSA) that can be used by the Project and its SIP to solve problems (Uhl, & Gollenia, 2012). The Research and Development Project (RDP) uses adapted HDT and BDD approaches, where both approaches resemble to empirical basic reasoning, or what can be called a learning process, like Action Research (AR) or Machine Learning (ML). This chapter's RDP for RE (RDP4RE), relies on various Project processes and components like, the DMS4RE, KMS4RE, and the RoREC (Trad & Kalpić, 2018a, 2018b). The RoREP is business driven and is agnostic to any specific business domain; and is founded on an avant-garde in-house framework that in turn is based on existing standards, like the Architecture Development Method for RE (ADM4RE) (The Open Group, 2011a, 2011b). Today, there are many methodologies, which can be used to develop Projects where the Project's team can use RoREP instances to synchronize ADM4RE cycles. RoREC's aim is to deliver recommendations for managing and aligning Project's processes like, RE's integration, ADM4RE and DevOps4RE with all other Project's resources. The RDP4RE is based on Literature Review Process for RE (LRP4RE) and a Qualitative Analysis for RE (QLA4RE). As shown in Figure 2, RoREP instances interact and are used by various Project's resources via an implemented DevOps4RE life management interface, which is used to synchronize and deploy versions of ICS components (CenturyLink, 2019).

Figure 2. DevOps4RE's Application Lifecycle Management concept (CenturyLink, 2019)

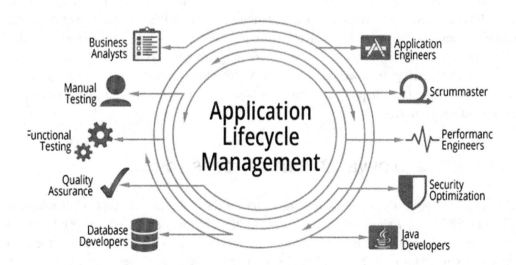

Synchronizing Project's resources, RE artefacts and modules, is a complex task and the main problem can arise due to the lack of the Entity's excessive and archaic RE related agility approach, that is why the RoREC will try to prove that it can be built on intelligent and coordinated mapping of all the Project's resources. Adapting archaic RE relates agilization methodologies for siloed ICS modules is not sufficient for dynamic Projects and is immature for the SIP (Trad, & Kalpić, 2022; Lindgren, & Münch, 2015). Due to Project stakeholders, who want to make excessive gains, there is immense pressure to implement such

Project*s* faster with the condition to keep ICS' high levels of robustness and availability. This implies that there is a need for mapping requirements to other Project*'s* resources (Kornilova, 2017), and that is this RDP4RE's and chapter's focus.

FOCUS OF THE CHAPTER

A digital transformation project supports the tracing and control of digital flows like in the domain of finance, where it can help the control of financial crimes (Trad, Nakitende, & Oke, 2021). Digitization is the process of converting information into a digital format which automates the Information and Communication Systems (ICS); and the result of an EA based transformation project is in fact the enterprise's digitization; that supports the enterprise's evolution (Gartner, 2016). EA methodologies, such as The Open Group's Architecture Framework's (TOGAF) and its dynamic ADM4RE, support essential activities are related to the specification of: 1) Project*'s* requirements' related documents; 2) Communications needed to verify the strategy and the defined goals; and 3) Synchronize DevOps4RE activities. Actual archaic RE techniques for Project*s* focus mainly (or even only) on prose-based documents, unsynchronized tools, noncentralized services/processes, and siloed ICS components, which generate Project*'s* volatility. The Manager and the Project team must ensure minimum support for RoREC's modelling, RoREP's implementations, and to define RE related strategic goals, like stakeholder concerns, which are uniquely financial. Unfortunately, such concerns are purely accountants' financial goals, and address only short-term commercial concerns, which are destructive for Project*s* in the long-term. As already mentioned, this chapter's focus is on the RoREC that supports the modelling and mapping of requirements to all Project*'s* resources, Microartefacts and modules. The RoREC is based on existing standards and frameworks, and for that goals, Microartefacts/Requirements (M/R) are aligned with the ADM4RE, atomic Building Blocks (aBB) concept, Unified Modelling Language (UML) or ArchiMate modelling languages. The aBB concept is based on TOGAF's Architecture Building Blocks (ABB) and Solution Building Blocks (SBB). This chapter also illustrates how Project*s* can benefit from the holistic RoREC and proposes an adequate RDP4RE.

THE RESEARCH AND DEVELOPMENT PROCESS FOR RE

The author's research project's keywords were introduced in the scholar engine (in Google's search portal) and the results have clearly shown the uniqueness and the absolute lead of the author's methodology, research and works. From this point of view and facts the author considers his works on the mentioned topics as successful and useful. This RDP's global topic is related to Project*s,* and in this continuous phase the Research Question (RQ) is: "Which RoREC features, characteristics, and which type of pattern should be used in the implementation phase of a Project?". Where the RDP4RE is based on a RE, ADM4RE, DMS4RE, CSFs, and CSAs. The major innovation in this chapter is linking of the Managers' beloved *Strengths (S), Weaknesses (W), Opportunities (O), and Threats* (T) (SWOT) basic method to CSAs, CSFs, and Key Performance Indicators (KPI).

Critical Success Areas, Factors Management

The SWOT analysis is one of the possible strategy views, supported by ArchiMate®. The SWOT analysis supports the evaluation of project's internal strengths and weaknesses, which includes needed resources and current capabilities, which can enable (or hinder) the implementation of transformation strategies. It can also reflect its external opportunities and threats for CSAs which enable (or hinder) project's performance. The SWOT analysis view supports the design of strategies that are aimed to exploit deduced Os and Ss, while proactively predicting possible Ts and to eventually improve Ws, and applying the following strategies (BizzDesign, 2022):

- WT strategy: minimize both Ws and Ts, to enable the achievement of defined goals.
- WO strategy: minimize Ws in order to be able to take advantage of new Os.
- ST strategy: maximize Ss in order to be able to deal with Ts.
- SO strategy: maximize Ss, using existing resources to promote new Os.

SWOT risk analysis is a basic technique to check Project's strategy or a Business Unit's (BU) viability; where the target domain can be local or global. SWOT checks Project's strengths, weaknesses, opportunities and threats or constraints, which can establish a link to the initial sets of CSAs, CSFs, and KPIs. The linked sets of factors can be external (O and T) and internal (S and W). SWOT is used for the Project's preparation activities, which is known as the AMD4RE's preliminary phase (Remawati, 2016; Wati, Ranggadara, Kurnianda, Irmawan, & Frizki, 2019). Managers take decisions for formulating a Project strategy based on the analysis of the external and internal CSAs and hence CSFs and KPIs. Strengths, weaknesses, opportunities, and threats (or SWOT) is an established concept for the categorization of highly important CSFs. An important CSF is how the Project can achieve and sustain a significant business competitive advantage, that can be done using value chain analysis. The Project's strategy shows how this value can be developed (Kitsios, Kyriakopoulou, & Kamariotou, 2022). So, to determine CSFs, there is a need to review SWOT items and they should reflect: 1) Project based on **S**; 2) Eliminate **W**; 3) Exploit **O** (by using **S**); and 4) Implement strategies to intercept **T**. CSFs are a key element in the Project and its business planning. A CSA is a category (or set) of CSFs where in turn a CSF is a set of KPI, where a KPI maps (or corresponds) to a single requirement and/or software feature, known as a Microartefact. As shown in Figure 3, for a given requirement and/or a Project problem, the Manager (or any Project team member) can identify the initial set of related CSAs, CSFs and KPIs, for the use in the SWOT analysis or DMS4RE, and they map to the sets of process solutions or recommendations (Trad, 2020). Hence, CSFs are important for the mapping between various types of RE artefacts, KMS4RE knowledge constructs, RoREP Microartefacts, Entity's organisational items, and the DMS4RE. Therefore, CSFs reflect areas that must meet the main strategic Project goals and predefined financial constraints. Measurement techniques, which are provided by the RDP4RE's framework (or simply the Framework), are used to evaluate the performance for each CSA, where CSFs can be internal or external (Trad & Kalpić, 2020a). Once the initial sets of CSFs and CSAs have been identified, then they can be used by the DMS4RE to generate solutions for a given RoREC problem and the gained knowledge/experience can be fed in the Entity's DMS4RE/KMS4RE; and that is how an Entity can build its own learning process. The CSF-based RDP4RE uses the DMS4RE in all the Framework's modules and all the Project's phases. RDP4RE's phase 1 (represented in pseudo-SWOT decision tables), and they form the empirical part of the RDP4RE. These tables check the following CSAs: 1) The RDP4RE, which is synthesized in Table 1; 2) The ICS, which is synthesized in Table 2; 3) The ADM4RE, which is synthesized in Table 3; 4) The KMS4RE/ DMS4RE, which is synthesized in Table 4; 5) The RoRE and RoREC Domain; which is synthesized in Table 5; and 6) Is the chapter's complete outcome, synthesized in Table 6. The RoRE chapter, delivers a set of recommendations and solutions for a Project, which uses an in-house Framework.

Figure 3. The relations between the preliminary phase and other components

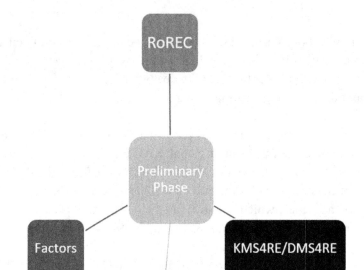

The Proposed Research Framework

RoREC's alignment strategy manages the Entity's *and* Project*'s* Microartefacts libraries and requirements that are implemented by the means of various types of ICS formalisms. The Framework supports various types of ICS and business formalisms, that includes the RoREC and its resultant RoREP, which is a fairly risky RE based process, that includes: 1) The role and implementation of requirements' link to Microartefacts and to the Project*'s* success rates, evaluated by factors; 2) Defining the granularity of the needed requirements and Microartefacts; 3) The management of RoREP instances, by using the ADM4RE coordinated DevOps4RE; 4) Applying a robust mapping mechanism; and 5) An iterative SIP' usage of RoREPs. The RoREC can be used to all types of Projects, and it is a part of Framework*'s* Implementation module (Im), and the Architecture module (Am). An RoREP instance is supported by ADM4RE's phases, where each RoREP Microartefact circulates through all its phases. RoREP Microartefacts contain their private set of CSFs and is how the overall Project*'s* status can be calculated; and this established a soft link between SWOT, factors and Microartefacts (Peterson, 2011). In this chapter, parts of previous works are reused for the better understanding of the RDP and RDP4RE; if everything needed were only referenced, it would be impossible to understand. This empirical engineering research model (simply the EEModel) can be considered as a non-conventional and innovative one, in the fields of complex transformations and EA.

The EEModel

RDP4RE is based on the EEModel which is optimal for Projects (Easterbrook, Singer, Storey & Damian, 2008), and the HDT is used to support Quantitative Analysis for RE (QNA4RE) and QLA4RE methods. EEModel's validity checks if the RQ is acceptable and valid as a contribution to existing scientific (and engineering) knowledge, proposed recommendations and the related PoC (or experiment). In engineering, an experiment is a design and prototyped using one or more CSFs (or independent variables, in theoretical research) and are processed to evaluate their influence on the EEModel's dependent variables. Experiments permit the evaluation of CSFs (SWOT and HDT) and checks if they are related. The RoREC is business centric and is founded on the Framework that in this chapter uses EA basics and standards (The Open Group, 2011a).

EA Basics and Standards

Using a just-enough EA (or the target architecture), helps the Project to align legacy architecture visions. The traditional architecture layers represent a silo concept, where it is complex to transform into an agile ICS. Using the RoREP, the Project transforms the legacy ICS into an agile classified directory of atomic business services (The Open Group, 2011a). The Project must have in depth knowledge of various Project's EA domains, which completes the profile of an Architect of Adaptive Business Information System (AofABIS, or the Manager), who uses a EEModel 1st approach (Trad, & Kalpić, 2013).

The EEModel First Approach - Top-Down Approach

The Framework uses a pseudo bottom-up concept that is mainly based on managing Microartefacts, by using the RoREC. It is an agile upstream approach that accommodates to legacy services environments. Architectures derived from standardized EA methodologies depend on the selected requirements' quality. In Projects architectures do not refer hard links to services, but they will include abstract services that map to requirements. In the SIP, the Framework supports the services to requirements mapping process. This ensures that Projects are well controlled and use the pool of RoREP. The RoREP supports the alignment between Microartefacts, requirements, organizational (re)structure, AMD4RE/governance phase(s), and the ICS. RoREPs are used and interfaced by using aBBs and their integration's status can be queried by using CSFs (Ylimäki, 2008). To avoid problems in the SIP, the bottom-up approach is highly recommended, where the first step would be to convert the legacy system into a structured ocean of Microartefacts. Nevertheless, this new structure needs an umbrella that is a high-level top-down concept. The "1:1" based bottom-up approach for a specific business requirement will look as follows: 1) It describes the used RoREPs and Use Cases (UC) for a specific business activity; 2) The EEModel the corresponding classes and diagrams; 3) Add the Microartefacts to the Project's architecture repository; and 4) Document and persist in directory for classification of the newly created Microartefacts. RoREP's holistic structure for the Project's SIP; it also offers: 1) A set of predefined requirements/Microartefacts development, operations, and intelligence templates to automate implementation processes; 2) Describes the RoREP development, operations and intelligence responsibilities and activities; and 3) Defines the major RoRE Microartefacts and processes. The RoREP is a set of idioms and activities, where an idiom is a basic automation activity that is generic and not specific to any RoREP implementation processes. That all depends on the role and the status of the Entity's ICS.

RDP4RE's CSFs

Based on the LRP4RE, the most important RDP4RE's CSFs that are used are:

Table 1. CSFs that have the rounded average of 9.20.

Critical Success Factors	KPIs	Weightings
CSF_RDP4RE_RoRE_Standards	Proven	From 1 to 10. **10 Selected**
CSF_RDP4RE_CSA_CSF_KPI_Integration	Proven	From 1 to 10. **10 Selected**
CSF_RDP4RE_Complexity	High	From 1 to 10. **08 Selected**
CSF_RDP4RE_Empirical_Engineering	Proven	From 1 to 10. **10 Selected**
CSF_RDP4RE_RoRE_Manager_Profile	Feasible	From 1 to 10. **09 Selected**
CSF_RDP4RE_RoRE_Components	Feasible	From 1 to 10. **09 Selected**
CSF_RDP4RE_Transformation_Framework	Possible	From 1 to 10. **09 Selected**
CSF_RDP4RE_LRP4RE	Feasible	From 1 to 10. **09 Selected**

valuation

THE ROLE OF INFORMATION AND COMMUNICATION SYSTEM

Complexity and Technology

The RoREC is related to the *business transformation architecture* category which has concepts for mapping designed business requirements to other Project's resources, and which satisfies the defined goals. The complexity of Projects and actual hyper-evolution of ICS technologies are challenging undertakings where Projects are very risky and their impacts on changes are hard to predict; that is why it is important to initiate SWOT and DMS4RE and the requirement phase. To successfully achieve this complex goal, there is a need first, to breakdown and unbundle the legacy ICS. For that goal Projects need RoREC and SIP tools, which comprise an important investment that can be wasted if badly used. Integrating such implementation and design tools requires the understanding also the vendor ranking in the market; and a tools roadmap is centralized and inter-operable across the entire Entity (Gartner, 2014a). At this time Manager and his team must ensure that these selected tools permit an anti-locked-in strategy. Such a *anti-locked-in strategy* has to be built on the role of standards, technologies, and EA methodologies.

The Roles of Standards, Avant-garde Technologies, and Methodologies

Today there are many business, EA and ICS related standards and they are to some degree applicable, like the following: TOGAF (and its ADM), Service Oriented Architectures (SOA), CMMi, COBIT, ITIL, UML, BPMN, BMM, SysML, … (Desfray, 2011). These standards, methodologies, and their implementation environments, support the Project*'s* breakdown unbundling of legacy ICS systems, using an empirical approach. An important CSF in the Project*s* is that changes done on the traditional ICS of Entities, to become agile innovative ones, should be based on the RoREC (The Open Group, 2011a), where the integration of related Microartefacts can be used by adopted standards. The RoREC englobes templates for the use of Project*'s* ABBs and SBBs. The theory and concept of reusable patterns, like the RoREP, suggests that RE implementers must be able to reuse proven components that emerge from the best *architecture & modelling* practices, in order to solve generic Project requests. Without the use of design RoREP, *the* Project would perform badly, applying architecture & modelling techniques, and that can result in the targeted business solution that: 1) Has bad performance; 2) Lacks scalabilities; 3) Brings human instabilities; and 4) Becomes un-usable and un-maintainable. Added to that, for practical reasons, many RE specialists have the tendency to *reinvent the wheel,* when attempting to implement RE templates. Therefore, the RoREC must apply: 1) Standardized tools and frameworks, like TOGAF and/or UML; 2) Standardized services' modelling methodology, like SOA Markup Language (SOAML); 3) Standardized Business Process Modelling (BPM); 4) Apply an adequate mapping model; and 5) Use the optimal agile concept for RE.

Agility Concepts and Requirements Engineering

Entity's agility is achieved by combining various synchronized business engineering concepts, ICS technologies and RoRE/EA related methodologies that promote global business automation schema to be implemented in various levels and parts of the Project. In order to unbundle and maintain the existing Entity's legacy ICS system and glue its innovated M/R mapping links in its dynamic ICS modules. ICS modules are made up of Microartefacts, where each Microartefact is a set of micro business services.

Atomic or Micro Business Services

There is no sure definition of business services architectural style, but there are common characteristics around the organization, business capability, business intelligence, and decentralized control of business environments. The transformed agile business system becomes coherently automated by unbundling of the legacy ICS system (Bebensee, & Hacks, 2019). This unbundling process delivers the needed sets of aBBs, where an ABB is a set of aBBs. This process starts with the classification of services/aBBs into CSAs. aBBs can be interfaced by using the Application Programming Interface (API) approach that is based on (Patni, 2017): 1) Modelling API's schema by creating a design document; 2) A schema model is a contract between the Entity and the clients; 3) A schema model is essentially a contract describing what the API is and how it works; and 4) Uses an agile strategy. The unbundling process is set up by the Project*'s* team who synchronize them with sets of requirements, by using implementation and tests procedures.

Implementation, Tests and Tools Diversity

Using many tools and gadgets, may cause major Project problems for RoREPs; many Project*s* presume that tools would solve all types of problems. Instead of using straightforward SIP, Project engineers spend most of their efforts in the search for libraries, scripts or gadgets which would shorten SIP time. Therefore the main focus must be set on: 1) RoREP's modelling; 2) Unit, aggregated and integration tests; 3) Continuous integration and deployment; 4) Change management concepts; 5) Performance and robustness estimations; and 6) And many others… Performance problems that cause a Project to be cancelled, are the performance and robustness problems that in general in Entities are interpreted as *human behavioural issues*; therefore there is a need for a granular agile DevOps4RE and M/R concept.

Agile Development, Operations and Mappings to Requirements

Project agility is achieved by combining synchronized domain, ICS and SIP methodologies that promote Entity's automation and business robustness. To unbundle, restructure and maintain the existing ICS and to glue its innovated Microartefacts in its choreography modules, and the DevOps4RE process, at various EA levels as shown in Figure 4. the DevOps4RE contains automated script to manage Microartefacts by applying a set of actions that coordinate and control SIP activities.

Figure 4. The requirements interaction with various ICS modules

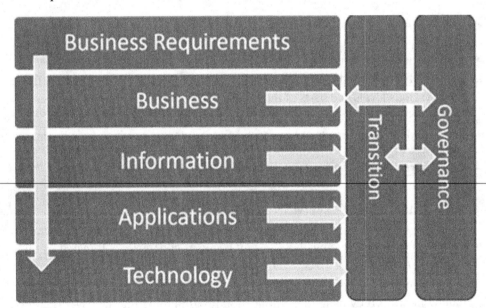

The DevOps4RE is based on a holistic systemic approach and its mechanics manage Microartefacts when it receives Project*'s* change requests. The DevOps4RE interacts with a multitude of Project resources, in a synchronized manner; using the ADM4RE to assist RoREP's integration activities (The Open Group, 2011a). The DevOps4RE supports mapping mechanisms that use the HDT to make the Project's integration flexible and to avoid major problems (Trad & Kalpić, 2017a). The RoREP supports a Project by offering requirements to Microartefacts mapping concept. Project's agility is supported by the following conditions: 1) Standardized and realistic agile methodologies; 2) Change management and integration procedures capacities; and 3) Efficient integration tests. Unfortunately, DevOps4RE processes cannot support an Entity wide robust agile concept that can drive dynamic business activities. Actual RoREC, development, operations, integration, and test tools are skeletons that use heterogenous scripting environments and methodologies, which do not offer a unified implementation strategy for Project. The Entity should use a Natural Language Programming (NLP), which includes DevOps4RE interactions and related transformation development, operations, integration, and development actions. Agile development environments use standards which: 1) Deliver a functional language that uses Microartefacts; 2) Uses the Entity's ICS components; 3) Delivers DMS4RE interfaces; and 4) Offers mechanisms for version management, deployment, and testing; and 4) Defines the levels of granularity.

The Granularity and the Unit of Work

Defining RE's, mapping and Microartefact granularities for a Project is a complex undertaking, added to that the "1:1" mapping and classification concept is a long process. Mapping of a requirement's UC(s) to Microartefact(s) in the form of a class diagram or communication diagram, can be done using a range of RoRE. This modelling and mapping concept is supported by a set of Microartefacts where its NLP module can evaluate compound expressions, according to AHMM4RE. DevOps4RE uses the AHMM4RE to evaluate the requirements, deliver solutions, where a requirement and Microartefact map to a class diagram (or communication diagram) and have a GUID RoREP's unit of work or a Project Microartefact, based on the alignment and classification of all the Project's requirements (and resources), using the "1:1" mapping concept.

ICS' CSFs

Based on the LRP4RE, the most important ICS's CSFs are:

Table 2. CSFs that have the rounded average of 9.60

Critical Success Factors	HMM enhances: KPIs	Weightings
CSF_ICS_Standards_Technologies_Methodologies	Stable	From 1 to 10. 09 Selected
CSF_ICS_RE_Agility_Concepts	Complex	From 1 to 10. 08 Selected
CSF_ICS_Atomic_Business_Services	Complex	From 1 to 10. 08 Selected
CSF_ICS_Implementation_Tests_Tools_Diversity	Stable	From 1 to 10. 09 Selected
CSF_ICS_Performance_Robustness	Stable	From 1 to 10. 09 Selected
CSF_ICS_DevOps4RE	Complex	From 1 to 10. 08 Selected
CSF_ICS_Granularity	Complex	From 1 to 10. 08 Selected
CSF_ICS_ADM4RE_Automation	Proven	From 1 to 10. 10 Selected

valuation

ARCHITECTURE DEVELOPMENT AND RoRE'S INTEGRATION

Integrating the RoREP

The advantage of using the RoREP is to optimize the use of RE across the Entity' processes (both manual and automated) and integrate its change and delivery strategy. To integrate the RoREP it is to: 1) Ensure architecture skills; 2) Unify dictionaries; 3) Use an NLP; 4) Integrate it in the development environments; 5) Define a business modelling strategy; 6) Define main UCs; 7) Use BPM; and 8) Define the sets of aBBs. The RoREP englobes all business requirements, information, and must facilitate Project's team in SIP. RoREP contain business models which are tuned through ADM4RE iterations. The RoREP offers multiple views that are based on methodology diagrams. The RoREP supports the integrity, modelling, and maintenance of requirements and its related Microartefacts; which assesses the readiness to transform the Entity.

Assess Readiness for Business Transformation

A *Business Transformation Readiness Assessment* evaluates and quantifies the Entity's readiness to change and start a Project. The Project*'s* assessment is based on the Framework*'s* readiness CSFs. The outcomes of the Framework*'s* readiness assessment is added to the *Capability Assessment*. These outcomes are used also to establish the ADM4RE and RoREP's interfaces, to support the Project, and to localize the risk CSFs. These CSFs are associated with the *Architecture Vision,* and are related to the initial level of risks, like catastrophic, critical, marginal, or negligible. These CSFs are to integrate and coordinate the ADM4RE.

ADM4RE's Coordination

RoREP's integration in Project*s* is done by using the ADM4RE, which supports it in the automation of DevOps4RE activities, especially to manage Microartefacts. Throughout ADM4RE phases, RoREP phases are created or improved. The ADM4RE encloses cyclic iterations, where all RoREP's instances actions are logged. RoREP is domain agnostic and technology independent. RoRE's integration with the ADM4RE has the following advantages: 1) Real-time RE, mapping and Microartefact management; 2) Improvement of ICS' performance, and robustness; 3) The RoRE enables the use of standard methodologies like UML or ArchiMate; and 4) The use of tests and an integration-driven developments approach (Visual Paradigm, 2019).

RoRE Enabled Tests

RoREPs must check if requirements respect (Tkachenko, 2015): 1) Completeness, where they must contain all needed information; 2) Clearity, where they should be transparent and clear; 3) Correctness, where all contents must be credible; 4) Consistency, where requirements should not mutually contradict; 5) Testability, validates that the SIP the requirements' sets. The ADM4RE controls, directs, and monitors RE activities; where mapping and Microartefacts' usage are verified by an adapted set of tests and integration-driven developments as shown in Figure 5. These tests are:

- Test Driven Developments for RE (TDD4RE): The actual standard for unit tests (or TDD) is an archaic semi-manual approach used in software development (known as the test first approach) which is related just to sets of code (Janzen & Saiedian, 2005); whereas the Design Driven Development for RE (DDD4RE) (or the Model first approach) is based on designing first the solution that starts with the modelling of Project requirements, mappings and Microartefacts by using UML or ArchiMate's UCs where each UC maps to a concrete set of diagrams. The class diagram maps to a Requirements and Microartefacts where the UC defines unit and integration tests. Automated tests evaluate the design for a given set of requirements and verifies their statuses. The DDD4RE or model first is optimal for the RoRE, that is present by the Acceptance Test Driven Development (ATDD) methodology (Design Patterns, 2015).

- Acceptance Test-Driven Development for RE (ATDD4RE): is applied in the case of collaborate business clients, Project testers and software engineers and to assist their communication (Koudelia, 2011). Based on standard TDD4RE, the standard ATDD4RE methodology is based on developing tests where tests represent the results of the requirement's behaviour of a set of Project Microartefacts. Business users contribute to define workable acceptance tests or use behaviour-driven development techniques (Koskela, 2007).

- Behaviour-Driven Development for RE (BDD4RE): The RoRE uses Framework's HDT and its NLP, which is the BDD4RE that includes unit, integration and acceptance tests, to be used in Projects (Bingham, Eisenhardt, & Furr, 2007; Soeken, Drechsler, & Wille, 2012). The BDD4RE has a pseudo-prose formalism that resembles to logical human scripts, so that business specialists implement RoRE scenarios and their corresponding tests. The BDD4RE includes a resources' mapping subsystem to link pseudo-prose keywords to Microartefacts, and to requirements. The HDT enables an automated requirements-based testing environment that is used with ADM4RE's iterations (Lazar, Motogna, Parv, 2010).

- Requirements' based Testing for RE (RbT4RE): The TDD4RE, ATDD4RE and BDD4RE are minimal and basic tests that confirm the coherence of the future system, but there is a need to do more profound domain tests, or the RbT4RE, to confirm that the actual sets of requirements are valid (Tutorialspoint, 2022). The RbT4RE is an approach in which test cases, pre-conditions and data flows are derived from RoRE; and it includes functional tests and non-functional ones, like, performance, reliability or usability. RbT4RE's phases are: 1) Completion criteria validation, which confirm when functional and non-functional validation is successful; 2) Design test cases; and 3) Execute and verify tests; 4) Track and manage defects. Another important CSF is that RbT4RE must: 1) Be carried out in real-time; 2) Add value to the DevOps4RE; 3) Provides the Project's overall status; and 4) Modelling languages can be used.

Figure 5. The Framework's tests sequence

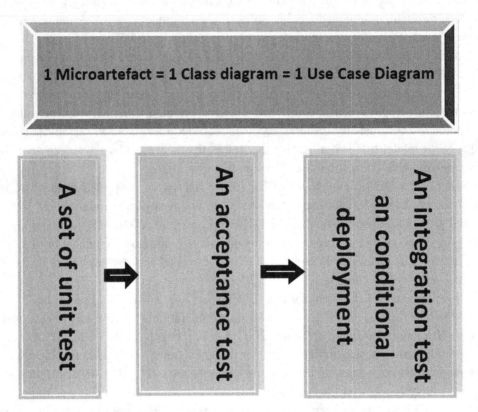

Modelling Language-ArchiMate

The *elicitation* of Project*'s* requirements is the first activity or step in the RE process in the context of ADM4RE. *Elicitation* refers to the capture of requirements, to avoid the assumption that Project*'s* requirements are ready to be simply collected, by using basic techniques. Information gathered from the RE process *elicitation* has to be interpreted, analysed, modelled and validated before RE specialists, who can confirm that a credible and coherent set of Project*'s* requirements has been located and collected. Therefore, Project*'s* requirements elicitation is directly related to all other RE activities. The *elicitation* discipline used is dependent on the used modelling scheme, and vice versa. Modelling schemes can imply the application of specific elicitation techniques, like the ones used with ArchiMate (Nuseibeh, & Easterbrook, 2000). As shown in Figure 5, RoREP's modelling using ArchiMate has the following characteristics (Hosiaisluoma, 2021; Bebensee, & Hacks, 2019): 1) It models behavioural and structural elements of a Project; 2) It enables EA modelling to support ICS infrastructure and landscapes; 3) RoREP related models implemented using ArchiMate can be stored in the Project*'s* repository; 4) It models data/information behaviours; 5) It uses an interchange format based on the Mark-up Language (XML) which can map to ArchiMate's Model Exchange File Format's XML schema(s); 6) Schema's properties are mapped to instances of ArchiMate property definitions; and 7) The used models should generate and map to ICS and its applications' cartography.

Figure 6. ArchiMate Framework (Hosiaisluoma, 2021)

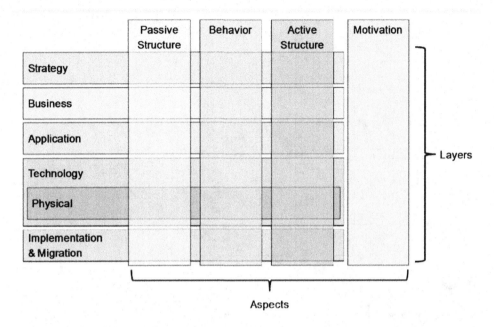

Application Cartography

The RoREP maps to Entity's *and/or* Project*'s* cartography of applications; Entity's applications are classified as follows (Togaf-Modeling, 2020): 1) Classifications can be done by EA capacities like TOGAF's Application Communication Diagram (ACD), which depicts its used models and mappings related to communications between applications and modules, in form an Entity's metamodel.

Figure 7. The architecture is layered (Togaf-Modeling, 2020)

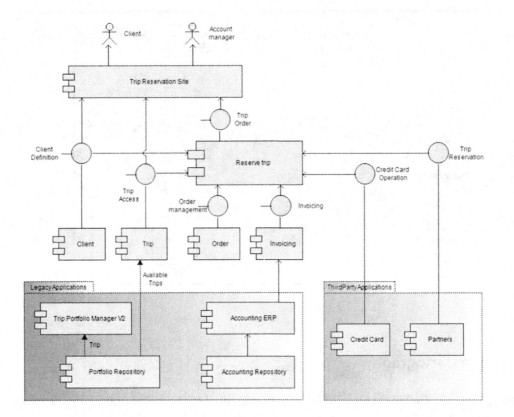

It presents applications, components, and interfaces (between various components); 2) Interfaces may be associated with data classes, applications can be related to Microartefacts; 3) Application communication diagrams can represent an existing applications' cartography, or a logical architecture of the transformed end-system. Microartefacts based EA is privileged; 4) Entities have hybrid (mixed) applications, repositories and new Microartefacts-based EA; 5) In the case of using Microartefacts, services based application components, should be structured according to their nature and their EA level; 6) Microartefacts based components are related by services, which use connectors; 7) A dimension of the applications' cartography should be dedicated to RoREP's integration; and 8) As shown in Figure 7, the EA is layered, where the interaction component layer is on top, process-based components in the middle, and entity components on the bottom.

Architecture Layers

The Framework*'s* architecture and modelling module (Am) helps in establishing EA principles that are defined in the Project's preliminary phase and it guides its vision. The Project*'s* EA superposes existing architecture standards, like TOGAF, as shown in Figure 8 (The Open Group, 2011a); the Framework is a tailored adoption of TOGAF and defines this approach as a just-enough EA, where the following layers RoRE are: 1) Business Architecture; 2) Data Architecture; 3) Application Architecture; and 4) Technology Architecture.

Figure 8. The architecture is layered (Togaf-Modelling, 2020)

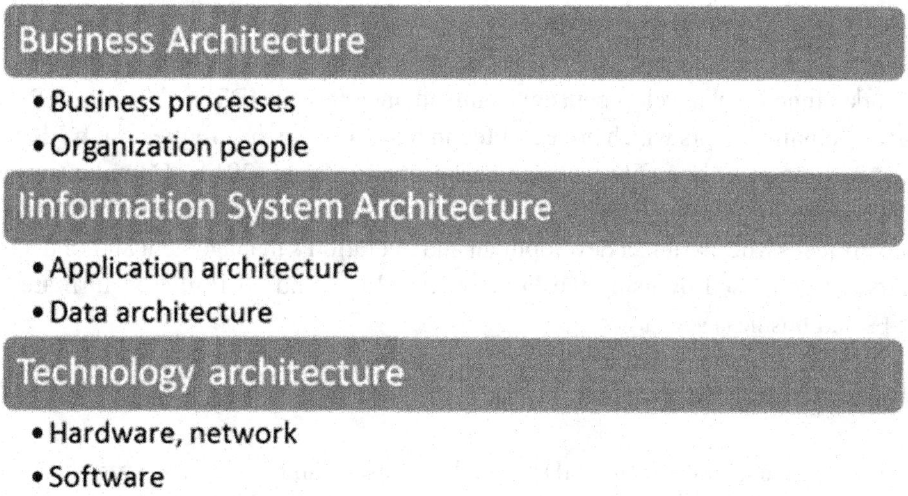

Business Architecture
- Business processes
- Organization people

Iinformation System Architecture
- Application architecture
- Data architecture

Technology architecture
- Hardware, network
- Software

Project*'s* ABBs and SBBs are used to solve assembled requirements and to support their principles, these blocks are a set of Project's deliverables. The dimensions of EA and RoREP are scoped to Project*'s* boundaries, which have to consider the heterogeneous types of architectures and legacy systems (The Open Group, 2011a); which could be supported by: 1) Defining deliverables in the form of templates; 2) Defining RoREP's and EA's interactions; 3) Applying Microartefacts integration; and 5) Applying a modelling strategy approach.

The ADM4RE and the RoRE CSFs

Based on the LRP4RE, the most important ADM4RE's CSFs are shown in Table 3.

Table 3. CSFs that have an average of 9.00

Critical Success Factors	HMM enhances: KPIs	Weightings
CSF_ADM4RE_IntegrationProcess	Complex	From 1 to 10. **08 Selected**
CSF_ADM4RE_Transformation_Readiness	Complex	From 1 to 10. **08 Selected**
CSF_ADM4RE_Tests	Proven	From 1 to 10. **10 Selected**
CSF_ADM4RE_ModellingLanguage	Proven	From 1 to 10. **10 Selected**
CSF_ADM4RE_Cartography	Possible	From 1 to 10. **09 Selected**

valuation

STRATEGIC INTELLIGENCE AND DECISION SYSTEMS FOR RE

Natural Functional Environments

The Framework's functional development environment, incudes the HDT; and the RoREP uses a NLP for RE (NLP4RE), to build scripts which are executed in real-time. Such activities can be done by business professionals with no prior computer science background (Moore, 2014). Complex systems management is an approach for building RoREP based end-system that is supported by heuristics models that automate the Project's Microartefact development and operation's management and/or decision making. Requirements are processed by using the Framework's HDT and NLP; that in turn are based on the selected CSFs and business scenarios.

Intelligent Business Scenarios

RoREP instances that use inter-related HDT based scenarios can be one of the types: 1) An interaction of Microartefacts; 2) A BPM instance; and 3) Other… The **How to read this chaptUUU** unbundling of the monolithic environment breaks down the actual monolithic environment into an automatized bank of Microartefacts to create scenarios like BPM (Richardson, 2014). The unbundling process upstreams (or refines) business scenarios, which are not altered just to integrate services and which align with the ADM4RE. When a Project starts the executive management, senior business analysts and business system designers, create the top-level organizational design artifact which is used to create the classification concept and that becomes a point of reference for the modelling process. This classification concept is used to classify the requirements, Microartefacts, and BPMs. RoRE's architectural style is intended to simplify the Project and the interoperation of ADM4RE and RE. By structuring and transforming capability, the authors refer to the notions of atomic, unique, meaningful, granular unbundling into business services; this enables the discovery of RE's functional capabilities, and to avoid duplicating similar business capabilities. From RE's perspective, Microartefacts focuses on transforming the legacy system, so that it enhances its flexibility and agility. Agility is crucial for dynamic systems, where Microartefacts modelling strategy's aim is to transform monolithic functions into portfolios of granular services. The unbundling of business activities and their decomposition in the form of M/R that can be filtered, traced, and queried; are stored in ICS' catalogues. Catalogues contain the following entities: 1) Organizational units' information and business function; and 2) M/Rs, their information service equivalence, and business activities (TOGAF, Directory, 2011).

Business Activities

Figure 9. Business Model and Atomic Business Service Interaction (TOGAF, Catalogue, 2011)

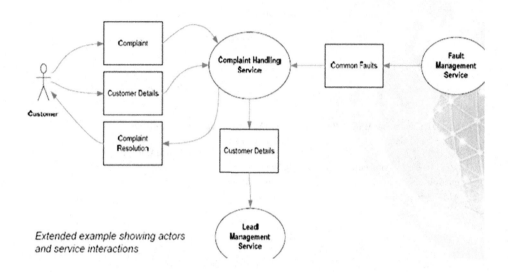

RoREPs' management improves the Entity's churn rate and would improve problems management. RE's modelling strategy delivers patterns to support modelling languages. RE and ADM4RE related patterns offer a generic approach, that makes the Project vendor independent, which recommends the following business activities: 1) Business cases or UCs modelling, which is a starting point for any new Microartefact/Requirement management; 2) The mapping concept supports the UCs; 3) A UC maps to a BPM, which links to a set of Microartefacts, as shown in Figure 30 (TOGAF, Catalogue, 2011); 4) RoREPs, represent the business architecture, focusing on business scenarios; 5) RoREs supports the design of new business activities and identifies data/knowledge Microartefacts.

Intelligent Data and Knowledge Microartefacts

The RoREP contains business data models and related modelling components and does not depend on the types of data sources. but Their diversity generates problems, especially in the SIP phase. Business data Microartefacts focuses primarily on the encapsulation of the data schema(s) (Pavel, 2011). The mapping concept is applied for business data models management and access, where the sets of requirements correspond to a data entity or a *business data view*, if the data can be encapsulated in a single class. A BPM Oriented Knowledge (BPMOK) management framework can be applied for knowledge management, which supports the implementation model.

RoREP's Implementation and Integration Model

Figure 10. The AHMM4RE model's nomenclature (Trad, & Kalpić, 2020a)

Basic Mathematical Model's (BMM) Nomenclature

Iteration	= An integer variable *"i"* that denotes a *Project/ADM iteration*	
microRequirement	= (maps to) KPI	(B1)
CSF	= Σ KPI	(B2)
Requirement	= (maps to) CSF = $\bigcup$ microRequirement	(B3)
CSA	= Σ CSF	(B4)
microMapping microArtefact/Req	= microArtefact + (maps to) microRequirement	(B5)
microKnowledgeArtefact	= $\bigcup$ knowledgeItem(s)	(B6)
neuron	= action->data + microKnowledgeArtefact	(B7)
microArtefact / neural network	= $\bigcup$ neurons	(B8)
microArtefactScenario	= $\bigcup$ microartefact	(B9)
AI/Decision Making	= $\bigcup$ microArtefactScenario	(B10)
microEntity	= $\bigcup$ microArtefact	(B11)
Entity or Enterprise	= $\bigcup$ microEntity	(B12)
EnityIntelligence	= $\bigcup$ AI/Decision Making	(B13)
BMM(*Iteration*) as an instance	= EnityIntelligence(*Iteration*)	(B14)

RoRE supports the configuration of M/Rs types to be used and mapped. M/Rs are orchestrated by using AHMM4RE's actions, which process the deployment and validation processes. The RoREP is implemented and used in all of Project's processes and the implementation of M/Rs, delivers final business components. Such set of actions can be modelled and orchestrated by the AHMM4RE (The Open Group, 2011a; Trad & Kalpić, 2020a). AHMM4RE's nomenclature supports the automated mapping model and is presented in a simplified form as shown in Figure 10. The symbol å indicates the summation of all the relevant named set members, while the indices and the set cardinality have been omitted. The summation should be understood in a broader sense, more like set unions. The AHMM4RE contains mathematical language, the NLP4RE, that can be used to script the behaviour of a system (Goikoetxea, 2004). The RoREP uses the AHMM4RE based DMS4RE, which is a part of the Framework.

Figure 11. The enterprise mathematical model

The Generic AHMM's Formulation

AHMM	= $\bigcup$ ADMs + BMMs	(B15)

AHMM's Application and Instantiation for RE and RoRE(P)

Domain	= RE	(B16)
AHMM4(*Domain*)	= $\bigcup$ ADMs + BMMs(*Domain*)	(B17)

The Enterprise AHMM4RE (EAHMM4RE) is the combination of an ADM4RE and the Framework. A transformation is the combination of an EA methodology like the TOGAF and the AHMM4RE that can be modelled after the following formula for the Transformational Model (TM):

TM = EA + AHMM4RE (B18).

(RoRE):

RoRE = å micro M/R + (synchronized) AHMM4RE (B19).

The RoREP is based on a concurrent and synchronized Framework which uses threads that can make various RE models run in parallel and exchange managing of SIP through the AHMM4RE.

DMS4RE's CSFs

Based on the LRP4RE, the most important DMS4RE CSFs that were used are:

Table 4. CSFs that have an average of 9.0

Critical Success Factors	HMM enhances: KPIs	Weightings
CSF_DMS4RE_Natural_Functional_Environments	Possible	From 1 to 10. **09 Selected**
CSF_DMS4RE_Intelligent_Business_Scenarios	Possible	From 1 to 10. **09 Selected**
CSF_DMS4RE_Business_Activities	Possible	From 1 to 10. **09 Selected**
CSF_DMS4RE_Data_Knowledge_Microartefacts	Proven	From 1 to 10. **10 Selected**
CSF_DMS4RE_Implementation_Integration_Model	Complex	From 1 to 10. **08 Selected**

valuation

RORE'S AND ROREP'S INTEGRATION

RE specialized environments exist, and are complex, due to various standards which are supposed to support an iterative SIP and DevOps4RE. Unfortunately, these environments are complex and hard to integrate, and cannot support Entity *wide* agile concepts. To simplify the integration there is a need to unbundle/restructure its M/R repository. This holistic integration must first define which of the architectures is optimal.

Defining the Optimal Architecture

A simplified and unified architecture must be developed for the Project, because there are many and the parallel use can cause creating parallel ICSs; such architecture approaches can be Cloud architecture, EA, SOA, JEE, UML… Entities need the possibility to customize the modelling process and include standards, notation and information relevant for its structure. This resulted in the extension of the modelling capabilities of the UML by the specification of extension mechanisms; and can be considered as a *lingua franca*. UML was the basis for evolution of many standard methodologies like System Modelling Language (SysML); in fact SysML is build on an UML Meta-Model.

SysML RE and EA Modelling

The optimal approach is to use common artefacts; and probably basis for such an approach is (Moreland, 2006)…

Figure 12. SysML's taxonomy of diagrams

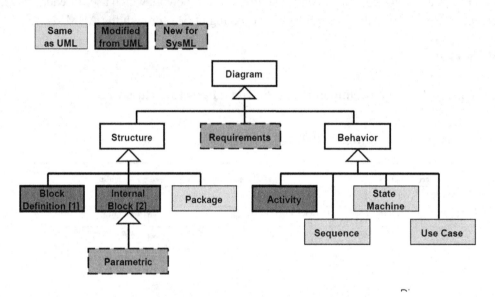

System ML (SysML): As shown in Figure 12, UML's evolution had brought significant capability for RE and was adapted by many vendors. And RE advantages were: 1) Define client's requests, goals, and expectations; 2) M/R is a problem which has a solution; 3) Definition of properties in a package; 4) Granularity; 5) Implementing traceability (derived and source); and 6) Constraints and verification test cases.

Figure 13. SysML's requirement's artefact

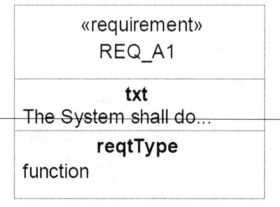

It also modified UML Class Diagram and enhanced UML Composite Structure Diagram. SysML reuses many of the major diagram types of UML. It is optimal to reuse UML diagrams like UC, sequence, state machine, and package diagram. SysML offers new diagrams: 1) Requirement diagram; and 2) Parametric diagram. The requirements diagram usage varies depending on the domain, Entity, Project, ... Its characteristics are: 1) To graphically represent requirements and their relations; 2) A requirement specifies a capability, function…; 3) Attributes contain a GUID and a description; 4) The methodology uses stereotypes and tags and adds specific properties. As shown in Figure 13, a requirement diagram shows requirements and their relations. Requirement properties are: 1) *Containment* (or *flowdown*) shows where requirements are decomposed into micro-requirements; 2) The *deriveReqt* and *copy* relationships exist between requirements only; 3) *trace, refine, satisfy, and verify,* exist between a requirement and other model attributes; 4) *verify* exists between a requirement and behaviour; 5) RoREC's element stereotyped as a *testcase*; and 6) *problem and rationale* comments can be used to depict problems and decisions.

Figure 14. SysML Taxonomy of Diagrams

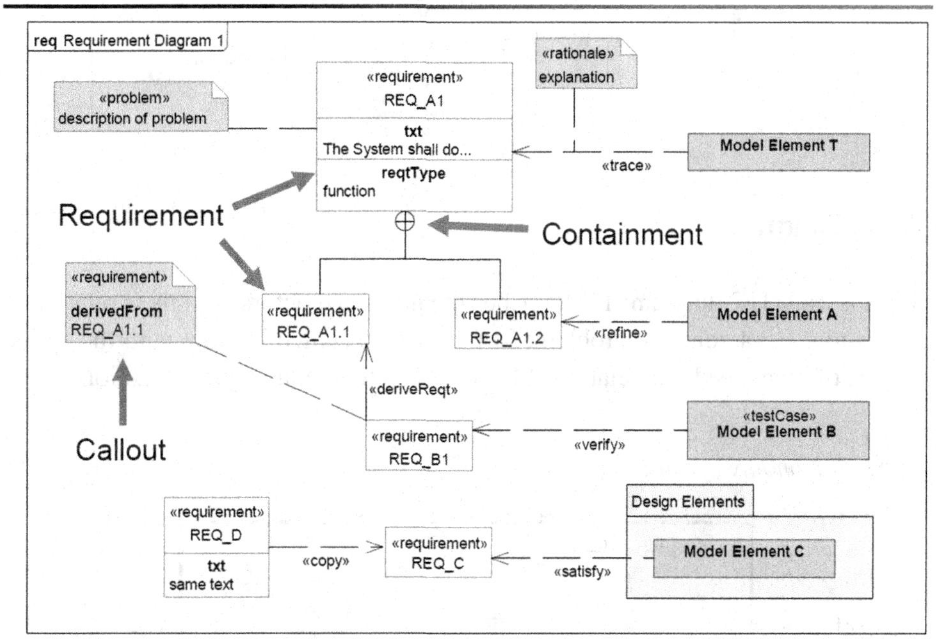

As shown in Figure 14, the package diagram visualizes the basic elements (like requirements) packages, which group structures of the model and define relationships between Project's groupings. These packages include namespaces and they have relations to sub-packages.

Figure 15. SysML's package diagram

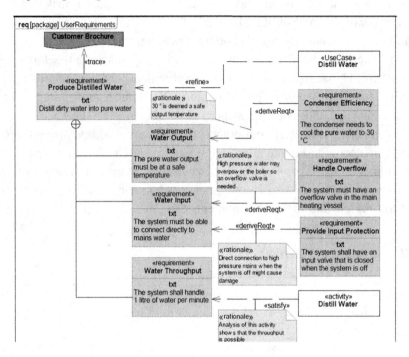

The Use Case Diagram

As shown in figure 16, a UC diagram: 1) Describes the goals and actors; and their relations to the end-system; 2) Needs to solve a concrete problem; 3) Uses actors which interact with the end-system, but they are not a part of it; and 4) Uses relations like: extend, include, and generalization.

Figure 16. SysML Taxonomy of Diagrams

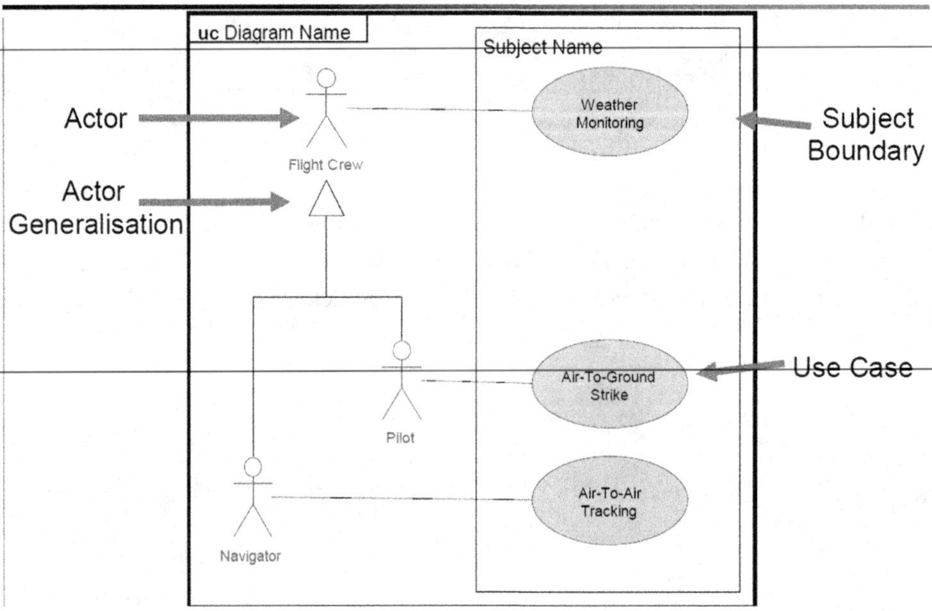

Another relevant methodology is DDD which is strongly related to a targeted domain and it has its own RE concept.

DDD RE and Modelling

DDD RE is based on (Manning-Franklin, 2022): 1) Strategic DDD patterns, which govern Project's RE, SIP and EA. It is agnostic to the used technology and environment; 2) Bounded context is a conceptual boundary in RE, where it depends on the Project's interpretation of a boundary; and 3) Functional DDD (FDDD) where data and behaviour are mono-directionally coupled rather than bi-directionally, which means that classes defined using NLP4RE cannot own their behaviour, but instead they couple functions to data in the form of parameter types. All functions can use classes, which match their type of signature. One of the oldest and most coherent RoRE modelling methodologies is BPM.

Architecture and BPM

RoREC manages BPM activities for Projects to simplify SIP, where BPMs rationalize business scenarios and they enable holistic development approach. BPMs can be used to enhance the Entity's KMS4RE by adopting a Business Process Oriented Knowledge Management approach (BPOKM) (Papavassiliou, Ntioudis, Mentzas & Abecker, 2001). BPM and other types of choreography, need mapping, development, operations, integration and tools that enclose a high-level interpreted NLP4RE environment, and its main characteristics are: 1) To deliver an in-house NLP4RE, with a BPM interface; 2) It supports existing standards (Beauvoir, & Sarrodie, 2018); 3) Using existing development environment(s); 4) It offers an adapted DevOps4RE; and 5) Supports M/R linking and mapping concepts.

Mapping Concepts

M/R mapping and granularities for Projects is a complex topic and classification of the discovered M/Rs respects a hierarchy like classes in Object Orientation (OO). An RoRE uses M/R through communication interfaces to interact with various ICS processes. Mapping of M/R to a UC uses a class diagram, communication diagram, and/or an Entity Relationship Model (ERM). The interaction is orchestrated by the NLP4RE. The M/R can dynamically evaluate compound expressions, according to the AHMM4RE's principles. The RoREP uses an AHMM4RE instance to evaluate the Project's requirements and to deliver solutions to problems (Neumann, 2002). The GUID identifies an M/R instance that contains information for the auto-generated mapping operations. Such an approach needs a specific M/R delivery strategy and a refinement concept.

M/R Delivery Strategy and Refinement

M/R delivery strategy supports the vision of contained intelligence which is a set of actions (or services) that coordinates mapping activities. RoREPs: 1) It generates mapping instances and receives requests like manage and change; 2) it uses the HDT to manage M/Rs on-request delivery basis; and 3) It offers a Refine Processes (RP) 4 RE (RP4RE). Projects *and* RoRE are complex and have limited business impacts. Optimal ADM4RE-based transformation iterations should be supported by RP4RE, which is a process that iteratively refactors ICS components and can regenerate RE artefacts. RPs' try to transform

Entity's BS components through their improvement, mapping, and interlinking. The RP4RE is used in the PoC, where automated tests/qualification activities are done on Project's components. The RP4RP can be an inhouse sub-system (or concept) which supports RE transformation activities and to refine and transform all Entity's resources (Koenig, Rustan, & Leino, 2016).The Project must select tools to support the refinement of M/R, resources and which can transform them into BBs. There are lots of tools having concepts like the triple conversion architecture. One of the main actions is to convert Project components that were developed using programming environments like, JEE, C++, PL/I, COBOL, Fortran, Assembly, … (MPSINC, 2021). RP4RE offers a set of methods for refining operations, which applies conversion to extract RE artefacts and CSFs (Wikipedia, 2021a).

RoREP's Development CSFs

Based on the LRP4RE, the most important RoREP CSFs that are used:

Table 5. CSFs that have an average of 9.00

Critical Success Factors	KPIs	Weightings
CSF_RE_Integration_Optimal_Architecture	Possible	From 1 to 10. 09 Selected
CSF_RE_Integration_Modelling	Possible	From 1 to 10. 09 Selected
CSF_RE_Integration_BPM	Possible	From 1 to 10. 09 Selected
CSF_RE_Integration_Mapping	complex	From 1 to 10. 08 Selected
CSF_RE_Integration_Delivery_Strategy_Refinement	Possible	From 1 to 10. 09 Selected

valuation

PoC'S IMPLEMENTATION

LRP4RE's Outcome

The LRP4RE (or Phase 1) outcome supports the PoC's background, using an archive of an important set of references and links that are analysed using a specific interface. After selecting, the CSA/CSFs tag is linked to M/Rs based scenarios; which are implemented as an item, in an Excel file; where all its details are defined; this concludes Phase 1. In this DMS4RE-related PoC (or Phase 2), the HDT processes a set of solutions. The empirical part is based on AHMM4RE's instance and M/Rs mechanics, which uses sets of CSFs that are used in phases 1 and 2; for the estimation of estimating risks.

Estimating Risk

ADM4RE's preliminary phase identifies the sets of Project problems. SWOT analysis results produced an opportunity for the Project. These results can make strategic decisions for developing the business with the standard references contained in the ICS (Remawati, 2016; Wati, Ranggadara, Kurnianda, Irmawan, & Frizki, 2019). For such activities, there is a need for optimal tools. The PoC used ArchiMate for the management dashboard based on the principles; this dashboard visualizes strategic and business CSFs and SWOT analysis. SWOT was used for the analysis, together with balanced scorecard method, goal setting, and resource, and capability analysis (Kitsios, Kyriakopoulou, & Kamariotou, 2022).

Setting Up SWOT and Analysing

As shown in Figure 17, the SWOT analysis was setup to support the evaluation of the Project's internal strengths and weaknesses, which includes needed resources and current capabilities to enable (or hinder) the implementation of transformation strategies.

Figure 17. The SWOT view

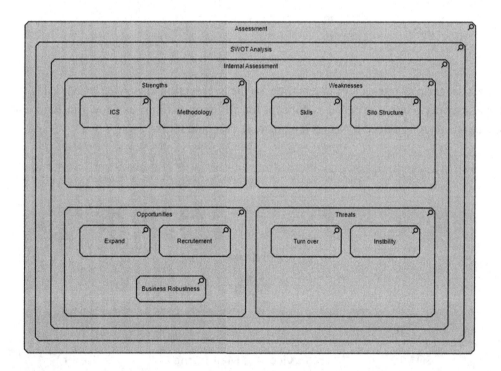

It can also reflect its external opportunities and threats within CSAs, which enable (or hinder) project's performance. The SWOT analysis view supports the design of optimal strategies that are aimed to exploit deduced Os and Ss, while proactively predicting possible Ts and to eventually improve Ws, while applying the following strategies (BizzDesign, 2022):

- WT strategy: minimize both Ws and Ts, to enable the achievement of defined goals.
- WO strategy: minimize Ws in order to be able to take advantage of new Os.
- ST strategy: maximize Ss in order to be able to deal with Ts.
- SO strategy: maximize Ss, using existing resources to promote new Os.

The first step ine the SWOT analysis is to link the defined set of CSAs, then CSAs are linked to CSFs; and finally CSFs are linked to KPIs. KPIs are charged in business scenarios.

Charging KPIs in Business Scenarios and Services

Figure 18. Charging a KPI in a business scenario

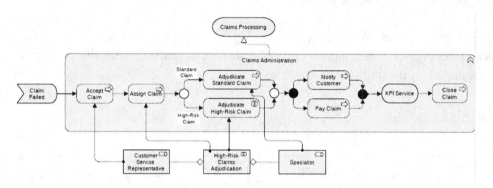

The business scenario charges a KPI in real time, so the aggregation also gives a CSF value in real time. The business scenario uses the services concept for all its operations, as shown in Figure 19. For all these operations there is a need to set up the business case and tools.

Figure 19. The use of Entity's services

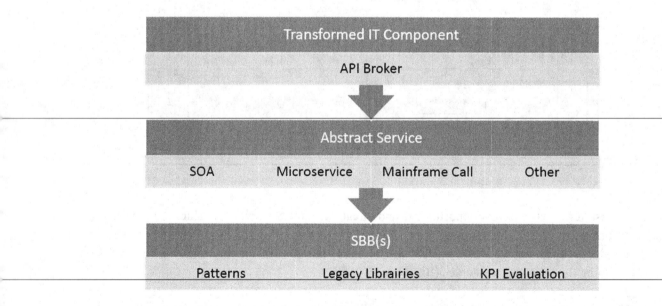

The Business Case and Tooling

The business case or UC has set the requirements of the future system from the business perspective; and that included high-level decisions related to Systems' Capabilities and Usage. The UC: 1) Identified business-level Constraints; 2) Described functionalities in terms of usages/goals of the system by actors; 3) Common functionalities were defined and used via the include relationship; and 4) Used are behavioural diagrams to describe business scenarios. An EA environment was responsible for the global concept that integrates standards, which models the Project's business vision and outcomes. The EA's transformation models business-outcome-driven M/Rs that support the DMS4RE needed for the evolution of the future-state EA paradigm that is required to deliver the desired goals that were defined in the SWOT table. The scope of business transformation architecture and modelling pattern includes: business cases, actors, BPMs, information models, business applications, business models and technology for the new transformed Entity. A differentiating characteristic of the discipline of business transformation architecture is that the selection of the right modelling pattern to deliver the expected business outcomes will reflect its business strategy and future-state vision. Worldwide groups, like Capgemini, what Gartner group's studies prove, rely on an EA tool, rather than on leading with traditional business outcomes. Managers analyse Project's vision, strategy and strengths in delivering business-outcome-driven EA based tooling environment. The next step is it to carry out phases 1 and 2.

From Phase 1 to Phase 2

The Project's enumeration of CSAs are: 1) The RDP4RE; 2) The ICS; 3) The Usage of the ADM4RE; 4) The ICS; 5) The KMS4RE/ DMS4RE; and 6) The RoREC and RoREP. Where Tables 1 to 6 were presented and evaluated in this chapter and they are the empirical part. That justifies PoC's feasibility (Trad, 2021).

The PoC

The RoREP's PoC was implemented using the research's Framework that had been developed using the Framework's NLP4RE, Microsoft Visual Studio .NET, C/C++ and Java. The PoC is based on the DMS4RE/AHMM4RE and the CSFs' binding, using a specific Project M/R and related resources, where the RoREP was designed using an UML and TOGAF methodologies.

Figure 20. The NLP4RE interface

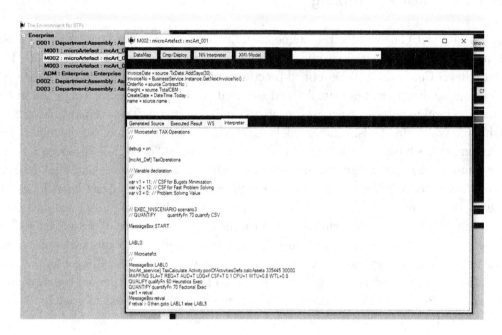

The RoREP represents the M/R, GUIDs and the CSFs. The Framework's frontend mapping/linking actions are activated by: 1) Selecting an HDT node that contains the M/R, and 2) Selecting the problem to be solved using NLP4RE. The Framework's frontend implements M/Rs. The RoREC uses a RoREP instances of knowledge (or learning) database that generates DevOps4RE actions and can make calls to KMS4RE/DMS4RE to solve a concrete Project problem. Once the Framework' development interface is activated, the NLP4RE interface can be launched to implement M/R scripts, as shown in Figure 20. These NLP4RE scripts that make up the KMS4RE/DMS4RE subsystem and RoREP instances relate to a set of services. RoREC-related CSFs were selected as demonstrated previously in this chapter's tables and the result of the processing of the DMS4RE, as illustrated in Table 6, shows that the RoRE is not an independent topic and is strongly bonded to the Project's overall risk management concept.

Table 6. The RoRE research's outcome

CSA Category of CSFs/KPIs	Transformation Capability	Average Result
The RDP4RE Integration	Transformable-Possible ▾	From 1 to 10. **9.20**
The Usage of the ADM4RE	Transformable-Possible ▾	From 1 to 10. **9.00**
The ICS Transformation	Transformable-Possible ▾	From 1 to 10. **9.00**
The Usage of the KMS4RE/DMS4RE	Transformable-Possible ▾	From 1 to 10. **9.00**
The RoRE's Integration	Implementable ▾	From 1 to 10. **9.00**

Evaluate First Phase

The model's main constraint is that CSAs having an average result below 8.5 will be ignored. As shown in Table 6 (which has a rounded average of 9.00), this fact keeps the CSAs (marked in green) that helps make this chapter's conclusion; and no CSA having a red colour. Which means that RoRE's integration will succeed and that the Project must be done in multiple transformation sub-Projects, using the ADM4RE, where the first one should try to transform the base systems, the ICS.

SOLUTION AND RECOMMENDATIONS

In this chapter that is related to the RoREC and RoREP, the authors propose a set of recommendations and table 6 shows that RoRE implementation is complex but possible; the resultant technical and managerial recommendations are:

- The absolute lead of the author's methodology, framework, research and works in the domain of digital transformation.
- The RoREC and RoREP are applicable and mature; and can be used with standard methodologies like, UML, SysML and TOGAF.
- The PoC confirmed RoREP's feasibility.
- A Project must build a global RoREC to integrate in all Entity's sub-systems.
- Efforts must be applied to integrate DevOps4RE concept, and the main issues are related to silos.
- The ADM4RE's integration with RoREPs enable the automated management of Microartefacts.
- The Project must be separated in multiple transformation Projects, where the first one should attempt to transform the ICS.
- Digitization transforms the *Entity's* legacy information into a digital format which automates the ICS.
- A digital transformation supports the tracing and control of digital flows like in the domain of finance, where it can help the control of financial crimes (Trad, Nakitende, & Oke, 2021).

FUTURE RESEARCH DIRECTIONS

The Framework future research efforts will focus on the role of software engineering and decision making.

CONCLUSION

This RDP4RE is part of a series of publications related to Project*s*, AI, and enterprise architectures; and it is based on mixed AR model; where CSAs and CSFs are offered to help Manager*s* and Project EA specialist to reduce the risks of failure when building Project*s* and its critical RE activities. In this chapter, the focus is on RoREC and RoREP, where its formalism defines a structured inter-relationship of RE activities, Microartefacts development and other resources. RoREP's development concept is an important CSF for the Project*'s* evolution. The most important managerial recommendation that was generated by the previous research phases was that the Manager must be an AofABIS. The PoC was based on the CSFs' binding to a specific research resources, and the internal reasoning model that represents the relationships between this research's concepts, requirements, Microartefacts and CSFs. The final result clearly implies that the RoREC supports Projec*ts*, is applicable. The authors recommend performing the Project operations through multiple independent sub-Projects; and always coordinating these activities with the RoREC.

ACKNOWLEDGMENT

In a work as large as this research project, technical, typographical, grammatical, or other kinds of errors are bound to be present. Ultimately, all mistakes are the authors' responsibility. Nevertheless, the authors encourage feedback from readers, identifying errors in addition to comments on the work in general. It was our great pleasure to prepare this work. Now our greater hopes are for readers to receive some small measure of that pleasure.

REFERENCES

Beauvoir, P., & Sarrodie, J.-B. (2018). *Archi-The Free Archimate Modelling Tool*. User Guide. The Open Group.

Bebensee, B., & Hacks, S. (2019). Applying Dynamic Bayesian Networks for Automated Modeling in ArchiMate-A Realization Study. In *2019 IEEE 23rd International Enterprise Distributed Object Computing Workshop (EDOCW)*. IEEE. DOI 10.1109/EDOCW.2019.0001

Bingham, Ch., Eisenhardt, K., & Furr, N. (2007). What makes a process a capability? Heuristics, strategy, and effective capture of opportunities. Strategic Entrepreneurship Journal. doi:10.1002ej.1

BizzDesign. (2022). *Modeling a SWOT analysis*. BizzDesign. https://support.bizzdesign.com/display/knowledge/Modeling+a+SWOT+analysis

CenturyLink. (2019). *Application Lifecycle Management (ALM)*. CenturyLink.

Desfray, Ph. (2011). *Using OMG Standards with TOGAF*. SOFTEAM – Modeliosoft. www.modeliosoft.com

Ebert, C., Gallardo, G., Hernantes, J., & Serrano, N. (2016). *DevOps4AI*. IEEE.

Easterbrook, S., Singer, J., Storey, M., & Damian, D. (2008). *Guide to Advanced Empirical Software Engineering-Selecting Empirical Methods for Software Engineering Research* (F. Shull, Ed.). Springer.

Gartner. (2014a). *What the Business Process Director Needs to Know About Enterprise Architecture*. Gartner.

Gartner. (2016). Gartner's 2016 Hype Cycle for ICT in India Reveals the Technologies that are Most Relevant to Digital Business in India Analysts to Explore Key Technologies and Trends at Gartner Symposium/ITxpo 2016, 15-18 November, in Goa, India. Retrieved April 3, 2018, from https://www.gartner.com/newsroom/id/3503417

Goikoetxea, A. (2004). A mathematical framework for enterprise architecture representation and design. *International Journal of Information Technology & Decision Making*, *3*(1), 5–32.

Hosiaisluoma, E. (2021). *ArchiMate Cookbook-Patterns & Examples*. Hosiaisluoma.

Janzen, D., & Saiedian, H. (2005). *Test-driven development concepts, taxonomy, and future direction*. IEEE.

Jonkers, H., Band, I., & Quartel, D. (2012a). *ArchiSurance Case Study*. The Open Group.

Kitsios, F., Kyriakopoulou, M., & Kamariotou, M. (2022). Exploring Business Strategy Modelling with ArchiMate: A Case Study Approach. MPDI. *Information (Basel)*, *2022*(13), 31. doi:10.3390/info13010031

Koenig, J., Rustan, K., & Leino, M. (2016). *Programming Language Features for Refinement*. Stanford University. doi:10.4204/EPTCS.209.7

Kornilova, I. (2017). *DevOps4AI is a culture, not a role!* Medium. Retrieved January 2, 2018, from https://medium.com/@neonrocket/devops-is-a-culture-not-a-role-be1bed149b0

Koskela, L. (2007). *Test driven: practical tdd and acceptance tdd for java developers*. Manning Publications Co.

Koudelia, N. (2011). *Acceptance test-driven development* [Master Thesis]. University of Jyväskylä, Department of Mathematical Information Technology, Jyväskylä, Finland.

Lazar, I., Motogna, S., & Parv, B. (2010). Behaviour-Driven Development of Foundational UML Components. Department of Computer Science, Babes-Bolyai University, Cluj-Napoca, Romania. doi:10.1016/j.entcs.2010.07.007

Lindgren, E., & Münch, J. (2015). Software Development as an Experiment System: A Qualitative Survey on the State of the Practice. XP 2015: Agile Processes in Software Engineering and Extreme Programming. *International Conference on Agile Software Development*.

Liu, A. (2022). *Rumbaugh, Booch and Jacobson Methodologies*. Opengenus. https://iq.opengenus.org/rumbaugh-booch-and-jacobson-methodologies/

Manning-Franklin, A. (2022). *Functional Domain Driven Design: Simplified*. Antman Writers Series. https://antman-does-software.com/functional-domain-driven-design-simplified

Moore, J. (2014). *Java programming with lambda expressions-A mathematical example demonstrates the power of lambdas in Java 8*. https://www.javaworld.com/chapter/2092260/java-se/java-programming-with-lambda-expressions.html

Moreland, C. (2006). *An Introduction to the OMG Systems Modeling Language (OMG SysML)*. Object Management Group. ARTiSAN Software Tools.

MPSINC. (2021). *Progamming Languages Conversion*. MPSINC. http://www.mpsinc.com/index.html

Neumann, G. (2002). Programming Languages in Artificial Intelligence. In Encyclopedia of Information Systems. Academic Press.

Nuseibeh, B., & Easterbrook, S. (2000). *Requirements Engineering: A Roadmap. Department of Computing Department of Computer Science*. Imperial College.

Patni, S. (2017). *Pro RESTful APIs Design, Build and Integrate with REST, JSON, XML and JAX-RS*. Apress.

Papavassiliou, G., Ntioudis, S., Mentzas, G., & Abecker, A. (2001). The DECOR approach to Business Process Oriented Knowledge Management (BPOKM). In *DEXA '01 Proceedings of the 12th International Workshop on Database and Expert Systems Applications*. IEEE Computer Society. https://astimen.wordpress.com/2009/10/28/the-decor-approach-to-business-process-oriented-knowledge-management-bpokm

Pavel, F. (2011). Grid Database—Management, OGSA and Integration. Academy of Economic Studies Romania.

Peterson, S. (2011). *Why it Worked: Critical Success Factors of a Financial Reform Project in Africa*. Faculty Research Working Paper Series. Harvard Kennedy School.

Remawati, D. (2016). Analisis SWOT Implementasi Green Computing Di Sekolah Kejuruan (Studi Kasus Pada SMK XYZ). *J. Ilm. SINUS*, 23–36.

Richardson, C. (2014). *Pattern: Microservices architecture*. https://microservices.io/patterns/microservices.html

Soeken, M., Drechsler, W., & Wille, R. (2012). Assisted Behavior Driven Development Using Natural Language Processing. In *International Conference on Modelling Techniques and Tools for Computer Performance Evaluation. Tools 2012: Objects, Models, Components, Patterns* (pp. 269-287). Springer-Verlag Berlin Heidelberg. 10.1007/978-3-642-30561-0_19

Tkachenko, E. (2015). *5 key attributes of requirements testing: Know before you code*. EPAM Systems. https://techbeacon.com/app-dev-testing/5-key-attributes-requirements-testing-know-you-code

Togaf-Modeling. (2020). *Application communication diagrams*. Togaf-Modeling.org. https://www.togaf-modeling.org/models/application-architecture/application-communication-diagrams.html

The Open Group. (2011a). *Architecture Development Method*. The Open Group. https://pubs.opengroup.org/architecture/togaf9-doc/arch/chap05.html

The Open Group. (2011b). *TOGAF 9.1*. The Open Group. https://www.opengroup.org/subjectareas/enterprise/togaf

TOGAF Skills. (2011). *Architecture skills framework*. http://pubs.opengroup.org/architecture/togaf9-doc/arch/chap52.html

TOGAF Catalogs. (2011). *Sample catalogs, matrices and diagrams.* http://www.opengroup.org/book-store/catalog/i093.htm

TOGAF Directory. (2011). *Using TOGAF to Define & Govern SOA.* https://pubs.opengroup.org/architecture/togaf9-doc/arch/chap 22.html

Tutorialspoint. (2022). *Requirement Based Testing.* Tutorialspoint. https://www.tutorialspoint.coM/Roftware_testing_dictionary/r equirements_based_testing.htm

Trad, A., & Kalpić, D. (2013). *The Selection, and Training Framework (STF) for Managers in Business Innovation Transformation Projects - The Background.* Conference on Information Technology Interfaces, Cavtat, Croatia.

Trad, A., & Kalpić, D. (2017a). *An Intelligent Neural Networks Micro Artefact Patterns' Based Enterprise Architecture Model.* IGI-Global.

Trad, A., & Kalpić, D. (2018a). *The Business Transformation Framework and Enterprise Architecture Framework for Managers in Business Innovation-Knowledge and Intelligence Driven Development (KIDD). In Encyclopedia of E-Commerce Development, Implementation, and Management.* IGI-Global.

Trad, A., & Kalpić, D. (2018b). *The Business Transformation Framework and Enterprise Architecture Framework for Managers in Business Innovation-Knowledge Management in Global Software Engineering (KMGSE). In Encyclopedia of E-Commerce Development, Implementation, and Management.* IGI-Global.

Trad, A., & Kalpić, D. (2020a). *Using Applied Mathematical Models for Business Transformation.* IGI Global. doi:10.4018/978-1-7998-1009-4

Trad, A., & Kalpić, D. (2020b). *The Business Transformation Framework and Enterprise Architecture Framework for Managers in Business Innovation-Intelligence Driven Development and Operations (IDDevOps).* IGI Global.

Trad, A. (2020). *Applied Mathematical Model for Business Transformation Projects: The Intelligent Strategic Decision-Making System (iSDMS). In Handbook of Research on IT Applications for Strategic Competitive Advantage and Decision Making.* IGI Global.

Trad, A., Nakitende, M., & Oke, T. (2021). *Tech-Based Enterprise Control and Audit for Financial Crimes: The Case of State-Owned Global Financial Predators (SOGFP). In Handbook of Research on Theory and Practice of Financial Crimes.* IGI Global.

Trad, A. (2021). An Applied Mathematical Model for Business Transformation and Enterprise Architecture: The Holistic Organizational Intelligence and Knowledge Management Pattern's Integration. *International Journal of Organizational and Collective Intelligence, 11*(1), 1–25. doi:10.4018/IJOCI.2021010101

Trad, A., & Kalpić, D. (2022). Business Transformation Project's Holistic Agile Management (BT-PHAM). *The Business & Management Review, 13*(1), 103–120. doi:10.24052/BMR/V13NU01/ART-12

Uhl, L., & Gollenia, L. A. (2012). *A Handbook of Business Transformation Management Methodology, Gower.* SAP.

Ylimäki, T. (2006). Potential critical success factors for enterprise architecture. *Journal of Enterprise Architecture*, 2(4), 29–40.

Wati, A., Ranggadara, I., Kurnianda, N., Irmawan, D., & Frizki, D. (2019). Enterprise Architecture for Designing Human Resources Application Standard Reference. *International Journal of Innovative Technology and Exploring Engineering, 8*(12).

Visual Paradigm. (2019). *TOGAF ADM4AI Tutorial. Visual Paradigm*. https://cdn.visual-paradigm.com/guide/togaf/togaf-adm-tutorial/02-togaf-adm.png

Wikipedia. (2021a). *Code refactoring*. https://en.wikipedia.org/wiki/Code_refactoring

Chapter 5
Digital Maturity Models:
A Holistic Framework for Digital Transformation

Kemal Özkan Yılmaz
https://orcid.org/0000-0003-1185-4397
Istanbul Kültür University, Turkey

ABSTRACT

This chapter aims to create a general framework on digital maturity models and reveal the highlights of the commonly used components of digital maturity models. The emphasis will be more on detailing human-related fundamental aspects of digital maturity models rather than the technological ones. Considering the incrementally increasing options to reach collaboration technologies and swift deployment of them to different sectors, the main challenge still remains as to manage the psychological asset related parts. Although technology is generally perceived as mechanical, whose boundaries can be drawn, tasks to be assigned can be defined; the technological elements within the concept of digital maturity also intersects with the human-related dimensions of digital transformation. The study aims to advise a systematic approach by offering a theoretical framework to manage human-related factors in a digital transformation journey with an inclusive perspective positioning technology as the core medium provider and contributes to the theoretical background in the field of study.

INTRODUCTION

The concepts of digitization, digitalization and digital transformation have become more and more audible in daily and corporate life, with the rapid development in the capabilities of technological opportunities and their presentation to larger masses in a more economical manner. This popularity usually ends up with the misunderstanding of these concepts even at the corporate world which highlights the need to assimilate the building blocks of the aforementioned concepts to ensure that the practitioners can effectively benefit from the ameliorating power of digital transformation in complex systems or in daily routines and operational practices.

DOI: 10.4018/978-1-6684-4102-2.ch005

The Covid-19 pandemic has acted as a directional impulse for many corporations, some of which have linked digital transformation as a savior to their sustainability (Agrawal, 2020; Bai et al., 2021), and majority of the others concentrated on their efforts to improve and make their digital processes perfect.

This chapter will introduce digital maturity models and highlight the important considerations of the commonly used components of featured digital maturity models to facilitate generation of a handy framework. The emphasis will be more on detailing human-related fundamental aspects of digital maturity models rather than the technological ones, since how human factor is utilized is regarded as the main source of differential value creating component in a digital transformation progress. On the other hand, such a content framing is necessary to handle such a broad topic in a chapter.

As it will be detailed throughout the chapter; although, technology is generally perceived as something mechanical, whose boundaries can be drawn, with what assigned tasks can be clearly defined, and even its course and journey to corporate goals are more easily manageable; the technological elements within the concept of digital maturity also intersect with the human-related dimensions of digital transformation. This intersection is a requirement of the role of human in an organizational setting, in other words, psychological capital that is integrated into ecosystems that are forming a meaningful whole. This situation creates a need to make some suggestions that will also be related to the technological elements of digital maturity throughout the chapter.

Digital maturity levels, on the other hand, are an inseparable measure of digital maturity concept that reveal the level and intensity of these concepts being applied within an organization, and in a way, they reveal the gap that a company has to cover in order to reach an acceptable maturity in digital transformation. Assessing digital maturity levels in an organizational setting also serves not only as a monitoring activity, but also as a starting point to reveal the current situation used as a check-up to offer the right prescription. Digital maturity level/stage assessment is to be mentioned to bind the efforts to an undeniable need to make progress, and adopt to the evolving needs of a business and an organization's requirements in time. The psychological asset involvement in the digital maturity journey brings a dynamic perspective and nature into existence.

While detailing these related dimensions, their roles in a digital transformation journey will also be explained, which will also include highlighting useful attributes of digital maturity dimensions. The need to sustain a progressive nature at digital transformation is also to be mentioned as products, services, technologies can become obsolete or they can change and get updated in a business environment; such a transformation will require organizations to catch up with contemporary ways of creating value (De Santos & Francisco, 2021).

The main purpose of this chapter is to offer an understanding of a digital maturity framework and detail the main aspects to take care of during a digital transformation remembering that digital maturity is not a destination, instead it is journey that would be applied in a holistic way.

The study can be considered as an attempt to make a remarkable theoretical contribution to the existing literature owing to its assertion that human related factors are not only facilitators but also requirements whose organizational and procedural links are to be bonded and strengthened by technological tools and platforms. The chapter only does not detail technological components in depth to focus on the main objective, but accepts and appreciates technology as a catalyst which in the first place requires the right environmental conditions. In other words, it foresees technology's interaction with the other dimensions in a digital maturity model such as strategy, leadership, culture and organization accepting it as a relational whole within a digital ecosystem by default.

The chapter is composed of seven sections. The first one after the introduction details the methodology, the second deals with theoretical background, as the third section gives background information, the fourth talks about the main focus of the chapter – digital maturity models, further sections detail solutions and recommendations, future research directions and conclusion of the study.

METHODOLOGY

The methodology of this chapter will be to start with brief definitions of digitization, digitalization, digital transformation and digital maturity concepts first. In order to sustain a common understanding, the term, digital maturity as a prominent context of digital transformation will be encapsulated. After setting a common background to facilitate further discussion, digital maturity models regarding digital transformation will be disclosed; first by highlighting prominent digital maturity models including the concepts and main dimensions of contemporary Industry 4.0 maturity models and/or indices. A general framework on digital maturity models will be depicted, and the highlights of the commonly used components of them will be explained. Furthermore, in the light of the detailed components, a theoretical framework that prioritizes applicability in business life will be proposed.

Industry 4.0 is also to be briefly defined since some of the organizations are industrial bodies whose processes and applications are highly related to industrial settings. In most of the cases companies' decisions to realize technological tools not only determine achieved and potential efficiency levels, but also provide organizations the opportunity to deal with more value-adding activities, and direct them to review (and even redesign) their value networks and inter-organizational relationships (Cennamo et al., 2020).

BACKGROUND

Prior to the coronavirus pandemic digital transformation was already a hot topic which has been utilized into daily practices. After experiencing shutdown periods due to the spread of the virus on a global scale, more companies appreciated the concept as a guiding opportunity to transform their businesses and the way they interact with their stakeholders to sustain their growth (Gavrila Gavrilo & De Lucas Ancillo, 2022), and more importantly their resilience (Organisation for Economic Co-operation and Development [OECD], 2020).

Actually, the real problem arises right at this stage, since most of the companies are unaware of success parameters in a digitalization process (Azizan et al., 2021), while some of them don't really have adequate idea about the fundamental concepts of digital transformation and their definitions (Eller et al., 2020). Especially, despite nearly ground-level of digitalization at large numbers of Small Medium Enterprises (SMEs) at the beginning of the Covid-19 pandemic, the challenge they faced was in affiliating related technologies and deploying them throughout their organizations (OECD, 2020) even in developed countries. SMEs has already started to perceive digitalization with an amplifying importance (Meffert et al., 2020). Therefore, digitization, digitalization, digital transformation and digital maturity concepts will be defined and termed to facilitate a mutual background before focusing on the digital maturity models and frameworks.

Digitization Concept

Digitization is defined as the basic binary process changing from analog to digital so that it can be stored in a database such as using a keyboard and a word processor for writing (Brennen & Kreiss, 2016; Bloomberg, 2018; Gobble, 2018; Savić, 2019; Gartner, 2022). As Ross (2017) puts forward, digitization standalone will not be able constitute a digitally operating organization. Although, some academicians like Ross (2017) define digitization as "standardizing business processes" (para. 1), the chapter will stick to the concept of digitalization when processes are to be involved; further justification will be supplied in the following paragraphs. It is also imperative to mention that, value generation becomes easier with digitization, such as experiencing its ability to reduce (Gobble, 2018) or avoid errors.

Digitalization Concept

Digitalization differs from digitization firstly because of its broader scope (Marushchak et al., 2021: 306), and secondly it has a potential to change the business since it connects digitized data with the processes in a system which Thomson (2021) names as an ecosystem. The main task in digitalization is generating autonomous business processes (Savić, 2019) contrary to the starting point of digitization which concentrates on data and its digitization. Digitalization is about utilizing digitized data/information in an effective way (Reis et al., 2020; Marushchak et al., 2021; Peranzo, 2021) to earn time to focus on value-adding activities such as decentralized decision making by autonomous systems, improvements in real-time data-driven analytics (Santos & Martinho, 2019). In other words, digitalization is changing the way data and/or information is processed; automation is the method to ignite this, but automation is not the final step of digitalization as its scope includes aggregating interrelated processes. For example, integration of software with required technical devices and creating a new set of processes for supplying service in medical services in an online way is one of the best implementation areas of digitalization, which is facilitated by digitization of data first; such a process not only enhances the performance of medical services but also creates new value for the consumer (Hilpert, 2021, p. 1591).

Social media is one of these systems to be integrated to sustain seamless connectivity and communication using digital platforms for information sharing about a company's products and services, social media mediums and satisfying customer service systems (Brennen & Kreiss, 2016) that are organized around the customer (VanBoskirk et al., 2017, p. 7; Gobble, 2018).

Digital Transformation Concept

According to Shallmo et al. (2017) and Teichert (2019), there is not a consensus on the definition of digital transformation. In order to explain the differences between the formerly defined terms just above this subtitle and digital transformation, the most conspicuous clue is integrating new digital technologies and tools into new organizational structures and reformulated ecosystem processes (Schwertner, 2017; Gobble, 2018; Marushchak, 2021). Shallmo and Williams (2021) explain digital transformation with the concept of strategy-oriented digital initiatives by giving digital strategy and digital business models as examples of these initiatives. They see strategy as an integral part of a digital transformation process. A comprehensive review of an organization's existing processes is a requirement throughout such a process which is utilized by Industry 4.0 concept in industrial settings to reconstitute, restructure and convert value chains taking advantage of digitization (Hilpert, 2021). A network structure, being interconnected and externally connected whenever necessary is embedded Industry 4.0 characteristics. Therewithal, as Hilpert (2021) states:

Industry 4.0 and digitization indicates how the context of innovation changes towards centralization of decision taking, dependencies of plants and locations as well as opportunities can be created based on policies to change from a supplier to a contributor to innovation (Hilpert, 2021, p. 1603).

The intended scope in a digital transformation is not limited with traditional organizational boundaries, the boundaries of an organization should be expanded by taking a broader perspective into play in order sustain an organization's competitiveness (Gong & Ribiere, 2021, p.14; Marushchak et al., 2021, p. 306). This broader perspective is about integrating different system into one; inevitably complexity of the ecosystem that has been created becomes one of the first issues to be managed in a digital transformation process highlighting collaboration (Wald et al., 2019) to enhance capabilities and assets to create more customer value (Ross, 2017; Gong & Ribiere, 2021, p.14).

The author aims to elaborate a digital transformation maturity framework by mentioning major components and manageable amount of digital maturity levels to highlight further development opportunities and a roadmap to progress with this perspective.

Digital Maturity Concept

Cambridge Dictionary (2022) defines maturity as "a very advanced or developed form or state".

Digital maturity is not only about modifying an organization's structures to effectively compete in digitalized systems (Kane et. al., 2017); apart from structures, it aims to reach an effective state that also coordinates people, culture and daily tasks and routines which are targeted to benefit from the provided technological advancements resulting from the internal or external environments (Rader, 2019). The concept also serves like a lighthouse to show the right roadmap and improvement areas when formulated as a model (Teichert et al., 2019). Psychological asset related endogenous components of digital maturity can mainly be grouped as cross-functional collaboration and corporate digital cultures (Teichert, 2019; Schallmo & Williams, 2021; Yılmaz, 2021) whose sub-dimensions will be detailed while exploiting the digital maturity models.

Digital tools such as Machine Learning (ML), Artificial Intelligence (AI), Virtual Reality (VR), Augmented Reality (AR), 5G, 6G, Internet of Things (IoT), blockchain technology, and automated decision support systems had already started to find application areas in business environments (Kane et al., 2017; Neugebauer, 2019; Harvard Business Review [HBR], 2020a; HBR, 2020b). Digital maturity level assessment of the technological sub-dimensions of a digital maturity model can be perceived to be easier due to their nature of being explicitly observed or from the productivity and efficiency generated by utilizing them.

Many digital maturity and Industry 4.0 models have been developed as digital transformation has always been in the agenda of corporations and the public institutions. Most of the popular models and the related resources of theirs can be named as follows: The Connected Enterprise Maturity Model - Rockwell Automation (Bradley, 2014), Digital Maturity Model 4.0. - Forrester (Gill & VanBoskirk, 2016), The ACATECH Industrie 4.0 Maturity Index (Schuh et al., 2017), Industry 4.0 Maturity Model (De Carolis et al., 2017), Industry 4.0 Maturity Model (Colli et al., 2018), Digital Maturity Model 5.0. – Forrester (VanBoskirk, et al., 2017), Industry 4.0 Maturity Model (Şener et al., 2018), Industry 4.0

Maturity Model for Manufacturing (Schumacher et al., 2019), Dimensions of Digital Transformation (Gurbaxani & Dunkle, 2019), Digital Readiness Index Assessment (Philipp, 2020), The New Elements of Digital Transformation (Bonnet & Westerman, 2021). Apart from these, many consulting companies and institutions such as Accenture, Booz & Company, CapGemini Consulting, Digital Transformation Group, DT Associates, KPMG, McKinsey & Company, PWC, MIT Center for Digital Business Strategy, and Transformation Consulting have also generated models to help companies transform themselves. The chapter will not try to cover every model that has been offered, but it will talk about the contemporary ones possessing detailed dimensions which can be applicable in corporate life according to the author's evaluation will be complied.

MAIN FOCUS OF THE CHAPTER

Technological tools and mediums developing day by day help organizations in the widening of their ecosystems of customers and partners bringing in new opportunities to improve the value they offer (Rader, 2019; Teichert, 2019; Thomson et al., 2021). Skylar et al. (2019) state how collaboration within ecosystem shareholders is enhanced and asserts that ecosystem stakeholders increase their capacity to transform organizational structures by influencing each other; in line with this thought, Azizan et al. (2021) name the outcome of this process as network value that has the potential enable companies to deliver new and more profitable value if managed in correctly. Consistent with this view, when a company can support the power of becoming better as a whole, the optimum task would be to start with assessing the digital transformation maturity at an organization or an ecosystem. Apart from a collective perspective, based on the mentioned characteristics digital maturity is also regarded as a value booster (Rader, 2019; de Santos & Francisco, 2021). The chapter will detail components of contemporary digital maturity structure and Industry 4.0 models/frameworks and their proposed levels to explain this perspective. In summary, digital transformation is a perpetual long-term journey which comprises a mentality to adapt to new conditions. Therefore, the important thing is to get started. In order to reach a sustainable result, digital maturity dimensions and a proposal to move the ladders forward will be discussed at this section.

Digital Maturity Models

A business model is the medium by which an organization portrays its provided benefits to the customers and business partners (Shallmo, et al., 2017). A digital maturity model is a systematic apparatus which first helps an organization to make an assessment of its current digitalization efforts and digital transformation phase (Santos & Martinho, 2021, p. 1026), and subsequently plan consequent steps based on the signals and indications gained from the assessment in a strategic way (Teichert, 2019; Barboza et al., 2022). Strömberg et al. (2020) describes digital maturity as something that becomes as usual generally coming into being like utilized technologies, processes, prevalent culture and ways of doing things. Fletcher and Griffiths (2020) suggest continuous improvement of digital maturity as a way of improvement in order to reduce losses and avoid potential losses encountered in the daily routines of enterprises, especially in turbulent times. Santos & Martinho (2021) support this idea, and state it as a future direction, as they call digital maturity as a forthcoming desired state and a goal to be reached towards a more sophisticated level. Williams et al. (2022), explain the term as an instrument to evaluate an organization's present and approaching capabilities in a systematic way by aiming to sort out the most critical dimensions and capabilities for a business setting.

Table 1 details complementary definitions about the proposed dimensions pillars and gives definition of key terms and clues about their utilization in an organization:

Table 1. Abbreviated Digital Maturity Dimensions

Digital Maturity Dimension	Definitions / Utilization	Definitions / Utilization - Industry 4.0 (I4.0) Focused
Strategy	Means of competitive advantage (Aversano et al., 2016; Strömberg et al., 2020; Cozzolino et al., 2021). Strategic vision and alignment (Gurbaxani & Dunkle, 2019). Cooperation decisions (Cozzolino et al., 2021).	Modification and/or adjustment of strategies such as R&D and product development; the fit (Hilpert, 2021). Enhancing digital capabilities (Eremina et al., 2019). The transformation of strategies to create and align value generation (Correani et al., 2020). Industry 4.0 implementation strategy and agility (Schuh et al., 2017).
Leadership	Merging traditional and new skills (Kane et al., 2019) and capabilities. Role and commitment of top executives. Vision (Strömberg et al., 2020). Ecosystem orchestrators and roles (Cozzolino et al., 2021). Leader should be in charge (Rader, 2019).	Willingness and competence leaders in management and coordination (Schumacher et al., 2016; Santos & Martinho, 2019). Digital competence of the leader (Sundberg et al., 2019). Situational factors and leader's personality (Ostmeier & Strobel, 2022).
Culture	Prominent success factor and core value creation capability (Kane, 2017; Kwiotkowska, 2022); corresponds to the organization. Role defining & cross-functional culture formation (Ubiparipović et al., 2022). Risk tolerant, collaborative and keen on learning (Rader, 2019); customer-centric (Teichert, 2019).	Adeptness to open-innovation and change; emphasis on knowledge sharing (Schumacher et al., 2016; Santos & Martinho, 2019). Cultural principles (Philipp, 2020). Deployed set of behavioral characteristics (Santos & Martinho, 2019).
Technology	Hardware and software, engagement with digital technologies (Cozzolino et al., 2021). Information and Communication Technologies (ICT) (Kwiotkowska, 2022). Behind the times technologies do not have the potential to sustain digital maturity (Rader, 2019; Kane, 2017)	Manufacturing and design technologies (Kwiotkowska, 2022), Internet of things (IoT), Cloud updates in the core technologies that shape related industries. Innovation capabilities brought by the preferred technologies and their rapport into competitiveness creating value chains (Hilpert, 2021).

Source: Compiled by the author (2022).

The holistic approach to manage digital transformation in new type of networks namely collaborative business ecosystems standalone does not assure prosperity (Barboza et al., 2022); especially at such interregnum states, assessment of maturity levels supplies a framework and a route to progress or maturing (Santos & Martinho, 2021) presented as the main objective of this chapter.

One of the most practical ways to formulate a digital maturity model is using four main pillars which are strategy, leadership, culture and technology (Strömberg et al., 2020); this is the most practical method; it is using a zipped version like sticking to a marketing mix composed of 4Ps instead of utilizing a 15Ps mix to manage a firm's marketing strategy and efforts.

Technology dimension will not be further extended since technological changes or shifts actually remain at the uncontrollable external environment of most of the companies (Kane, 2017) that are keen reaching higher levels of digital maturity. At this stage, it will be significant to recall the importance and interconnectedness of innovation (Schumacher et al., 2016; Santos & Martinho, 2019) and digital transformation which foresees culture as the starting point of becoming successful and concentrating on the customer (Teichert, 2019) and on competitiveness creating value creation Kwiotkowska, 2022).

First three dimensions which are strategy, leadership and culture given in Table 1 are called psychological asset (human) related dimensions. The most practical way to formulate a digital maturity model is using four main dimensions which are strategy, leadership, culture and technology (Strömberg et al., 2020) as also given at Table 1. As already mentioned for a more detailed dimension structure, Table 2 will be used to breakdown the so-called psychological asset (human) related dimensions into more detail to indicate embedded concepts that are meant to be included in the abbreviated model version that is proposed to be used to avoid complexity while its application. Table 2 details the extended version of the so-called psychological asset related digital maturity dimensions in a digital transformation continuum:

Table 2. Digital Maturity Dimensions - Extended Version of Psychological Asset Related Ones

Type	Digital Maturity Dimension	Definitions / Utilization Including Industry 4.0 Perspective
Psychological Asset Related Dimensions – Extended Version	(Digital) Culture	Please refer to Table 1.
	Strategy	Please refer to Table 1.
	Leadership	Please refer to Table 1.
	Organization	Organizational structure and its flexibility; leadership; information systems (Brusakova, 2022); intangible and tangible resources (Brusakova, 2022; Kwiotkowska, 2022). Organizational/ structural capabilities evolving in Industry 4.0 efforts (Şener et al., 2018).
	Digital Skills	Self-initiated/proactive & career opportunity facilitating skill development (Ostmeier & Strobel, 2022). Their crucial role in the transformation process (Kane et al., 2019; Eller et al., 2020). Education and development for the right skills (Hilpert, 2021).
	Innovation	In search of digital innovation for business model transformation opportunities (Azizan et al., 2021). Innovation is derived by ecosystems and networks; Industry 4.0's interconnectedness with innovation, processes; innovation arising from technology & digitization; innovation chains (Hilpert, 2021).
	Customer Insight & Experience	Ecosystems bridging processes with customers and stakeholders (Thomson et al., 2021). The power of customer insights & customer related big data in updating customer experience (Correani et al., 2020; Marushchak et al., 2021). Artificial intelligence administered digital experiences (Correani et al., 2020). Technology driven augmentation in value creation and customer centricity (Barboza et al., 2022).
	Vision	The need for a digital mindset and entrepreneurship for a real change and transformation (Warner & Wäger, 2019). The required technologies to design, bring to life and prosper intelligent systems.
	Digital Ecosystem	Platform based ecosystem; digital advertising ecosystem; materialization of digital ecosystems (Cozzolino et al., 2021). Organizational boundaries getting indistinctive (Skylar et al., 2019). The communication and relationship management within the ecosystems can generate upgrading the set to enable autonomous quick fixes (Thomson et al., 2021).

Source: Compiled by the author (2022).

Table 2 offers a more disassembled outlook of Table 1; for example: vision dimension can be studied under Table 1's strategy dimension, but if an organization needs to express some sort of special emphasis on vision, extended maturity levels can be more suitable for such purposes. There is a similar relationship between culture and innovation; actually, innovation is a subset of culture which is vital for any organization. Since innovation finds ground by facilitating a risk tolerant, collaborative and open to change mindset such a culture has first to be created in any organizational setting to open room for digital transformation efforts. An employee's motivation to build digital skills can also be related with the characteristics of the culture that is tried to be built within an ecosystem (Ostmeier & Strobel, 2022);

so that, this extended dimension can be taken as set resulting from culture and digital ecosystems or having being facilitated by them concurrently.

Table 2 also utilizes digital ecosystem dimension to emphasize that the boundaries of the organization is no longer limited to its employees and customers and most of the change drivers result from an extended external environment. Platform based ecosystems or network of companies has evolved as a standard business execution tool helping companies gather focused insights about the customer bringing about effective transactions and collaboration between network members (Cozzolino et al., 2021). To underline customer insight and experience the above dimensions of the human related part of the model exploits these concepts as a separate dimension.

Theoretical Framework

The chapter proposes a holistic theoretical framework, detailed in Figure 1, in which technology plays the role of a conductive fluid like role in a pool of experiments, thereby permeating all other dimensions and connecting them as well.

Figure 1. The proposed digital maturity model. Source: Compiled by the author (2022).

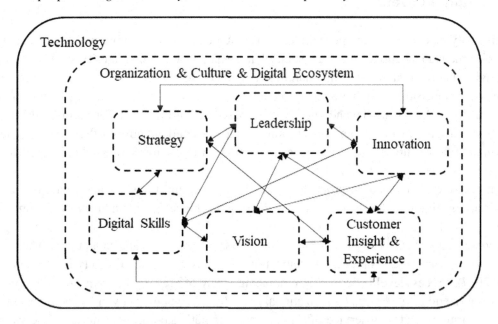

Instead of using technology as a separate dimension the author prefers to depict it as a fluidic and uniting element. Figure 1 aims to illustrate that all the other dimensions exist and get functional with technology. There is a mutual relationship between digital maturity dimensions; for example, technology sustains a fertile medium for spreading an organization's culture along its ecosystem, and also an organization's culture facilitates necessary background for the deployment of technology. In the proposed model, technology surrounds all the other dimensions or in other words, others are soaked into technology like test objects in an experiment basin ensuring technology's penetration into them resulting in a commune. Xie and Chang's (2022) points of view supports this idea as they elaborate that an organizations readiness for digital innovation affects its business model's innovation capacity. Williams et al. (2022) also highlight the capacity of digital maturity models to facilitate innovation, thus Figure 1 uses dashed boxes to depict and demonstrate the digital maturity dimensions' interaction both with each other and with technology, organization, culture and digital ecosystems as habitat facilitators and providers.

The findings of Perera et al. (2022) support the author's perspective; they indicate that digital maturity ameliorate innovations as it supplies better cohesion within an organization's digital ecosystem. Another supporting evidence is the existence of 150 maturity models advised to manage innovation even in 2019 (Schallmo & Williams, 2021). Furthermore, some researchers often use the terms Industry 4.0 and innovation interchangeably (Santos & Martinho, 2019; Hilpert, 2021; Tutak & Brodny, 2022; Xie & Chang, 2022), and the chapter's proposed framework adopts a holistic framework that also encompasses this idea.

The other elements of the of the proposed digital maturity model; namely strategy, leadership, innovation, customer insight & experience, digital skills and vision all are thought to have mutual effects on each other. This situation is also used to reveal the interdependence of the variables in the proposed model in line with the ideas of Phuyal et al. (2020), and the more effective management of a model component will have a significant positive effect on the others. Similarly, if a variable is not well-managed or insufficient, it has the potential to have a negative impact on other variables.

Digital Maturity Levels

Digital maturity models aim to propose dimensions that are to be further improved which brings in the idea of assessment and identifying current status that are named as digital maturity levels. Barboza et al. (2022) advise that both qualitative and quantitative methods should be used in the assessment process of a maturity model. Such a depth is a requirement because of the variety of aforementioned digital maturity dimensions. Organizational and/or ecosystem-based evolution or transformation is the main goal of a maturing progress, where we need a certain level to start and quantity further plans and activities which can be followed up using definite measurement criteria (Santos & Martinho, 2021) which aims to avoid uncertainties (Kane et al., 2019).

Shallmo and Williams (2021) propose digital strategy, digital transformation, and digital implementation as constitutional levels to be used to gauge and achieve digital maturity. According to Gong and Ribiere (2021, p. 15) digital transformation is an inevitable collective issue that aims to generate constitutional change in an ecosystem which requires a set of resources and capabilities. These changes occur at diverse levels such as process, organization, business domain and society level (Parviainen & Tihinen, 2017). Process level is defined as embracing newly offered digital tools and redesigning processes by digitization first design. Organizational level digital level maturity talks about designing new approaches, bringing out new service offerings, and eliminating outdated patterns in the value flow. Business domain level carries the discussion to ecosystems that an organization is involved and mainly focuses on value generation chains and reorganizing roles within the value flow. Society level details structures (Parviainen & Tihinen, 2017, p. 64) in an ecosystem.

Another perspective, the integrated approach consists of several fundamental levels for strategically oriented digitalization. These include digital strategy, digital transformation, and digital implementation. The approach also consists of three relevant steps: an initial analysis of digital maturity, an in-depth analysis, and the derivation of measurements (Schallmo & Williams, 2021, p. 1). Table 3 depicts five staged two proposals for digital maturity level determination from current literature, whose application will be practical and can be used to reach results frictionlessly:

Table 3. Digital Maturity Levels/Stages

Level/Stage	Digital Maturity Level Description
Source: Azizan et al., 2021	
Initialized	Lack of process monitoring. The organization is a reactive mechanism, digitalization efforts need to be activated.
Managed	Processes can be partly carried out due the absence of correct enabling technologies. Digitalization is partly realized; further progress is required.
Defined	Robustly architectured processes that are equipped with comprehensive digitalization instruments. Level of digitalization has the potential visualize areas to be further improved.
Integrated & Interoperable	Present systems and processes are upgraded and developed. Technology is integrated to processes to maintain good quality outputs.
Digitally Oriented	Company utilizes digital technologies and has already transformed its processes. Although, fruits of digital transformation are enjoyed and improved results (reliable & swift information, etc.) is experienced, the need for better internal cooperation in decision making is observed.
Source: Barboza et al., 2022	
Infancy	Processes are mainly managed using conventional instruments. ICT is regarded as a cost generating set of tools rather than an efficiency medium. Short-term focus and closed loop boundaries dominate the organization.
Developing	Awareness on ICT utilization is a useful tool to progress business. Getting started with ICT without eliminating objectivity. Dependence on specific individuals is still dominant in conducting daily routines and tasks, so that no consistency in results and limited concern for customer's engagement.
Transforming	ICT utilization is realized as a must have to progress business and finds prominent formal place in corporate strategies to enable co-creation amongst ecosystem members.
Optimized	Dominance of ICT in company's footprints. Innovativeness is encouraged and becomes possible as a result of coordinated process flow within ecosystems. ICT is embedded in processes; co-creation within ecosystem members is applied and a desired method.
Digital Maturity	The form of planning and actualizing co-created value as a natural outcome of a transformed culture and mindset. Collaborating and making use of technology within ecosystems.

Source: Compiled by the author (2022).

Shallmo and Williams (2021) assert a simpler digital maturity assessment model than given at Table 2, which is focusing on the outcomes of digital transformation as emerged elements at the organizational structures (and ways of doing business). Both of the results and structures are to be evaluated with three main pillars are digital strategy, digital transformation and digital implementation. Each of those divided into 4 stages which formulates an "initial analysis" (p. 10) reference as a starting point. The proposed maturity levels are ranked from one to four, in a hierarchical order they are listed as: not having started or existing; achieved some progress; achieved considerable progress and fully achieved.

Rather than the number of digital maturity levels presented in a model, how these levels are detailed is more important to assess current situation and make further plans for a company. The depth of existence of digitalization evidence varies from infancy to digitally mature based on the assessment of the dimensions in the utilized digital maturity frameworks given at Table 3. The suggested digital maturity levels to be used to assess an organization's status are: infancy, developing, transforming, optimized and digitally mature. Although, Barboza et al. (2022) emphasize the outstanding value of culture and mindset transformation or their restatement as a natural outcome of a successful digital transformation, the author finds digitally transformed processes (Azizan et al., 2021) an imperative set of evidence guaranteeing digital maturity.

The details summarized in each level's description given at Table 3 can be used to work on must haves to reach higher levels of digital maturity and to identify a position for further development plans and realize the (full/better) advantage of their digitalization potency (Santos & Martinho, 2021; Schallmo & Williams, 2021; Thomson et al., 2021; Ostmeier & Strobel, 2022).

Another important contribution of digital transformation to organizations is the atmosphere it creates for the engagement of leaders and employees (Azizan et al., 2021). Cozzolino et al., (2020) extends this idea as a trait to be sustained between ecosystems members that is aimed to generate cooperation and even competition. Digital maturity state of an organization should be crowned by the deployment of collaboration throughout its ecosystem using technology as a facilitator.

SOLUTIONS AND RECOMMENDATIONS

Digital transformation progress assessment facilitates being or becoming innovative over time, based on systematic efforts to develop employees, which encourage them to act in an innovative way to update or redesign organizational structures (Hilpert, 2021) and behavior – in other words cogitating culture (De Santos & Francisco, 2021). Digital transformation efforts would target to involve customers (Ross, 2017); with the help of technology, and the interaction capability of the mediums offered, new ways of engagement and more touch points with the customers would be designed in the transformation journey. Enhancing engagement by default extends organizational boundaries resulting in a broader span of culture (De Santos & Francisco, 2021); such an effect conjures up stereotyping, monitoring and alignment of culture for value offerings matching the desires and expectations of a company's customers.

Determining the current digital maturity level of an organization serves both as a starting point and a roadmap formulating instrument. Of course, the time required to see remarkable changes can take more time than proposed (Ross, 2017), so that the key element in a transformation process is to incessantly striving, sticking to the plan and also rechecking digital maturity levels in order to not to be late for reviewing and taking corrective action where necessary.

For a manufacturing company understanding ecosystem dynamics from research & development to cost structure (Hilpert, 2021) and nonlinearities within its network in search of understanding where value is generated or should be generated can be used as a starting point before trying to determine a roadmap for digital transformation which is to be backed with adequate amount of digitization (Ross, 2017). Scenario planning (having alternative roadmaps) can also be used to maintain agility and cope with ambiguities present in ecosystem complexities and uncontrollable factors.

FUTURE RESEARCH DIRECTIONS

The level of maturity to be achieved or in other words what is an optimum level in a digital maturity roadmap has been under discussion (Rader, 2019; Kane 2017); it would be useful to conduct sector specific research in some specific industries such as retail and services in order to offer a sequence of activities to reach digital maturity on a proven track that covers many of the companies that had to digitize themselves and have to transform digitally.

It is also worth noting that, when an organization digitally transforms; this also transforms other companies and stakeholders since interconnectedness or being in an ecosystem is a default characteristic of digitalization. The power of influence or potency of digital transformation and becoming mature is not only limited to a single organization, it can be deployed to partners and stakeholders or they can be encouraged to transform which will possibly results in upheaval of the society as a whole. This concept puts forward collaboration and re-thinking organizational boundaries which will also foster customers to get involved in the transformation process. In such an efficacy, digital maturity levels can also serve as objective assessment criteria.

Research focused on dynamic capabilities and digitally developed ecosystem dynamic capabilities to be related with maturity states would give more insight and provide recommendations to business life. On the other hand, Phuyal et al. (2020) point out the interconnected structures in industrial systems. Concentrating on industrial organizations and their Industry 4.0 specific characteristics can also be a good area of study to understand interconnected systems and data management-oriented opportunities.

Digital maturity finds extensive coverage in white papers (Strömberg et al., 2020) so that it is a good idea to also follow up consulting powerhouses to keep updated about practical implications to compare them with academic studies to reveal potential research areas; especially in case of chasing sector specific roadmaps.

CONCLUSION

Considering that digital transformation is a sort of an activity and a concept that will continue progressively due to the dynamic nature of an ecosystem's stakeholders and their fostering maturity levels, it is essential to develop qualifications, to monitor environmental changes and to reflect them to the organization and its ecosystem structures. The requirement of developing new capabilities amongst people and ecosystems so that they can respond to continuously developing business environment, digital maturity models serve as a tool to pinpoint the areas to be improved firstly (Ubiparipović et al., 2022).

The prominent theoretical contribution of this chapter is proposing a practical framework to manage digital transformation efforts which it coins and asserts the concept of "the psychological asset related elements" (organization, culture, digital ecosystems, strategy, leadership, innovation, customer insight & experience, digital skills and vision) in a digital maturity model. The study also claims the cohesion of these elements and technology within ecosystem, and pinpoint the interconnectedness of each pillar.

The advancement of technology and the new capabilities brought by it is an incessant reality even in today's business environment. This fact can be seen as a major theoretical implication of the study that still necessitates the adaptation of digital maturity models as a tool for advancement. This implication is also compatible with the progressive realities that evolve in digital ecosystems that an organization takes part.

As detailed throughout the chapter the intent is to clearly state that there is not a single recipe or correct way of enhancing digital maturity levels and climbing the ladder. Although digital transformation and related efforts to reach higher digital maturity levels are expansionist due to competition and being in a relationship with an ecosystem, it should always be taken into account that there is not a single correct or advised method to transform, for every organization considering its culture, properties of its psychological assets, dynamic capabilities and resources are unique (Kwiotkowska, 2022). These complexities should be taken into account to not to pursue a single correct method while becoming mature. Digital maturity model dimensions detailed at Table 1 (strategy, leadership, culture and technology) and Table 2 offer a framework to be coordinated; a company can start with the abbreviated model and further proceed with refracted human related dimensions given at Table 2 – organization, digital skills, innovation, customer insight & experience, vision and digital ecosystems - to make sure that alignment activities cover adequate span of concepts. Furthermore, the chapter asserts that the success of technological tools is also contingent upon their well-defined interaction with the psychological assets and the congruity of defined process flow and its revision where necessary.

The fundamental limitation of the study is a natural outcome that is generally encountered in broad topics which can be named as framing it to a more manageable size; the author preferred to describe technology as a catalyzer to show its undeniable importance while concentrating on the human-related dimensions in digital maturity models.

Technology is not detailed into its dimensions throughout this chapter; further research can be conducted to investigate correlations between tech-related dimensions and human-related dimensions of digital maturity model in specific sectors. Potential studies to cover also the technology dimension is a promising future research topic. Further breakdown of technology dimension can be used as another study's focal point. Quantitative research using the proposed theoretical framework of the chapter can be further developed to prioritize and rank given dimensions where the level of interconnectedness can be also be tested in different sectors and also shed light to internal dynamics of some sectors possessing the ability to formulate practical business life implications. Further academic studies can be used to compare and contrast different sectors or in a broader perspective; industrial corporations versus others can also offer a promising potential to differentiate manufacturing sectors from services industry. Employing both qualitative and quantitative methods in business settings have the potential to contribute to the existing literature in the field of study; at least such studies can be used to test whether practical conclusions can be reached or not.

Operating medium of an organization is in dynamic nature considering the pace of technology and also as a trait of its presence in a digital ecosystem; besides an organization generally finds its place in sets of ecosystems that are forming a bigger whole which increases the level of complexity and non-linearity. Digital maturity models sustain a framework to manage these complexities and form an agile structure by adopting maturing as an ongoing process or even a utopic state of becoming. This chapter per the author advises an abbreviated digital maturity model composed of strategy, leadership, culture and technology. The first four of the five dimensions are named as human (psychological asset)-related ones and a more comprehensive in other extended digital maturity model covering this termed portion is also detailed to express the concepts included in the abbreviated version in a more comprehensive way.

While the author depicts the intertwined structure of the extended dimensions with each other to support the idea to employ a simplified model to communicate details in an effective way also supplies an extended model with the addition of other dimensions which are organization, digital skills, innovation, customer insight & experience, vision and digital ecosystems. Since becoming digitally mature is a perpetual process the author offers the assessment of digital maturity levels as the starting point of such a peregrination. Value flow or customer value generation is the main goal behind utilizing a digital maturity framework; separating digital maturity into stages provides an organization with the segmentation that will enable the ecosystem to recognize and comprehend the existing customer value embedded within its processes. One of the advised digital maturity levels/stages in an ascending hierarchical order is composed of infancy, developing, transforming, optimized and digital maturity

This chapter backs up the perspective that the progress achieved in digital maturity stages is a result of the interaction between endogenous and exogenous nonlinear determinants. Digital maturity levels come to the fore as a tool that provides direction and planned development, instead of the randomness that adapts to the flow of these constraints and the winds of enablers, resembling a ship without an uncertain destination port and route. The term "holistic" used at the chapter headline is selected to highlight an extended span of factors which are to be considered based on the concentrated sector or industry. Creating both internal and external collaborations structurally can be used as the main method for rapid progress at digital maturity levels. Bettering the level of digital maturity is by nature a difference making performance improvement tool.

ACKNOWLEDGMENT

Funding: This research received no specific grant from any funding agency in the public, commercial, or not-for-profit sectors.

REFERENCES

Agrawal, A. (2020). COVID-19 – Driving Digitization, Digitalisation & Digital Transformation in Healthcare. *Science Reporter*, 20–21.

Azizan, S., Ismail, R., Baharum, A., & Hidayah Mat Zain, N. (2021). Exploring The Factors That Influence The Success Of Digitalization In An Organization's IT Department. *2021 6th IEEE International Conference on Recent Advances and Innovations in Engineering (ICRAIE)*, 1–6. 10.1109/ICRAIE52900.2021.9704018

Bai, C., Quayson, M., & Sarkis, J. (2021). COVID-19 pandemic digitization lessons for sustainable development of micro-and small- enterprises. *Sustainable Production and Consumption*, *27*, 1989–2001. doi:10.1016/j.spc.2021.04.035 PMID:34722843

Barboza, R. J. G., Michalke, S., & Siemon, D. (2022). Towards the Design of a Digital Business Ecosystem Maturity Model for Personal Service Firms: A Pre-Evaluation Strategy. *PACIS 2022 Proceedings*, 86. https://aisel.aisnet.org/pacis2022/86

Bloomberg. (2018). *Digitization, Digitalization, And Digital Transformation: Confuse Them At Your Peril.* https://www.forbes.com/sites/jasonbloomberg/2018/04/29/digitization-digitalization-and-digital-transformation-confuse-them-at-your-peril/?sh=6f9ca6142f2c

Bonnet, D., & Westerman, G. (2021). The New Elements of Digital Transformation. *MIT Sloan Management Review*, *62*(2), 83–89. https://sloanreview.mit.edu/article/the-new-elements-of-digital-transformation/

Bradley, A. (2014). *The Connected Enterprise Maturity Model*. Rockwell Automation.

Brennen, J. S., & Kreiss, D. (2016). Digitalization. In K. B. Jensen, E. W. Rothenbuhler, J. D. Pooley, & R. T. Craig (Eds.), *The International Encyclopedia of Communication Theory and 239 International Journal for Modern Trends in Science and Technology Philosophy* (pp. 556–566). Wiley-Blackwell.

Brusakova, I. A. (2022). Comparative Analysis of Models for Assessing the Digital Maturity of the Transformation Infrastructure. *2022 XXV International Conference on Soft Computing and Measurements (SCM)*, 209–211. 10.1109/SCM55405.2022.9794829

Cambridge Dictionary. (2022). *Cambridge Dictionary: Maturity.* https://dictionary.cambridge.org/dictionary/english/maturity

Cennamo, C., Dagnino, G. B., Di Minin, A., & Lanzolla, G. (2020). Managing Digital Transformation: Scope of Transformation and Modalities of Value Co-Generation and Delivery. *California Management Review, 62*(4), 5–16. doi:10.1177/0008125620942136

Colli, M., Madsen, O., Berger, U., Møller, C., Wæhrens, B. V., & Bockholt, M. (2018). Contextualizing the outcome of a maturity assessment for Industry 4.0. *IFAC-PapersOnLine, 51*(11), 1347–1352. doi:10.1016/j.ifacol.2018.08.343

Correani, A., Massis, A. D., Frattini, F., Petruzzelli, A. M., & Natalicchio, A. (2020). Implementing a Digital Strategy: Learning from the Experience of Three Digital Transformation Projects. *California Management Review, 62*(4), 37–56. doi:10.1177/0008125620934864

Cozzolino, A., Corbo, L., & Aversa, P. (2021). Digital platform-based ecosystems: The evolution of collaboration and competition between incumbent producers and entrant platforms. *Journal of Business Research, 126*, 385–400. doi:10.1016/j.jbusres.2020.12.058

De Carolis, A., Macchi, M., Negri, E., & Terzi, S. (2017). A Maturity Model for Assessing the Digital Readiness of Manufacturing Companies. In H. Lödding, R. Riedel, K. D. Thoben, G. von Cieminski, & D. Kiritsis (Eds.), *Advances in Production Management Systems. The Path to Intelligent, Collaborative and Sustainable Manufacturing. APMS 2017. IFIP Advances in Information and Communication Technology, 513*. Springer. doi:10.1007/978-3-319-66923-6_2

De Santos, J. A., & Francisco, E. de R. (2021). Digital Maturity Level of a B2B Company: Case Study of a Brazilian Complex Manufacturing Company. *International Conference on Information Resources Management (CONF-IRM)*, 12. https://aisel.aisnet.org/confirm2021/22/

Eller, R., Alford, P., Kallmünzer, A., & Peters, M. (2020). Antecedents, consequences, and challenges of small and medium-sized enterprise digitalization. *Journal of Business Research, 112*, 119–127. doi:10.1016/j.jbusres.2020.03.004

Eremina, Y., Lace, N., & Bistrova, J. (2019). Digital Maturity and Corporate Performance: The Case of the Baltic States. *Journal of Open Innovation, 5*(3), 54. doi:10.3390/joitmc5030054

Fletcher, G., & Griffiths, M. (2020). Digital Transformation During a Lockdown. *International Journal of Information Management, 55*, 102185. doi:10.1016/j.ijinfomgt.2020.102185 PMID:32836642

Gartner. (2022). *Gartner Information Technology Glossary: Digitization.* https://www.gartner.com/en/information-technology/glossary/digitization

Gavrila Gavrila, S., & De Lucas Ancillo, A. (2022). Entrepreneurship, innovation, digitization and digital transformation toward a sustainable growth within the pandemic environment. *International Journal of Entrepreneurial Behaviour & Research, 28*(1), 45–66. doi:10.1108/IJEBR-05-2021-0395

Gill, M., & VanBoskirk, S. (2016). *The Digital Maturity Model 4.0 Benchmarks: Digital Business Transformation Playbook.* Forrester. https://www.forrester.com/report/The-Digital-Maturity-Model-40/RES130881

Gobble, M. M. (2018). Digitalization, Digitization, and Innovation. *Research Technology Management, 61*(4), 56–59. doi:10.1080/08956308.2018.1471280

Gong, C., & Ribiere, V. (2021). Developing a unified definition of digital transformation. *Technovation, 102*, 102217. doi:10.1016/j.technovation.2020.102217

Gurbaxani, V., & Dunkle, D. (2019). Gearing Up for Successful Digital Transformation. *MIS Quarterly Executive, 18*(3), 209–220. doi:10.17705/2msqe.00017

Harvard Business Review. (2020a). *Reevaluating Digital Transformation During Covid-19* [Research Report]. Harvard Business Review Analytic Services. https://hbr.org/sponsored/2020/11/reevaluating-digital-transformation-during-covid-19

Harvard Business Review. (2020b). *Reconciling Cultural and Digital Transformation to Design the Future of Work* [White Paper]. Harvard Business Review Analytic Services. https://hbr.org/sponsored/2020/10/reconciling-cultural-and-digital-transformation-to-design-the-future-of-work

Hilpert, U. (2021). Regional selectivity of innovative progress: Industry 4.0 and digitization ahead. *European Planning Studies, 29*(9), 1589–1605. doi:10.1080/09654313.2021.1963047

Kane, G. C. (2017). 'Digital Transformation' Is a Misnomer. *MIT Sloan Management Review.* https://sloanreview.mit.edu/article/digital-transformation-is-a-misnomer/?og=Digital+Leadership+Tiled

Kane, G. C., Palmer, D., Phillips, A. N., Kiron, D., & Buckley, N. (2017). *Achieving Digital Maturity.* MIT Sloan Management Review and Deloitte University Press. https://sloanreview.mit.edu/projects/achieving-digital-maturity/

Kane, G. C., Phillips, A. N., Copulsky, J., & Andrus, G. (2019). How Digital Leadership Is(n't) Different. *MIT Sloan Management Review, 60*(3), 34–39. https://www.proquest.com/scholarly-journals/how-digital-leadership-is-nt-different/docview/2207927776/se-2

Kwiotkowska, A. (2022). The interplay of resources, dynamic capabilities and technological uncertainty on digital maturity. *Dynamic Capabilities*, 1-17. doi:10.29119/1641-3466.2022.155.15

Marushchak, L., Pavlykivska, O., Khrapunova, Y., Kostiuk, V., & Berezovska, L. (2021). The Economy of Digitalization and Digital Transformation: Necessity and Payback. *2021 11th International Conference on Advanced Computer Information Technologies (ACIT)*, 305–308. 10.1109/ACIT52158.2021.9548529

Meffert, J., Mohr, N., & Richter, G. (2020). *How German "Mittelstand" copes with COVID-19 challenges.* https://www.mckinsey.com/business-functions/mckinsey-digital/our-insights/how-the-german-mittelstand-is-mastering-the-covid-19-crisis

Neugebauer, R. (2019). *Digital Transformation* (1st ed.). Springer-Verlag GmbH Germany. doi:10.1007/978-3-662-58134-6

OECD. (2020). *Coronavirus (COVID-19): SME policy responses.* OECD Policy Responses to Coronavirus (COVID-19). https://www.oecd.org/coronavirus/policy-responses/coronavirus-covid-19-sme-policy-responses-04440101/

Ostmeier, E., & Strobel, M. (2022). Building skills in the context of digital transformation: How industry digital maturity drives proactive skill development. *Journal of Business Research, 139,* 718–730. doi:10.1016/j.jbusres.2021.09.020

Parviainen, P., Kääriäinen, J., Tihinen, M., & Teppola, S. (2017). Tackling the digitalization challenge: How to benefit from digitalization in practice. *International Journal of Information Systems and Project Management, 5*(1), 63–77. doi:10.12821/ijispm050104

Peranzo, P. (2021). What is digital transformation and why it is important for business. *Imaginovation.* https://imaginovation.net/blog/what-is-digital-transformation-importance-for-businesses/

Perera, U. T., Heeney, C., & Sheikh, A. (2022). Policy parameters for optimising hospital ePrescribing: An exploratory literature review of selected countries of the Organisation for Economic Co-operation and Development. *Digital Health, 8.* doi:10.1177/20552076221085074 PMID:35340903

Philipp, R. (2020). Digital readiness index assessment towards smart port development. *Sustainability Management Forum | Nachhaltigkeits Management Forum, 28*(1–2), 49–60. doi:10.1007/s00550-020-00501-5

Rader, D. (2019). Digital maturity – the new competitive goal. *Strategy and Leadership, 47*(5), 28–35. doi:10.1108/SL-06-2019-0084

Reis, J., Amorim, M., Melão, N., Cohen, Y., & Rodrigues, M. (2020). Digitalization: A Literature Review and Research Agenda. In Z. Anisic, B. Lalic, & D. Gracanin (Eds.), *Proceedings on 25th International Joint Conference on Industrial Engineering and Operations Management – IJCIEOM* (pp. 443–456). Springer International Publishing. 10.1007/978-3-030-43616-2_47

Ross, J. (2017). Don't confuse digital with digitization. *MIT Sloan Management Review.* https://sloanreview.mit.edu/article/dont-confuse-digital-with-digitization/

Santos, R. C., & Martinho, J. L. (2019). An Industry 4.0 maturity model proposal. *Journal of Manufacturing Technology Management, 31*(5), 1023–1043. doi:10.1108/JMTM-09-2018-0284

Savić, D. (2019). From Digitization, Through Digitalization, to Digital Transformation. *Online Searher, 43*(1). https://www.infotoday.com/OnlineSearcher/Articles/Features/From-Digitization-Through-Digitalization-to-Digital-Transform ation-129664.shtml?PageNum=2

D. Schallmo, & C. A. Williams (Eds.). (2021). Integrated Approach for Digital Maturity: Levels, Procedure, and In-Depth Analysis. In *The ISPIM Innovation Conference – Innovating Our Common Future*. LUT Scientific and Expertise Publications.

Schallmo, D., Williams, C. A., & Boardman, L. (2017). Digital transformation of business models—Best practice, enablers, and roadmap. *International Journal of Innovation Management, 21*(08), 1740014. doi:10.1142/S136391961740014X

Schuh, G., Anderl, R., Gausemeier, J., ten Hompel, M., & Wahlster, W. (Eds.). (2017). Industrie 4.0 Maturity Index. Managing the Digital Transformation of Companies (acatech STUDY). Herbert Utz Verlag.

Schumacher, A., Nemeth, T., & Sihn, W. (2019). Roadmapping towards industrial digitalization based on an Industry 4.0 maturity model for manufacturing enterprises. *Procedia CIRP, 79*, 409–414. doi:10.1016/j.procir.2019.02.110

Schwertner, K. (2017). Digital transformation of business. *Trakia Journal of Sciences, 15*(Suppl.1), 388–393. doi:10.15547/tjs.2017.s.01.065

Şener, U., Gökalp, E., & Eren, P. E. (2018). Towards a maturity model for industry 4.0: A systematic literature review and a model proposal. In Industry 4.0 From the Management Information Systems Perspectives (pp. 291–303). Academic Press.

Sklyar, A., Kowalkowski, C., Tronvoll, B., & Sörhammar, D. (2019). Organizing for digital servitization: A service ecosystem perspective. *Journal of Business Research, 104*, 450–460. doi:10.1016/j.jbusres.2019.02.012

Strömberg, J., Sundberg, L., & Hasselblad, A. (2020). Digital Maturity in Theory and Practice: A Case Study of a Swedish Smart-Built Environment Firm. *2020 IEEE International Conference on Industrial Engineering and Engineering Management (IEEM)*, 1344–1348. 10.1109/IEEM45057.2020.9309760

Sundberg, L., Gidlund, K. L., & Olsson, L. (2019). Towards Industry 4.0? Digital Maturity of the Manufacturing Industry in a Swedish Region. *2019 IEEE International Conference on Industrial Engineering and Engineering Management (IEEM)*, 731–735. 10.1109/IEEM44572.2019.8978681

Teichert, R. (2019). Digital Transformation Maturity: A Systematic Review of Literature. *Acta Universitatis Agriculturae et Silviculturae Mendelianae Brunensis, 67*(6), 1673–1687. doi:10.11118/actaun201967061673

Thomson, L., Kamalaldin, A., Sjödin, D., & Parida, V. (2021). A maturity framework for autonomous solutions in manufacturing firms: The interplay of technology, ecosystem, and business model. *The International Entrepreneurship and Management Journal*. Advance online publication. doi:10.100711365-020-00717-3

Tutak, M., & Brodny, J. (2022). Business Digital Maturity in Europe and Its Implication for Open Innovation. *Journal of Open Innovation*, *8*(1), 27. doi:10.3390/joitmc8010027

Ubiparipović, B., Matković, P., & Pavlićević, V. (2022). Key activities of digital business transformation process. *Strategic Management*, *00*(00), 1–8. doi:10.5937/StraMan2200016U

VanBoskirk, S., Gill, M., Green, D., Berman, A., Swire, J., & Birrel, R. (2017). The Digital Maturity Model 5.0. *Forrester*. https://www.forrester.com/report/The+Digital+Maturity+Model+50/-/E-RES137561

Wald, D., de Laubier, R., & Charanya, T. (2019). *The Five Rules of Digital Strategy*. Boston Consulting Group. https://web-assets.bcg.com/img-src/BCG-The-Five-Rules-for-Digital-Strategy-May-2019_tcm9-220981.pdf

Warner, K. S. R., & Wäger, M. (2019). Building dynamic capabilities for digital transformation: An ongoing process of strategic renewal. *Long Range Planning*, *52*(3), 326–349. doi:10.1016/j.lrp.2018.12.001

Williams, C. A., Krumay, B., Schallmo, D., & Scornavacca, E. (2022). An interaction-based Digital Maturity Model for SMEs. *PACIS 2022 Proceedings*, 254. https://aisel.aisnet.org/pacis2022

Xie, X., Zhang, H., & Blanco, C. (2022). How organizational readiness for digital innovation shapes digital business model innovation in family businesses. *International Journal of Entrepreneurial Behaviour & Research*. Advance online publication. doi:10.1108/IJEBR-03-2022-0243

Yılmaz, K. Ö. (2021). Mind the Gap: It's About Digital Maturity, Not Technology. In T. Esakki (Ed.), *Managerial Issues in Digital Transformation of Global Modern Corporations* (pp. 222–243). IGI Global. doi:10.4018/978-1-7998-2402-2.ch015

ADDITIONAL READING

Al Amoush, A. B., & Sandhu, K. (2020). Digital Transformation of Learning Management Systems at Universities: Case Analysis for Student Perspectives. *Digital Transformation and Innovative Services for Business and Learning*, 41-61. doi:10.4018/978-1-7998-5175-2.ch003

Berghaus, S., & Back, A. (2016). Stages in Digital Business Transformation: Results of an Empirical Maturity Study. *MCIS 2016 Proceedings*. 22. Retrieved from https://aisel.aisnet.org/mcis2016/22

Chanias, S., Myers, M. D., & Hess, T. (2019). Digital transformation strategy making in pre-digital organizations: The case of a financial services provider. *The Journal of Strategic Information Systems*, *28*(1), 17–33. doi:10.1016/j.jsis.2018.11.003

Fischer, E. A. (2016). Cybersecurity Issues and Challenges: In Brief. *Congressional Research Service*, 1–12. Retrieved from https://fas.org/sgp/crs/misc/R43831.pdf

Hess, T., Benlian, A., Matt, C., & Wiesböck, F. (2016). Options for formulating a digital transformation strategy. *MIS Quarterly Executive*, *15*(2), 123–139. doi:10.4324/9780429286797-7

Li, F. (2020). The digital transformation of business models in the creative industries: A holistic framework and emerging trends. *Technovation*, *92–93*(January), 102012. doi:10.1016/j.technovation.2017.12.004

Mergel, I., Edelmann, N., & Haug, N. (2019). Defining digital transformation: Results from expert interviews. *Government Information Quarterly*, *36*(4), 101385. doi:10.1016/j.giq.2019.06.002

Singh, A., & Hess, T. (2017). How chief digital officers promote the digital transformation of their companies. *MIS Quarterly Executive*, *16*(1), 1–17. doi:10.4324/9780429286797-9

Vial, G. (2019). Understanding digital transformation: A review and a research agenda. *The Journal of Strategic Information Systems*, *28*(2), 118–144. doi:10.1016/j.jsis.2019.01.003

Weill, P., & Woerner, S. L. (2018). Is your company ready for a digital future? *MIT Sloan Management Review*, *59*(2), 21–25. https://sloanreview.mit.edu/article/is-your-company-ready-for-a-digital-future/

KEY TERMS AND DEFINITIONS

Collaboration: Collective engagement of contributors in a coordinated effort to solve an issue, a problem, or an incident together.

Digital Ecosystem: A distribute, cooperative, open socio-technical system with properties of self-adaptive organizational flexibility, scalability, extended organizational boundaries and resilience.

Digital Maturity: Adapting an organization to challenge effectively in increasingly digitalized ecosystems.

Digital Strategy: Using digital resources to fulfil customer expectations and to digitally transform business processes with an open-minded mentality to triumph in business models.

Digital Transformation: Incorporating digital technologies into daily corporate life to run routine or new to the earth activities by forming new processes or by updating them making use of digital technologies and tools to fulfill and adapt to market, customer, and business demands.

Digitalization: Utilization of digital tools and technologies to modify a business model and generate chance and enable new opportunities for deciding about actions to be taken.

Digitization: Process of converting analog data to digital form. Its target is to become paperless, transforming into bytes.

Industry 4.0: Indicates the 4th Industrial Revolution; it is a form of swift digital transformation where systems, machines and industries are integrated using smart automation and digitized connectivity tools.

Chapter 6
Process Mining for Social and Economic Needs:
An Introduction

Sibanjan Das
ⓘ https://orcid.org/0000-0002-2437-0482
Indian Institute of Management, Ranchi, India

Pradip Kumar Bala
ⓘ https://orcid.org/0000-0002-9028-4902
Indian Institute of Management, Ranchi, India

ABSTRACT

Process mining is a paradigm shift from traditional process understanding methodologies like interviews and surveys to a data-driven understanding of the actual digital processes. It analyzes business processes by applying algorithms to the event data generated by digital systems. The chapter provides insight into various uses of process mining in different social and economic processes, with examples from past works demonstrating how practical process mining is in detecting and mitigating bottlenecks in these sectors. Then the chapter further delves into the details of process mining algorithms, key features, and metrics that can help practitioners and researchers evaluate process mining for their work. It also highlights some data quality issues in the event log that can inhibit obtaining fair results from process models. Additionally, some current limitations and concerns are described for creating awareness and building over the body of knowledge in the process and sequential mining techniques.

INTRODUCTION

Connected Machines, digital transformation, and process automation have become ubiquitous. These terms are appearing more frequently than what it was a decade ago. The Covid 19 pandemic further made the world physically distant and virtually connected. Organizations were forced to evaluate their manual process and re-think digitalizing their business processes. This digital transformation of busi-

DOI: 10.4018/978-1-6684-4102-2.ch006

ness processes led to multiple workflows generating a new train of rich data for almost every event in the workflows.

Additionally, digitalization leads to the generation of various logs in different digital systems (Law et al. 2021). Analyzing these logs with event data can provide rich information in governing or further digital process transformation. However, traditional data mining or machine learning algorithms are unsuitable for crunching these event logs and providing process-related insights. This is where the process mining algorithms excel. Process mining algorithms can process millions of event data points to tap into these datasets and generate the much-needed intelligence to help discover the process divergence and loopholes in the workflows. Digital transformation solutions are geared toward automating business processes, and process mining provides intelligence to drive these digital workflows to deliver the desired outcome. Every process is a chain of events made up of specific steps that do some activity and passes on to the next stage in the process. This continues until the goal of the process is met or achieved. So, a process has a definite start and end. Most of the time, standard processes are well known based on which the workflows are laid out. However, in real life, there are chances of deviations. For example, let's say an Organization implements an ERP system to streamline its Order to Cash process. The Organization is quite aware of the standard process it follows, from capturing an order to delivering the final product to its customers. Based on this knowledge, they lay out the workflows in the new ERP system. However, it being a new system, adopting these ERP systems might not be an easy job for their employees. So, they might think of ways to bypass the process, for example, continue using spreadsheets for order entry and upload the details at the end of the day. Process mining can help analyze the process and find the bottlenecks in these digital workflows. As stated, process mining is the "Analytics of Processes". The main goal of Process Mining is to analyze how the process emerges, how they deviate from the standard/ideal process, and where the bottlenecks are to optimize and improve the processes.

Process mining finds its application in various industries and domains. Jans et al. (2011) used process mining as a tool for mitigating internal transaction fraud. It is also helpful as a tool to analyze business processes post implementation of any process automation or enterprise business systems. Mahendrawathi, Zayin, and Pamungkas (2017) used process mining for a post-implementation review of an ERP system that was used to improve the procurement process and discovered bottlenecks related to cycle time for procurement activities. Dogan, Fernandez-Llatas, and Oztaysi (2019) used process mining to evaluate differences among the customer visits in a shopping mall. Process mining also aids in bringing a scientific approach to improving sales process and performance management (Bernard et al. 2016). It is also employed for management of the software development process and proves an effective tool for improving the maturity level of the software engineering process (Lemos et al. 2011). These are a few examples that cited the enormous contribution potential that process mining has in improving different digital workflows.

This chapter aims to provide an introduction and enable discussion on the current state of process mining and identify different opportunities in further research or application of process mining in their field of work. Moreover, the chapter provides information professionals and researchers with examples of how others apply process mining, especially in healthcare, education, and energy industries. These examples of process mining work are contextualized within social and economic sectors to provide another possible research methodology for exploring the digital workflows within these sectors.

LITERATURE REVIEW

In a literature review conducted by Dakic et al. (2018), the literature related to process mining in healthcare, education, energy, and agriculture collectively accounts for 50% of all process mining publications reviewed. This is because process mining can identify various critical issues in social and economic processes. He et al. (2018) explored the process mining approach to improve emergency rescue processes of fatal gas explosion accidents in Chinese coal mines and were intrigued by their findings. It is known that gas explosion is a hazard in Chinese coal mines, and researchers actively scout for techniques to address this irregularity. Experts looked at process mining for answers. Research revealed that the average length of accident cases is 23%, and 44% of emergency operations that lasted more than four days. The accident grade affects the rescue performance compared to other factors of the site. Besides this finding, the process mining technique research revealed the significant role of ownership issues in accidents in the coal mine region. Researchers delved into the usefulness of process mining to introduce it as a mechanism to understand rescue operations and bring organizational change. This is one example where process mining strategies helped uncover bottlenecks and gas position deviations as being responsible for the explosion. In exploring its relevance, it was described as a computationally reconstructed algorithm. This means that the program can be adjusted to visualize and analyze events data logs. Process Mining is usually linked to improving business management. However, it can be argued that process mining can be applied as a method to facilitate organizational change (Grisold et al. 2020). Agronomists and crop producers can use process mining to visualize and predict crop rotations. Dupuis, Dadouchi, and Agard (2022) used Process Mining and Directly-Follows Graphs(DFG) to understand the frequent behavior patterns of crop producers.

The use cases of process mining go beyond these micro-economic growth drivers and are useful to improve some of the long-term issues related to education, healthcare (Kaur et al. 2021), and other aspects of human capital such as Sustainable Development Goals(SDG) to eliminate poverty and protect the environment (Chopra et al. 2022), which are also a determinant of economic growth (Barro 2001). Awatef et al. (2015) used process mining to analyze the interactions between training providers, courses, and other resources involved in a student's training journey. With the growth of EdTech digital platforms, there has been a revolution in the field of digital education, enabling courses and study materials to be accessible from anywhere and fostering better collaborations in discussion forums. The relationship between self-learning and formative assessment is an active line of research in the educational community. Dom´ınguez et al. (2021) used process mining to discover that increasing the number of self-assessments cannot compensate for the lack of effort in formative assessment. Still, activities promoting self-assessments improved students inclination toward formative assessments. There are better benefits when process mining is combined with other traditional analytic approaches. Silva et al. (2020) demonstrated this benefit by using the analytical techniques to correlate students' grades with their behavior while watching the video lectures and then used process mining to analyze the dynamic student behavior. They further stated that using only descriptive analytics would not provide such analysis results as these methods don't consider the process perspective in their algorithms.

In studies related to healthcare, researchers used process mining to discover care delays and bottlenecks in a patient's journey to correlate the patient pathways and outcomes before and after COVID-19 during stroke care. Process mining helped uncover the worsening of patients' health status, delay in receiving therapies, and rise in hospital referrals instead of emergency services during and after COVID-19

(Leandro et al. (2022)). Similarly, Augusto et al. (2022) used process mining techniques to analyze the impact of COVID-19 on vaccination patterns.

The use cases for process mining in discovering process deviations and bottlenecks are not just limited to explicit process events data. It can be used on any event data, including social media. It is a known fact that social media has become a powerful platform for generating and analyzing user data. Various data mining techniques are applied to create multiple machine learning models for recommending different actions in the user or consumer journey. While these techniques are helpful, it must be noted that there is room for further exploration regarding human behavior on these platforms (Li and M. de Carvalho 2019). In today's world of rising social media usage, process mining can be used to analyze a user's journey to uncover deep insights into their behavior, and any behavioral deviations can be investigated further for potential criminal activity. Process mining helps discover processes and is also a valuable tool for analyzing social structures from event logs. This is considered when analyzing tweets, evaluating comments, and assessing the content of posts themselves. While many data mining or machine learning techniques exist to get insights from this vast data, there is a scope to uncover more profound meaning to data extracted in terms of event data. These event data provides an opportunity to find insights related to important behavioral patterns and operational process on social media sites. Process mining equips us with the algorithms and framework to do this analysis efficiently and cost-effectively.

Process mining is a beneficial tool in the security domain as well. In this age of the digital economy, securing digital applications plays a very crucial role. Analyzing the trail of all events across all systems and networks is a daunting task ((Stergiou et al. 2018), (Zhou et al. 2021)). Aalst and Medeiros (2005) advocates using process mining techniques to analyze the audit trail for security violations. Auditing the authenticity of transactions is essential for smart contracts as well. Process mining aids in validating the observed transaction sequence against the expected ones, making it easier for auditors to check if the smart contracts fit as per the designer's expectation Corradini et al. (2019). Alrahili (2021) in his article perceive process mining as a strategy for analyzing security integrity, unlike social media human behavior patterns. The primary focus here is on Role-Based Access Control (RBAC). However, the difficulty lies in not having enough data to explore what this phenomenon is all about. Consequently, a systematic literature review was conducted to obtain the relevant data. It was discovered that even though 27 publications were evaluated, data extraction was inconclusive. This was despite the fact that 40 approaches had been discussed. It was concluded that utilizing process mining for RBAC analysis could be a worthwhile idea. However, much more research must be conducted to determine its feasibility in making it operational.

PROCESS MINING METHODOLOGY

Process mining focuses on discovering the behavior aspects of processes from log data. Weijters, Aalst, and Medeiros (2006) defined three different perspectives related to process, Organization, and case for carrying out the process mining.

- Process perspective refers to the activities or steps taken within a process. This perspective is known as "Control-flow," which helps teams identify dependencies between the activities based on their ordering.

- Organization perspective captures the involvement of various actors/performers in the processes. It focuses on the performance of people or roles in performing activities in an organizational unit and finding the relationship between these performers.
- Case perspective focuses on analyzing the characteristics of every case by specific attributes or values. This perspective is used to discover various anomalies in the process outcomes.
- A time-based perspective is also proposed by Zerbino, Stefanini, and Aloini (2021), which is to extract and analyze the time and frequency of the events in a process.

The starting point of any process mining activity is the availability of an event log, a collection of various events taking place over a period of time for multiple cases. Process mining algorithms expect an event log to have a Case ID/Transaction ID, which can be used to uniquely identify a case record, an activity that has a description of a task accomplished in that case, and a timestamp of the activity execution. Table 1 illustrates a sample of patient treatment event log data with all data elements for successfully carrying out process mining (shoot2kill gitlab n.d.).

Table 1. Sample Event Log

patient	action	org:resource	DateTime
patient 0	First consult	Dr. Anna	2017-01-02 11:40:11
patient 0	Blood test	Lab	2017-01-02 12:47:33
patient 0	Physical test	Nurse Jesse	2017-01-02 12:53:50
patient 0	Second consult	Dr. Anna	2017-01-02 16:21:06
patient 0	Surgery	Dr. Charlie	2017-01-05 13:23:09
patient 0	Final consult	Dr. Ben	2017-01-09 08:29:28
patient 1	First consult	Dr. Anna	2017-01-02 12:50:35
patient 1	Physical test	Nurse Jesse	2017-01-02 13:59:14
patient 1	Blood test	Lab	2017-01-02 14:20:19
patient 1	X-ray scan	Team 1	2017-01-06 09:13:40
patient 1	Second consult	Dr. Anna	2017-01-06 10:38:04
patient 1	Medicine	Pharmacy	2017-01-06 11:47:36
patient 1	Final consult	Dr. Anna	2017-01-06 16:49:21
patient 2	First consult	Dr. Anna	2017-01-04 10:02:49

Once the event log is available, it can be used to conduct three types of analysis for deriving insights from the processes.

Process Discovery

Process discovery is the starting point for using process mining. This is a crucial step in synthesizing the process model from raw event data without an already available process model. The idea is to generate process models from event logs without any a-priori information. Process models provide a graphical way to visualize the workflows for process visibility and understanding. They are represented as graphs such as Petri-nets whose nodes represent the activities and arcs are transitions between the activities. Additionally, Petri-nets is a design language for process models that provide semantics which is essential to represent the behaviour of a given model (Aalst 1998). It is an easy to understand graphical representation that the practitioners can easily interpret as well (Hee, Sidorova, and Van der Werf 2013). Adam, Atluri, and Huang (1998) demonstrated the use of Petri-net as an effective tool for identifying inconsistent

Figure 1. Process Model

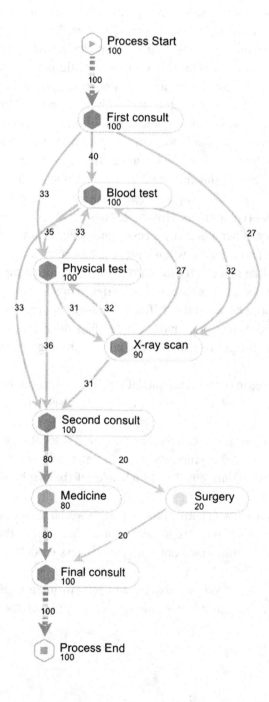

dependencies specifications in a workflow and checking the feasibility of a workflow execution when temporal constraints are present. Various discussions have been on improving Petri-nets, such as a new adaptive Petri-Net to model complex, distributed, and real-time systems (Baldellon, Fabre, and Roy 2011). Bause and Kritzinger (2013) proposed the concept of stochastic Petri-Nets, which can be used for performance analysis of systems using Markovian techniques. Czerwinski et al. (2020) worked on improving the reachability problem in Petri-nets. Aalst et al. (2011) introduced the concept of workflow nets, a sub-class of Petri-Net that provides a structure to express workflows. They have a definite start and end in a place that can be used to express workflows. A workflow net has properties such as no dead transitions, proper terminations, and options to complete so that there are no unboundedness, deadlocks, or dead transitions, which makes it suitable for applying to a workflow. A process model created using Celonis process mining tool using the sample patient treatment dataset is show in Figure 1

The Alpha algorithm is one of the first process mining algorithms that can discover workflow nets(an extended version of Petri-nets). It reconstructs the workflow from an event log, and the output is a process model in the form of a Petri-net. It can adequately deal with concurrency. However, it has problems with noise, infrequent/incomplete behaviour, and complex routing constructs. In the absence of the actual "As-is" process model, this can act as a standard model or compare with the already available process model. One major drawback of the Alpha algorithm is the under-fitting problem (Diamantini et al. 2016). The Heuristic miner algorithm addresses this drawback and makes it suitable to use in practice (Prathama, Nugroho Yahya, and Lee 2021). It is also a noise-tolerant algorithm that can deal with data quality issues such as incompleteness and data entry errors. However, it has a problem dealing with many activities and unstructured behaviour. Fuzzy miner, another process mining algorithm, can provide process models at a desired level of abstraction and is useful when a process has many activities or exhibits unstructured behaviour, such as spaghetti-like models. It uses correlation metrics to simplify the complex workflows to concise process models by hiding less essential activities in clusters or leaving out those activities.

It is often difficult to choose between the different kinds of algorithms for process modelling. The authors in Buijs, Dongen, and Aalst (2012) have adopted four different evaluation metrics that can be used to measure and compare the performance of the process mining algorithms.

- Fitness(Recall) - This metric captures the model's ability to accurately reproduce the cases present in the event log. If the process model can reproduce each sequence in the event log, it's a 100% fitted model.
- Precision - It quantifies how much behaviour a process model allows, which was not observed in the event log. A precise model captures each case as a separate path in model. If the reproduced logs from the process model are equal to or a subset of the original log, the model has a 100% precision.
- Generalization - This measure quantifies the extent to which a process model can generalize behaviour observed in the processes to reproduce future behaviour of the processes.
- Simplicity - This metric captures the complexity of process models.

These four model quality dimensions are also used to get information related to differences between the model, and an event log in the conformance checking activity as well (Naderifar, Sahran, and Shukur 2019).

Conformance Checking

Conformance checking is a technique to compare an existing reference process model to the event logs of the same process. This technique determines whether the actual process conforms with the defined standard process. The reconstructed process model reveals various key information about the process and helps understand if the process is running as expected. It compares event logs of an actual process case-by-case with the existing reference model and determines if there are discrepancies and deviations in the model behaviour, such as any skip in a planned step, duplicates, or unplanned activity. In short, conformance checking is defined as checking the alignment of the process model with the log and execution of processes within the boundaries defined by management and Organization (Burattin 2015). There are a few metrics that can be used in the conformance checking process to get some critical insights related to the processes.

- Percentage of non-conforming cases- This metric provides insights into the volume of non-conforming cases that can alert the process owners when there is an increase in nonconformant cases.
- Top violations - Metric is associated with detecting compliance violations in the process. For example, violations of mandatory activities, breaches in the approval levels, or any compliance rules set in the process.
- Top variations - Metric to provide the most deviated paths in the process. For example, Activity A to Activity D is the most occurring deviation in a process.
- Throughput time - Provides the amount of time taken to run a process. Measuring throughput time can provide insight into the time taken to execute every activity in a process as well as the complete process. This metric helps measure a process's effectiveness and helps identify any delays associated with any activity in the process. The throughput time for each connection between activities for the sample patient treatment dataset is shown in Table 2. The median time to X-ray scan from first consultation and blood test is more than 3 days. If this step is optimized, it might help reduce the overall throughput time of the process.

Table 2. Throughput time per connection

Source	Target	# Occurences	Total Throughput Time
Second consult	Surgery	20	4.10 days
First consult	X-ray scan	27	3.97 days
X-ray scan	Second consult	31	3.93 days
Blood test	X-ray scan	32	3.14 days
Physical test	Second consult	36	2.97 days
Blood test	Second consult	33	2.31 days
Surgery	Final consult	20	2.05 days
Physical test	X-ray scan	31	1.96 days
Medicine	Final consult	80	1.92 days
First consult	Physical test	33	1.02 days
Blood test	Physical test	35	0.67 days
Second consult	Medicine	80	0.10 days
First consult	Blood test	40	0.05 days
Physical test	Blood test	33	0.03 days
X-ray scan	Physical test	32	0.03 days
X-ray scan	Blood test	27	0.03 days

The conformance checking step acts as a starting point for process enhancement.

Process Enhancement

Here the apriori model is extended with a new aspect or perspective (Aalst 2016). If the process model during the conformance checking phase reflects some discrepancies or deviations and is not reflecting reality, then the process can be either extended or repaired. This is done by removing bottlenecks, and deadlocks, correcting the deviations, or improving the through put Yasmin, Bukhsh, and De Alencar Silva (2018) discusses various perspectives of process enhancement as per Table 3.

It is essential to have a good event log fit for process mining modelling. The event log sometimes has issues related to missing event data, incorrect timestamps, and duplicate activities. (R.P., Mans, and Aalst 2013) identified four categories of process characteristics and 27 classes of data issues that might be present in an event log.

Table 3. Process Enhancement Perspectives

Type	Perspective	Defintion
Repair	Control flow	Repairing the flow by replaying each trace in the log and finding closet possible process to reflect the reality with minimum number of changes. The final choice is the reconstructed model that best balances simplicity, precision and generalisation. (de Leoni 2022)
Extension	Organisation/Resource	Focuses on the resources to anything or anyone involved in performance certain activities, such as employees, system or equipment. The aim is to model how resources are grouped, and how they interact with each other.
	Time	Focuses on amount of time taken to accomplish a certain activity such as
		arrival of items, availability of parts etc.
	Case	Focuses of the properties of cases such as its attributes, path taken and flow times

Process characteristics issues stem from high volume processes, very granular activities, or process variability. The 27 classes of data issues might arise in the quality of data logged in the event logs. They identified four categories of these problems.

- Missing Data - This scenario is where data related to activities, timestamps, and the relationship between these values are missing in the event log.
- Incorrect Data - These are the cases where although complete data is available in the event log, they are out of context. For example, the relationships or values with no association with the process are incorrectly logged.
- Imprecise Data - These are the cases where the data is not fit for analysis. For example, the timestamps are at a significantly higher order than the required unit of measurement for the analysis.
- Irrelevant Data - The logged data is not in a form that can be directly used for analysis. Instead, some data transformation like filters or aggregation is required to bring the log data to the desired format.

Process mining heavily depends on the sequence of events based on the timestamps recorded in the log. The contemporary process mining assumes the process to be in a steady state (Bose et al. 2011). However, this might not be the case due to the concept drift, which means the process might change over time while the process mining activity is still in progress. But this might not be a severe problem considering it is still an offline analysis. The assessment of concept drift in process mining is complex considering the lack of common metrics, datasets, and protocols (Sato et al. 2021). These are some of the challenges related to data quality issues that might require significant effort to resolve these issues. Martin (2020) presented a framework for potentially identifying and cleaning the data quality issues in process mining. Early assessment of some of these well-known data issues might help save efforts in correcting these challenges at an initial stage, thereby saving some time and effort.

Various tools can be used to conduct process mining. Berti, van Zelst, and Aalst (2019) presented a process mining library in Python known as PM4Py(Process Mining for Python) that has integration with the start of art data science and machine learning libraries in Python such as NumPy, Pandas, Scikit-Learn, and Scipy. Similarly, bupaR Janssenswillen et al. (2018) is an R package for process mining in R systems. These two packages in Python and R respectively support process mining scientists in a largescale experimental setting and also with an ability to customize the algorithms and metrics. However, easy-to-use graphical tools like ProM (Van der Aalst et al. 2009), Celonis, or Disco can be used to analyze processes by business analysts or managers without much coding experience. Kebede and Dumas (2015) provided a comparison of various process mining tools based on a framework they proposed to compare the tools. The analysis showed that the tools differ in terms of how functionalities are provided in the process mining tools.

CONCLUSION

This chapter highlights how process mining is an essential data mining technique in this world where processes and workflows are rapidly getting digitized. Understanding processes is always a priority to enhance and fix issues related to a process. Interviews, anecdotal feedback, or surveys were preferred methods to understand the processes. Process mining is a paradigm shift toward understanding how work is being done. It has provided a way to build a data-driven view of how work is getting done by applying models on top of the data points in a process. It is essential to think about the unit of analysis and perspectives at the outset of carrying out the process mining activity. This lays a foundation for selecting the type of algorithms and metrics that can be utilized to arrive at the outcome that we want to drive using process mining. There are several situations where the event log might suffer from data quality issues. Some of the well-known data quality issues are described in this chapter and can be used to assess the event log's data quality before starting the mining task. Moreover, this chapter highlighted several benefits and prior applications of process mining in different industries and domains, providing an insight into the broad area of work that process mining techniques can be applied. These publications can be further studied to build upon the body of knowledge already available in process mining and sequential process analysis.

FUTURE WORK

The current process mining techniques focus on the process discovery of historical or offline data. It does a great job of identifying inefficiencies and bottlenecks in the processes, thereby helping in enhancing or fixing the issues in the processes in case of any deviations or violations. Process mining algorithms can be enriched further to function on ongoing case data in real-time that can predict certain situations like the time to complete a running workflow, possibility of any deviations or violations in the future. There is some current work in this area, such as path prediction using process mining (Bernard and Andritsos 2019), prediction of medical expenses for gastric cancer based on process mining (Cao et al. 2021) and time prediction in manufacturing systems using process mining (Choueiri et al. 2020).

However, few such works provide an opportunity for developing generalized process mining algorithms with predictive power.

Another area of improvement and further research is in event data extraction and transformation. In a study conducted by IEEE Task Force on Process mining, it was discovered that data preparation for process mining is highly time-consuming, and there are significant challenges with complex events data with inconsistent relationships and incomplete and missing data (Wynn et al. 2022). Participants indicated a need for automatic data quality assessment and data preparation modules for efficient data transformations on events data for process mining.

REFERENCES

Aalst, W. (1998). The Application of Petri Nets to Workflow Management. *Journal of Circuits, Systems, and Computers*, *8*(01), 21–66. doi:10.1142/S0218126698000043

Aalst, W. (2016). *Process Mining: Data Science in Action*. Academic Press.

Aalst, W., Hee, K., Ter, A., & Sidorova, N. (2011). Soundness of workflow nets: Classification, decidability, and analysis. *Formal Aspects of Computing*, *23*(3), 333–363. doi:10.100700165-010-0161-4

Aalst, W., & Medeiros, A. K. A. (2005). Process Mining and Security: Detecting Anomalous Process Executions and Checking Process Conformance. *Electronic Notes in Theoretical Computer Science*, *121*, 3–21. doi:10.1016/j.entcs.2004.10.013

Adam, N., Atluri, V., & Huang, W. (1998). Modeling and Analysis of Workflows Using Petri Nets. *Journal of Intelligent Information Systems*, *10*(2), 131–158. doi:10.1023/A:1008656726700

Alrahili, R. (2021). *Towards Employing Process Mining for Role Based Access Control Analysis: A Systematic Literature Review*. Academic Press.

Augusto, A., Deitz, T., Faux, N., Manski-Nankervis, J.-A., & Capurro, D. (2022). Process mining-driven analysis of COVID-19's impact on vaccination patterns. *Journal of Biomedical Informatics*, *130*, 104081. doi:10.1016/j.jbi.2022.104081 PMID:35525400

Awatef, Gueni, Fhima, Cairns, & David. (2015). *Process Mining in the Education Domain*. Academic Press.

Baldellon, Fabre, & Roy. (2011). Modeling distributed realtime systems using adaptive Petri nets. *Actes de la 1re journ´ee 3SL*, 10.

Barro, R. J. (2001). Education and economic growth. *The contribution of human and social capital to sustained economic growth and well-being, 79*, 13–41.

Bause & Kritzinger. (2013). *Stochastic Petri Nets -An Introduction to the Theory*. Academic Press.

Bernard, G¨., & Andritsos, P. (2019). Accurate and transparent path prediction using process mining. In *European Conference on Advances in Databases and Information Systems* (pp. 235–250). Springer. 10.1007/978-3-030-28730-6_15

Bernard, G¨., Boillat, T., Legner, C., & Andritsos, P. (2016). When sales meet process mining: A scientific approach to sales process and performance management. *Proceedings of the 37th International Conference on Information Systems (ICIS 2016).*

Berti, van Zelst, & Aalst. (2019). *Process Mining for Python (PM4Py): Bridging the Gap Between Process- and Data Science.* Academic Press.

Bose, van der Aalst, Zliobaite, & Pechenizkiy. (2011). Handling concept drift in process mining. In *International Conference on Advanced Information Systems Engineering* (pp. 391–405). Springer.

Buijs, J., Dongen, B., & Aalst, W. (2012). On the Role of Fitness, Precision. *Generalization and Simplicity in Process Discovery.*, *7565*(09), 305–322.

Burattin, A. (2015). *Process Mining Techniques in Business Environments* (Vol. 207). doi:10.1007/978-3-319-17482-2

Cao, Y., Guo, Y., She, Q., Zhu, J., & Li, B. (2021). Prediction of medical expenses for gastric cancer based on process mining. *Concurrency and Computation*, *33*(15), e5694. doi:10.1002/cpe.5694

Chopra, M., Singh, D. S. K., Gupta, A., Aggarwal, K., Gupta, B. B., & Colace, F. (2022). Analysis & prognosis of sustainable development goals using big data-based approach during COVID-19 pandemic. *Sustainable Technology and Entrepreneurship*, *1*(2), 100012. doi:10.1016/j.stae.2022.100012

Choueiri, A. C., Denise, M. V. S., Scalabrin, E. E., & Eduardo, A. P. S. (2020). An extended model for remaining time prediction in manufacturing systems using process mining. *Journal of Manufacturing Systems*, *56*, 188–201. doi:10.1016/j.jmsy.2020.06.003

Corradini, F., Marcantoni, F., Morichetta, A., Polini, A., Re, B., & Sampaolo, M. (2019). Enabling auditing of smart contracts through process mining. In *From Software Engineering to Formal Methods and Tools, and Back* (pp. 467–480). Springer. doi:10.1007/978-3-030-30985-5_27

Czerwinski, Lasota, Lazi´c, Leroux, & Mazowiecki. (2020). The Reachability Problem for Petri Nets Is Not Elementary. *Journal of the ACM, 68*, 1–28.

Dakic, Stefanovic, Cosic, Lolic, & Medojevic. (2018). Business process mining application: A literature review. *Annals of DAAAM & Proceedings*, 29.

de Leoni, M. (2022). Foundations of Process Enhancement. *Springer International Publishing.*, *8*. Advance online publication. doi:10.1007/978-3-031-08848-3

Diamantini, C., Genga, L., Potena, D., & Aalst, W. (2016). Building Instance Graphs for Highly Variable Processes. *Expert Systems with Applications*, *59*, 59. doi:10.1016/j.eswa.2016.04.021

Dogan, O., Fernandez-Llatas, C., & Oztaysi, B. (2019). Process mining application for analysis of customer's different visits in a shopping mall. In *International Conference on Intelligent and Fuzzy Systems* (pp. 151–159). Springer.

Dominguez, C., Garcia-Izquierdo, F. J., Jaime, A., Perez, B., Rubio, A. L., & Zapata, M. A. (2021). Using Process Mining to Analyze Time Distribution of Self-Assessment and Formative Assessment Exercises on an Online Learning Tool. *IEEE Transactions on Learning Technologies, 14*(5), 709–722. doi:10.1109/TLT.2021.3119224

Dupuis, A., Cam'elia, D., & Agard, B. (2022). Predicting crop rotations using process mining techniques and Markov principals. *Computers and Electronics in Agriculture, 194*, 106686. doi:10.1016/j.compag.2022.106686

Grisold, Wurm, Mendling, & vom Brocke. (2020). *Using Process Mining to Support Theorizing About Change in Organizations*. Academic Press.

He, Z., Wu, Q., Wen, L., & Fu, G. (2018). A process mining approach to improve emergency rescue processes of fatal gas explosion accidents in Chinese coal mines. *Safety Science*, 111.

Hee, Sidorova, & Van der Werf. (2013). *Business Process Modeling Using Petri Nets*. Academic Press.

Jans, M., Van Der Werf, J. M., Lybaert, N., & Vanhoof, K. (2011). A business process mining application for internal transaction fraud mitigation. *Expert Systems with Applications, 38*(10), 13351–13359. doi:10.1016/j.eswa.2011.04.159

Janssenswillen, G., Depaire, B^., Swennen, M., Jans, M., & Vanhoof, K. (2018). bupaR: Enabling reproducible business process analysis. *Knowledge-Based Systems*, 163.

Kaur, Singh, Kumar, Gupta, & El-Latif. (2021). Secure and energy efficient-based E-health care framework for green internet of things. *IEEE Transactions on Green Communications and Networking, 5*(3), 1223–1231.

Kebede, M., & Dumas, M. (2015). *Comparative evaluation of process mining tools*. University of Tartu.

Law, Ip, Gupta, & Geng. (2021). *Managing IoT and Mobile Technologies with Innovation, Trust, and Sustainable Computing*. CRC Press.

Leandro, G., Miura, D., Safanelli, J., Borges, R., & Cl'audia, M. (2022). *Analysis of Stroke Assistance in Covid-19 Pandemic by Process Mining Techniques* (Vol. 294). doi:10.3233/SHTI220394

Lemos, Sabino, Lima, & Oliveira. (2011). Using process mining in software development process management: A case study. In *2011 IEEE International Conference on Systems, Man, and Cybernetics* (pp. 1181–1186). IEEE. 10.1109/ICSMC.2011.6083858

Li & de Carvalho. (2019). Process Mining in Social Media: Applying Object-Centric Behavioral Constraint Models. *IEEE Access*, 1–1.

Mahendrawathi, E. R., Zayin, S. O., & Pamungkas, F. J. (2017). ERP post implementation review with process mining: A case of procurement process. *Procedia Computer Science, 124*, 216–223. doi:10.1016/j.procs.2017.12.149

Martin, N. (2020). *Data Quality in Process Mining*. Academic Press.

Naderifar, V., Sahran, S., & Shukur, Z. (2019). A review on conformance checking technique for the evaluation of process mining algorithms. *TEM Journal, 8*(4), 1232.

Prathama, F., Yahya, B. N., & Lee, S.-L. (2021). A Multi-case Perspective Analytical Framework for Discovering Human Daily Behavior from Sensors using Process Mining. *2021 IEEE 45th Annual Computers, Software, and Applications Conference (COMPSAC)*, 638–644. 10.1109/COMPSAC51774.2021.00093

R.P., Bose, Mans, & Aalst. (2013). *Wanna improve process mining results?* Academic Press.

Sato, D. M. V., De Freitas, S. C., Barddal, J. P., & Scalabrin, E. E. (2021). A survey on concept drift in process mining. *ACM Computing Surveys, 54*(9), 1–38.

Silva, F. G., Reis da Silva, T., Alan de Oliveira, S., & Aranha, E. (2020). Behavior analysis of students in video classes. In *2020 IEEE Frontiers in Education Conference (FIE)* (pp. 1–8). IEEE. doi:10.1109/FIE44824.2020.9274274

Stergiou, C., Psannis, K. E., Gupta, B. B., & Ishibashi, Y. (2018). Security, privacy & efficiency of sustainable cloud computing for big data & IoT. *Sustainable Computing: Informatics and Systems, 19*, 174–184. doi:10.1016/j.suscom.2018.06.003

Van der Aalst, van Dongen, Gu¨nther, Rozinat, Verbeek, & Weijters. (2009). ProM: The process mining toolkit. *BPM, 489*(31), 2.

Weijters, A. (2006). Process Mining with the Heuristics Mineralgorithm. Academic Press.

Wynn, M. T., & Lebherz, J. (2022). Rethinking the input for process mining: insights from the XES survey and workshop. In *International Conference on Process Mining* (pp. 3–16). Springer. 10.1007/978-3-030-98581-3_1

Yasmin, F. A., Bukhsh, F. A., & Patricio, D. A. S. (2018). Process enhancement in process mining: A literature review. *CEUR Workshop Proceedings, 2270*, 65–72. http://simpda2018.di.unimi.it/

Zerbino, P., Stefanini, A., & Aloini, D. (2021). Process Science in Action: A Literature Review on Process Mining in Business Management. *Technological Forecasting and Social Change, 172*, 121021. doi:10.1016/j.techfore.2021.121021

Zhou, Z., Wang, M., Ni, Z., Xia, Z., & Gupta, B. B. (2021). Reliable and Sustainable Product Evaluation Management System Based on Blockchain. *IEEE Transactions on Engineering Management.*

Chapter 7
The Concept of Modularity in the Context of IS/IT Project Outsourcing:
Analyzing the Role of Interface From the Findings of Empirical Studies

Shahzada Benazeer
University of Antwerp, Belgium

Jan Verelst
University of Antwerp, Belgium

Philip Huysmans
University of Antwerp, Belgium

ABSTRACT

Information systems and/or information technology (IS/IT) outsourcing became a very common practice in developed and emerging economies. Despite IS/IT outsourcing's importance, reports on outsourcing initiatives indicate problematic situations. A large number of IS/IT outsourcing projects are being renegotiated or prematurely terminated, and many IS/IT outsourcing failures are not even publicly reported due to the fear of negative responses from the market and stakeholders. Literature suggests that the IS/IT project outsourcing is a complex maneuver. The concept of modularity has been applied in many other fields in order to manage complexity and enhance agility/flexibility; hence, four cases were analyzed using the lens of modularity in order to understand and identify the relationship between the concept of modularity and IS/IT project outsourcing. The interface aspect of modularity has emerged as the most relevant and identified in all four cases. It implies that the interface aspect should get greater attention when designing/planning a new IS/IT outsourcing project.

DOI: 10.4018/978-1-6684-4102-2.ch007

INTRODUCTION

Digital technologies have advanced more rapidly than any innovation in the history of mankind (UN report, 2019) prompting redesigning of every aspects people interact with, be it connectivity, healthcare, business processes, financial services, economic policies, trade and public services, etc. Digitalization has a profound impact on organizational strategies which revolutionized connectivity and as a result, the prevalence of IS/IT project outsourcing has become a norm in contemporary organizations. Findings from many empirical studies suggest that despite prevalence of IS/IT project outsourcing the failure rate is high. Premature contract terminations and frequent dissatisfaction with IS/IT outsourcing results are commonly encountered. Literature also suggests that many IS/IT outsourcing projects are being re-negotiated and/or prematurely terminated.

Table 1. IS/IT outsourcing success and failure in literature

	Description	Authors
40%	findings from a Dutch field study of a representative sample of 30 IS/IT outsourcing deals totalling to more than 100 million Euro. Of the 30 deals, 18 (60%) were successful.	Delens, Peters, Verhoef, & Van Vlijmen (2016)
50%	almost half of the IS/IT outsourcing projects failed to realize intended benefits or targets	Jabangwe, Smite, & Hesbo (2016)
60%	60% of customer organizations were not able to meet their pre-defined targets	Schmidt, Zoller, & Rosenkranz (2016)
71%	71% projects were considered failures or challenged	Wojewoda & Hastie, (2015)
78%	78% of projects discontinued either by switching vendors or terminating the projects	Gorla & Lau (2010)
78%	In the long term, the relationship between customer and vendor reaches the point of failure in the 78% of the projects	Mehta & Mehta (2010)
44%	44% of the projects failed: Cancelled prior to completion or delivered and never used	Ciric & Rakovic (2010)
32%	32% of the projects challenged: Late, over budget, and/or with less than the required features and functions	
24%	24% of the projects brought back in-house (back-sourced)	Tadelis (2007)
35%	35% projects failed	Gay & Essinger (2000)

Empirical data found in literature from 2000 until 2016 clearly illustrates that the failure rates in IS/IT outsourcing projects remain persistently high (table 1). Empirical research has attempted to quantify the high probability of IS/IT outsourcing project failures. For instance, table 1 illustrates the findings of a joint longitudinal study conducted in 2000 by Oxford University's Institute of Information Management and the University of Missouri. This study tracked 29 major IS/IT project outsourcing over eight years and reported that more than 35 percent of the projects failed (Gay & Essinger, 2000). Similarly, a Dutch field study, representing a sample of 30 outsourcing deals concludes that out of the 30 deals 18 were successful (Delens, Peters, Verhoef, &, van Vlijmen, 2016). Another study suggests that 60% of customer organizations involved in IS/IT project outsourcing were not able to meet their pre-defined targets (Schmidt, Zoller, & Rosenkranz, 2016). Wojewoda and Hastie (2015) suggest in their findings that 71% of IS/IT outsourcing projects were considered failures or challenged.

A pertinent question this paper deals with how IS/IT project outsourcing failure may be addressed. So far, the literature includes many suggestions offered by both scholars and practitioners. Peterson and Carco (1998) suggested to streamline operations and 'fix the problem' before outsourcing IS/IT services. Various suggestions were introduced: the interested reader is referred to (1) Lambert, Emmelhainz, and Gardner (1999) who introduced their '*Partnership Model*'; (2) Greaver (1999) who formulated '*seven steps to successful outsourcing*'; (3) Logan (2000) who proposed two solutions in order to avoid failure in IS/IT project outsourcing. She suggests firstly, diagnosing the relationship from both sides of the contract and secondly, engaging agency theory to help design the types of contracts and relationships necessary to provide and support an environment of trust; (4) Lee (2001) who suggested knowledge sharing; (5) Rottman (2008) who elaborates on the importance of 'knowledge transfer'; (6) Harris, Herron and Iwanicki (2008) who stressed the importance of a high quality 'service level agreement' (SLA); (7) Karimi-Alaghehband and Rivard (2012) who proposed a model of IS/IT outsourcing success grounded in dynamic capabilities perspective; (8) Ishizaka & Blakiston, (2012) who proposed the "*18 C's model*" for a successful long-term outsourcing arrangement; and (9) Zheng and Abbott (2013) who argued that reconfiguration of organizational resources is vital to be successful in outsourcing. Despite the introduction of such remedies, the empirical research referred to above (table 1) continue to attest to the high failure rate of IS/IT project outsourcing. It seems that these remedies, if used, turned out to be partially successful at best.

Indeed, several authors already pointed out inherent complexity and weaknesses in the current IS/IT project outsourcing approach. Although IS/IT project outsourcing is considered by many scholars as a complex business strategy (e.g., Beulen & Ribbers, 2003; Jacques, 2006), many customer organizations do not even fully consider the risks associated with IS/IT project outsourcing and often fail to make decisions systematically and rigorously (Oshri, Kotlarsky, & Willcocks, 2015). Aron, Clemons, and Reddi (2005) suggest that the complexity of processes plays a significant role in IS/IT project outsourcing decisions. Cohen and Young (2006) argue that *ad hoc* sourcing approaches of yesteryears are ineffective in today's complex world. Findings from the research of *British Computer Society* indicates that complexity is the most common attribute to the failed outsourced IS/IT projects (Nauman, Aziz, & Ishaq, 2009). The findings from aforesaid studies further strengthen the argument in literature that IS/IT project outsourcing is a complex maneuver and inherent complexity is one of the main reasons for high failure rate of IS/IT outsourced projects. Moreover, the findings of a survey conducted by Tadelis (2007) focused on 25 world class firms having outsourcing contracts suggest that *"many companies learned that 'unexpected complexity', 'lack of flexibility' among outsourcing providers, and other 'unforeseen problems' added costs as well as friction, ultimately translating into higher total costs than anticipated. One-quarter of the outsourced transactions were brought back in-house"* (p. 261-262). The problems highlighted by Tadelis (e.g., "unexpected complexity", "lack of flexibility" and "unforeseen problems") might be addressed by applying the concept of modularity. In the following section, the concept of modularity is discussed briefly. In order to understand modularity in organizational structure better, one of the important aspects of modularity the "interface", and how SLA (Service Level Agreement) acts as an interface are explained briefly in the next section.

This paper is structured in the following fourteen sections. The current section is a brief introduction of the most important concepts in the research, i.e., IS/IT outsourcing as well as modularity, and a brief discussion of the background, the issues, problems, research objectives and importance of the study.

In the following section the main concepts in this study, i.e., modularity and one of the most important aspects of modularity, the "interface" is discussed. Next, in literature review modularity and outsourcing are discussed in detail. After literature review, a discussion of research methodology and data collection process adopted for this study. In the fifth section, theoretical framework is discussed and based on this discussion research questions are formulated. The sixth section briefly introduce the four cases, and in the seventh section findings are briefly discussed. Next, a cross-analysis using replication logic approach of four cases is done and the findings presented in the eighth section. In the next four sections, four cases are analyzed. In the thirteenth section, a conclusion has been drawn from the findings of the analyses and finally, contributions, limitations, and future research directions are discussed.

THE CONCEPT OF MODULARITY

The concept of modularity originated in systems science and has been applied to a wide range of fields that deal with complex systems (Schilling 2000). Modularity is defined as a property of a complex system, whereby the system is decomposed into several subsystems (i.e., modules). Simon (1962, p. 474) explained modularity as *"nearly decomposable systems, in which the interactions among the subsystems are weak, but not negligible"*. Due to the seamless connectivity and digitalization, the use of modular architecture is exponentially increased. The modularity concept is generally applied, but not limited to control or manage complexity; provide agility and flexibility. Besides domain-specific theories, also general, domain-independent research has been conducted, for example by Baldwin and Clark at Harvard Business School (Baldwin & Clark, 2000). The modularity literature prescribes some important aspects of modularity which are crucial in designing a modular system. In the context of these four cases, relevant and important modularity aspects are "interface" (Baldwin & Clark, 2000; Langlois, 2002; Sako, 2005; Sanchez & Shibata, 2021); "modular architecture" – refers to a predefined set of prescriptive rules which all modules of the system need to adhere to (Baldwin & Clark, 2000; Mannaert, Verelst, & De Bruyn 2016; Sanchez & Shibata, 2021); "cohesion" – refers to the degree of similarity between the parts of a single module (Chidamber & Kemerer, 1994); "modular operator" – refers to the actions that facilitates change to existing structure in order to improve a complex system (Baldwin & Clark, 2000; Sanchez & Shibata, 2021); "coupling" – refers to the degree of connection, relationship, and dependencies between modules (Orton & Weick 1990; Van der Linden, Mannaert, & De Bruyn, 2012); "dependencies" – refers to the degree to which a module relies on other modules in order to function (Ethiraj, Levinthal, & Roy, 2008); "separation of concerns" – refers to a design principle that advocates a single module should be allocated a single task or responsibility (Dijkstra, 1974; Mannaert, Verelst, & De Bruyn 2016); "design rules" – refers to an architecture that ensures the modules remain compatible to each other so that they work as a whole (Baldwin & Clark, 2000; Langlois, 2002; Sanchez & Shibata, 2021); "standards" – refers to adhering to some uniform rules in designing a modular system because uniformity in 'design rules' facilitates mass production (Sako, 2005) and "encapsulation" – refers to hiding information within a module so that modules can interact without requiring or being able to have full knowledge of the contents of each other (Parnas, 1972). This paper mainly focuses on the interface aspect of modularity. It is because, the role of interface aspect observed in all four cases (table 2). In the following, the interface aspect is described briefly.

Interface

Modules should communicate with each other through the interfaces (Langlois, 2002). The decomposition of a larger complex system into modules allows breaking apart or splitting up the complexities into smaller pieces (modules). The split modules are still mutually compatible as they work together as a whole towards a common goal. The modules' compatibility logically follows the adoption of specific 'design rules' (architecture) using an interface as a connector (Baldwin & Clark, 2000; Langlois, 2002). In a modular architecture, interfaces between modules must be fully specified explicitly and unambiguously, and the interfaces must be adhered to throughout the development process (Huysmans et al., 2014; Sako, 2005; Sanchez & Shibata, 2021). A modular architecture can be described as *"a predefined set of prescriptive rules which all modules of the system need to adhere to"* (Huysmans et al., 2014, p. 4418). An interface is a common boundary where direct contact between two or more modules occurs and where these modules communicate with each other. The interface can be a virtual or physical document where the rules of interaction (dependencies, conditions) among modules are exhaustively and unambiguously documented. The interface describes the inputs required by a module to perform its part of the functionality, and the output it will provide to its external environment (which includes other modules in the system). An interface is one of the most important aspects of the concept of modularity which describes how different modules can interact and work as a whole. Organisations are complex systems and the concept of modularity may be applied to manage the complexity, to enhance agility and flexibility, by developing well-defined interfaces between organisation units, and a clear task to-organisation unit mapping at various levels in organisational hierarchies (Sako 2005). Interfaces for organisation architecture are more difficult to specify, as compared to those for physical product architecture (Sako, 2005) due to the involvement of human actors and they are often considered as units of a modular architecture. For instance, Terlouw (2011, p. viii) states, *"modules can comprise humans and/or software systems"* and in addition, Dietz (2006, p. 81) proposed a method to identify modular actor role structures and thereby asserts that *"an enterprise is constituted by the activities of actor roles, which are elementary chunks of authority and responsibility, fulfilled by subjects"*.

A Simple Example of an Interface

Let's consider an SD (secure digital) card and a laptop as two modules. The SD card connects to the laptop through a slot (port). This slot (port) can be considered as an interface between SD card and laptop. The different manufacturers can make different versions of SD cards, with different characteristics (e.g., class 2, 4, 6, 10, etc.), if they adhere to the same interface.

Service Level Agreements (SLAs) as Interface

In general, an IS/IT outsourcing deal concerns an agreement or contract between (mostly two) parties in which one party (the vendor) agrees to deliver certain services to another party (the customer). The outsourcing contracts (SLA's) need to be managed by good arrangements stipulating the roles and responsibilities of each of the involved actors as these deals are often highly complex and of crucial importance for both parties. In an abstract way, the SLA can be considered by some people (especially by practitioners) as a means to measure the performance of the vendor (i.e., KPI's) and level of services expected from the service provider. The SLA is also used in measuring the KPI's, for example in detailing the compensation which has to be paid for failures in delivering the committed service within specific budget and within a specific time. At the industry level, these interfaces often consist of regulatory frameworks, rules, standards, and technical specifications that allow different players to connect (Jacobides, Knudsen, & Augier, 2006). Interfaces in services can include people, information, and rules governing the flow of information (Voss & Hsuan, 2009). The team-leads representing customer organization and the vendor organization can be conceived as interfaces between these two organizations. The SLA between the customer organization and the vendor organization can also be conceived as an interface. For instance, Sako (2005) explains that interfaces between modular organization units must be well defined and standardization is one of the several ways in which interfaces can be well defined. Moreover, some scholars suggest that the contract procedures of outsourcing can also be regarded as modular (e.g., Blair, O'Connor, & Kirchhoefer, 2011; Miguel, 2005). Analyzing seven outsourcing contracts (6 from IS/IT projects and 1 from manufacturing project) in a law firm in the U.S., evidence emerges that all these contracts have modular structures with features designed to reduce interdependencies and mechanisms applied for managing the interface (Blair, O'Connor, & Kirchhoefer, 2011). More recently, Sanchez and Shibata (2021) proposed a set of Modularity Design Rules (MDRs) for governing the organizational and managerial processes essential in developing modular architectures. They proposed total ten rules for MDRs in three phases of development, namely, "before component development" (seven proposed rules); "during component development" (two proposed rules) and; "after initial component development" (1 proposed rule). Interestingly, they considered the importance of interface aspect in all three phases of development. An IS/IT outsourcing project can be considered as a modular structure consisting of two (or more) organizations, i.e., the vendor and customer organization, transferring a number of responsibilities for IS/IT-systems under a collaboration defined in an SLA. The SLA can then, in terms of modularity, be considered as the interface of the modules. Even though the importance of an SLA is universally recognized in the literature, the point of view of modularity taken in this study resulted in interesting observations. As it is depicted in figure 1, at organizational level the SLA signed (the top layer of figure 1) between customer and vendor organizations may be considered as an interface. Similarly, two team leads (the bottom layer of figure 1) from two organizations (customer and vendor) may also be considered as an interface. The mid-layer illustrates interface at technical level where interface connects two devices or two modules. As it is explained above, the slot that connects the SD card to the laptop, may also be considered as an interface. The following section describes how modularity relates to outsourcing.

Figure 1. Different types of interface between two modules

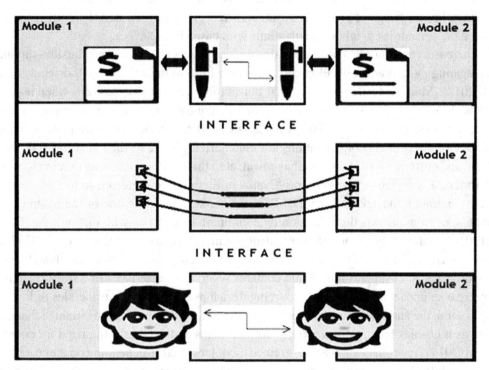

LITERATURE REVIEW: MODULARITY AND OUTSOURCING

This section reviews the literature on the role of modularity in the context of outsourcing. The concept of modularity was discussed by Simon in 1962 but it is observed that of late, the concept is drawing a great deal of attention from scholars and practitioners of varied domains (Campagnolo & Camuffo, 2010). The literature suggests that, in the past, modularity and outsourcing were predominantly studied by separate research communities (Fixson, Ro, & Liker, 2005). But recently a new stream of research has emerged that combines the scholars from the communities of Economics, Management, and Organizational Sciences and the scholars from the communities of Systems Sciences, Design Sciences, and Engineering. As modularization and outsourcing are becoming increasingly inseparable (Camuffo, 2004), the scholars from the new research stream are exploring some principles of the concept of modularity in addressing the complexities in an ever-changing business and economic environment. When different parts of a system require different types of knowledge, it is better to split those different parts by following the rules described by the concept of modularity. Thus, each split part (module) can be allocated to experts either in-house or to an external team (outsource) whichever option is more attractive. Splitting of a complex system into many modules creates many options including cost advantage through outsourcing, mass-customization, and mass-production. Voss and Hsuan (2009) argue that in products (e.g., software, business applications, etc.), modular architecture enables a firm to consider outsourcing. The mass customization also requires creating modular architectures that allow plug-and-play, mix-and-match compatibility of modules to configure product variations to meet specific preferences of customers (Sanchez & Mahoney, 2013). The production systems of complex products are decomposed (split) into separate process modules. In a modular production system, interfaces of each module are defined precisely and unambiguously, as a result each module becomes an autonomous part of a loosely coupled system. Such a design of production systems offers greater agility and flexibility and producers can outsource these modules to other organizations with ease (Tee, 2009).

There is increasing evidence that producers of complex systems (e.g.., software, business applications, etc.) are delegating more product development responsibilities to the suppliers (Rodrigues, Carnevalli, & Miguel, 2014). Mikkola (2003) suggests that outsourcing can only be realized when a system can be decomposed in such a way that interfaces of the components are well specified and standardized, which is one of the important factors of the modularization strategy. Due to the trend of high digitalization in the service sector, many organizations are modularizing some of their services (i.e., 'logistics', supply-chain, and customer relationship management, etc.) that may facilitate outsourcing to specialized external partners. For instance, contemporary banks in literature are referred to as *"modular banks"* (Wackerbeck, Helmuth, Skritek, & Putz, 2017) as the banking sector is one of the leading sectors in taking advantages from process digitalization by combining modularity with outsourcing. Nagpal and Lyytinen (2010) analyzed data from 55 respondents (senior executives) in the U.S. The findings suggest that modularity has a significant effect on IS/IT outsourcing success. Kramer, Heinzl, and Neben (2017) propose a decision support model that contains several characteristics of a software component, which may have an impact on software outsourcing decisions. Applying the principles of low coupling and high cohesion, the authors suggest, a fine-grained modular structure can be obtained that facilitates outsourcing as it enables handing over strictly defined items and decreases the need for coordination. Cunningham (2014) argues in favor of a fine-grained modular structure depending on the cost involved. A fine-grained modular designed project can have a competitive advantage if the costs of integrating those fine-grained modules are lower than the value it adds to the system. Sako (2014, p. 8) also contends that, *"Outsourcing/offshoring requires a certain degree of disaggregation, standardization, and modularization of tasks before services can be delivered from one legal entity to another at a geographic distance"*. As the economic environment is constantly evolving with new challenges and opportunities and the digital revolution is reshaping customers behaviours and demands, both the vendor and the customer organizations are constantly redesigning their business model. In this hostile environment, in order to adapt quickly to the emerging situations, organizations are fast-tracking digitalization which may help to create a fine-grained modular architecture. The literature suggests that modular architecture is one of the factors of of IS/IT outsourcing decision (Dedrick, Carmel, & Kraemer, 2017). Moreover, many scholars consider the organization as a complex system and proposed to modularize organizations to achieve agility and flexibility (e.g., Daft & Lewin, 1993; Sanchez & Mahoney, 1996; Sanchez & Shibata, 2021; Sako, 2005; Campagnolo & Camuffo, 2010; Schilling & Steensma, 2001; Nadler & Tushman, 1999; Op't Land, 2008; Baldwin & Clark, 1997). Many scholars suggest that modular organization structure may enhance the opportunity to outsource with ease (e.g., Anand & Daft, 2007; Wu & Park, 2009; Schilling & Steensma, 2001; Benazeer, 2018). It is interesting to note that the literature on modularity in IS/IT outsourcing remains at a high level of abstraction. There is limited attention for the intricate details of modularity, more specifically the aspects of modularity such as the interface, coupling, cohesion, etc. in the context of complex projects. Therefore, using a systematic approach, this paper offers an in-depth analysis of four cases in order to address the interface aspect of modularity in more detail.

METHODOLOGY AND DATA COLLECTION PROCESS

An interview-based descriptive, qualitative, case study research approach has been adopted as this study is intended to get better insights of a new phenomenon using the lens of the concept of modularity. This qualitative approach was deemed more suitable, as in this context, the goal is to gain an in-depth understanding of the manifestations (the "how") of modularity (Yin, 2009; 2014). Moreover, little theoretical knowledge on modularity within outsourcing is currently available, making a more descriptive case study appealing. This paper is focused on the interface aspect of modularity in four cases. It is however important to note that these four cases were embedded as a larger research project, in which insights into all aspects of modularity was considered. This has allowed applying the perspective within various situations and within different contexts that reflect on the generalizability of the findings in a more informed way. In order to enhance the external validity, in addition to a thematic analysis, a cross-case analysis is also conducted (described in detail in a following section). The first two cases were international IS/IT outsourcing cases, which have been documented extensively and have been re-analyzed based on secondary data. Additionally, two Belgian IS/IT outsourcing cases have been analyzed in this study, based on primary data. The primary data were collected through open-ended, semi-structured and exploratory interviews from senior executive of two Belgian customer organizations. According to Myers and Newman (2007), the qualitative interview has been used extensively in IS research, is a powerful research tool and an excellent means of gathering data. Furthermore, Mintzberg (1979) asserts that *"semi-structured interviews provide a controlled framework which facilitates analysis but also allows for the collection of 'soft' anecdotal data"* (p. 587). The data was collected from multiple sources, for instance, semi-structured onsite interviews of six sessions in each case (with each session lasting for two hours), direct observations during four onsite visits by three investigators, online archival records, documentations and presentations by the informants, and media outlets. Data triangulation was performed by comparing coded data from different sources (i.e., interview, direct observations, online archival records, documentations and presentations by the informants, and media outlets). The presence and active participation of three investigators assured investigators triangulation. The next section describes the theoretical framework in order to formulate the research questions.

THEORETICAL FRAMEWORK AND RESEARCH QUESTION

The findings from the literature review suggest that in software development and/or business applications the use of modularity is prevalent. For instance, some modularity aspects such as 'interface', 'encapsulation' or 'information hiding', 'separation of concerns', and loose coupling, etc. are widely used in software development projects. The IS/IT project outsourcing is often dealing with the software and/or business application developments. These relationships between complexity, modularity, and IS/IT project outsourcing and the insights from the literature guided the authors to think about the potential use of the concept of modularity as an ideal theoretical lens in the context of IS/IT project outsourcing. In order to study the application of the modularity concept in the context of IS/IT project outsourcing, a broad, exploratory investigation has been conducted to look for the phenomena in the context of IS/IT project outsourcing that can be interpreted as modular structures. These phenomena could be instances, examples or counter-examples of modularity in a wide-variety of aspects of IS/IT project outsourcing, both product- and process-oriented, both at the technical and at the non-technical (organizational) level. The technical level refers to modular structures in software (from specifications to the programming code), whereas the non-technical (organizational) level refers to the possibility of interpreting 'IS/IT outsourcing' as *'organizational modularity'* in the sense that two organizations collaborate and communicate based on an SLA (interface). In addition to identifying instances, examples and counter-examples of modularity, this study strives in obtaining indications of their relevance or importance in terms of the IS/IT outsourcing project. This relevance or importance can be derived in multiple ways, including: first, instances/examples/counterexamples could be unimportant in the sense that they have little or no impact on the efficiency, effectivity, success or failure of the project. On the other hand, they could be linked to success/failure factors in the project, which makes their relevance or importance more likely. Second, if instances/examples/counterexamples are related to design rules, design principles or theories regarding modularity, they could derive relevance from these theoretical foundations. For example, a known violation of a modularity design rule is likely to have, based on its theoretical grounding, a priori negative impact on the modularity aspects of the products and processes that it is a part of. In this sense, the theoretical grounding establishes a certain measure of relevance or importance of the instance/example/counter-example. In order to pursue the abovementioned research goal, the research questions are formulated as follows:

RQ1: *Which instances (examples, counter-examples) of the use of modularity in the context of IS/IT outsourcing can be identified?*

RQ2: *How can the relevance and/or importance of these instances (examples, counter examples) for IS/IT outsourcing project be assessed?*

IS/IT PROJECT OUTSOURCING: INTRODUCTION TO CASES

In the following, four cases are introduced briefly. The first case concerns the BSkyB vs. EDS outsourcing project. This case was of significant scale and complexity and ended in failure. The subsequent court case ended in 2010 and provided an exceptional amount of documentation. The re-analyses and results of this case study have been presented at the '*47th Hawaii International Conference on System Sciences*' (Huysmans et al., 2014) and a further elaborated analysis has been published in the '*International Journal of IT/Business Alignment and Governance*' (Huysmans et al., 2014). The second case concerns an IS/IT outsourcing project in a public university in a developing country in Asia. This IS/IT outsourcing project was small in budget and scope, simple and not complex. Nevertheless, this project encountered significant issues and ended in failure. The findings of re-analysis have been published as a book chapter in the '*Encyclopedia of Information Science and Technology*' (Benazeer, Huysmans, De Bruyn & Verelst, 2018). The third case concerns one of the biggest service sector companies in Belgium. During data collection, this organization was actively involved in an IS/IT project with an international service provider (single-vendor). The analysis of this case has been published in the '*International Journal of Information System Modeling and Design*' (Benazeer, Verelst, & Huysmans, 2020). Finally, the fourth case concerns a Belgian Financial Institution. This case is important because the financial service sector in Europe is becoming highly digitalized, hence it is indispensable to get some insights into this sector. Gewald and Dibbern (2005, p. 2) assert that *"One industry where digitization has dramatically altered the way in which business processes are carried out is the Banking Industry. Almost the entire portfolio of banking products is available in digital form and many services are now provided through the internet. The balance in a current account, an international payment, or the purchase of mutual funds is nowadays merely an electronic transaction which takes place in bits and bytes on a storage system within a corporate data center. Associated business processes like trade settlement or execution control are of an electronic nature as well".*

During data collection, this organization was involved in a multi-vendor IS/IT outsourcing project. The analysis of this case has been presented at the *'Enterprise and organizational Modeling and Simulation' (EOMAS 2017)* conference and later published in the *'Lecture Notes in Business Information processing: Enterprise and Organizational Modeling and Simulation'* (Benazeer, De Bruyn & Verelst, 2017). In the following section, a brief discussion elaborates the findings of the analysis that highlights the relevance of "interface" aspect of modularity in each of the four cases.

FINDINGS

The findings of four analyzed cases answers the research questions. The findings illustrate many aspects of modularity (e.g. interface, modular architecture, cohesion, modular operator, coupling, dependencies, separation of concerns, design rules, standards, and encapsulation) are relevant to the IS/IT outsourcing projects. Interestingly, the relevance of "interface" aspect found in all four cases which was not the case for any other aspects of modularity (table 2). More recently in an empirical study, Sanchez and Shibata (2021) proposed a set of ten Modularity Design Rules (MDRs) which are essential in developing modular architectures and should be considered during three proposed development phases. Sanchez and Shibata (2021) suggest the interface aspect as an essential aspect in all three phases of developing modular architectures. Moreover, in these four cases, an indication of a positive correlation between the failure or difficulties in the projects and failure to meet the modularity requirements are also observed. In other words, while analyzing the cases applying the lens of modularity, violations in prescribed modularity requirements (as it is suggested by the modularity literature) were observed when a project was heading towards failure or going through difficulties. However, the authors by no means consider modularity a panacea and do not claim that the failure or difficulties are only due to these violations, but it is definitely a matter of interest to look at. Scholars and practitioners alike will find good value in the insights uncovered and the lens of modularity might complement other tools/theories in order to avoid failures or difficulties in future IS/IT outsourcing projects. In the following, using cross-case analysis, the relevance of "interface" aspect in all four cases briefly discussed.

CROSS CASE ANALYSIS

In the following table (2), a cross-analysis using replication logic approach of four cases has been done. In this analysis it has been observed that some modularity aspects have emerged multiple times illustrating the level of relevance in analyzed four cases. For instance, 'Interface/SLA' as the most relevant modularity aspect has emerged in all four cases.

Table 2. Cross case analysis of the findings

Modularity aspects	Case 1	Case 2	Case 3	Case 4	Result
Interface / SLA	X	X	X	X	4
Modular architecture	X	X		X	3
Cohesion	X			X	2
Modular operator		X	X		2
Coupling			X		1
Dependencies		X			1
Separation of concerns	X				1
Design rules		X			1
Standards					0
Encapsulaion					0

The second most relevant modularity aspect 'modular architecture' has emerged in three cases. The third most relevant modularity aspects 'cohesion' and 'modular operator' both have emerged in two cases and the least relevant modularity aspects 'design rule', 'dependencies', 'coupling', and 'separation of concerns' each have emerged in one case. The authors further like to clarify that the level of relevancy is a context specific factor. The above illustrated results in table 2, are relevant in the context of these four analyzed cases. However, the order of relevance may differ in different context. The contribution of this cross-case analysis is at least it facilitates a deeper insight and understanding about the weight attributed to each aspect. In the following sections, each of the four cases are discussed using the following systematic approach: first, the modular structure of the problem domain is made explicit. Most importantly, the identification of modules is addressed. Second, the relevant modularity aspects are selected. Third, the resulting modularity requirements are listed, and fourth, the absence of modularity characteristics will be discussed in the context of violation or non-conformance of the modularity requirements.

CASE #1: INTRODUCTION

The project described in this case concerns an outsourcing deal to develop a Customer Relationship Management (CRM) system by Electronic Data Systems (EDS) for the British Sky Broadcasting (BSkyB) group. BSkyB appointed EDS in 2000, estimating that the project would take 18 months to complete and will cost £48 million. However, the project was considered a failure, and the contract was terminated in December 2002. By then, BSkyB had already spent £170 million. Eventually, the project would take six years to complete and cost £265 million. Conflicts regarding the proper project execution resulted in the filing of a case at the London Technology and Construction Court. Due to the complexity of the case, the judgment procedure lasted very long, and the case ended in July 2008, but the judgment was finalized in January 2010. BSkyB was awarded £318 million. The court proceedings documents with all the arguments and counter-arguments are publicly available. An earlier discussion of the risks involved in this IS/IT outsourcing project was provided by Verner and Abdullah (2012). This analysis mainly focuses on '*proven technology representation*' issue because this selection guarantees the analysis of IS/IT outsourcing risk factors which were highly relevant in this case. Consequently, conclusions were drawn by interpreting the excerpts and making explicit links between the IS/IT outsourcing risk factors and the aspects of modularity.

Analysis Case #1: Interface specification consisting of undefined or hidden dependencies

From a technical perspective, the context in which the project was embedded was complex. Before the project was initiated, several legacy applications handled the Customer Relationship Management functionality at Sky which is depicted in figure 2 as "initial applications" (BSkyB v. EDS, 2010, para. 9-10).

Figure 2. The modular structure of the application portfolio

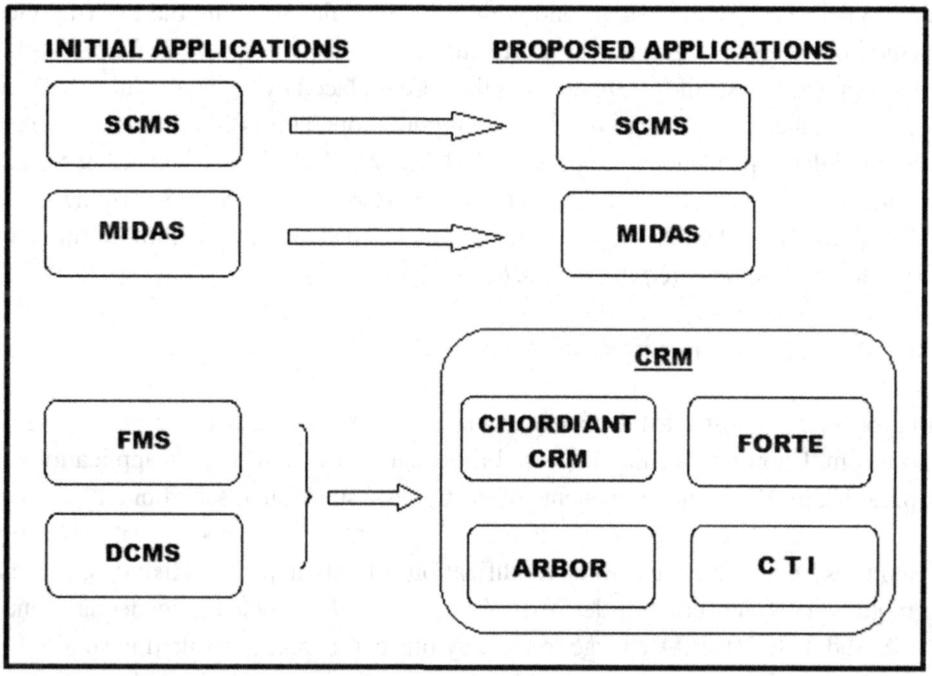

The following legacy applications handled the Customer Relationship Management of BSkyB:

(1)   'DCMS', the Digital Customer Management System,
(2)   'SCMS', the Subscriber Card Management System,
(3)   'MIDAS', the Management of Information for Digital and Analogue Systems, and
(4)   'FMS', the Field Management System.

The EDS proposed a new CRM to be constituted using new four different software technologies (i.e., Chodriant CRM, Forte, Abror, and CTI). This is depicted in figure 2 as "proposed applications". The new CRM system needed to replace some of the legacy applications (i.e., DCMS and FMS) while another application remained operational and needed to interface with the new CRM application (i.e., SCMS) (BSkyB v. EDS, 2010, para. 11 - 12). Moreover, different software technologies that were used by the vendor to build the new CRM system would have following functions: 'Chordiant CRM'; 'Arbor' as billing software; 'CTI' for the call centers; 'Forte' as a development framework and a middleware product (BSkyB v. EDS, 2010, para. 49 - 50). The alleged failure to deliver a promised seamless integration of all these (new) technologies was referred to as the *"proven technology representation"* in the court proceedings.

i. Identifying the modular structure and requirements

From a modularity point of view, the applications represent separate modules at a coarse-grained level. In figure 2, the application modules are represented as rounded boxes. The IS/IT application portfolio of BSkyB is the system in scope and within this modular structure the IS/IT applications are conceived as modules. This configuration is referred to as *modular structure MS1*. The modular operator 'substitution' is applied to a set of two modules as they are replaced by a new module. According to the modularity literature, the successful application of a modular operator is conditional on the assumption that no hidden modular dependencies are present (Baldwin & Clark, 2000; Sanchez & Shibata, 2021). Therefore, besides dependencies listed in the interface specification of the SCMS system in the Invitation to Tender (ITT) (BSkyB v. EDS, 2010, para. 11), no other dependencies are allowed for. This requirement is referred to as *modularity requirement MR1*.

ii. Assessing the modularity requirements

The court proceedings attested to the presence of numerous hidden dependencies. For example, a new log-on method from 'Chordiant' resulted in instability issues in various legacy applications (BSkyB v. EDS, 2010, para. 1236). Different components of the CRM system (such as Arbor and CTI) and legacy applications (such as SCMS) make log-on calls to 'Chordiant'. Consider now a modification to how 'Chordiant' requires a call to be made. Such modification is likely in projects like this, especially when the project's requirements were not well-defined upfront and labeled 'unclear', 'inadequate' and 'ambiguous' (Verner & Abdullah, 2012). A change in the way this call is made, resulted in (possibly multiple) changes in each application, implying that modular dependencies exist besides dependencies listed in the interface specification. Moreover, given the hidden nature of these dependencies, the impact of this change is unknown upfront. Hence, the absence of an explicit specification of all modular dependencies in the structure of the system shows that *modularity requirement MR1* was not met.

CASE #2: INTRODUCTION

In this case, the vendor organization is referred to as *'Aries'* and the customer organization is referred to as *'Taurus'*. *'Aries'* was a very competent and well-reputed provider as it was one of the leading independent companies working as a business unit of a large and reputed international company. *'Taurus'* was a big public-sector university in a developing country in Asia. In order to embrace digitalization, *'Taurus'* outsourced an IS/IT project to *'Aries'*. The task was to create a web-based portal for academic records management. *'Taurus'* assigned a team of experts referred to as the 'focal team' who were responsible for communicating with vendor *'Aries'* and were responsible in supervising each and every aspect of this project. The project became complex although it was small and simple, which involved low technical and functional complexities. Moreover, the project benefits were tangible and measurable; nevertheless, the project failed even though an extra 12 months period was extended after expiring the original project completion time (Nauman, Aziz, & Ishaq, 2009). The project was unanimously termed as a failure. *'Taurus'* was dissatisfied with the solution and was not using it and *'Aries'* was asking for more time and resources to complete the project.

Analysis Case #2: SLA consisting of undefined or hidden dependencies

The modularity literature suggest that modules should interact with one another through the interface (Langlois, 2002). To function adequately, the interaction between modules should be exhaustively and unambiguously documented in the interface. In the context of IS/IT project outsourcing, the SLA essentially provides an interface between the vendor and the customer (Blair, O'Connor, & Kirchhoefer, 2011; Miguel, 2005; Sako, 2005). As far as the SLA is concerned, responsibilities of each module, rights of each module, and the relationships between modules are to be described in detail.

i.   Identifying the modular structure and requirements

The outsourcing collaboration is the system in scope and within this modular structure, the focal team of organization *'Taurus'* and the organization *'Aries'*, are conceived as modules. This configuration is referred to as *modular structure MS2*. The SLA serves as the interface connecting both parties (modules), thus the focal team of organization *'Taurus'* is only allowed to ask organization *'Aries'* for services which are described in the SLA. Any service asked for which is not described in the SLA constitutes a hidden dependency, which may result in unwanted outcomes, namely coupling or ripple effects. Hence, in a good modular design, besides dependencies listed in the interface (SLA) specification, no other (hidden) dependencies are allowed for and this requirement is referred to as *modularity requirement MR2*.

ii.   Assessing the modularity requirements

After many hurdles and delays, the outsourced IS/IT project completed the 'testing phase'. As agreed in the SLA, after the testing phase, the team leader of *'Taurus'* would start to lead the implementation of the IS/IT project. But within three months *'Taurus'* discontinued implementing the IS/IT project and asked *'Aries'* to appoint an expert who would supervise the entire implementation process, and to train the end users in using the new IS/IT system. As such services were not part of the SLA, which was agreed upon by both parties, *'Aries'* declined to offer such services without getting a financial benefit. In turn, *'Taurus'* was inflexible and insisted on receiving the services for free. As a result, both teams eventually turned away from the IS/IT project. The following excerpt shows that the SLA consisted of hidden (i.e., undefined) dependencies:

"Head of Department of Computer Science started to lead the team to implement the project. However, the project implementation came to a standstill when the client organization desired deputation of full-time experts by the vendor organization to supervise the implementation which included training of the end users to use the system and subsequently adopt it. Vendor expressed their inability to depute an expert without charging a further expenditure to the customer" (Nauman, Aziz, & Ishaq, 2009, p.271).

The unanticipated support asked by the focal team of organization *'Taurus'* to organization *'Aries'* was an indication of the existence of undefined inter-modular dependencies. The SLA specifying the dependencies between *'Taurus'* and *'Aries'* was, therefore, said to be poorly defined and contained hidden dependencies, as such *modularity requirement MR2* was not met.

CASE #3: INTRODUCTION

The selected case deals with a vendor organization referred to as *'Alpha'*, and a customer organization referred to as *'Omega'*. *'Alpha'* was regarded as a competent service provider. *'Omega'* was one of the biggest service sector companies in Belgium. The IS/IT outsourcing project involved managing and maintaining the entire IS/IT systems of *'Omega'*. The main motivation of the IS/IT outsourcing was cost reduction. As part of this outsourcing contract, almost all of the IS/IT headcounts were transferred from the *'Omega'* organization to the *'Alpha'* organization with job guarantees for a certain period. Those people were highly skilled IS/IT experts and were well paid due to their long experience. The contract period was of medium terms and at the time of the interviews a 2nd year was running.

Analysis Case #3: SLA was vague, ambiguous, incomprehensive, inexplicit, and not well defined

Modules should communicate with one another through interfaces (Langlois, 2002). An interface is a common boundary where direct contact between two modules occurs and through which these two modules communicate with each other. The interface is a virtual or physical document where the rules of interaction among modules are exhaustively and unambiguously documented. The interface describes the inputs required by a module to perform its part of the functionality, and the output it will provide to its external environment (which includes other modules in the system). In the context of IS/IT project outsourcing, the SLA can be considered as an interface between two modules (Blair, O'Connor, & Kirchhoefer, 2011; Miguel, 2005; Sako, 2005). In this context the vendor 'Alpha' and the customer 'Omega' are conceived as two modules. The importance of an SLA relating to the success of the IS/IT project is recognized and understood by the informant.

i.   Identifying the modular structure and requirements

The outsourcing collaboration is the system in scope and within this modular structure, the organization *'Alpha'* and the organization *'Omega'* are conceived as modules. This configuration is referred to as *modular structure MS3*. The SLA serves as the interface connecting both organizations. To function adequately, the interaction between modules *'Alpha'* and *'Omega'* should be exhaustively and unambiguously documented in the interface. Hence, all the interactions and settlements between modules

'*Alpha*' and '*Omega*' should be conducted through the interface (SLA) and this requirement is referred to as *modularity requirement MR3*.

ii.   Assessing the modularity requirements

As long as the highly skilled former employees of '*Omega*' were working for '*Alpha*', no major problems were reported. But '*Alpha*' started replacing those highly skilled people out and problems started to surface. Although it was stated in the SLA that the '*Omega*' would get similar services as it was used to get from the in-house team, the actual situation seems to be different. The following excerpts are highlighting the actual situation:

"It was stated (in the SLA) that we would get similar services". The informant further said that:

"Probably, there is something behind. Why they are not delivering, why? Are they not capable or is it something financially not interesting for them to deliver in time?".

The '*Omega*' team did not include several items in the SLA and as a result, they have to ask for extra services from the '*Alpha*' team for which the '*Alpha*' team charges them extra. As a result, the cost reduction motivation was overshadowed. An example can be given about the incomplete SLA from the following excerpt:

"We have to ask for extra things (services), it was not calculated in the predicted cost reduction".

The service delivery situation became so uncertain that the service managers from the '*Omega*' team had to travel regularly to the site of the '*Alpha*' organization in order to explain the priorities of '*Omega*' team, and to explain the '*Alpha*' team what they needed to do in order to deliver in time. At some point, it seems that the urgency and frustration triggered to ignore the SLA which is reflected in the following excerpt:

"Our service managers are physically traveling 2 3 times a week to the vendor in order to explain to them what the priorities are and what they need to do, jamais-vu".

Later the informant added that: *"I don't think that the SLA is important right now, it just has to work".*
Although the importance of a well-defined SLA is recognized by the informant, probably, this realization came too late. The above excerpts illustrate that the interface (SLA) was weak, vague, ambiguous, incomprehensive, inexplicit, and not well defined; therefore, the *modularity requirement MR3* was not met.

CASE #4: INTRODUCTION

This case concerns a Belgian banking organization (further referred to as '*AB bank*', a fictitious name in order to guarantee anonymity and confidentiality). *AB bank* focuses on private banking activities, implying that compared to traditional retail bankers, their customer base is smaller but wealthier. Within the Belgian financial services industry, the organization can be considered as medium-sized in terms of the number of employees, number of clients, turnover, etc. While being a private bank in its core, the bank also welcomes investment clients with smaller budgets which can be served via an online investment portal. The portfolio management activities for bigger clients are offered through personal advice. Due to its relatively limited headcount consisting of 140 full-time employees in total, the IS/IT department of *'AB bank'* is rather small as well, consisting of ten full-time employees. The bank considers its IS/IT activities as operational and necessary but not as a strategic issue to obtain a competitive advantage. In that context, the modus operandi of multi-vendor outsourcing was chosen over the years. This means that taking advantage of digitalization, multiple and different external suppliers were used to provide different types of services. First, most of the development and maintenance work of the IS/IT infrastructure was outsourced to external parties. Additionally, the IS/IT department was dealing with many applications from different vendors. Given its relatively small internal team and the focus on outsourcing, *'AB bank'* had only developed three core applications internally: CRM, client on-boarding (registering information of newly acquired customers), and an order management system (Figure 3, grey ovals inside the big box). This aligns with the ambition of *'AB Bank'* of attempting to limit the number of customized products but instead giving preference to the use of package solutions. Therefore, the main activity of the in-house IS/IT team of *'AB Bank'* was concerned with the integration of all outsourced activities as well as its general management (package selection, vendor negotiations, etc.). Modularity is inherently a recursive concept that can be applied at different levels. This analysis revealed two major levels at which modularity could be clearly applied to the case at hand (i.e., inter-organizational level and intra-organizational level). Due to space limitations, this analysis will focus only on the intra-organizational level (the internal organization of *'AB Bank'*, such as the architecture of its different IS/IT applications and their integration). The intra-organizational level in the figure 3, (big box on the right-hand side) depicts a general overview of the IS/IT system modules present within the case organization; the grey ovals indicate the internally developed and maintained applications; the arrows within the big box at intra-organizational level, depict the most important interactions between the systems. At the inter-organizational level, It has been observed that a large majority of the IS/IT applications (i.e., the white ovals) were outsourced to external parties and for these outsourced applications, a set of SLAs (small rectangular boxes with letters "SLA") was agreed upon with a set of external IS/IT service providers (big rectangular boxes A to F on the left-hand side).

Figure 3. Modular structures identified: 'AB Bank' case

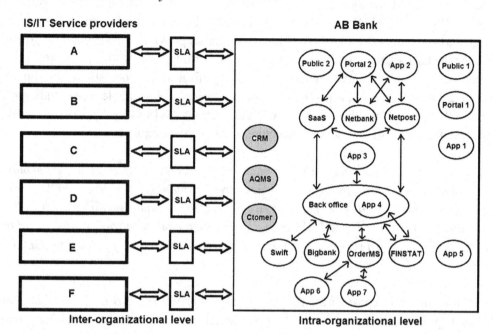

The informant is the head of the IS/IT department (CIO). While sketching the current situation of his department as well as the outlook for the future, the integration of the different (often externally acquired) applications was already indicated as a major concern. The IS/IT department of *'AB Bank'* was mostly busy with systems integration due to multi-vendor IS/IT outsourcing practice. The informant identifies systems integration as a challenge he was dealing with. This study intends to get some more insights using the lens of the concept of modularity, why the integration issue was so difficult to resolve.

Analysis 1 - Case #4: Lack of exhaustively documented inter-application interfaces at the <u>intra-organizational level</u>

At a fine-grained level, IS/IT project outsourcing concerns transferring certain responsibilities regarding a (set of) IS/IT applications from the client to the vendor. Clearly, within the client organization, those externally developed IS/IT applications should be integrated (both with internal systems and with systems from other vendors) so that they can collaborate with one another if required. As it has been aforementioned and illustrated (figure 3), *'AB Bank'* has adopted a multi-vendor outsourcing strategy encompassing numerous medium-sized applications and only a limited amount of applications developed and maintained by themselves. Given the importance to *'AB Bank'* of managing this set of applications, during the interview, the conversation went in-depth on how the organization dealt with this particular configuration. More specifically, questions were asked about how the integration between these applications was established. Was this easy or problematic? Whose responsibility was this? And how was this taken into account during the different phases of the IS/IT outsourcing project (e.g., initiation, start-up, execution, etc.)?

i.   Identifying the modular structure and requirements

The modular structure to study the communication and integration within the IS/IT systems of *'AB Bank'* can be easily identified. That is, each individual IS/IT application is a module per se. The informant discussed and elaborated the IS/IT application landscape within *'AB Bank'* (e.g., the different applications for the back office, front office, customer onboarding, etc.), with some applications being internally managed and some of them externally. As the organization did not distinguish subparts within each application, the application level is the lowest granularity level available when studying the integration issue in this case. When the informant was asked about the IS/IT systems of *'AB Bank'* which seems to be a configuration of the modular structure, the informant replied:

"Yes, it is like granularity. Indeed, we have different applications that work together. In the outsourcing part we have 'SaaS' type of outsourcing, we then have 'remote managed services'. 'ECM' is managing the hardware of operational stuff of 'X' machine (which runs the private banking system), and software is indeed managed and delivered by 'BLU'. The back office of the online portal is 'SaaS', and front services are running on 'managed services outsourcing'. The hardware is run by 'SR' but 'GL' developed the software".

Therefore, it is logical to consider the IS/IT application portfolio of *'AB Bank'* as the system, with every individual application being a module. This configuration is referred to as *modular structure MS4*. Theoretical knowledge regarding modularity allows to formulate several requirements for this structure. A well-designed modular system should have a clear and well-established modular architecture. This means that the set of modules (here: IS/IT applications) in the system should be identified and the dependencies between the IS/IT applications (i.e., the interfaces) should be exhaustively documented. Based on this information, a set of design rules can be created which form boundary conditions with which the IS/IT applications have to comply with (i.e., it describes a set of required inputs and outputs). Within these limitations, each IS/IT application can freely choose its specific implementation. Therefore, the existence of exhaustively documented inter-application interfaces and the adherence to centrally defined design rules referred to as *modularity requirement MR4*.

ii.   Assessing the modularity requirements

During the interview and in-depth discussions about IS/IT applications and its integrations, the initial feeling was that the architecture looked rather complicated and questions came to the author's mind how integrations were managed. It was immediately noted by the informant that integration was an important challenge within *'AB Bank'* as it was straightforward for him to enumerate a set of pertinent issues in this area:

"The integration is a challenge we have. We explored many issues in 'AB Bank' which are linked to the integration of applications. The integration challenges that we have are not really resolved".

It was an easy task for the informant to enumerate a set of examples of related issues. For instance, the informant stated that if a new customer is coming to open an account, the administrative employee needs to enter data manually in 7 different systems and that in some cases this number of systems can go up to 15. Or, if a customer likes to order a particular equity, the portfolio manager first has to look at the equity offers on a system 'A' and then needs to go to system 'B' to execute the order as no direct links between these two systems were established. Stated otherwise, not all systems which can or should automatically interact were properly connected in the case of *'AB Bank'*. Furthermore, this did not even seem to be a real priority when asking about the process of vendor and application selection:

"When we select an application, the first things we look at are the functional requirements. Do they match with our business requirements? Then we look at the non-functional requirements. We look at things like, are we able to manage the operating systems, the database systems? But indeed, we don't look at the requirements in terms of what kind of interfaces do we want [....]".

Finally, it was interesting to note that the integration problem was not only technical or on a syntactical level, but equally semantic:

"Different systems use different concepts. The most difficult part is to match different concepts to each other. For me, the pain is in the interfacing part".

From the above excerpts, it becomes clear that the interfaces between the different applications within the IS/IT portfolio of *'AB Bank'* were often not exhaustive if they existed at all. Therefore, it can be concluded that *modularity requirement MR4* was not met.

DISCUSSION AND CONCLUSION

The instances or examples of modularity in the context of IS/IT project outsourcing that were observed in four analyzed cases are listed on table 2. It is not surprising to find that in all four cases, the interface/ SLA related issues were observed. Firstly, in case 1 the 'IS/IT Application Portfolio' issue related to the BSkyB v. EDS (2010) court proceedings pointed out that the original bid team had under-estimated size and complexity. It is interesting that the court did not attribute this under-estimation on incorrect budgeting procedures or other managerial issues, but on technical issues related to size and complexity. This complexity can be related to multiple hidden dependencies in the interface specification of the SCMS system and violations of design principles (such as separation of concerns) which were observed in the case and were reported as resulting in unknown and large change impacts in the project. Secondly, in cases 2, 3 and 4, issues with incompleteness of the SLA were observed. Even though the importance of an SLA is universally recognized in the literature, the point of view of modularity taken in this study resulted in interesting observations. For instance, in case 2, examples of incompleteness in the SLA were observed which led to additional financial claims by the vendor organization. In case 3, the replacement of formerly in-house IS/IT staff by new hiring's, led to a loss of knowledge. This also exposed incompleteness in the SLA with the customer organization asking for supposedly additional services from the vendor organization which led to additional financial claims, as well as frequent and costly visits to the vendor organization in order to clarify uncertainties in the priorities and setup of the IS/IT outsourcing project. In case 4, it has been emerged from the findings that even an important issue like "systems integration" is directly related to the interface. An incomplete and vague interface or an interface consisting undeclared (hidden) dependencies may hinder the smooth progress of an IS/IT outsourcing project and even it may lead to a failure.

Literature in management- and IT management-research stress the importance of completeness of an SLA and the advantages it provides. However, it remains interesting nonetheless that these recommendations can be related to a (technical or systems theoretic) concept such as modularity, which prescribes that interfaces should not contain hidden dependencies, which implies completeness. The fact that the insights based on modularity correlate with insights from other sources, still constitute an indication of the relevance of the role of modularity in the context of IS/IT project outsourcing, even if they are not new. Many studies point to the importance of drafting an appropriate SLA at the start of an IS/IT outsourcing project, and of monitoring its execution and updating the SLA if and when required.

Answering the first research question, this study provides instances or examples of the role of modularity in the context of IS/IT outsourcing, at two levels; technical and organizational. In one hand, analyses of cases 1 and 4 clearly illustrates of how the concept of modularity is relevant at the technical levels of IS/IT outsourcing projects. On the other hand, analyses of cases 2 and 3 clearly illustrates how the concept of modularity is relevant at the organizational levels of IS/IT outsourcing projects. Answering the second research question, these instances or examples can be linked to several theoretical frameworks, design principles and success/failure factors, providing further indications of their importance in IS/IT outsourcing projects. These instances or examples in the analyzed cases can be interpreted as violations of, or at least insufficient attention to, well-known design principles on modularity, thereby providing indications that they likely negatively impact aspects such as complexity and project success. The global impression resulting from this study, is that in a highly digitalized environment, IS/IT outsourcing projects deal with several types of modular structures (technical, organizational). This implies that the aspects of these modular structures actually do play a role in IS/IT project outsourcing, and could be studied not only at the high, abstract level that is already dealt with in management- and IT management-research. The current cases have provided indications to this extent in the form of instances or examples. The analyzed cases indeed contain many violations against modularity (as well-known modularity design principles are not applied), or a lack of 'attention', which can be linked to theoretical frameworks, principles or success factors to illustrate their importance. In conclusion, the results of the analyses indicate that the concept of modularity seems to be certainly relevant in the context of IS/IT project outsourcing. This finding can be considered consistent with the literature on modularity in general and domain-independent frameworks such as Baldwin and Clark's modularity design rules (Baldwin & Clark, 2000), but this relevance seems to be underemphasized in most current research on IS/IT project outsourcing.

Finally, the authors would like to mention that in no way, this study is implying that modularity is the only or dominant factor determining the success of IS/IT outsourcing projects. Instead, the aim of this study was to explore the role of a factor that is in author's opinion often underexposed in the context of IS/IT project outsourcing, which in no way minimizes the role of other factors. In the next section, first theoretical contributions and contributions to practice of this study are discussed followed by the limitations and future research directions.

CONTRIBUTIONS, LIMITATIONS AND FUTURE RESEARCH DIRECTIONS

In this section, first discussion is about the theoretical contributions of this study, followed by the discussion about the contributions to practice. Next, limitations of this study are discussed and finally, some suggestions about future research directions are provided.

Theoretical Contributions

This study provides interpretations of phenomena in IS/IT outsourcing projects based on the concept of modularity, as instances or examples of modular structures with indications of their importance based on links to theoretical frameworks, design principles or success/failure factors. On the one hand, for researchers into IS/IT project outsourcing, these interpretations can contribute to a richer understanding of IS/IT outsourcing projects, and the reasons why they are (not) successful. Aspects of this contribution include the wide range of areas where modularity can be applied and the variety of ways in which modularity influences IS/IT outsourcing projects (including organizational and technical), as well as the significant influence technical modularity issues seems to play, even though IS/IT project outsourcing is often considered from a non-technical, IT-management point of view. On the other hand, for researchers into modularity these interpretations show that modularity can be applied in a concrete way to IS/IT project outsourcing. These contributions provide further support for the increasing interest in studying IS/IT project outsourcing from the point of view of modularity especially, when digital revolution is constantly reshaping the boundaries of the firm.

Contributions to Practice

This study also offers several guidelines for practitioners in IS/IT project outsourcing highlighting the complexity involved in IS/IT outsourcing projects, based on the concept of modularity. The analysis shows that, even in IS/IT outsourcing projects of moderate size, not only complex modular structures can be present at the technical level (i.e., software), but also at the organizational level (i.e., teams, departments, SLA's, business processes, documents, etc.). The combination of these modular structures and their interaction (coupling, dependencies) can result in highly complex situations, especially considering the amounts and size of the modular structures present, combined with the width of the effects of modularity. Moreover, the complexity is persistently increasing due to the digital transformation with many moving parts and underlying integrations and processes that all need to fit together. In the following paragraphs first limitations of this study is discussed and then the next paragraph provides some suggestions for future research directions.

Limitations

Currently, three main limitations are present in the study. The first limitation concerns the limited numbers of cases. As the current findings are based on four cases, no valid generalizations can be made yet, but the literature suggests that *"[...] case study without any attempt to generalize can certainly be of value in this process and has often helped cut a path toward scientific innovation"* (Flyvbjerg, 2006, p.10). The second limitation is that this study is unintentionally biased towards medium and large organizations because in the context of IS/IT project outsourcing, mostly medium and large organizations are seeking help from an external provider. Third, conducting descriptive qualitative case study research is necessary in the context of modularity and outsourcing because the objective is to get in-depth information about a contemporary phenomenon.

Future Research Directions

This study has provided indications of the role that modularity plays in IS/IT project outsourcing, in a wide range of domains, from organizational to technical. In order to build on these indications and maximize the insights that can be gained from modularity in this context, the authors call for future research to provide more detail on the role of modularity and its potential to address the issues that are currently causing IS/IT outsourcing projects to fail. In other words, the authors call for the addition of modularity aspects as a complement to the current management-approaches to IS/IT project outsourcing, providing a combination of more management-oriented and more structure-oriented (i.e., modularity-oriented) approaches to provide a richer view of factors influencing the success of IS/IT outsourcing projects. Some additional case studies may be conducted in order to validate the extent to which the modularity perspective (and the identified modularity aspects) may help in obtaining a better ex-ante understanding of IS/IT outsourcing project risks leading to project failure. Moreover, as additional cases are analyzed, a clear set of hypotheses can result, proposing modularity aspects and requirements as instruments to identify IS/IT outsourcing project risk factors, of which the relationship may be tested quantitatively. The confirmation of (several of) these hypotheses may provide a sound basis to identify and understand IS/IT outsourcing risk factors more confidently.

REFERENCES

Anand, N., & Daft, R. L. (2007). What is the right organization design? *Organizational Dynamics*, *36*(4), 329–344. doi:10.1016/j.orgdyn.2007.06.001

Aron, R., Clemons, E. K., & Reddi, S. (2005). Just right outsourcing: Understanding and managing strategic risk. *Journal of Management Information Systems*, *22*(2), 37–35. doi:10.1080/07421222.200 5.11045852

Baldwin, C. Y., & Clark, K. B. (1997). Managing in an age of modularity. *Harvard Business Review*, *75*(5), 84–93. PMID:10170333

Baldwin, C. Y., & Clark, K. B. (2000). *Design rules: The power of modularity*. The MIT Press. doi:10.7551/mitpress/2366.001.0001

Benazeer, S. (2018). *On the feasibility of applying the concept of modularity in the context of IS/IT outsourcing* [Doctoral dissertation, University of Antwerp]. University of Antwerp Research Repository. https://repository.uantwerpen.be/docman/irua/199d4e/153326.pdf

Benazeer, S., De Bruyn, P., & Verelst, J. (2017). Applying the concept of modularity to IT outsourcing: A financial services case. In R. Pergl, R. Lock, E. Babkin, & M. Molhanec (Eds.), *Enterprise and organizational modeling and simulation* (pp. 68–82). Springer. doi:10.1007/978-3-319-68185-6_5

Benazeer, S., De Bruyn, P., & Verelst, J. (2020). The concept of modularity in the context of IS/IT project outsourcing: An empirical case study of a Belgian technology services company. *International Journal of Information System Modeling and Design*, *11*(4), 1–17. doi:10.4018/IJISMD.2020100101

Benazeer, S., Huysmans, P., De Bruyn, P., & Verelst, J. (2018). The concept of modularity and normalized systems theory in the context of IS outsourcing. In M. Khosrow-Pour (Ed.), *Encyclopedia of information science and technology* (pp. 5317–5326). IGI Publications.

Beulen, E. J. J., & Ribbers, P. M. A. (2003). International examples of large-scale systems - theory and practice: A case study of managing IT outsourcing partnerships in Asia. *Communications of the AIS*, *11*, 357–376. doi:10.17705/1CAIS.01121

Blair, M. M., O'Connor, E. O., & Kirchhoefer, G. (2011). Outsourcing, modularity, and the theory of the firm. *BYU Law Review*, *2011*(2), 262–314.

BSkyB v. EDS, High Court of Justice, UK (2010).

Campagnolo, D., & Camuffo, A. (2010). The concept of modularity in management studies: A literature review. *International Journal of Management Reviews*, *12*(3), 259–283.

Camuffo, A. (2004). Rolling out a world car: Globalization, outsourcing, and modularity in the auto industry. *Korean Journal of Political Economy*, *2*(1), 183–224.

Chidamber, S. R., & Kemerer, C. F. (1994). A metrics suite for object-oriented design. *IEEE Transactions on Software Engineering*, *20*(6), 476–493. doi:10.1109/32.295895

Ciric, Z., & Rakovic, L. (2010). Change management in information system development and implementation projects. *Management Information Systems*, *5*(2), 23–28.

Cohen, L., & Young, A. (2006). *Multisourcing: Moving beyond outsourcing to achieve growth and agility*. Harvard Business School Press.

Cunningham, R. (2014). *Information environmentalism: A governance framework for intellectual property rights*. Edward Elgar Publishing. doi:10.4337/9780857938442

Daft, R. L., & Lewin, A. Y. (1993). Where are the theories for the new organizational forms? *Organization Science*, *4*(4), i–vi.

Dedrick, Carmel, & Kraemer (2017). A dynamic model of offshore software development. In L. P. Wilcocks, M. C. Lacity, & C. Sauer (Eds), *Outsourcing and offshoring business services* (pp. 281-320). Springer.

Delens, G. P. A. J., Peters, R. J., Verhoef, C., & van Vlijmen, S. F. M. (2016). Lessons from Dutch IT-outsourcing success and failure. *Science of Computer Programming*, *130*(32), 37–68. doi:10.1016/j.scico.2016.04.001

Dietz, J. L. G. (2006). *Enterprise ontology: Theory and methodology*. Springer. doi:10.1007/3-540-33149-2

Dijkstra, E. W. (1974). *On the role of scientific thought. E.W. Dijkstra Archive (EWD447), Center for American History*. The University of Texas at Austin.

Ethiraj, S. K., Levinthal, D., & Roy, R. R. (2008). The dual role of modularity: Innovation and imitation. *Management Science*, *54*(5), 939–955. doi:10.1287/mnsc.1070.0775

Fixson, S. K., Ro, Y., & Liker, J. (2005). Modularity and outsourcing: Who drives whom? A study of generational sequences in the U.S. automotive cockpit industry. *International Journal of Automotive Technology and Management, 5*(2), 166–183. doi:10.1504/IJATM.2005.007181

Flyvbjerg, B. (2006). Five misunderstandings about case study research. *Qualitative Inquiry, 12*(2), 219–245. doi:10.1177/1077800405284363

Gay, C. L., & Essinger, J. (2000). *Inside outsourcing: The insider's guide to managing strategic sourcing*. Brealey.

Gewald, H., & Dibbern, J. (2005). *The influential role of perceived risks versus perceived benefits in the acceptance of business process outsourcing: Empirical evidence from the German banking industry* (Working Paper Nr. 2005-9). J. W. Goethe University, Germany: E-Finance Lab.

Gorla, N., & Lau, M. B. (2010). Will negative experiences impact future outsourcing? *Journal of Computer Information Systems, 50*(3), 91–101.

Greaver, M. F. II. (1999). *Strategic outsourcing: A structured approach to outsourcing decisions and initiatives*. Amacom.

Harris, M. D. S., Herron, D., & Iwanicki, S. (2008). *The business value of IT: Managing risks, optimizing performance and measuring results*. CRC Press. doi:10.1201/9781420064759

Huysmans, P., De Bruyn, P., Benazeer, S., De Beuckelaer, A., De Haes, S., & Verelst, J. (2014). On the relevance of the modularity concept for understanding outsourcing risk factors. *Proceedings of the 47th Hawaii International Conference on System Sciences*.

Huysmans, P., De Bruyn, P., Benazeer, S., De Beuckelaer, A., De Haes, S., & Verelst, J. (2014). Understanding outsourcing risk factors based on modularity: The BSkyB case. *International Journal of IT/Business Alignment and Governance, 5*(1), 50–66. doi:10.4018/ijitbag.2014010104

Ishizaka, A., & Blakiston, R. (2012). The 18C's model for a successful long-term outsourcing arrangement. *Industrial Marketing Management, 41*(7), 1071–1080. doi:10.1016/j.indmarman.2012.02.006

Jabangwe, R., Smite, D., & Hesbo, E. (2016). Distributed software development in an offshore outsourcing project: A case study of source code evolution and quality. *Information and Software Technology, 72*, 125–136. doi:10.1016/j.infsof.2015.12.005

Jacobides, M. G., Knudsen, T. R., & Augier, M. (2006). Benefiting from innovation: Value creation, value appropriation, and the role of industry architectures. *Research Policy, 35*(8), 1200–1221. doi:10.1016/j.respol.2006.09.005

Jacques, V. (2006). *International outsourcing strategy and competitiveness: Study on current outsourcing trends: IT, business processes, contact centers*. Publibook.

Karimi-Alaghehband, F., & Rivard, S. (2012, December). Information technology outsourcing success: A model of dynamic, operational, and learning capabilities. *33rd International Conference on Information Systems (ICIS)*.

Kramer, T., Heinzl, A., & Neben, T. (2017, January). Cross-organizational software development: Design and evaluation of a decision support system for software component outsourcing. *Proceedings of the 50th Hawaii International Conference on System Sciences*. 10.24251/HICSS.2017.041

Lambert, D. M., Emmelhainz, M. A., & Gardner, J. T. (1999). Building successful logistics partnerships. *Journal of Business Logistics, 20*(1), 165–181.

Langlois, R. N. (2002). Modularity in technology and organization. *Journal of Economic Behavior & Organization, 49*(1), 19–37. doi:10.1016/S0167-2681(02)00056-2

Lee, J.-N. (2001). The impact of knowledge sharing, organizational capability and partnership quality on IS outsourcing success. *Information & Management, 38*(5), 323–335. doi:10.1016/S0378-7206(00)00074-4

Logan, M. S. (2000). Using agency theory to design successful outsourcing relationships. *International Journal of Logistics Management, 11*(2), 21–32. doi:10.1108/09574090010806137

Mannaert, H., Verelst, J., & De Bruyn, P. (2016). *Normalized systems theory: From foundations for evolvable software toward a general theory for evolvable design*. Koppa.

Mehta, N., & Mehta, A. (2010). It takes two to tango: How relational investments improve IT outsourcing partnerships. *Communications of the ACM, 53*(2), 160–164. doi:10.1145/1646353.1646393

Miguel, P. A. C. (2005). Modularity in product development: A literature review towards a research agenda. *Product: Management & Development, 3*(2), 165–174.

Mikkola, J. H. (2003). Modularity, component outsourcing, and inter-firm learning. *R & D Management, 33*(4), 439–454. doi:10.1111/1467-9310.00309

Mintzberg, H. (1979). *The structuring of organizations*. Prentice-Hall.

Myers, M. D., & Newman, M. (2007). The qualitative interview in IS research: Examining the craft. *Information and Organization, 17*(1), 2–26. doi:10.1016/j.infoandorg.2006.11.001

Nadler, D. A., & Tushman, M. L. (1999). The organization of the future: Strategic imperatives and core competencies for the 21st century. *Organizational Dynamics, 28*(1), 45–60. doi:10.1016/S0090-2616(00)80006-6

Nagpal, P., & Lyytinen, K. (2010, December). Modularity, information technology outsourcing success, and business performance. *International Conference on Information Systems*.

Nauman, A. B., Aziz, R., & Ishaq, A. F. M. (2009). Information system development failure and complexity: A case study. In M. G. Hinter (Ed.), *Selected reading on strategic information systems* (pp. 251–275). IGI Publications. doi:10.4018/978-1-60566-090-5.ch017

Op't Land, M. (2008). *Applying architecture and ontology to the splitting and allying of enterprises* [Doctoral dissertation, Delft University of Technology]. Gildeprint.

Orton, J. D., & Weick, K. E. (1990). Loosely coupled systems: A reconceptualization. *Academy of Management Review, 15*(2), 203–223. doi:10.2307/258154

Oshri, I., Kotlarsky, J., & Willcocks, L. (2015). *The handbook of global outsourcing and offshoring*. Palgrave Macmillan. doi:10.1057/9781137437440

Parnas, D. L. (1972). On the criteria to be used in decomposing systems into modules. *Communications of the ACM*, *15*(12), 1053–1058. doi:10.1145/361598.361623

Peterson, B. L., & Carco, D. M. (1998). *The smart way to buy information technology: How to maximize value and avoid costly pitfalls*. Amacom.

Rodrigues, E. A., Carnevalli, J. A., & Miguel, P. A. C. (2014, January). Modular design and production: An investigation on practices in an assembler and two first-tier suppliers. *Proceedings of the International Conference on Industrial Engineering and Operations Management*.

Rottman, J. W. (2008). Successful knowledge transfer within offshore supplier networks: A case study exploring social capital in strategic alliances. *Journal of Information Technology*, *23*(1), 31–43. doi:10.1057/palgrave.jit.2000127

Sako, M. (2005). Modularity and outsourcing. In A. Principe, A. Davies, & M. Hobday (Eds.), *The business of system integration* (pp. 229–253). Oxford University Press. doi:10.1093/acprof:oso/9780199263233.003.0012

Sako, M. (2014). Outsourcing and offshoring of professional services. In L. Empson, D. Muzio, J. Broschak, & B. Hinings (Eds.), *The oxford handbook of professional service firms* (pp. 327–347). Oxford University Press.

Sanchez, R., & Mahoney, J. T. (1996). Modularity, flexibility, and knowledge management in product and organization design. *Strategic Management Journal*, *17*(S2), 63–76. doi:10.1002mj.4250171107

Sanchez, R., & Mahoney, J. T. (2013). Modularity and economic organization: Concepts, theory, observations, and predictions. In A. Grandori (Ed.), *Handbook of economic organization* (pp. 383–399). Edward Elgar Publishing. doi:10.4337/9781782548225.00031

Sanchez, R., & Shibata, T. (2021). Modularity design rules for architecture development: Theory, implementation, and evidence from the development of the Renault–Nissan alliance "common module family" architecture. *Journal of Open Innovation*, *7*(4), 1–22. doi:10.3390/joitmc7040242

Schilling, M. A. (2000). Towards a general modular systems theory and its application to inter-firm product modularity. *Academy of Management Review*, *25*(2), 312–334. doi:10.2307/259016

Schilling, M. A., & Steensma, K. (2001). The use of modular organizational forms: An industry-level analysis. *Academy of Management Journal*, *44*(6), 1149–1168. doi:10.2307/3069394

Schmidt, N., Zoller, B., & Rosenkranz, C. (2016). The clash of cultures in information technology outsourcing relationships: An institutional logics perspective. In J. Kotlarsky, I. Oshri, & L. P. Willcocks (Eds.), *Shared services and outsourcing: A contemporary outlook* (pp. 97–117). Springer. doi:10.1007/978-3-319-47009-2_6

Simon, H. A. (1962). The architecture of complexity. *Proceedings of the American Philosophical Society*, *106*(6), 467–482.

Tadelis, S. (2007). The innovative organization: Creating value through outsourcing. *California Management Review, 50*(1), 261–277. doi:10.2307/41166427

Tee, R. (2009, January). Product, organization, and industry modularity exploring interactions across levels. *DRUID-DIME Academy Winter Ph.D. Conference on Economics and Management of Innovation, Technology, and Organizational Change.*

Terlouw, L. I. (2011). *Modularization and specification of service-oriented systems* [Doctoral dissertation, Delft University of Technology]. Gildeprint.

UN Report. (2019). *The impact of digital technologies.* Retrieved August 25, 2022, from https://www.un.org/sites/un2.un.org/files/2019/10/un75_new_technologies.pdf

Van der Linden, D., Mannaert, H., & De Bruyn, P. (2012, February). Towards the explicitation of hidden dependencies in the module interface. *Proceedings of the Seventh International Conference on Systems (ICONS).*

Verner, J. M., & Abdullah, L. M. (2012). Exploratory case study research: Outsourced project failure. *Information and Software Technology, 54*(8), 866–886. doi:10.1016/j.infsof.2011.11.001

Voss, C. A., & Hsuan, J. (2009). Service architecture and modularity. *Decision Sciences, 40*(3), 541–569. doi:10.1111/j.1540-5915.2009.00241.x

Wackerbeck, P., Helmuth, U., Skritek, B., & Putz, A. (2017). *Building the modular bank: Sourcing strategies in the age of digitization.* Strategy & PWC Consulting. Retrieved August 24, 2022, from https://www.strategyand.pwc.com/reports/building-modular-bank

Wojewoda, S., & Hastie, S. (2015). *Standish group 2015 chaos report - Q&A with Jennifer Lynch.* Infoq. Retrieved August 20, 2022, from https://www.infoq.com/articles/standish-chaos-2015

Wu, L., & Park, D. (2009). Dynamic outsourcing through process modularization. *Business Process Management Journal, 15*(2), 244–255. doi:10.1108/14637150910949461

Yin, R. K. (2009). *Case study research: Design and methods.* Sage Publications.

Yin, R. K. (2014). *Case study research: Design and methods.* Sage Publications.

Zheng, Y., & Abbott, P. (2013, June) Moving up the value chain or reconfiguring the value network? An organizational learning perspective on born global outsourcing vendors. *Proceedings of the 21st European Conference on Information Systems.*

Chapter 8
Research, Development, and Innovation Capability Maturity Model Reference for European Projects

Cozmiuc Claudia Diana
West University of Timisoara, Romania

Andreea Bozesan
Ioan Slavici University, Timisoara, Romania

Liviu Herman
Ioan Slavici University, Timisoara, Romania

Sinel Galceava
Ioan Slavici University, Timisoara, Romania

Cristian Pitic
Ioan Slavici University, Timisoara, Romania

ABSTRACT

Capability maturity and capability readiness models are designated management tools in scholarly literature. One of their applications is in shaping roadmaps, projects, and programs. Typically, individual articles tackle the two topics separately and the way the two management tools are intertwined. Scholarly literature shows the need to link the two tools. These tools tend to refer to digital transformation or include digitalization in their construction. A specific example is the TRL1-TRL9 capability maturity model used by NASA and by the European Union. This is the reference for conducting research, development, and innovation activities at the European Union. It models roadmaps and all project management techniques. A specific case study is considered in this chapter. Findings show the reference to manage research, development, and innovation activities at the European Union and NASA is a capability maturity model, a project management tool, and includes digitalization business information systems as the tools for business activity and business process management.

DOI: 10.4018/978-1-6684-4102-2.ch008

INTRODUCTION

Capability Maturity Models CMM have been used since 1986, in a wide variety of forms. In 2006, the Software Engineering Institute at Carnegie Mellon University developed the Capability Maturity Model Integration CMMI, which has largely superseded the CMM and addresses some of its drawbacks. A maturity model can be viewed as a set of structured levels that describe how well the behaviors, practices and processes of an organization can reliably and sustainably produce required outcomes. The basic constituents of capability maturity models are capability maturity levels and organizational dimensions. Organizational dimensions achieve the capability maturity levels. Business processes are one such organizational dimension. On capability maturity model is the TRL 1 to TRL 9 model, which is used for research, development and innovation activities. The TRL model is used to describe the activities and results or deliverables. An activity or process is completed when certain deliverables have been attained as outcome of the activities. The TRL1-9 capability maturity model is used by Nasa and European Union as the benchmark for research, development and innovation activities. The model assumes digital simulation technologies, such as Computer Aided Technologies or Product Lifecycle Management software, underpin these activities and their deliverables. The model is, by now, harmonized with digital simulation technologies and will be analyzed as such. Typically scientific literature review alots individual articles to capability maturity models. Another topic that is designed individual articles is roadmaps. Literature review shows articles tie roadmaps to maturity indices and show them as the next management tool to be used after the roadmap. Roadmaps thereby overarch programs and projects. The goal of this article is to describe the TRL 1-9 capability maturity model and the way it is used to derive roadmaps, programs and projects. The topic as such is prevalent in scholarly articles focused on maturity indices and roadmaps. The methodology is a descriptive case study. The subject of the case study is the use of the TRL 1 to 9 model in European project calls for funding on a specific European project case. The case study illustrates how the capability maturity model becomes a requirement for the project activities and deliverables and how this can easily be organized in milestones. Findings are that the TRL 1-9 model is a simple capability maturity model and that its deliverables and milestones are recommended to be used to schedule projects. These deliverables include proof of concept which may be realized via digital means, Computer Aided Technologies or Product Lifecycle Management Software. Findings from the empirical data analysis are that the Technological Readiness Model is easy to apply in technologically advanced agriculture. A reference example in classic theory is thereby confirmed. The case study is a useful example for practitioners.

LITERATURE REVIEW

The Definition and Role of Capability Maturity Models and Readiness Models

In 1986, the capability maturity model was created by the US Department of Defense. In the capability maturity model, maturity levels are given by progressive capabilities that describe how the behaviors, practices and processes of an organization can reliably and sustainably produce required outcomes (Nayab, 2010). The maturity levels are: initial; repeatable; defined; capable; efficient. In digital transformation, maturity indexes refer to processes and other organizational dimensions and have several capability maturity levels. The goal tends to be value. According to Porter and Heppellmann (2014), smart connected products have several maturity stages given by their capabilities: monitoring, control, optimization, autonomy. They are to transform several organizational dimensions: business model, value chain, information technology architecture, corporate functions, relationships, processes, organizational structures, value. In Industry 4.0, maturity models are organizational capabilities, include a stream of objectives and sequential levels or stages (Sener & Gokalp, 2018). Maturity models function as improvement along with maturity progress (Sener & Gokalp, 2018). Maturity models are used to determine the current maturity model and generate a roadmap to move to the next maturity model (Sener & Gokalp, 2018). Maturity models represent a theory of staged based evolution and its basic purpose consists of describing stages and maturation paths through a scale of maturity (Bertolini et al, 2019). Mettler (2009, 2011) defined maturity as a development of a specific ability or reaching to a targeted success from an initial to an anticipated stage. In the view the most popular Industry 4.0 maturity model proponents, Schumacher, Eroll and Sihn (2016) maturity models are a tool for comparing current maturity level to the desired maturity level of an organization or process, by conceptualizing and measuring. Erol, Schumacher and Sihn, in 2016, relate maturity models to digital transformation and digital transformation in manufacturing, industry 4.0. Maturity systems increase their capabilities over time regarding the achievement of some future state. This definition is shared by other popular authors such as Proença and Borbinha (2016), Mittal et al (2018). According to Proença and Borbinha (2016), maturity models can be used as evaluation criteria and described as complete, perfect or ready. Maturity models may be used from progression from basic state to a more advanced final state. The role of maturity models is complex: audit; benchmark; process appraisal; organization appraisal; progress tracking; diagnostic (Proença and Borbinha, 2016). Maturity models are models that help an individual or entity to reach a more sophisticated maturity level, in people/ culture, processes/ structures, and or objects/ technologies following a step by step continuous improvement process (Mittal et al, 2018). Maturity models are questionnaires with several options to choose from for each question (Akdil et al, 2017). Nickkhou et al (2016) defined maturity as guidance to correct or prevent problems, evidence of an achievement or a perfect state to be reached. Maturity models enable organizations to audit and benchmark regarding to assessment results; to track process towards a desired level; to evaluate strengths, weaknesses, threats, opportunities; to sequence stages from basic to advanced. Duffy (2001) teach organizations which actions should be considered in order to reach an advanced maturity level, and when and why to decide these actions. Tarhan at al (2016) describe maturity models as desired logical path for processes in several business fields which include discrete levels of maturity. According to Backlund et al (2014), maturity models are extremely important tools to appraise organizations.

Maturity indices are defined as radar charts, where the X axis represents digital transformation capability maturity levels and the Y axis the organizational dimensions impacted by digital transformation. Scales are used to measure how organizational dimensions are digitally transformed in order to score in sum at the level for each progressive level of digital transformation capability maturity.

Capability Maturity Models CMM have been used since 1986, in a wide variety of forms. In 2006, the Software Engineering Institute at Carnegie Mellon University developed the Capability Maturity Model Integration CMMI, which has largely superseded the CMM and addresses some of its drawbacks. A maturity model can be viewed as a set of structured levels that describe how well the behaviors, practices and processes of an organization can reliably and sustainably produce required outcomes. Maturity models involve the following: maturity levels, key process areas, goals, common features, and key practices. There are different capability maturity levels, and they involve: initial; repeatable; defined; capable; efficient. One of these CMMIs is the Technological Readiness Model TRL, which is the reference at Nasa and European Union. This capability maturity model is the object of this research, as one of several capability maturity models in scientific literature review.

Readiness models have the goal to capture the starting point and allow for initializing the development process (Schumacher et al, 2016). Readiness Assessments (Benedict et al, 2017; Mittal et al, 2018) are evaluation tools to analyze and determine the level of preparedness of the conditions, attitudes, and resources, at all levels of a system, needed for achieving its goal(s). Readiness assessments are recognized by top authors (De Carolis, 2017; Leyh, 2016; Schumacher et al, 2016).

Maturity model (Mittal et al, 2018) scales and readiness scales are prepared by consultants, by the original proponents of the Industry 4.0 vision, by scientific research, by technology proponents or analysts. Mittal et al (2018, p. 28, 29) noted that a large body of literature concerning Industry 4.0 offers to small and medium enterprises in scientific journals. Reading the works of consultants, it was noted many work with large corporations in information technology or customers as well as small to medium enterprises. Schumacher et al (2019) categorize maturity models into two groups: holistic approaches and specific approaches. Holistic approaches aim to assess and utilize elements of Industry 4.0 from all possible angles. They may be used to derive encompassing success factors. Specific approaches focus on a limited number of aspects relevant in Industry 4.0 with greater detail.

THE TECHNOLOGY READINESS 1-9 MODEL

The Nasa capability maturity model has several technological readiness stages: TRL1 means basic principles are observed and reported; TRL2 means technology concept and/or application formulated; TRL3 means analytical and experimental critical function and/or characteristic proof of concept; TRL4 means component and/or breadboard validation in laboratory environment; TRL5 means component and/or breadboard validation in relevant environment; TRL6 means system of subsystem model or prototype demonstration in a relevant environment (ground or space); TRL7 is system prototype demonstration in a space environment; TRL8 is actual system completed and flight qualified through test and demonstration (ground or space); TRL9 is actual system flight proven through successful proven mission operations. This technology maturity model has been in place at Nasa for many years. It is used to diagnoze the stages of research and development maturity of various Nasa projects (Mankins, 1995). This capability maturity model has been used as reference for scientific research in several industries.

The research covers all points TRL1, TRL2 and TRL3. Research begins with TRL1, discovering and formulating basic principles. At this stage, theoretical / fundamental scientific research begins to migrate to research and development. The results can be found in publications that present the basic principles of technology, or observations of physical reality. The following is a brief review of the general literature on Computer Aided Technologies, which include Computer Aided Engineering, Computer Aided Engineering. These are the technology used to produce the granulate complex presented here. Many of the Computer Aided Technologies modules are integrated into Product Lifecycle Management. These technologies are a reference in research and development in multiple industries, perfectly compatible with the TRL1-TRL9 technological maturity degree model used at NASA.

The company makes equipment and production lines using technologies for a traditional industrial field, the agricultural one, using innovative computer-assisted technologies, the so-called Computer Aided Technologies, which includes Computer Aided Engineering. The research starts with the father of computer-assisted technology, Dr. Grieves at NASA. This technology is the basic definition and the first example of fully virtual research, development and innovation.

The review of the scientific literature and market reports globally shows the research-development supported by computer-assisted technology in individual modules or Product Lifecycle Management Software. The use of technologies for virtual or cyber simulation of physical systems began at NASA through Dr. Grieves and incorporates several technologies that virtually manage research, development and innovation projects with degrees of technological maturity TRL type. This is the simulation technology in Industry 4.0. PLM technology is sold in the high-tech technical sector such as aeronautics, automobiles, industrial equipment, medicine - customer industries. This project uses computer-assisted technology in a traditional field that is rarely present in global sales, which is agriculture. The technology is not used as an end in itself, but competing with existing technologies through performance indicators that can be quantified and measured by this ratio. The research hypotheses are that Computer-Aided Technology such as Computer-Aided Design and the like could design and make machines capable of more efficient production processes than traditional ones. In this sense, the research is supported by specific studies in the field of agriculture and forestry that address existing technology and the ability to improve it through new technological means and adapted to customer needs.

OPERATIONS MANAGEMENT TOOLS: THE LEAN START-UP, ROADMAPS, PROGRAMS, PROJECTS, OPERATIONS

Digital transformation roadmap literature reviews (Zaoui and Souissi, 2020) show roadmaps are used for the evaluation of the digital transformation, by evaluating the dimensions and stages of digital transformation. Management consultants (Accenture, 2017, 2019; Deloitte, 2018, EY 2020a, 2020b, IDC 2015, 2020; KPMG, 2019; Mc Kinsey 2017, 2018, 2020; PWC, 2018) define roadmaps as the management instruments that fill the gap between the initial level of digital maturity and the desired level of digital maturity. Roadmaps may be represented as a chart, where the horizontal axis shows milestones, timeline and several vertical axes the dimensions of digital transformation. Deriving the roadmap from the maturity index is the core of the original digital transformation proposal and Acatech Industry 4.0 methodology. The Acatech Industry 4.0 methodology involves a gap analysis of existing Industry 4.0 capabilities to needed Industry 4.0 capabilities. This is used to derive a value based planning of the digital transformation. After the maturity analysis, the roadmap approach follows. This understanding of roadmaps is shared by scientific research (Albukhitan, 2020, Daim and Oliver, 2008, De Carolis et al, 2017, Dombrowski et al, 2015, 2018, Issa et al, 2018, Kostoff et al, 2001, Mielli and Bulanda, 2019, Mourtzis, 2020; Peter et al, 2020, Vishnevskiy et al, 2016, Project Management Institute, 2013, Roland Berger, 2018; Rohrbeck et al, 2013, Santos et al, 2017; Sarvari et al, 2018, Schumacher et al, 2019; Siedler et al, 2019; Zaoui and Souissi, 2020). Management consultants and scientific resources refer to roadmaps as instrumental to the strategic orientation of the company and the setting of strategic goals. According to Project Management Bodies of Knowledge by the Project Management Institute in 2017, roadmaps are defined as a project management tool (Project Management Institute, 2017). Roadmaps describe the overview and strategy (Project Management Institute, 2017). Roadmaps define the goals and milestones (Project Management Institute, 2017). According to the project management standard setting body, roadmaps are used to oversee programs and projects (Project Management Institute, 2017). A project roadmap is a graphical, high level overview of the project's goals (milestones), deliverables and risks presented on a timeline. Programs are defined as a group of related projects managed in a coordinated way, in order to obtain benefits and control not available from managing projects individually (Project Management Institute, 2017). Program offices are hierchically superior to projects and oversee project management. Projects are defined as a temporary endeavor undertaken to create a unique product, service, or result. Examples are product design, production planning, construction. Operations, by contrast, represent the ongoing production of goods and/or services. Roadmaps may be represented as a chart, where the horizontal axis shows milestones, timeline and several vertical axes the dimensions of digital transformation.

Furthermore, project management techniques are supported by Product Lifecycle Management Software according to practioners and scientific resources (Binder, 2007, Project Management Institute, 2017, Ruchkin and Trofimova, 2017, Stark, 2015, Stewart, 1997). Management accounting has designated methods for research and development intensive manufacturing, lifecycle costing and target costing (Ruchkin and Trofimova, 2017). Roadmaps are tied to complex cause and effect relationships and simulations. Liebrecht et al in 2018, 2020, 2021 rate roadmaps as instrumental to implementing strategy, to the simulation of implementation sequences, to system dynamics simulation. Roadmaps shape the procedure for deriving the beneficial implementation scenario. Roadmaps are preceeded by strategic assessment, monetary evaluation, procedure for deriving beneficial implementation scenarios with prioritized method order. The results are implementation scenarios with prioritized method trees. The general toolbox for Industry 4.0 is derivation of cause-effect relationships and general method characteristics. This takes the form of a decision tree with several scenarios. The result is the recommended Industry 4.0 roadmap. In the next stage, cause and effect relationships are used to derive the roadmap (Liebrecht et al, 2018, 2020, 2021).

THE USE OF COMPUTER AIDED TECHNOLOGIES FOR RESEARCH, DEVELOPMENT AND INOVATION

The review of the scientific literature and market reports globally shows the research-development supported by computer-assisted technology in individual modules or Product Lifecycle Management Software. The use of technologies for virtual or cyber simulation of physical systems began at NASA through Dr. Grieves and incorporates several technologies that virtually manage research, development and innovation projects with degrees of technological maturity TRL type. This is the simulation technology in Industry 4.0. PLM technology is sold in the high-tech technical sector such as aeronautics, automobiles, industrial equipment, medicine - customer industries. The digital twin is defined as "…virtual products are rich representations of products that are virtually indistinguishable from their physical counterparts…". The software that unifies all Computer Aided Technologies, Product Lifecycle Management software, comprises: "physical products in real space", "virtual products in virtual space" and "the connections of data and information that ties the virtual and real products together". The links between product lifecycle management and manufacturing are the object of research for digital manufacturing (Kutin, Dolgov & Sedykh, 2016). Product Lifecycle Management software is defined as the technology to design products and plan production in virtual copies of physical properties (Bergsjo, 2009; Draghici & Draghici, 2009; Fasterholdt et al, 2018; Goldbeck and Simperler, 2019; Morris, 2009; Stark, 2015; Ulrich and Eppinger, 1995). Scientific references show Product Lifecycle Management software cumulates several Computer Aided Technologies: Computer Aided Design, Computer Aided Engineering, Computer Aided Quality

Assurance and others. The integration of these technologies shapes research, development and innovation activities and business processes. They are not technologies per se, but business project and business process management solutions. Market reports from CIMdata (2016, 2022) and Statista (2021a, 2021b) show these markets have proeminent customers in the automotive, aerospace, manufacturing equipment and medical industries. Furthermore, OECD (2018) data shows the following industries have the highest research, development and innovation intensity: electronic and optical products; pharmaceuticals; air and spacecraft; total manufacturing; mining and quarrying; total services; utilities; agriculture, forestry and fishing; construction.

METHODOLOGY

This case study is descriptive and instrumental. The data therein is based on European projects A descriptive, analytical and instrumental case study is deployed on European project "Performance and excellence in the field of environment and renewable energy through modern cluster entities", SMIS number 138692, funded by the Romanian Ministry of Research, Innovation and Digitisation through the Operational Competitiveness Program (POC). The empirical data in the case study comes from this project. The case describes and analyzes the activities pertinent to innovation ecosystems of start-ups, as they have been subjected in the funding request. The focus of the case study are European project requirements, meaning the documents required by the European funding agency to approve the project. The instrumental nature of the case study comes from its utility to other European project calls.

Empirical Data Analysis

The Trl 1 To Trl9 Capability Maturity Index And Its Use In European Funds

European projects are required to follow the TRL1-TRL9 research and development capability maturity classification. This means research and development activities are to be classified by the following stages:

TRL 1 – basic principles observed

TRL 2 – technology concept formulated

TRL 3 – experimental proof of concept

TRL 4 – technology validated in lab

TRL 5 – technology validated in relevant environment (industrially relevant environment in the case of key enabling technologies)

TRL 6 – technology demonstrated in relevant environment (industrially relevant environment in the case of key enabling technologies)

TRL 7 – system prototype demonstration in operational environment

TRL 8 – system complete and qualified

TRL 9 – actual system proven in operational environment (competitive manufacturing in the case of key enabling technologies; or in space) These stages are consistent with the European Union Methodology for Classification and Market Readiness (2018).

European project authorities give grades for the identified connections between the TRL model and the project scheduling. There is no specified way to do this. Amongst European Union programs, the Operational Competitiveness Program typically involves the use of stages TRL1 to TRL9. The program has deliverables demanded after each TRL stage, and the passage to the next stage is conditioned by the attainment of these deliverables. The deliverables inherent to the capability maturity stages are inherent to each stage of capability maturity.

At stage TRL1, the activity is that theoretical/fundamental scientific research begins to migrate to R&D. At this stage, deliverables are publications presenting the basic principles of technology, or observations of physical reality (including "discovery experiments"). The ways to test the attainment of this deliverable are answering the following questions: are the working hypotheses underlying the innovative conceptual model presented and to be tested and validated in subsequent stages of concept development? and can the existence of an innovative conceptual model relevant to the sector in which it is to be developed be proven?.

At stage TRL2, the activity is the transition from TRL 1 to TRL 2 and represents the transfer of purely theoretical ideas to applied research. Publications or other documents that present and argue the conceptual model including from the perspective of the feasibility of its implementation. The tests to meet are: are experiments being designed to corroborate the fundamental results from TRL 1?; are the results obtained in TRL 1 better/completely understood?; are there practical applications for this innovative conceptual model?.

At stage TRL3, the activity is the demonstration of the functionality of the concept and validation of the individual components of the technology. The approach must demonstrate the transition from the theoretical phase ("on paper") to the experimental one, in order to verify the functionality of the concept. Numerical modeling and simulation can be used to complement the physical experiment. The deliverables are: "Report on the Presentation of Research Hypotheses and Expected Results"; "Analytical and Experimental Studies at Laboratory Scale"; "Laboratory Test Results to Determine Parameters of Interest and Compare With Theoretical Predictions for Critical Sub-Systems". The tests to meet are: analytical and experimental studies place the proposed conceptual model in an appropriate functional and laboratory context; laboratory experiments are designed with repeatability and quality control in mind to quantitatively verify that the concept works as expected; laboratory demonstrations lead to the validation of the presented working hypotheses.

At stage TRL4, the activity is the laboratory validation of components and/or assembly/system. The main components of the technology are integrated and tested in the laboratory under conditions similar to those of operation in order to establish the functionality of the assembly. The deliverables are: "Report Describing How the Demonstration Was Carried Out" (e.g. the main stages of the laboratory demonstration, the reference values considered, etc.); "Report On the Functionality of the Components as a System; Test Results for the Set of Components, Highlighting the Similarity (or Differences) in Relation to the Expected Functionality and Performance". The tests to meet are: determining whether the individual components of a technology are interoperable and can be integrated as a process; it is aimed that in the laboratory, in simulated conditions as close as possible to real environments, the components of the process are individually validated and subsequently integrated in an ad hoc manner.

At stage TRL5, the activity is the validation of the laboratory model, on a reduced or increased scale, as appropriate, with the reproduction by similarity of the real operating conditions. All technology components are assembled so that the system configuration is similar to the final application in almost all aspects. The main difference between TRL 4 and TRL 5 is the increased fidelity of laboratory tests compared to the real system, respectively with the actual operating conditions. The tested system is very close to the prototype. The deliverables are: "Laboratory Test Results", "Difference Analysis Report Between Laboratory Conditions, Analysis of the Significance of Laboratory Tests for the Operation of the Real System". The tests to meet are: it is proven that the main components of the technology are integrated with the functional ones and are tested in a simulated operational environment with realistic elements; it describes how the level of development of the concept meets market requirements under real operating conditions.

In the POC request, each stage of capability maturity comprises research, development and innovation activities and the required deliverables. There are standards on European Union level (European Union, 2022a, 2022b, 2022c). The attainment of these deliverables conditions the passage from one stage to the other. Stages TRL4 and TRL5, development activities, are indeed to be organized as projects with activities, milestones and deliverables. These are to be defined by the TRL capability model, where each TRL stage may be organized as work package in projects. The activities in the TRL model are to become the work package activities. The deliverables are to become the milestones. The attainment of one stage of capability maturity is conditioned on the deliverables. Only when the deliverables are met, the passage to the next stage is allowed. It follows as argument that the deliverables are also the milestones to each work package the TRL stages are.

The Reports That Attest the Attainment of Stages Trl 1 To Trl3

A European project has been initiated to prove the attainment of stages TRL 1 to TRL3. One of three documents proves the level of technological development TRL3: "Report About the Research Hypotheses and the Results Expected to be Attained"; "Analytical and Experimental Studies at Laboratory Scale"; "The Results of Laboratory Tests for the Determination of Interest Parameters and the Comparison to Theoretical Predictions for Critical Systems". One of the documents intended to prove the concept is presented as it was submitted to the European Union funding office.

The first part of the document shows the attainment of technology readiness model TRL1 and TRL2. The content of the document presented to the European Union is enclosed herein as follows. Specific research in the field includes the doctoral thesis of engineer Mihai Chițoiu, focused on the process of crushing biomass with the help of hammer mills, completed in 2018 at the Polytechnic University of Bucharest. The research refers to the process of pelletizing and obtaining biofuel. The doctoral thesis is completed with the selective bibliography related to the subject. Another thesis about the production process of pellets and briquettes belongs to Simina Mariș, published at the Polytechnic University of Timișoara. The market segment refers to unconventional fuels, used on the energy market in the form of pellets or briquettes which are a cheap source of heating. The doctoral thesis studies the production process of pellets and briquettes, using artificial intelligence and statistics. The thesis is supported by a production line at Cena, made for multiple academic research. Simina Maris's thesis is supported by other research carried out on the Cena production line, respectively: the present project refers to a granulation line as a workstation for the pellet and briquette production line from Cena. The project is thus integrated into research on the production of pellets and briquettes through new and advanced technologies, such as computer-assisted technologies and artificial intelligence.

The following sections of the document deal with the attainment of technology maturity level TRL3. The granular technological complex is at the research stage TRL3, which aims to prove the concept (provide proof of concept) which refers to an innovative process. The concept is the architecture of solutions for production equipment and will shape future production processes. The research included several stages, which start from the existing technology and lead to an innovative process. The technical drawings of both research stages, made through the Autocad program, represent the transition from the initial research stage to the final research stage. Virtual simulation programs of product design and production processes can be called Computer Aided Engineering as a whole and reference computer systems in research and development worldwide. Autocad is a Computer Aided Engineering program.

The production process cumulates production activities, which use production resources: production equipment and labor. The production process has a certain level of energy consumption and emits a certain level of carbon dioxide. Through the production process, raw materials and materials are converted into finished products and sold products. The object of the present research and innovation is the improvement of the existing granulation technology at the beginning through investments in new equipment that lead to the final stage of the production process. The solution architecture at the beginning and at the end of the research process is presented therein. At the starting point, the innovative process refers to a production line.

The research hypotheses are the following: solution architecture for pellet and briquette production concept is feasible; the use of innovative machines will lead to quantifiable process innovations, such as specific improvements in the performance of production machines and improvements that concern the production processes as a whole. The improvements are as follows: processing three bales instead of one; the reduction of energy consumption by 50%; the elimination of noxes. The results of the research will confirm the realization of the hypotheses. Autocad research leads to improvements in the original production process by replacing production equipment with new ones, digitally simulated by Autocad. The original document has enclosed the solution architecture before and after the Autocad based improvements, based on technical drawings and quantifiable key performance indicators. The drawings prove the concept exists.

Incremental innovation of the production process leads to improved operational performance indicators. The first such indicator is the quantity produced, which migrates from processing one bale to processing three bales. By processing triple the amount of raw material, the amount produced is also tripled. Energy consumption is reduced by 50%, reaching 30 kw per hour. Noxious emissions are reduced to zero.

The Project that Schedules Activities to Capability Maturity Levels Trl 4 and Trl5

European projects require proof of concept, stages TRL1 to TRL3, to have funding approved for the subsequent stages. Such documentation demands the attainment of stages TRL1 to TRL3. Some European funds cover stages TRL4 to TRL5, requiring the concept to be tested and validated in laboratory conditions and in relevant environment.

The European Union allocates public funds via several types of programs that may be available on EU basis or on national basis. One of these project calls is POC/62/1/3 Stimulating the enterprise demand for innovation via research, development and innovation projects or in partnership with research and development institutes and universities, to the goal of product or process innovation in the economic sectors with high growth potential. A project has been awarded, innovation and economic and functional optimization in the energy production for thermal energy materials. The project has been financed using the net present value of discounted cash flow and internal rate or return. The European Union has offered a file to compute these indicators, based on operational, investment and financial cash flow.

The project is a construction which will act as a start-up incubator in the field of renewable energy. The construction will host office space, conference rooms, other type of rooms designed to promote renewable energy and its vendors. Whereas the costs refer to the building itself and have several sources of financing, the income comes from using this space for business activities in several scenarios.

These activities should achieve the building in a timeframe of three years. The following years offset income from rent, royalties and similar with expenses needed to operate the start-up incubator which the building will be. The future cluster, now in construction, will support several activities, amongst which the following: activities related to obtaining, validating and protecting patents and other intangible assets or activities of realization of expenses related to obtaining, validating and protecting patents and other intangible assets belonging to the cluster; consulting activities in the field of innovation - activities for making expenses for the purchase of consulting services in the field of innovation; the innovation support activity – the activity of making expenses for support services an innovation.

One of the activities via which the green energy cluster supports start-ups is the consulting activity. This activity contains several subactivities. One is the sub-activity of purchases of general consulting services in the field of innovation - renewable energies. It involves the purchasing of consulting services in the field of innovation. Consulting services will be purchased from legal entities that have already implemented innovation in the field of renewable energies. The result is the general consultancy insurance contract in the field of innovation. Consulting may be provided for the validation of technologies in laboratory and simulated conditions, stages TRL4 and TRL5. Thereby universities support proof of concept, stages TRL1 to TRL3 and technology validation, stages TRL4 to TRL5. European projects may fund the attainment of stages TRL4 and TRL5 for start-ups. They require consulting services from universities and research institutes.

Acknowledgment: This research was funded through the grant" Performance and excellence in the field of environment and renewable energy through modern cluster entities", SMIS number 138692, funded by the Romanian Ministry of Research, Innovation and Digitisation through the Operational Competitiveness Program (POC).

CONCLUSION

Scholarly literature review presented earlier on in this article shows capability maturity indices are a management tool that shows the development of a specific ability or reaching to a targeted success from an initial to an anticipated stage. Therein, definitions show maturity levels are given by progressive capabilities that describe how the behaviors, practices and processes of an organization can reliably and sustainably produce required outcomes. Maturity models enable organizations to audit and benchmark regarding to assessment results; to track process towards a desired level; to evaluate strengths, weaknesses, threats, opportunities; to sequence stages from basic to advanced. The TRL 1-9 capability maturity model oversees research, development and innovation activities. The activities are to have a linear sequence, where each stage of capability maturity is required to pass to the next. Stages TRL1 to TRL9 comprise activities and their deliverables in an ideal state which organizations that implement research, development and innovation must reach. The adepts of this standard include Nasa and the European Union. The former created the model. The latter uses it for all project funding calls. The TRL 1-9 requirements meet the definition of capability maturity models in that they have capability maturity levels for the organizational dimension research, development and innovation activities. This shows a classic understanding of capability maturity levels that is staple in theory and practice.

An individual section in the literature review shows maturity models are used to determine the current maturity model and generate a roadmap to move to the next maturity model. Maturity models are used to generate roadmaps and to schedule programs, projects and their activities. Maturity indices may be graphically represented as radar charts, where the X axis represents digital transformation capability maturity levels and the Y axis the organizational dimensions impacted by digital transformation. Roadmaps may also be conceived in chart form; in this case, the horizontal axis shows milestones, timeline and several vertical axes show the dimensions of digital transformation. Deriving roadmaps from maturity indices involves switching the radar chart to a line chart. The X axis in the radar chart becomes the X axis in the line chart, which comprises capability maturity levels which become milestones and are assigned timeline. In the maturity index radar chart, the Y axis comprises several organizational dimensions which become several vertical axes in the roadmap chart. These will take different values at each stage of capability maturity on the X axis. Empirical data evidence shows the TRL 1 to TRL 9 capability maturity model may be used the same way. Capability maturity levels become milestones in projects and programs. The TRL 1-9 capability maturity model has one dimension: activities, basically research, development and innovation activities. These activities are matched with the capability maturity model as each level of capability maturity is assigned inherent activities. The functionality of capability maturity models is thereby used to schedule projects and programs.

In the TRL1-TRL9 capability maturity model, proof of concept argues the transition to stages TRL4 and TRL5, which correspond to the development stage and are managed via projects. This corresponds to a Proof of Concept funding at the European Union, where the existence of the concept may be argued by technical drawings for the future prototype and product. In literature review, Computer Aided Technologies and Product Lifecycle Management Software are argued to create digital twins or virtual copies of the physical devices and the processes they are capable of. They are the business information systems used to shape business projects and business processes in theory and practice alike. These software systems manage product design and production planning. Empirical data analysis shows Autocad is such software and provides proof the product concept exist in virtual form. This will be later on built physically and tested in laboratory and simulated conditions. Digital simulation technologies are thereby included in managing research, development and innovation activities via the capability maturity index.

REFERENCES

Acatech. (2017). *Industrie 4.0 Maturity Index*. Managing the Digital Transformation of Companies. Retrieved April 12th, 2020, from https://www.acatech.de/wp-content/uploads/2018/03/acatech_STUDIE_Maturity_Index_eng_WEB.pdf

Accenture. (2017). *Digital Transformation of Industries. Demystifying Digital and Securing $100 Trillion for Society and Industry by 2025*. Retrieved on June 30th, 2021, from: https://www.accenture.com/t00010101t000000z__w__/ru-ru/_acnmedia/accenture/conversion-assets/dotcom/documents/local/ru-ru/pdf/accenture-digital-transformation.pdf

Accenture. (2019). *An AI roadmap to maximize the value of AI*. Retrieved on June 30th, 2021, from: https://www.accenture.com/us-en/insights/artificial-intelligence/ai-roadmap

Accenture. (2019). *Rethink, reinvent, realize*. Retrieved April 12th, 2020, from https://www.accenture.com/_acnmedia/thought-leadership-assets/pdf/accenture-ixo-industry-insights-hightech.pdf

Accenture. (2020). *Industry X.0*. Retrieved April 12th, 2020, from https://www.accenture.com/us-en/services/industryx0-index

Akdil, K. Y., Ustundag, A., & Cevikcan, E. (2018). *Maturity and Readiness Model for Industry 4.0 Strategy. In Industry 4.0: Managing the Digital Transformation*. Springer.

Albukhitan, S. (2020). Developing Digital Transformation Strategy for Manufacturing. *Procedia Computer Science*, *170*, 672–679. doi:10.1016/j.procs.2020.03.173

Alcayaga, A., Wiener, M., & Hansen, E. G. (2019). Towards a Framework of Smart Circular Systems: An Integrative Literature Review. *Journal of Cleaner Production*, *221*, 622–634. doi:10.1016/j.jclepro.2019.02.085

Anderl, R., Picard, A., Wang, Y., Fleischer, J., Dosch, S., Klee, B., & Bauer, J. (2015). *Guideline Industrie 4.0 – Guiding Principles for the Implementation of Industrie 4.0 in Small and Medium Sized Businesses*. In VDMA Forum Industrie, Frankfurt, Germany.

Backlund, F., Chroneer, D., & Sundqvist, E. (2014). Project management maturity models – A critical review: A case study wish Swedish engineering and construction organizations. *Procedia: Social and Behavioral Sciences*, *119*, 837–846.

Barata, J., & da Cunha. (2017). Climbing the Maturity Ladder in Industry 4.0: A Framework for Diagnosis and Action that Combines National and Sectorial Strategies. *Twenty-third Americas Conference on Information Systems*, Boston, MA.

BCG. (2015). *Industry 4.0: The Future of Productivity and Growth in Manufacturing Industries*. Retrieved April 14th, 2020, from https://www.bcgperspectives.com/content/articles/engineered_products_project_business_industry_40_future_productivity_growth_manufacturing_industries/?chapter=2

BCG. (2020a). *Digital Acceleration Index*. Retrieved April 14th, 2020, from https://www.bcg.com/capabilities/technology-digital/digital-acceleration-index.aspx

BCG. (2020b). *Digital Transformation*. Retrieved April 10th, 2020, from https://www.bcg.com/digital-bcg/digital-transformation/overview.aspx

Benedict, N., Smithburger, P., Donihi, A. C., Empey, P., Kobulinsky, L., Seybert, A., Waters, T., Drab, S., Lutz, J., Farkas, D., & Meyer, S. (2017). Blended Simulation Progress Testing for Assessment of Practice Readiness. *American Journal of Pharmaceutical Education*, *81*(1), 14.

Benedict, N., Smithburger, P., Donihi, A. C., Empey, P., Kobulinsky, L., Seybert, A., Waters, T., Drab, S., Lutz, J., Farkas, D., & Meyer, S. (2017). Blended Simulation Progress Testing for Assessment of Practice Readiness. *American Journal of Pharmaceutical Education, 81*(1), 14. doi:10.5696/ajpe81114

Bergsjo, D. (2009). *Product Lifecycle Management: Architectural and Organizational Perspectives* [Doctoral Dissertation]. Chalmers University of Technology, Goteborg, Sweden.

Bertolini, M., Esposito, G., Neroni, M., & Romagnoli, G. (2019). Maturity models in industrial internet: A review. *Procedia Manufacturing, 39*. Advance online publication. doi:10.1016/j.promfg.2020.01.253

Bertolini, M., Esposito, G., Neroni, M., & Romagnoli, G. (2019). Maturity models in industrial internet: A review. *Procedia Manufacturing, 39*, 1854–1863. doi:10.1016/j.promfg.2020.01.253

Bibby, L., & Dehe, B. (2018). *Defining and Assessing Industry 4.0 Maturity levels - Case of the Defence sector.* Production Planning and Control. Retrieved April 14th, 2020, from https://pure.hud.ac.uk/ws/files/14152037/0_final_accepted_manuscript_plain_text_with_T_F_pure.pdf

Binder, J. (2007). *Global Project Management*, Communication, Collaboration and Management Across Borders. Routledge Taylor and Francis Group.

CapGemeni. (2011). *Digital transformation: a roadmap for billion-dollar organizations.* Retrieved April 12th, 2020, from https://www.capgemini.com/wp-content/uploads/2017/07/Digital_Transformation__A_Road-Map_for_Billion-Dollar_Organizations.pdf

CapGemeni. (2014). *Digital transformation review. Crafting a Compelling Customer Experience.* Retrieved April 12th, 2020, from https://www.capgemini.com/wp-content/uploads/2017/07/digital-transformation-review-6_3.pdf

CapGemeni. (2017). *The Digital Advantage: How digital leaders outperform their peers in every industry.* Retrieved April 15th, 2020, from https://www.capgemini.com/wp-content/uploads/2017/07/The_Digital_Advantage__How_Digital_Leaders_Outperform_their_Peers_in_Every_Industry.pdf

CapGemeni. (2018). *Industry 4.0 Maturity Model – Mirroring today to sprint into the future.* Retrieved April 15th, 2020, from https://www.capgemini.com/fi-en/2018/09/industry-4-0-maturity-model-mirroring-today-to-sprint-into-the-future/

CIMdata. (2016). *Free Resources.* Retrieved June 2022, from: https://www.cimdata.com/en/resources/about-plm

CIMdata. (2022). *CIMdata Publishes PLM Market and Solution Provider Report.* Retrieved June 2022, from: https://www.cimdata.com/en/news/item/6459-cimdata-publishes-plm-market-and-solution-provider-report

CMMI Product Team. (2007). *CMMI for Acquisition.* Technical Report. Retrieved June 2022, from: ftp://ftp.sei.cmu.edu/pub/documents/07.reports/07tr017.pdf

Colli, M., Bergerb, U., Bockholta, M., Madsena, O., Møllera, C., & Vejrum Wæhrens, B. (2019). A maturity assessment approach for conceiving context-specific roadmaps in the Industry 4.0 era. *Annual Reviews in Control, 48,* 165–177. doi:10.1016/j.arcontrol.2019.06.001

Cooper, R. G. (2008). Perspective: The Stage-Gate®Idea-to-Launch Process-Update, What's New, and Nex Gen Systems. *Journal of Product Innovation Management, 25,* 213–232.

Daum, J. H. (2003). *Intangible Assets and Value Creation.* Wiley.

De Carolis, A., Macchi, M., Negri, E., & Terzi, S. (2017), Guiding manufacturing companies towards digitalization a methodology for supporting manufacturing companies in defining their digitalization roadmap. *IEEExplore International Conference on Engineering, Technology and Innovation.* https://doi.org/10.1145/3477911.3477924

De Carolis, A., Macchi, M., Negri, E., & Terzi, S. (2017). A maturity model for assessing the digital readiness of manufacturing companies. In *IFIP International Conference on Advances in Production Management Systems.* Springer.

Deloitte. (2018a). *Digital Maturity Model. Achieving digital maturity to drive growth.* Retrieved April 12th, 2020, from https://www2.deloitte.com/content/dam/Deloitte/global/Documents/Technology-Media-Telecommunications/deloitte-digital-mat urity-model.pdf

Deloitte. (2018b). *Digital Transformation with new SAP Technologies.* Retrieved April 13th, 2020, from: https://www2.deloitte.com/content/dam/Deloitte/lu/Documents/technology/lu-digital-transformation-sap.pdf

Deloitte. (2019). *Pivoting to digital maturity. Seven capabilities central to digital transformation.* Retrieved April 12th, 2020, from https://www2.deloitte.com/us/en/insights/focus/digital-maturity/digital-maturity-pivot-model.html

Deloitte. (2020). *Uncovering the connection between digital maturity and financial performance. How digital transformation can lead to sustainable high performance.* Retrieved April 12th, 2020, from https://www2.deloitte.com/us/en/insights/topics/digital-transformation/digital-transformation-survey.html

Dombrowski, U., Krenkel, P., Falkner, A., Placzek, F., & Hoffmann, T. (2018). Potenzialanalyse von Industrie 4.0-Technologien: Zielorientiertes Auswahlverfahren. ZWF - Zeitschrift für wirtschaftlichen Fabrikbetrieb, 107–111.

Dombrowski, U., Richter, T., & Ebentreich, D. (2015). *Auf dem Weg in die vierte industrielle Revolution: Ganzheitliche Produktionssysteme zur Gestaltung der Industrie-4.0.* Architektur. zfo Zeitschrift Führung +Organisation.

Draghici, G., & Draghici, A. (2009). Collaborative Multisite PLM Platform. *CENTERIS 2009, Conference on Enterprise Information Systems, Conference: CENTERIS 2009, Conference on Enterprise Information Systems.* https://www.researchgate.net/publication/235673601_Collaborative_Multisite_PLM_Platform

Duffy, J. (2001). Maturity models: Blueprints for evolution. *Strategy and Leadership*, *29*(6), 19–26.

Essig, M., Glas, A. H., Selviaridis, K., & Roehrich, J. K. (2016). Performance-based contracting in business markets. *Industrial Marketing Management*, *59*, 5–11. doi:10.1016/j.indmarman.2016.10.007

European Commission. (2018). *EU businesses go digital: Opportunities, outcomes and uptake*. Retrieved on August 12th, 2020, from https://ec.europa.eu/growth/tools-databases/dem/monitor/site s/default/files/Digital%20Transformation%20Scoreboard%202018 _0.pdf

European Commission. (2020a). *Parteneriat în exploatarea Tehnologiilor Generice Esenţiale (TGE), utilizând o PLATformă de interacţiune cu întreprinderile competitive TGE-PLAT*. Accesat în iunie 2022: https://www.imt.ro/TGE-PLAT/e-news/Concept%20adus%20la%20TRL %207-8.pdf

European Commission. (2020b). *The European Commission's science and knowledge service*. Joint Research Centre, Accesat în iunie 2022: https://s3platform-legacy.jrc.ec.europa.eu/documents/20182/4 43954/M.+Ranga+-+Day+3+TRLs+and+tech+transfer.pdf/f9214fc2-7 514-4b67-a18f-962d45cdd5ab

European Commission. (2022a). *Documents download module*. Retrieved April 13th, 2020, from: https://ec.europa.eu/research/participants/documents/downloa dPublic?documentIds=080166e5cce7b21b&appId=PPGMS

European Union. (2022a). *H20201 Model for ERC Proof of Concept Grants 2 (H2020 ERC PoC — Multi)*. Retrieved April 13th, 2020, from: file:///D:/Desktop/h2020-mga-erc-poc-multi_v2.0_en.pdf

European Union. (2022b). *Access 2EIC National Contact Points for Innovation*. Retrieved April 13th, 2020, from: https://access2eic.eu/wp-content/uploads/2020/04/Access2EIC_ How-to-apply-successfully_template.pdf

EY. (2018). *Digital Readiness Assessment. Does your business strategy work in a digital world?* Retrieved April 14th, 2020, from https://digitalreadiness.ey.com/

EY. (2020a). *Digital Strategy and Transformation*. Retrieved on June 30th, 2021, from: https://www. ey.com/en_gl/digital/transformation

EY. (2020b). *Strategic Roadmap*. Retrieved on June 30th, 2021, from: https://www.ey.com/en_se/advi sory/strategic-roadmap

Fasterholdt, I., Lee, A., Kidholm, K., Yderstræde, K., Møller, B., & Pedersen, K. (2018). A qualitative exploration of early assessment of innovative medical technologies. *BMC Health Services Research*, *18*(1). Advance online publication. doi:10.118612913-018-3647-z

Global Center for Digital Business Transformation. (2015). *Digital Vortex. How Digital Disruption Is Redefining Industries*. Retrieved April 13th, 2020, from https://www.cisco.com/c/dam/en/us/solutions/collateral/indus try-solutions/digital-vortex-report.pdf

GoldbeckG.SimperlerA. (2019). *Business Models and Sustainability for Materials Modelling Software, Projects*. European Materials Modelling Council, EMMC - European Materials Modelling Council. Doi:10.5281/zenodo.2541722

Grieves, M. (2006). *Product Lifecycle Management. Driving the Next Generation of Lean Thinking*. Mc Graw Hill.

Groza, I. V., & Balint, R. (2022). *Realizarea și planificarea modelului științific pentru produsele noi. Proiect: INOvări și optimizări economice și funcționale în producția industrială de MATeriale pentru energie termică, „INO-MAT", cod SMIS 119412. Realizarea și planificarea modelului științific pentru produsele noi*. Titus Industries SRL.

Hou, J., & Neely, A. (2018). Investigating risks of outcome-based service contracts from a provider's perspective. *International Journal of Production Research, 56*(6), 2103–2115. doi:10.1080/00207543 .2017.1319089

Hou, J., & Neely, A. (2018). Investigating risks of outcome-based service contracts from a provider's perspective. *International Journal of Production Research, 56*(6), 2103–2115. doi:10.1080/00207543 .2017.1319089

IBM. (2014). *Big Data & Analytics Maturity Model*. Retrieved April 13th, 2020, from https://www.ibmbigdatahub.com/blog/big-data-analytics-maturi ty-model

IBM. (2019). *Industry 4.0 and Cognitive Manufacturing. Architecture Patterns, Use Cases and IBM Solutions*. Retrieved April 13th, 2020, from https://www.ibm.com/downloads/cas/M8J5BA6R

IDC. (2015). *A Digital Transformation Maturity Model and Your Digital Roadmap*. Retrieved April 13th, 2020, from https://pdf4pro.com/amp/download?data_id=3c6d16&slug=a-digit al-transformation-maturity-model-and-your-digital

IDC. (2020). *Worldwide Digital Transformation Strategies*. Retrieved April 13th, 2020, from https://www.idc.com/getdoc.jsp?containerId=IDC_P32570

IEEE. (2019). *Platforms, Present and Future XIV*. Retrieved April 13th, 2020, from https://cmte.ieee.org/futuredirections/2019/11/13/platforms-present-and-future-xiv/

IMD. (2020). *The digital business agility perspective*. Retrieved April 13th, 2020, from https://www.imd.org/contentassets/929fc49598cc47ba888277438b d963b6/tc044-16---digital-business-agility-pdf.pdf

Issa, A., Hatiboglu, B., Bilstein, A., & Buernhansl, T. (2018). Industrie 4.0 roadmap: Framework for digital transformation based on the concepts of capability maturity and alignment. *Procedia CIRP, 72*. Doi:10.1016/j.procir.2018.03.151

Johansson, C. (2009). *Knowledge Maturity as Decision Support in Stage-Gate Product Development: A Case From the Aerospace Industry* [Doctoral Thesis]. Department of Applied Physics and Mechanical Engineering, Luleå University of Technology.

Johansson, P., Christian, J., Ola, I., Ola, T., & Larsson, T. (2020). *Take the knowledge path to support knowledge management in product/service systems.* Academic Press.

Jorgensen, M., Mohagheghi, P., & Grimstad, S. (2017). Direct and indirect type of connection between type of contract and software project outcome. *International Journal of Project Management, 35,* 1573–1586.

Kamal, M. M., Sivarajah, U., Bigdeli, A. Z., Missi, F., & Koliousis, Y. (2020). Servitization implementation in the manufacturing organisations: Classification of strategies, definitions, benefits and challenges. *International Journal of Information Management, 50.* Advance online publication. doi:10.1016/j.ijinfomgt.2020.102206

Keith, B., Vitasek, K., Manrodt, K., & Kling, J. (2016). *Strategic Sourcing in the New Economy: Harnessing the Potential of Sourcing Business Models for Modern Procurement.* Springer.

Knowledge Exchange and Fraunhofer. (2017). *Industry 4.0 maturity assessment.* Retrieved April 15th, 2020, from https://www.researchgate.net/profile/Alexander_Kermer-Meyer/publication/317720108_Industry_40_Maturity_Assessment/links/594a68b8aca2723195de5ed1/Industry-40-Maturity-Assessment.pdf

Kostoff, R. N., & Schaller, R. R. (2001). Science and technology roadmaps. *IEEE Transactions on Engineering Management, 48*(2), 132–143. doi:10.1109/17.922473

Kowalkowski, C., & Kindström, D. (2014). Service innovation in product-centric firms: A multidimensional business model perspective. *Journal of Business and Industrial Marketing, 29*(2), 96–111. doi:10.1108/JBIM-08-2013-0165

KPMG. (2016). *Digital Readiness Assessment.* Retrieved April 13th, 2020, from https://www.future.consulting/en/foresight/trend-analyses/analyses/article/new-digital-readiness-assessment-by-2b-ahead-and-kpmg/

KPMG. (2019). *Strategic roadmap.* Retrieved April 13th, 2020, from: https://home.kpmg/xx/en/home/insights/2019/04/strategic-road-map.html

Kutin, A., Dolgov, V., & Sedykh, M. (2016). Information links between product life cycles and production system management in designing of digital manufacturing. *Procedia CIRP, 41,* 423–426. doi:10.1016/j.procir.2015.12.126

Leyh, C., Bley, K., Schäffer, T., & Forstenhäusler, S. (2016). SIMMI 4.0 - a maturity model for classifying the enterprise-wide it and software landscape focusing on Industry 4.0. In *2016 Federated Conference on Computer Science and Information Systems (FedCSIS).* IEEE.

Li, D., & Mishra, N. (2020). Engaging Suppliers for Reliability Improvement under Outcome based Compensations. *Omega.* Advance online publication. doi:10.1016/j.omega.2020.102343

Liebrecht, C. (2020). *Entscheidungsunterstutzung fur den Industrie 4.0-Methodeneinsatz: Strukturierung, Bewertung and Ablietung von Implementierungsreihenfolgen. Zugl: Karlsruhe, Diss, 2020.* Shaker Verlag G.

Liebrecht, C., Kandler, M., Lang, M., Schaumann, S., Stricker, N., Wuest, T., & Lanz, G. (2021). Decision support for the implementation of Industry 4.0 methods: Toolbox, Assessment and Implementation Sequences for Industry 4.0. *Journal of Manufacturing Systems, 58*(3), 412–430. doi:10.1016/j.jmsy.2020.12.008

Liebrecht, C., Schaumann, S., Zeranski, D., Antonszkiewicz, A., & Lanza, G. (2018). Analysis of Interactions and Support of Decision Making for the Implementation of Manufacturing Systems 4.0 Methods. *10th CIRP Conference on Industrial Product-Service Systems, Procedia CIRP*, 161-166. DOI: 10.1016/j.procir.2018.04.005

Liinamaa, J., Viljanen, M., Hurmerinta, A., Ivanova-Gonge, M., Luotola, H., & Gustafsson, M. (2016). Performance-Based and Functional Contracting in Value Based Solution Selling. *Industrial Marketing Management, 59*, 37–49. doi:10.1016/j.indmarman.2016.05.032

Mankins, J. C. (1995). *Technology Readiness Levels*. NASA Advanced Concepts Office.

McKinsey. (2016). Industry 4.0 at McKinsey's model factories. Retrieved April 14th, 2020, from-http://sf-eu.net/wp-content/uploads/2016/08/mckinsey-2016-industry-4.0-at-mckinseys-model-factories-en.pdf

McKinsey. (2017). *A roadmap for a digital transformation*. Retrieved on June 30th, 2021, from: https://www.mckinsey.com/industries/financial-services/our-insights/a-roadmap-for-a-digital-transformation

McKinsey. (2018). *Unlocking success in digital transformations*. Retrieved April 13th, 2020, from https://www.mckinsey.com/business-functions/organization/our-insights/unlocking-success-in-digital-transformations

McKinsey. (2019). *Hannover Messe 2019: The 3 P's of Industry 4.0*. Retrieved April 13th, 2020, from https://www.mckinsey.com/business-functions/operations/our-insights/operations-blog/hannover-messe-2019-the-3-ps-of-industry-40

McKinsey. (2020). *Digital 20/20*. Retrieved on June 30th, 2021, from: https://www.mckinsey.com/business-functions/mckinsey-digital/how-we-help-clients/digital-2020/overview

Methodology for Classification and Market Readiness. (2018). *Methodology for the Classification of Projects/Services and Market Readiness*. https://cyberwatching.eu/sites/default/files/D2.3%20Methodology%20for%20the%20classification%20of%20projects%20and%20market%20readiness_0.pdf

Mettler, T. (2009). *A Design Science Research Perspective on Maturity Models in Information Systems*. Working Paper. Institute of Information Management, University of St. Gallen.

Mettler, T. (2011). Maturity Assessment Models: A Design Science Research Approach. *International Journal of Society Systems Science, 3*(1-2).

Mielli, F., & Bulanda, N. (2019). Digital Transformation: Why Projects Fail, Potential Best Practices and Successful Initiatives. *2019 IEEE-IAS/PCA Cement Industry Conference (IAS/PCA),* 1-6. doi:10.1109/CITCON.2019.8729105

Mittal, S., Romero, D., & Wuest, T. (2018). Towards a Smart Manufacturing Model for SMEs. AICT, 536(2), 155-163.

Morris, R. (2009). *The fundamentals of product design.* AVA Publishing.

Mourtzis, D. (2020). Simulation in the design and operation of manufacturing systems: State of the art and new trends. *Internaional Journal of Production Resources, 58,* 1927–1949. doi:10.1080/00207543.2019.1636321

NASA. (2022). *Appendix G: Technology Assessment/Insertion.* Accesat în iunie 2022: https://www.nasa.gov/seh/appendix-g-technology-assessmentinsertion

Nayab, N. (2010). *The Difference Between CMMI vs CMM. Bright Hub PM.* Retrieved April 13th, 2020, from https://www.brighthubpm.com/certification/69744-cmmi-vs-cmm-which-is-better/

Nickkhou, S., Taghizadeh, K., & Hajiyakhali, S. (2016). Designing a portfolio management maturity model. *Procedia: Social and Behavioral Sciences, 226,* 318–325.

OECD. (2018a). *Oslo Manual 2018. Guidelines for Collecting, Reporting and Using Data on Innovation.* Accesat în iunie 2022, de la https://www.oecd-ilibrary.org/docserver/9789264304604-en.pdf?expires=1656233036&id=id&accname=guest&checksum=4CCC30AA4DCCE4B7811A0F98517C9292

OECD. (2018b). *R&D intensity by industry: Business enterprise expenditure on R&D as a share of gross valued added, 2018 (or nearest year).* https://www.oecd-ilibrary.org/social-issues-migration-health/r-d-intensity-by-industry-business-enterprise-expenditure-on-r-d-as-a-share-of-gross-valued-added-2018-or-nearest-year_f9722f58-en

OECD. (2020). *Research and development statistics.* Retrieved April 13th, 2020, from: https://www.oecd.org/sti/inno/researchanddevelopmentstatisticsrds.html

OECD (2022), Gross domestic spending on R&D (indicator). doi:10.1787/d8b068b4-en

Paulk, M. C. (1993). Capability Maturity Model SM for Software, Version 1.1. Software Engineering Institute, Carnegie Mellon University.

Peter, M. K., Kraft, C., & Lindeque, J. (2020). Strategic action fields of digital transformation: An exploration of the strategic action fields of Swiss SMEs and large enterprises. *Journal of Strategy and Management, 13*(1), 160–180. doi:10.1108/JSMA-05-2019-0070

Pezzotta, G., Sassanelli, C., Pirola, F., Sala, R., Rossi, M., Fotia, S., Koutoupes, A., Terzi, S., & Mourtzis, D. (2018). The Product Service System Lean Design Methodology (PSSLDM): Integrating product and service components along the whole PSS lifecycle. *Journal of Manufacturing Technology Management*, 29(8), 1270–1295. doi:10.1108/JMTM-06-2017-0132

Proença, D., & Borbinha, J. (2016). Maturity models for information systems - A state of the art. *Procedia Computer Science*, *100*, 1042–1049. doi:10.1016/j.procs.2016.09.279

Project Management Institute. (2017). PMBOK® Guide – Sixth Edition. Author.

PWC. (2015). *2015 Global Digital IQ® Survey. Lessons from digital leaders 10 attributes driving stronger performance.* Retrieved April 13th, 2020, from https://www.pwc.com/gx/en/advisory-services/digital-iq-survey-2015/campaign-site/digital-iq-survey-2015.pdf

PWC. (2016). *Industry 4.0: Building the digital enterprise*. Retrieved April 13th, 2020, from https://www.pwc.com/gx/en/industries/industries-4.0/landing-page/industry-4.0-building-your-digital-enterprise-april-201 6.pdf

PWC. (2018). *Delivering digital change*. Retrieved April 13th, 2020, from: https://pwc.blogs.com/fsrr/2018/11/delivering-digital-change.html

PWC. (2020). *Accelerate digital transformation*. Retrieved April 13th, 2020, from https://www.pwc.co.uk/services/consulting/accelerate-digital.html

Raddats, C., Kowalkowski, C., Benedettini, O., Burton, J., & Gebauer, H. (2019). Servitization: A Contemporary Thematic Review of Major Research Streams. *Industrial Marketing Management*, *2019*, 207–223. doi:10.1016/j.indmarman.2019.03.015

Rohrbeck, R., & Schwarz, J. O. (2013). *The value contribution of strategic foresight: insights from an empirical study of large European companies*. Retrieved on June 30th, 2021, from: https://www.researchgate.net/publication/236977761_The_Value _Contribution_of_Strategic_Foresight_Insights_From_an_Empiri cal_Study_of_Large_European_Companies

Roland Berger. (2015). *The digital transformation of industry*. Retrieved April 13th, 2020, from www.rolandberger.com › publications › publication_pdf

Roland Berger. (2018). *Industrie 4.0? Step this way!* Retrieved April 13th, 2020, from www.rolandberger_coo_insights_e

Ruchkin, A., V., & Trofimova, O., M. (2017), Project Management: Basic Definitions and Approaches. *Management Issues / Voprosy Upravleniâ, 46*, 1-10.

Santos, C., Mehrsai, A., Barrosaa, C., Araújob, M., & Ares, E. (2017). Towards Industry 4.0: An overview of European strategic roadmaps. *Procedia Manufacturing*, *13*, 972–979. doi:10.1016/j.promfg.2017.09.093

Sarvari, P. A., Unstundag, A., Cevikcan, E., & Kaya, I. (2018). Technology Roadmap for Industry 4.0. In Industry 4.0: Managing The Digital Transformation (pp. 95-103). Springer.

Schumacher, A., Erol, S., & Sihn, W. (2016). A maturity model for assessing Industry 4.0 readiness and maturity of manufacturing enterprises. *Procedia CIRP - Changeable, Agile. Reconfigurable & Virtual Production*, *52*(1), 161–166.

Schumacher, A., Nemeth, T., & Sihn, W. (2019). Roadmapping towards Industrial Digitalization based on an Industry 4.0 Maturity Model for Manufacturing Enterprises. *Procedia CIRP*, *79*, 2019. doi:10.1016/j.procir.2019.02.110

Sener, U., Gokalp, E., & Eren, P., E. (2018). Towards a Maturity Model for Industry 4.0: a Systematic Literature Review and a Model Proposal. *Industry 4.0 from the MIS*, 221

Siedler, C., Sadaune, S., Zavareh, M. T., Eigner, M., Zink, K. J., & Aurich, J. C. (2019). Categorizing and selecting digitization technologies for their implementation within different product lifecycle phases. *12th CIRP Conference on Intelligent Computation in Manufacturing Engineering*, 274–279. DOI: 10.1016/j.procir.2019.02.066

Sîrbu, C., Groza, I. V., Mnerie, D., & Mnerie, G. V. (2022). *Proiectarea și optimizarea arhitecturii instalației pilot de producere peleți-bricheți. Proiect: Inovări și optimizări economice și funcționale în producția industrială de Materiale pentru energie termică, „INO-MAT", cod SMIS 119412*. Titus Industries SRL.

Stark, J. (2015). *Product Lifecycle Management: The Devil Is in the Details* (Vol. 2). Springer International Publishing.

Statista. (2021). *Product lifecycle management (PLM) & engineering software market revenues worldwide from 2019 to 2025(in million U.S. dollars)*. Accesat în iunie 2022, de la https://www.statista.com/statistics/796151/worldwide-product-lifecycle-management-engineering-software-market/

Statista. (2021). *Size of computer-aided engineering (CAE) market worldwide, from 2015 to 2021*. Accesat în iunie 2022, de la https://www.statista.com/statistics/732384/worldwide-computer-aided-engineering-market-revenues/

Stewart, G. (1997). Supply-chain operations reference model (SCOR): The first cross-industry framework for integrated supply-chain management. *Emerald*, *10*(2), 62–67.

Storbacka, K. (2011). A solution business model: Capabilities and management practices for integrated solutions. *Industrial Marketing Management*, *40*(5), 707–711. doi:10.1016/j.indmarman.2011.05.003

Șuta, A., I., Mariș, Ș., A., Gomoi, V., S., Molnar, M., & C., F. S. (2022). *Conceperea și depunerea documentației pentru obținerea unui brevet. Proiect: Inovări și optimizări economice și funcționale în producția industrială de Materiale pentru energie termică, „INO-MAT", cod SMIS 119412*. Titus Industries SRL.

Şută, A., Mnerie, A. V., & Holotescu, C. (2022). *Evaluarea industrială și testarea indicatorilor tehnici a produselor obținute. Proiect: Inovări și optimizări economice și funcționale în producția industrială de Materiale pentru energie termică, „INO-MAT", cod SMIS 119412.* Titus Industries SRL.

Şuta, A. I., & Cazan, A. C. (2022). *Realizarea fizică a eșantioanelor de peleți/bricheți. Proiect: INOvări și optimizări economice și funcționale în producția industrială de MATeriale pentru energie termică, „INO-MAT", cod SMIS 119412.* Titus Industries SRL.

Tarhan, A., Turetken, O., & Reijers, H. A. (2016). Business process maturity models: A systematic literature review. *Information and Software Technology, 75,* 122–134.

Valencia, A., Mugge, R., Schoormans, P. L., & Schifferstein, H. N. J. (2015). The Design of Smart Product-Service Systems (PSSs). An Exploration of Design Characteristics. *International Journal of Design, 9,* 13–28.

VDMA. (2017). *Industrie 4.0 Readiness study.* Retrieved April 11th, 2020, from https://industrie40.vdma.org/en/viewer/-/v2article/render/15525817

Vishnevskiy, K., Karasev, O., & Meissner, D. (2016). Integrated roadmaps for strategic management and planning. *Technological Forecasting and Social Change.* Advance online publication. doi:10.1016/j.techfore.2015.10.020

Visnjic, I., Jovanovic, M., Neely, A., & Engwall, M. (2017). What brings the value to outcome-based contract providers? Value drivers in outcome business models. *International Journal of Production Economics, 192,* 169–181. doi:10.1016/j.ijpe.2016.12.008

Visnjic, I., Neely, A., & Jovanovic, M. (2018). The Path to Outcome Delivery: Interplay of Service Market Strategy and Open Business Models. *Technovation, 72-73,* 46–59. doi:10.1016/j.technovation.2018.02.003

Wendler, R. (2012). The maturity of maturity model research: A systematic mapping study. *Information and Software Technology, 54*(12), 1317–1339. doi:10.1016/j.infsof.2012.07.007

Yasseri, S., & Bahai, H. (2018). System Readiness Level Estimation of Oil and Gas Production Systems. *International Journal of Coastal and Offshore Engineering, 2*(2). Doi:10.29252/ijcoe.2.2.31

Zaoui, F., & Souissi, N. (2020). *Roadmap for digital transformation: A literature review.* The 7th International Conference on Emerging Inter-networks, Communication and Mobility (EICM), Leuven, Belgium. DOI: 10.1016/j.procs.2020.07.090

Chapter 9
Renewable Energy Ecosystems Financed by European Union Public Funds

Cozmiuc Claudia Diana
Ioan Slavici University, Timisoara, Romania

Cosmina Carmen Florica
Polytechnics University, Timisoara, Romania

Delia Albu
Ioan Slavici University, Timisoara, Romania

Avram Greti
Ioan Slavici University, Timisoara, Romania

Octavian Dondera
Ioan Slavici University, Timisoara, Romania

ABSTRACT

The past decades have seen the emergence of business ecosystems. Their historic evolution is important, as ecosystems are deemed new business models to progressively replace the old ones. Ecosystems are broadly and vaguely defined in literature review, where they represent new network business models. Empirical data about European project calls show precise definitions of ecosystem types and activities that are subject to European project funding calls. These definitions apply to new commercial entities recently created and funded by such European project calls. The chapter aims to build theory with a case study about emerging business practices with high instrumental value to other businesses, which may apply for European project calls. The methodology is a descriptive case study. Findings are the level of detail and focus on business practices exceed that of general and vague theory, thereby allowing business practice to contribute to theory and enrich it. Definitions of ecosystem entities and ecosystems management activities to be used across the European Union are this addition.

DOI: 10.4018/978-1-6684-4102-2.ch009

INTRODUCTION

This article takes a historic view at innovation ecosystems as they have emerged since 1997. Recently, innovation ecosystems have evolved to green innovation ecosystems. The issue of managing innovation ecosystems has arisen, highlighting the activities involved in managing green ecosystems. In the past years billion Euros have been invested by the European Union to finance innovation ecosystems and define several types of them: technological transfer centers, technological and business incubators, technological information centers, liaison offices with industry. Ecosystems are pivotal to the new Horizon programme. New entities emerge to manage innovation ecosystems, which the European Union calls centers of technological transfer. One of these entities are innovation and technological transfer clusters. The goal of this article is to build theory via the activities provisioned by European funds as compatible with innovation clusters. These activities are predefined by the European Union for the entities designated to manage innovation ecosystems, which innovation ecosystems should target digitalization, health, renewable energy and intelligent buildings (Agentia pentru Dezvoltare Regionala Vest, 2022). One of these entities is the innovation cluster, which is predefined in structure on all European Union level by provisions which include funding resources and activities. European Union funded innovation clusters are predefined, identified and included in public databases (Ministerul Educatiei si Cercetarii, 2021). This article is a descriptive and analytical case study on the activities in the new management entities, as they are provisioned by European Funds and adopted by a cluster in Timisoara, Romania which is financed via these funds. The empirical data section describes and analyzes these activities based on European Union funding provisions about activities and based on their adoption and use for all management planning and analysis documents by the management cluster. The data comes from European Union funding provisions and European Union project funding call documents. The paper juxtaposes scholarly literature review about management ecosystems and the activities therein, which are deemed novel status 2012. Findings show managing ecosystems may hold predefined activities to be adopted by all such entities. They are used for project planning, financial planning and analysis, funding requests in the local innovation cluster. This offers first definitions for the activities to manage innovation ecosystems. It also shows how these activities may be used to conduct financial planning, that is forecast the net present value of discounted cash flow. Empirical findings are more advanced than theory and the case study may therefore be used to build theory based on them. The contribution to scholarly literature is to bring new theory to scientific literature and support this and further research about the activities to manage ecosystems. From a practical perspective, the article creates attention on new practices from the European Union with designated activities to be used by all European funding adepts. In one of the European Union programs, Horizon 2020, innovation ecosystems are financed billion Euros on European Union level. Knowing how to plan business based on these funding opportunities will increase enterprise adoption of these funds. The propensity to be used by practioners gives the instrumental case value of this case study.

BACKGROUND

The Use of Business Models and Network Business Models Therein

By the late 20th century, authors like Mc Grath (McGrath and MacMillan, 1995, 2009) note the emergence of the uncertain environment. The business plan is replaced by the business model, which may be defined as the hypotheses to build a business upon according to Drucker (Ovans, 2015). In a new look at management, business models become the main management tools (Hamel, 2000) and are made up of: customer interface, customer benefits, core strategy, configuration, strategic resources, company boundaries and value network. There are several definitions of business models. It is around 1997 that the concept of ecosystems has been introduced in management literature (Corallo et al, 2019). The organization and description of a comprehensive ecosystem model useful to ecosystem management is necessary. In this article, we propose the human ecosystem as an organizing concept for ecosystem management. First, we describe the history of the human ecosystem idea; both biological ecology and mainstream social theories provide useful guidance. Next we present the key elements of a human ecosystem model: critical resources (natural, socioeconomic, and cultural), social institutions, social cycles, and social order (identities, norms, and hierarchies). In each element, we provide a general definition and description, suggest ways that the variable can be measured, and give selected examples of how it may influence other components of the human ecosystem. The article concludes with specific suggestions as to how the human ecosystem model can play an organizing role in ecosystem management.

The main stream definition is the business model canvas from Osterwalder, Pigneur and Smith (2010) and includes the following elements: the value proposition; customer relationships; customer segments; channels; revenue streams; key activities; key resources; key partners; cost structure. This is the most popular approach. Essentially, business models are divided into pipeline business models and network based business models. Traditional business models are pipeline business models. New Economy business models are network business models. They may be described as ecosystems (Ben Letaifa, 2014), where all stakeholders thrive together. Ecosystems create more value than their constituent components, Deloitte argue. There are two basic types of business ecosystem that can be observed in practice: solution ecosystems, which create and/or deliver a product or service by coordinating various contributors, and transaction ecosystems, which match or link participants in a two-sided market through a (digital) platform (BCG, 2019). Similarly Roland Berger notes there are ecosystems of value networks. One of these ecosystems is ecosystems of innovation. They have been noted by Chesbrough (2002, 2012), where open innovation in ecosystems is regarded as the contemporary form of innovation. By moving from the corporate laboratory to the open network of innovation, innovation is to be boosted.

New business models involve open networks. According to Van Alstyne (2016), open innovation is part of the new, network based, business model logic. The logic of the New Economy underpins the generation of invention: invention tends to be a knowledge intensive act (Daum, 2003), involving intellectual capital, where knowledge is managed in open networks. Knowledge is created in networks, and economics theory poses the value of knowledge grows as the network and the use of knowledge grow too. The proponent of open innovation asserts the new strategy accelerates innovation via inflows and outflows of knowledge, via creating markets for the knowledge. Open innovation is a radical paradigm shift to laboratory innovation (Chesbrough, 2002, 2012; Open Innovation Community, 2017). Van der Borgh et al. (2012) note that organizations moved from closed enterprises to eco-systems in order to exploit open innovation. In this view, open innovation is placed eminently in start-ups (Chesbrough,

2002, 2012). Open innovation involves the following techniques: external partnerships, crowdsourcing, idea contests, co-creation, social media (blogging, wikis). Social media, Web 2.0 and Enterprise 2.0 (Mc Afee, 2006) have facilitated open innovation, collaboration, innovation eco-systems. Velu et al (2013) note that innovation opportunities are stimulated, captured and exploited in eco-systems. Innovation opportunities are stimulated, captured and exploited in eco-systems (Ben Letaifa, 2014). Managing eco-systems is a novel concept.

Ecosystem Typology

Scholarly literature review shows ecosystems have evolved years 1993 to 2016, shifting from business ecosystems to innovation ecosystems (Gomes et al, 2018; Haller et al, 2012). Business ecosystem relates mainly to value capture, while innovation ecosystem relates mainly to value creation. Literature review articles conclude by describing six research streams in innovation ecosystem (Gomes et al, 2018): industry platform x innovation ecosystem; innovation ecosystem strategy, strategic management, value creation and business model; innovation management; managing partners; the innovation ecosystem lifecycle; innovation ecosystem and new venture creation. These streams lead us to propose opportunities for further research to solidify the innovation ecosystem concept. Whereas the dynamics of ecosystems have changed, scholarly literature review shows the latest trend in ecosystems are green innovation ecosystems (Fan et al, 2022).

Scholarly literature review shows that green innovation ecosystems have emerged recently and are expected to have high future potential. Empirical data analysis also shows green innovation ecosystems have high future potential. The United Nations Sustainable Development Goals have been decided in 2015 and are intended to be achieved until 2030. The decision was made by the United Nations General Assembly. The Sustainable Development Goals were developed in the Post-2015 Development Agenda as the future global development framework to succeed the Millennium Development Goals which were ended in 2015. There are seventeen United Nations Sustainable Development Goals: no poverty, zero hunger, good health and well-being, quality education, gender equality, clean water and sanitation, affordable and clean energy, decent work and economic growth, industry, innovation and infrastructure, reduced inequality, sustainable cities and communities, responsible consumption and production, climate action, life below water, life on land, peace, justice, and strong institutions, partnerships for the goals. Affordable and clean energy is one of these United Nations Sustainable Development Goals. Horizon Europe is a European Commission funding program 2021-2027 which succeeds the previous program, Horizon 2020. Horizon 2020 is known to have given digitalization in manufacturing in the preceding program. The current Horizon program: health, industrial technologies, climate change and environment, energy, fundamental research, society, food and natural resources, transport and mobility, security, space. Horizon 2020 sponsors renewable energy via the Life Programme, a search for available translations of the preceding link is the European Union's financial instrument supporting environmental, nature conservation and climate action projects throughout the European Union. Since 1992, LIFE has co-financed more than 4500 projects. Horizon 2022 is part of Horizon Europe, where the European Union aims to create more connected and efficient innovation ecosystems to support the scaling of companies, encourage innovation and stimulate cooperation among national, regional and local innovation actors. Innovation ecosystems bring together people or organizations whose goal is innovation, and include the links between resources (such as funds, equipment, and facilities), organizations (such as higher education institutions, research and technology organizations, companies, venture capitalists and financial intermediaries), investors and policymakers. The actions supported by European Innovation Ecosystems complement the actions carried out by the European Innovation Ecosystems and the European Institute of Innovation and Technology, activities across Horizon Europe, initiatives at national, regional and local level as well as private and third sector initiatives.

Literature review shows attempts to address entrepreneurial ecosystems (Kshetri, 2014; Mason and Brown, 2014; Stam, 2015; Zahra and Nambisan, 2011, 2012). Entrepreneurial ecosystems have several dimensions: the first dimension is government and leadership; the availability of human capital; the financial capital and funding available for entrepreneurs; the market where the entrepreneurs work; the embedded culture by which the society tolerates the honest mistakes and honorable failure by the entrepreneurs and accepts the attitude of contrarian thinking and risk taking; fifth dimension concerns the embedded culture by which the society tolerates the honest mistakes and honorable failure by the entrepreneurs and accepts the attitude of contrarian thinking and risk taking; the sixth dimension comprises the required supports for the ecosystem from nongovernment institutions, infrastructure and professions. The first dimension, government and leadership, advocates entrepreneurship concept and creates stimulating institutions that directly relate to entrepreneurs such as research institutions, centers for public–private dialog, overseas liaisons. The third dimension is the financial capital and funding available for entrepreneurs. The financial capital required to initiate an infant business requires private equity funds, venture capital funds, public capital markets, micro loans, angel investors, debt financing, which could be available at a presale stage (zero stage). The fourth dimension relates to the market where the entrepreneurs work. The fifth dimension concerns the embedded culture by which the society tolerates the honest mistakes and honorable failure by the entrepreneurs and accepts the attitude of contrarian thinking and risk taking. The final dimension pertains to the required supports for the ecosystem from nongovernment institutions, infrastructure and professions.

On the energy market, the value chain comprises activities power generation, transmission and distribution (Mahmud et al, 2020). This value chain may replace the automation hierarchy with the digitalization network (Aarikka-Stenroos, L., and Ritala, P., 2017, Inderwildi et al, 2020, Shuaiyin, et al, 2019). In the automation hierarchy, a centralized source manages supply chain management to all consumers, who use the energy from the centralized source. In the digitalization hierarchy, power is generated via renewable energy sources by various prosumers. It will be distributed by microgrids to consumers, where cyber-physical systems as microgrids will use artificial intelligence to decide energy consumption in peer-to-peer decisions and networks. Cyber-physical systems have the capability to make and negotiate decisions on their own and in the systems of systems they form across the Internet of Things. The decisions concern the level of energy to be distributed amongst prosumers. Payments are managed via blockchain. The digitalized technologies create the network or ecosystem supply business model. Status 2022, these are one of the energy management technology solutions that are abundant and have zero carbon foot print. The technologies have a competitive advantage compared to fossil fuels which are responsible for 60% of the global carbon footprint. Global initiatives aim at zeroing the carbon footprint and one of the technologies involved are distributed energy management solutions, which create ecosystem business models (Inderwildi et al, 2020).

The Intellectual Property Index IPI aims to evaluate the intellectual property ecosystem (Pugatch et al, 2018). The index has eight specific categories that can be divided into 40 indicators. According to Global IP Center (2018), a proper and legal intellectual property architecture, which encourages and protects creators, may positively affect creativity and innovation. Innovation results mean the effective profusion of protected creations. We use the IPI assuming that the intellectual property ecosystem is closely related to the innovation ecosystem, as the efforts of creative and inventive human production tend to seek legal protection for their creations (Global IP Center, 2018). We thus test the hypothesis that the IPI positively affects innovation output. Innovation involves: patents, related right and limitations; copyrights, related rights and limitations; trademarks, related rights and limitations; trade secrets and relate rights; commercialization of IP assets; enforcement; systemic efficiency; membership in and ratification of international treaties. Innovation results mean the effective profusion of protected creations. We use the IPI assuming that the intellectual property ecosystem is closely related to the in-novation ecosystem, as the efforts of creative and inventive human production tend to seek legal protection for their creations (Global IP Center, 2018). Innovation results mean the effective profusion of protected creations. We use the IPI assuming that the intellectual property ecosystem is closely related to the in-novation ecosystem, as the efforts of creative and inventive human production tend to seek legal protection for their creations (Global IP Center, 2018). Since 2012, the IPI has observed the performance of intellectual property ecosystems in several countries (Pugatch et al., 2018). Developed by the Global Innovation Policy Center, the index has measured several dimensions of intellectual property (Pugatch et al., 2018). The innovation ecosystem construct has emerged as a promising approach in the literature on strategy, innovation and entrepreneurship (Gomes et al, 2018). This study performs a literature review years 1993 to 2016 and notices the transition from business ecosystems to innovation ecosystems (Gomes et al, 2018).

Entities in Innovation Ecosystems

Start-ups in innovation ecosystems are financed by the lean start-up technique. As Mc Grath (McGrath and MacMillan, 1995, 2009) and Magretta (2002) point out, the lean start-up technique is suitable for an uncertain business environment, in which financing is based on hypotheses rather than a predicted business plan. Business plans are the classic financial management tools, suitable for a certain business environment. They are financed by equity and debt. The logic of the financing process is consistent with the Capital Asset Pricing Model. By contrast, business models are financed by venture capital. This has given the success of Sillicon Valley. Venture capital and other forms of risk capital are now available on an international basis, as a different type of capital employed than equity. Equity financing requires the complete forecast of the net present value of discounted cash flow. This involves a business plan completed with a five years' forecast (Blank, 2009, 2013, 2014). Lean start-ups, in contrast, begin by searching for a business model understood as hypotheses to be tested, revised and discarded. Lean start-ups continually gather customer feedback and develop their products in agile mode. This approach asserts business plans are too rigid and challenged or contested by the contact with the customer. Business models involve design, test, pivot (Osterwalder, 2011). They define the stages in a start-ups life (Ries, 2010). The first stage is customer development, and involves seeking a business model, ending when the business model has been found. Customer building entails the following stages: customer development, via customer discovery, customer validation, customer creation, company building (Ries, 2010). This stage involves hypotheses. By the customer validation stage, a scalable business model has been found. Agile development is the next stage, where both the problem and the solution are known (Blank, 2009, 2013, 2014). The Build–Measure–Learn loop emphasizes speed as a critical ingredient to product development. A team or company's effectiveness is determined by its ability to ideate, quickly build a minimum viable product of that idea, measure its effectiveness in the market, and learn from that experiment. In other words, it's a learning cycle of turning ideas into products, measuring customers' reactions and behaviors against built products, and then deciding whether to persevere or pivot the idea; this process is repeated as many times as necessary. The phases of the loop are: Ideas ® Build ® Product ® Measure ® Data ® Learn (Ries, 2011a, 1011b; Maurya, 2012). Most of the experimenting is realized via minimum pivot products, which has the goal to test hypotheses about the product, strategy, and engine of growth ideas that are used to build a minimum viable product.

Start-ups involve high risk and require risk capital, from institutions such as angel investors, family offices, venture capitalists and hedge funds (Blank, 2009, 2013, 2014). Venture capital is a type of private equity, a form of financing that is provided by firms or funds to small, early-stage, emerging firms that are deemed to have high growth potential, or which have demonstrated high growth (in terms of number of employees, annual revenue, or both). The open innovation model is tied to venture capital (Chesbrough, 2002). According to Chesbrough (2002), a corporate venture capital investment is defined by two characteristics: its objective and the degree to which the operations of the investing company and the start-up are linked. Valuation via venture capital means several opportunities are invested in, with just a percentage deemed for long-term success. This allows for high risk inherent to start-ups. Innovation will be successfully diffused in agile organizations, for example start-ups.

MAIN FOCUS OF THE CHAPTER[1]

A descriptive, analytical and instrumental case study is deployed on European project "Performance and excellence in the field of environment and renewable energy through modern cluster entities", SMIS number 138692, funded by the Romanian Ministry of Research, Innovation and Digitisation through the Operational Competitiveness Program (POC). The empirical data in the case study comes from this project. The case describes and analyzes the activities pertinent to innovation ecosystems of start-ups, as they have been provisioned by the European Union and subjected in the funding request. The focus of the case study are European project funding requirements, meaning the documents required by the European funding agency to approve the project. The instrumental nature of the case study comes from its utility to other European project calls.

Issues, Controversies, Problems

The European Union has organized several types of entities that manage innovation and technological transfer. The European Union funds several types of innovation entities: technological transfer centers – CTT, technological and business incubators-ITA, technological information centers - CIT, liaison offices with industry-OLI. These types of innovation entities are prescribed via European Union law, in order to facilitate European Union funding. These entities are defined by European funds as follows. CTT is an infrastructure entity whose activity consists in stimulating ITT in order to introduce research results into the economic circuit, transformed into new or improved products, processes and services. ITA represents an infrastructure entity whose activity is mainly oriented towards facilitating the initiation and development of new innovative enterprises based on advanced technology. CIT represents an infrastructure entity that carries out information dissemination activities related to CI results, technological documentation and training of economic agents, in order to stimulate the valorization of the results, the creation and development of innovative behavior of the socioeconomic environment. OLI, assimilated as competence centers, including hubs and regional centers, is an infrastructure entity whose object of activity consists in establishing, maintaining and expanding the links between the providers of

CI results and the socio-economic environment/economic agencies, in order to facilitating technological transfer. OLI is intended to support internationalization measures (e.g. adaptation to new business models, adaptation of technological production processes to certification and standardization systems specific to export markets, etc.) for SMEs operating in the fields of smart specialization, with application in competitive fields.

There are several domains of intelligent specialization that match the new types of commercial entities: bioeconomy, which includes agriculture and food industry, forest and wood engineering, biotechnology; information and communications technologies, which includes space and security, information society, cyber security and Industry 4.0; energy, environment and climate change, which include environment and climate change, energy; eco technologies and advanced materials, which include textiles and new materials; health and tourism for a healthy lifestyle.

In accordance with legal requirements, the initiative of building an innovation and technological transfer entity may belong to the central or local public administration authorities, research units, universities, chambers of commerce and industry, other commercial entities. The innovation and technological transfer entity may be built by public institutions or private entities.

This document represents the terms to access European funds for the call for projects POR/2018/1/1.1.A./1 – Priority Axis 1 – Promoting technological transfer. These type of European funds are intended to promote investment in research and development, developing connections and synergies amongst companies, technological transfer, social innovation, ecoinnovation and applications of public services, stimulating demand, creating networks and groups of open innovation via intelligent specialization, supporting activities of applied technological research, pilot lines, actions to validate products early on, advanced production capacity and prime production, especially in the field of essential generic technologies and diffusing general use technologies. The goal is to enhance company innovation via supporting innovation entities and technological transfer in fields of intelligent specialization. The attainment of this goal happens via supporting the entities of innovation and technological transfer.

The activities to be performed in an innovation and technological transfer entity are as follows. The first activity is "the development of databases with information on research - development - innovation (CDI) offers and the demand in the field from the economic environment, the portfolio of patents and licenses of the CI centers, the existing facilities in research-demonstration-trial laboratories that can be marketed". The second activity is "inventory/monitoring of research works developed in the fields of ITT completed with technologies, patents, products, prototypes, experimental models; the evaluation and selection of technologies with the potential for capitalization through technological transfer and the creation of a specialized database". The third activity is "the dissemination of information regarding local, regional and national priorities in the areas of interest targeted by ITT". The fourth activity is "the permanent identification of national and international scientific events to ensure the increase in the visibility of the potential and scientific results". The fifth activity is "the dissemination of legislation regarding industrial property rights. The sixth activity is the elaboration of analyses, studies, market research, forecasts on topics related to RDI or technological transfer at the request of the private sector". The seventh activity is "organizing and participating in events to increase awareness of innovation such as fairs, conferences, study visits, meetings between actors in the field (companies and research organizations)". The eighth activity is "granting support to innovative small medium enterprises for the recruitment of qualified personnel; placement of students and those in vocational education in the business environment; technological investment studies". The nineth activity is "providing business assistance for innovation and technological transfer, which involves investigating market needs and developing market studies upon request and elaboration of reports, analyzes and information studies and technological forecasting". Tenth, "training activities and development of the innovative culture of industrial partners, especially SMEs, through seminars, courses, editing of promotional and informative materials, scientific events, fairs and exhibitions, etc".. The eleventh activity is "specialized technical assistance and consultancy for the application / purchase of technologies". The twelvth activity is "the technological evaluation and technological audit; technological vigilance, technological information, assistance in the retechnology of economic agents)". The thirteenth activity is "the verification of pilot plans consisting of: evaluation of assumptions, development of new production formulas, establishment of new production specifications, design of special equipment and structures required by new processes, preparation of operating instructions or process manuals provided that they do not be used for commercial purposes; development, realization, experimentation of the new model or solution for the product/ method/system/technology/service, etc". The fourteenth activity is "the assistance and consultancy for the creation of experimental models and prototypes". Activity fifteen is "the assistance and consultancy for the exploitation of intellectual property rights". Activity sixteen is "assistance and consulting services, including legislative, at the European and international level, for patenting activity and for the exploitation of industrial property rights".

The European Union allocates public funds via several types of programs that may be available on EU basis or on national basis. One of these project calls is POC/62/1/3 Stimulating the enterprise demand for innovation via research, development and innovation projects or in partnership with research and development institutes and universities, to the goal of product or process innovation in the economic sectors with high growth potential. A project has been awarded, innovation and economic and functional optimization in the energy production for thermal energy materials. The project has been financed using the net present value of discounted cash flow and internal rate or return. The European Union has offered a file to compute these indicators, based on operational, investment and financial cash flow.

The project is a construction which will act as a start-up incubator in the field of renewable energy. The construction will host office space, conference rooms, other type of rooms designed to promote renewable energy and its vendors. Whereas the costs refer to the building itself and have several sources of financing, the income comes from using this space for business activities in several scenarios.

In order to have the construction, the first step has been financing cash inflow, in the form of equity from the capital providers. This has created cash flow inflow from financing activities, the starting point of the construction project. Next, this income has been used for cash outflow from investing activities, mainly: tangible assets; intangible assets; experts and other services. The cash outflow from investment activities has been achieved at the beginning of the forecast period. Equity has been matched with the fixed assets invested in, meaning equity covers all investment in the first three years fully. Equity has been used to finance the fixed assets invested in.

Managing the innovation cluster involves adopting some of these activities and building the business plan based on them. The business plan submitted to the European Union involves the adoption of some of these activities and their being tied to business results and the computation of net discounted cash flow. The activities provisioned by the law are the activities in the innovation cluster and have been adopted in the business plan as such, with the exact words provisioned by European law. These activities are to shape the cash inflow and cash outflows of the future innovation cluster. The proposal for the European Union project elaborates these activities and ties them to economic results from the cluster. The chosen activities shape the documents to be submitted to the European Union, including the request for financing. The chosen activities also shape the Gantt chart of activities, their content and sequence. The chosen activities are also to shape the budget for European projects, the expenses inherent to each activity type. All these documents are intended to be correlated and shape the future project.

The income from these activities will generate the cash inflows. The building intends to be a start-up incubator for renewable energy, and will generate income from the following sources: income from royalty payments, as a result of technology transfer; income from consulting and specialty technical assistance; income from renting office space or conference rooms; income from renting technology equipment; income from transferring and exploiting intellectual property rights; other operational income. As the rooms to be rented or otherwise used will be rented on a repetitive basis involving identical or similar activities, renting the office space has been treated as ongoing operations. Expenses needed to operate the start-up incubator include: salaries, consulting and social insurance; expenditure on materials and tools; utility costs; maintenance; administrative expenses; other operational expenses. These sources of income and expenditure refer to operational income and give operational cash flow. The financial forecast assumes zero working capital and chances to working capital to be reflected in cash flow. In summary, the start-up incubator begins with financing cash inflow, which is consumed by investment activities cash outflow. After three years this creates a building space for a start-up incubator for scientific research in renewable energy and related activities, which are repetitive and rated as operations. Whereas equity covers assets in full plus a safety margin, once operating activities begin, cash inflows for the duration of the project. This creates cash and cash equivalents at the end of each reporting period, and the positive nature of these cash and cash equivalents is measured by European funds. Whereas the net present value of discounted cash flow is the only indicator for investment valuation, other strategic criteria are considered in the form of creating start-up ecosystems; incentivizing innovation and technological transfer; innovating in terms of renewable energy.

In the choice of the innovation cluster, activity sixteen has been deemed the most important. The first of the activities in the innovation cluster is consulting: activities for making expenses for the purchase of consulting services in the field of innovation. This activity contains the sub-activity of purchasing general consulting services in the field of innovation - renewable energies. The suppliers will be national research institutes, universities in this country and legal entities that have already implemented innovation in the field of renewable energies. Examples of suppliers are: entities that have parks of photovoltaic cells, parks of wind power plants, producers of pellets and briquettes, producers of energy willow. This activity is measured by key performance indicator or intended result: the general consultancy insurance contract in the field of innovation.

Acquisition of consulting activities also involves the acquisition of intangible assets and the capitalization of the intangible assets within the cluster (Maza-Ortega et al, 2017). The cluster provides several subactivities: general consulting services in the field of innovation, consulting services in the field of innovation - renewable energies, consulting services in the field of innovation regarding the acquisition of intangible assets for the cluster and capitalization of the intangible assets of cluster, consulting services regarding the use of standards and regulations that contain them, consulting services in the field of innovations, assistance and professional training regarding the transfer of acquaintances. Innovation concerns the field of innovation and renewable energies. An important part is the acquisition of know-how, namely knowledge from the field of databases from the field of ISI Thomson, Springer Ferlag, Scopus, etc. The result is the insurance contract for consultancy in the field of innovation as regards acquisition of intangible assets; to see the opportunity to purchase some assets intangibles, existing offers on the national and European market. The key performance indicator for this activity is the consulting contract.

The next sub-activity is the acquisition of consulting services regarding the use standards and regulations that contain them. Within this subactivity, specialists in the field of standards will be consulted regulations in the field of renewable energies: solar energy; kinetic energy of flowing river water; wind energy; biomass; other forms. The result is the insurance contract for consulting in the field of innovation regarding the use the standards and regulations they contain. Thus, the standards in the field will be detailed pellets and briquettes, wind energy, solar energy, biomass, etc. Result are achieved through the following activities: the sub-activity of procurement of services consulting in the field of innovation regarding the use of standards and regulations that contain conclusion on arguments; contract. The document that proves the receipt is the consulting contact.

Activity sixteen is assistance and consulting services, including legislative, at the European and international level, for patenting activity and for the exploitation of industrial property rights. Other consulting services to be acquired are in the field innovations, assistance and professional training regarding knowledge transfer. The applicant (beneficiary/applicant) will purchase these services. This involves professional training consultancy in what concerns the transfer of knowledge, with mainly elements of learning pedagogy, general and special psycho-pedagogy. The focus of this knowledge transfer is green skills. The particularities of this knowledge, namely "green skills", will be emphasized.

Green skills is a topic on the agenda of European research centers several years, and now the subject has become particularly important for employers as well, concerned with developing their businesses sustainably. In these countries the ILO – Institute International Labor Office and CEDEFOP – EU Research Center on Training the professional have already carried out studies at the country level for the estimated needs of green skills. Findings show green skills refer to all economy sectors: agriculture, energy, manufacturing industry, services, etc. Employees' competencies are to be enhanced. This is particularly focused on the issue of sludge. The issue of sludge from the perspective of the circular economy follows an ambitious way of valorization of sludge regarding sustainable development through a correct management of sludges regarding the reduction (by drying), reuse (of the biosolid) and recycling (of the fraction Wet). Romanian legislation provides by Order no. 591/2017 solutions for the recovery of sludge and clearly identify the types of sewage sludge: 1 - sludge from wastewater treatment plants in localities and from other treatment plants wastewater treatment, with a composition similar to urban wastewater; 2 - sludge from septic tanks and other similar installations, for purification waste water; 3 - sludge from sewage treatment plants, other than those mentioned in points 1 and 2; 4 - sludge treated by a biological, chemical or thermal process, by long-term storage or by any other appropriate procedure, which would significantly reduce their power of fermentation and the health risks resulting from their use.

Waste-sludge management from the perspective of the circular economy follows the procedures of valorization: technological valorization (recovery of industrial products), energy valorization (secondary and renewable energy resources) and capitalization in agriculture and animal husbandry (fertilizer, feed, etc.). The result is the consulting insurance contract in the field of innovation for assistance and professional training in knowledge transfer. By courses are held, meetings with specialists, printing a professional training course and a guide in the field renewable energies, training courses within renewable energy entities. Result is again conclusion of contract.

The next major support activity is the activity of making expenses for support services a innovation. This contains subactivities: innovation services sub-activity through data banks, libraries; innovation services sub-activity through labs; innovation services sub-activity through more effective processes and services includes services subactivity to support innovation - testing costs and quality certification for development purposes of more products, processes or services effective. Results are: benefiting from access to the libraries of prestigious institutions, access to doctoral theses, access to indexed research journals, scientific papers published in various journals and conferences and classified in different type databases: ISI Thomson, SCOPUS, Copernicous, etc.; patents of inventions / innovations, international doctorates, other intellectual property rights, studies of feasibility developed by prestigious institutions, studies at prestigious universities, consultancy and last generation services. Result achieved through the following activities is: the sub-activity of innovation services by data banks. The document proving the receipt is the contact/contracts. The sub-activity of innovation services through labs involves the applicant (beneficiary/applicant) will purchase these services. Within this subactivity will appeal to the specialized laboratories in the field belonging to the national institutes of research and universities in the country, laboratories in the following fields: a. Solar energy: a1. Solar systems with the concentration of solar rays; a2. Photovoltaic systems; b. Kinetic energy of flowing river water; c. Wind energy; d. Biomass; e. Other forms. Access to high-performance equipment is also proposed for: measuring calorific power using special calorimeters; determination of the chemical compositions of the various raw materials and the residues resulting afterwarded combustion; determination of noxes following combustion. Benefiting from access to the specialized laboratories in the field belonging to the institutes national research institutes and universities in the country, laboratories in the following fields: a. Solar energy; b. Kinetic energy of flowing river water; c. Wind energy; d. Biomass; e. Other forms. Access to high-performance equipment is also proposed. Result achieved through the following activities: the sub-activity of innovation services by laboratories. Conclusion are contact/contracts, result corresponding to this activity (according to applicant guide). The document proving the receipt is the contact/contracts. The sub-activity of innovation services through more effective processes and services – services subactivity to support innovation - testing costs and quality certification for development purposes of more products, processes or services effective. Within this sub-activity, entities with the latest generation services will be called upon to provide them incorporate the current know-how in the products and processes offered. Examples: creation of the latest generation know-how, quality testing of pellet type products and biomass, etc. Predicted results are: within this sub-activity, entities with the latest generation services will be called upon to provide them incorporate the current know-how in the products and processes offered. Examples: creation of the latest generation know-how, quality testing of pellet type products and biomass, etc.

SOLUTIONS AND RECOMMENDATIONS

Green ecosystems and ecosystems of renewable energy sources are one future direction.

FUTURE RESEARCH DIRECTIONS

Green ecosystems and ecosystems of renewable energy sources have recently emerged in scholarly literature. It is also in the past years they have been pivotal to European Union programs like Horizon 2020 and other adjacent. There are several types of innovation and technological transfer entities provisioned by the European Union. They are designed to manage digitalization and renewable energy systems that may be digitalized. The European Union provisions the funding, structure and activities within such management entities. The activities are predefined in full name. Applicants will use these activities to plan and organize the future innovation and technological transfer management entities. These activities will plan the project and its structure and will be tied to the projected elements of financial statements. The activities are tied to key performance indicators, which measure their attainment or not. The activities will be used to plan cash inflow and outflow and thereby compute the net present value of discounted cash flow. The activities are complex and deserve to be studied in detail. Such scholarly research does not exist at the date. Whereas innovation and green innovation ecosystems have recently emerged, definitions tend to be general and vague, in contrast with EU provisions which refer to specific activities for all funding calls and specific key performance indicators to measure the success of these activities. The article is a case study about an innovation cluster now in construction. It is one in a series of case studies about the renewable energy cluster and the various aspects of its planning and functionality. It shows how this cluster has been planned and funded, in close connection with several other innovation clusters of the same type. This may be treated as an empirical management innovation that brings new empirical data to theory and creates new research directions. One of them refers to the newly created management entities that manage innovation and technological transfer that have just been created. This provides first empirical evidence about green energy innovation clusters. The issue of distributed energy ecosystems of prosumers also arises for some of the renewable energy sources which will be managed and operated by digitalized technologies. As a simple case study about projected future operations, this case study is descriptive of business practices and limited in its capacity to study more than the documents submitted by the future naturae of the operations which the project tackles.

CONCLUSION

Ecosystems of start-ups are incentivized by minimum documentation requirements for funding and by European funds. Literature review and empirical data analysis via European funding practices show that value creation has shifted from individual companies towards steering the ecosystem in all. A frequent from of ecosystems refer to innovation ecosystems, and open innovation via venture capital remains primer. Science needs to be enriched with research about steering ecosystems and the type of activities to be conducted therein.

ACKNOWLEDGMENT

Acknowledgment: This research was funded through the grant "Performance and excellence in the field of environment and renewable energy through modern cluster entities", SMIS number 138692, funded by the Romanian Ministry of Research, Innovation and Digitisation through the Operational Competitiveness Program (POC).

REFERENCES

Aarikka-Stenroos, L., & Ritala, P. (2017). Network management in the era of ecosystems: Systematic review and management framework. *Industrial Marketing Management, 67*, 23–36. doi:10.1016/j.indmarman.2017.08.010

Acharya, V., Kumar, S., Sunand, S., & Gupta, K. (2018). Analyzing the factors in industrial automation using analytic hierarchy process. *Computers & Electrical Engineering, 71*, 877–886. doi:10.1016/j.compeleceng.2017.08.015

Adner, R., & Kapoor, R. (2010). Value Creation In Innovation Ecosystems: How the Structure of Technological Interdependence Affects Firm Performance in New Technology Generations. *Strategic Management Journal, 31*(3), 306–333. doi:10.1002mj.821

Agentia pentru Dezvoltare Regionala Vest. (2022). *1.1.A Entitati de inovare...* Retrieved on October 26th, 2022, from https://adrvest.ro/ghidul-specific-pi-1-1-a/

Apitz, S. E., Elliott, M., Fountain, M., & Galloway, T. S. (2006). European Environmental Management: Moving to an Ecosystem Approach. *Integrated Environmental Management: Moving to an Ecosystem Approach, 2*(1), 80–85. doi:10.1002/ieam.5630020114 PMID:16640322

BCG. (2019). *What Is Your Business Ecosystem Strategy?* Retrieved on August 20th, 2022, from https://www.bcg.com/publications/2022/what-is-your-business-ecosystem-strategy

Ben Letaifa, S. (2014). *The Uneasy Transition from Supply Chains to Ecosystems.* Academic Press.

Blank, S. (2009). *Customer Development at Startup2Startup.* Retrieved on August 20th, 2022, from https://www.slideshare.net/sblank/customer-development-at-startup2startup/28-Thanks_Startup_Lessons_

Blank, S. (2013). Why the Lean Start-Up Changes Everything. *Harvard Business Review.*

Blank, S. (2014). *What Founders Need to Know: You Were Funded for a Liquidity Event – Start Looking.* Retrieved on August 20th, 2022, from https://steveblank.com/category/venture-capital/

Block, Z., & MacMillan, I. C. (1985). Milestones for Successful Venture Planning. *Harvard Business Review, 63*(5), 84–90.

Bonchek, M., & Choudary, S. P. (2013). Three Elements of a Successful Platform Strategy. *Harvard Business Review.*

Boston Consulting Group. (2012). *The Most Innovative Companies 2012.* Retrieved on August 20th, 2022, from https://www.bcgperspectives.com/Images/The_Most_Innovative_Companies_2012_Dec_2012_ tcm80-125210.pdf

Boston Consulting Group. (2013). *The Most Innovative Companies 2013.* Retrieved on August 20th, 2022, from https://www.bcgperspectives.com/Images/Most-Innovative-Companies-2013_tcm80-186913.pdf

Boston Consulting Group. (2014). *The Most Innovative Companies 2014.* Retrieved on August 20th, 2022, from https://www.bcgperspectives.com/Images/Most_Innovative_Companies_2014_Oct_2014_tcm80-174313.pdf

Boston Consulting Group. (2015). *The Most Innovative Companies 2015*. Retrieved on August 20th, 2022, from https://www.bcgperspectives.com/Images/BCG-Most-Innovative-Companies-2015-Dec-2015_ tcm80-203388.pdf

Boston Consulting Group. (2016). *The Most Innovative Companies 2016*. Retrieved on August 20th, 2022, from https://media-publications.bcg.com/MIC/BCG-The-Most-Innovative-Companies-2016-Jan-2017.pdf

CapGemeni. (2014). *Digital Transformation Review, Crafting a Compelling Customer Experience*. Retrieved on August 20th, 2022, from https://www.capgemini-consulting.com/digital-transformation-review-6

CapGemeni. (2016a). *The Digital Strategy Imperative: Steady Long-Term Vision, Nimble Execution*. Retrieved on August 20th, 2022, from https://www.capgemini-consulting.com/dti/digital-strategy-review-9

CapGemeni. (2016b). *The New Innovation Paradigm for the Digital Age: Faster, Cheaper and Open*. Retrieved on August 20th, 2022, from https://www.capgemini-consulting.com/digital-transformation-review-8

CapGemeni. (2016c). *Strategies for the Age of Digital Disruption*. Retrieved on August 20th, 2022, from https://www.capgemini-consulting.com/resource-file-access/resource/pdf/digital_transformation_review_ 7_1.pdf

CapGemeni. (2016d). *Gearing Up for Digital Operations*. Retrieved on August 20th, 2022, from https://www.capgemini-consulting.com/digital-transformation-review-5

Casadesus-Masanell, R., & Ricart, J. E. (2011). How to Design a Winning Business Model. *Harvard Business Review*.

Chesbrough, H. (2002). Making Sense of Corporate Venture Capital. *Harvard Business Review*. PMID:11894386

Chesbrough, H. (2012). *Open Innovation. Where We've Been and Where We're Going*. Research Technology Management.

Christensen, C., Kaufman, S., & Shih, W. (2008). Innovation Killers: How Financial Tools Destroy Your Capacity to Do New Things. *Harvard Business Review*, *86*(1), 98–105, 137. PMID:18271321

Corallo, A., Errico, F., Latina, M. E., & Menegoli, M. (2019). Dynamic Business Models: A Proposed Framework to Overcome the Death Valley, Springer. *Journal of the Knowledge Economy*, *10*(1). Advance online publication. doi:10.100713132-018-0529-x

Daum, J. H. (2003). *Intangible Assets and Value Creation*. Wiley.

Drucker. (1994). What Is a Business Model? *Harvard Business Review*.

Fan, X. Y., Shan, X. S., Day, S., & Shou, Y. Y. (2022). Toward Green Innovation Ecosystems: Past Research on Green Innovation and Future Opportunities from an Ecosystem Perspective. *Industrial Management & Data Systems, 122*(9), 2012–2044. Advance online publication. doi:10.1108/IMDS-12-2021-0798

Gallo, A. (2017). A Refresher on Discovery Driven Planning. *Harvard Business Review*.

Gioratra, K., & Netessine, S. (2014). Four Paths to Business Model Innovation. *Harvard Business Review*.

Global I. P. Center. (2022). *US Chamber of Commerce*. https://www.theglobalipcenter.com/

Godin. (2005). *The Linear Model of Innovation: The Historical Construction of an Analytical Framework*. Retrieved on August 20th, 2022, from http://www.csiic.ca/PDF/Godin_30.pdf

Gomes, L. A. D., Facin, A. L. F., Salerno, M. S., & Ikenami, R. K. (2018). Unpacking the Innovation Ecosystem Construct: Evolution, Gaps and Trends. *Technological Forecasting and Social Change, 136*, 30–48. doi:10.1016/j.techfore.2016.11.009

Grossman, R. (2016). The Industries That Are Being Disrupted the Most by Digital. *Harvard Business Review*.

Gupta, S. K. S., Mukherjee, T., Varsamopoulos, G., & Banerjee, A. (2011), Research directions in energy-sustainable cyber–physical systems. *Sustainable Computing: Informatics and Systems, 1*(1), 57-74.

Haller, M., Ludig, S., & Bauer, N. (2012). Bridging the scales: A conceptual model for coordinated expansion of renewable power generation, transmission and storage. *Renewable & Sustainable Energy Reviews, 16*(5), 2687–2695. doi:10.1016/j.rser.2012.01.080

Hamel, G., & Prahalad, C. K. (2000, July). Competing for the Future. *Harvard Business Review*.

Horizon Europe. (2022). https://cordis.europa.eu/

Horizon Europe. (2022). https://eismea.ec.europa.eu/programmes/european-innovation-ecosystems_en

Horizon Europe. (2022a). https://research-and-innovation.ec.europa.eu/funding/funding-opportunities/funding-programmes-and-open-calls/horizon-europe_en

Iansiti, M., & Levien, R. (2004). The Keystone Advantage. Harvard Business School Press.

Inderwildi, O., Zhang, X., Wang, X., & Kraft, M. (2020). The Impact of Intelligent Cyber-Physical Systems on the Decarbonization of Energy. *Energy and Environmental Science, 3*.

Kavadias, S., Ladas, K., & Loch, C. (2016). The Transformative Business Model. *Harvard Business Review*.

Kothandaraman, P., & Wilson, D. T. (2001). The Future of Competition: Value-Creating Networks. *Industrial Marketing Management, 30*(4), 379–389. doi:10.1016/S0019-8501(00)00152-8

Kshetri, N. (2014). Developing Successful Entrepreneurial Ecosystems: Lessons from a Comparison of an Asian Tiger and a Baltic Tiger. *Baltic Journal of Management, 9*(3), 330–356. doi:10.1108/BJM-09-2013-0146

Lackey, R. T. (1998). Seven pillars of ecosystem management. *Landscape and Urban Planning*, *40*(1-3), Page21–30. doi:10.1016/S0169-2046(97)00095-9

Liu, K., & Chen, M. H. (2017). Research on Innovation Ecosystem Based on Green Management. *Proceedings of International Symposium on Green Management and Local Government's Responsibility*, 174-178.

Machlis, G., Force, J. E., & Burch, W. R. Jr. (1997). The human ecosystem. 1. The human ecosystem as an organizing concept in ecosystem management. *Society & Natural Resources*, *10*(4), 347–367. doi:10.1080/08941929709381034

Magretta. (2002). Why Business Models Matter. *Harvard Business Review*.

Mahmud, K., Sahoo, A., K., Fernandez, E., Sanjeevikumar, P., & Holm-Nielsen, J. (2020). Computational Tools for Modeling and Analysis of Power Generation and Transmission Systems of the Smart Grid. *IEEE Systems Journal, 14*(3). doi:10.1109/JSYST.2020.2964436

Maroufkhani, P., & Ralf Wagner, R. (2018). Entrepreneurial ecosystems: A systematic review. *Journal of Enterprising Communities: People and Places in the Global Economy*.

Mason, C., & Brown, R. (2014). Entrepreneurial Ecosystems and Growth Oriented Entrepreneurship, Conference: Entrepreneurial Ecosystems and Growth Oriented Entrepreneurship. OECD LEED Programme and the Dutch Ministry of Economic Affairs on Entrepreneurial Ecosystems and Growth Oriented Entrepreneurship, The Hague, Netherlands.

Maurya, A. (2012). *Running Lean: Iterate from Plan A to a Plan That Works*. Retrieved on August 20th, 2022, from https://books.google.ro/books?id=j4hXPn233UYC&redir_esc=y

Maza-Ortega, J., M., Acha, E., Garcia, S., & Gómez-Expósito, A. (2017). Overview of power electronics technology and applications in power generation transmission and distribution. *Journal of Modern Power Systems and Clean Energy, 5*(4), 499 – 514. doi:10.1007/s40565-017-0308-x

McAfee. (2006). *Enterprise 2.0, Version 2.0*. Retrieved on August 20th, 2022, https://andrewmcafee.org/2006/05/enterprise_20_version_20/

McGrath, R. G., & MacMillan, I. C. (1995). Discovery Driven Planning. *Harvard Business Review*, *73*(4), 44–54.

McGrath, R. G., & MacMillan, I. C. (2009). *Discovery Driven Growth: a Breakthrough Process to Reduce Risk and Seize Opportunity*. Harvard Business Publishing.

Ministerul Educatiei si Cercetarii. (2021). *Registrul Entidin Infrastructurtatilor acreditate si Autorizate Provizoriu din Infrastructura de Inovare si Transfer Tehnologic*. Retrieved on October 26th, 2022, from https://www.research.gov.ro/uploads/sistemul-de-cercetare/infrastructuri-de-cercetare/infrastructura-de-inovare-si-transfer-tehnologic/2021/registru-entitati-de-inovare-si-transfer-tehnologic-ianuarie-2021.pdf

Open Innovation Community. (2017). *Open Innovation Community*. Retrieved on August 20th, 2022, http://openinnovation.net/

Osterwalder, A. (2010). *Business Model Canvas.* Retrieved on August 20th, 2022, https://strategyzer. com/canvas/business-model-canvas

Osterwalder, A. (2011). *Burn Your Business Plan.* Retrieved on August 20th, 2022, https://www. slideshare.net/Alex.Osterwalder/creativity-worl d-forum-belgium/undefined

Osterwalder, A., & Pigneur, Y. (2013). *Business Model Generation: A Handbook for Visionaries, Game Changers, and Challengers.* Wiley.

Osterwalder, A., Pigneur, Y., & Smith, A. (2010). *Business Model Generation.* Independently Published.

Ries, E. (2010). *Introduction to Customer Development at the Lean Startup Intensive at Web 2.0 Expo by Steve Blank.* Retrieved on August 20th, 2022, https://www.slideshare.net/startuplessonslearned/

Ries, E. (2011a). *The Lean Startup: How Today's Entrepreneurs Use Continuous Innovation to Create Radically Successful Businesses.* Retrieved on August 20th, 2022, from https://books.google.ro/ books?id=tvfyz-4JILwC&redir_esc=y

Ries, E. (2011b). *The Lean Startup - Google Tech Talk.* Retrieved on August 20th, 2022, from https://www.slideshare.net/startuplessonslearned/eric-ries-t he-lean-startup-google-tech-talk

Schmidtab, M., & Åhlundb, C. (2018). Smart buildings as Cyber-Physical Systems: Data-driven predictive control strategies for energy efficiency. *Renewable & Sustainable Energy Reviews*, *90*, 742–756. https://doi.org/10.1016/j.rser.2018.04.013

Shuaiyin, M., Yingfeng, Z., Jingxiang, L., Haidong, Y., & Jianzhong, W. (2019). Energy-cyber-physical system enabled management for energy-intensive manufacturing industries. *Journal of Cleaner Production*, *226*, 892–903. doi:10.1016/j.jclepro.2019.04.134

Stam, E. (2015). Entrepreneurial Ecosystems and Regional Policy: A Sympathetic Critique September 2015. *European Planning Studies*, *23*(9). Advance online publication. doi:10.1080/09654313.2015.10 61484

UN Sustainable Development Goals. (2022). *Do you know all 17 SDGs?* https://sdgs.un.org/goals

Van Alstyne, M. W., Parker, G. G., & Choudary, S. P. (2016). Pipelines, Platforms, and the New Rules of Strategy. *Harvard Business Review*.

Van der Borgh, M., Cloodt, M., & Romme, A. G. L. (2012). Value Creation by Knowledge-Based Ecosystems: Evidence from a Field Study. *R & D Management*, *42*, 150–169.

Velu, C., Barrett, M., Kohli, R., & Salge, T. L. (2013). *Thriving in Open Innovation Ecosystems: Towards a Collaborative Market Orientation.* Cambridge Service Alliance Working Paper.

Westerman, G., Bonnet, D., & McAfee, A. (2011). *Digital Transformation: A Roadmap for Billion-Dollar Organizations.* Harvard Business School Press.

Westerman, G., Bonnet, D., & McAfee, A. (2014). *Leading Digital: Turning Technology Into Business Transformation.* Harvard Business Review Press.

World Economic Forum. (2016). *Digital Disruption Has Only Just Begun*. Retrieved on August 20th, 2022, from https://www.weforum.org/agenda/2016/01/digital-disruption-has-only-just-begun/

Zahra, S. A., & Nambisan, S. (2010). Entrepreneurship in Global Innovation Ecosystems, March 2011. *Academy of Marketing Science Review*, *1*(1), 4–17. doi:10.100713162-011-0004-3

Zahra, S. A., & Nambisan, S. (2012). Entrepreneurship and Strategic Thinking in Business Ecosystems. *Business Horizons*, *55*(3), 219–229. doi:10.1016/j.bushor.2011.12.004

Zahra, S. A., & Nambisan, S. (2012). Entrepreneurship and Strategic Thinking in Business Ecosystems. *Business Horizons*, *55*(3), 219–229.

Zhipeng, C., & Zheng, X. (2018). *A Private and Efficient Mechanism for Data Uploading in Smart Cyber-Physical Systems*. IEEE. doi:10.1109/TNSE.2018.2830307

Chapter 10
Digital Transformation of Museum Conservation Practices:
A Value Chain Analysis of Public Museums in Hong Kong

Athena Kin-kam Wong

 https://orcid.org/0000-0003-2053-8724
The University of Hong Kong, Hong Kong

Dickson K. W. Chiu

 https://orcid.org/0000-0002-7926-9568
The University of Hong Kong, Hong Kong

ABSTRACT

Museums offer education and enjoyment to the public through exhibitions and public programs, but what happens behind the museums can be a mystery. Thus, this study uses the visit to conservation laboratories as a behind-the-scenes tour to illustrate the conservation practices at public museums in Hong Kong and thus potential digital transformation. The value chain analysis was used to systematically exanime the environment and operations of the conservation process in depth, focusing on museum education and extension activities. Lacking human resources, safety, and access constraints often limited the capacity to offer such tours. Some suggestions on digital transformation using contemporary information and communication technologies (ICTs) were proposed for engagement improvement and expanding the audience. Scant studies research how such conservation tours facilitate learning and engagement with visitors or analyze museum operations with value-chain analysis for digital transformation, especially in East Asia.

DOI: 10.4018/978-1-6684-4102-2.ch010

INTRODUCTION

There are fourteen public museums and three cultural spaces under the purview of the Leisure and Cultural Services Department (LCSD) in Hong Kong (LCSD, 2021) to promote local culture as well as knowledge of arts, history, and science to the public (LCSD, 2017; Chen et al., 2018; Lo et al., 2019; Deng et al., 2022; Meng et al., 2022). The public museums offer free admission to permanent exhibitions, outreach programs, and annual museum pass to encourage more visits to build a broader audience base (LCSD, 2017). There are over 1.5 million items stored in Hong Kong public museums (LCSD, 2017). However, only a tiny fraction of the collection is on display.

Behind-the-scene tours such as visits to collection storage and museum conservation facilities offer new ways to engage the public on what happened behind the museum to learn more about the museum's roles and functions. Thus, this study chose to visit conservation laboratories as the behind-the-scene tour. It aims to show the making of an exhibition in displaying and mounting on exhibits (van Saaze, 2011) and collection care in the museum storage (Parowicz, 2019), where conservation practices are often a mystery to the public. The tour brings conservation activities inside Hong Kong public museums to the public. However, these tours are held several times yearly with limited participation capacity. Value chain analysis is applied to systematically exanimate the environment and operation of a public museum in delivering conservation laboratory tours. Based on the results, we propose adopting information and communication technologies (ICTs) to help deliver conservation laboratories tours to a broader audience and improve engagement ().

BACKGROUND

Cabinets of curiosity, also known as *Wunderkammers* in German, where collectors categorize and assemble their treasures to the collection for display (Amsel-Arieli, 2012), are the early formation of the museum. The development of museums is open to all and represents the cultures and values of a wide variety of the public (Bennett, 1995). According to the International Council of Museums (ICOM, 2007), the museum is an institution in the service of society and its development, open to the public, which acquires, conserves, researches, communicates, and exhibits the tangible and intangible heritage of humanity and its environment for education, study, and enjoyment. Museum has a professional role in the stewardship and long-term preservation of its collection.

The Conservation Office is responsible for the long-term preservation of public museum collections, with three conservation laboratories located at the Hong Kong Museum of Art, the Hong Kong Museum of History, and the Hong Kong Heritage Museum under LCSD (2020a). The Office's mission is to provide competent and professional excellence to collection care, promote conservation awareness through educational and extension programs to engage with the community, and generate public support for conservation endeavors (Conservation Office, 2021). The Office's extension activities include visiting the conservation laboratories, virtual reality games held at Museum Festival, and do-it-yourself workshops under the School Cultural Day Scheme.

For example, the extension activity "Visit the conservation laboratories" is held annually at the International Museum Day and the Museum Festival in summer to enhance public interest in the museum scene through various activities (LCSD, 2017). Compared to the figure where approximately 3.67 million visitors patronized public museums in 2019, the behind-the-scenes tours through the conservation laboratories only attracted about 100 participants (LCSD, 2020b). Unfortunately, with a limited quota, only a small number of visitors can interact with the conservators and observe their work. The lack of staff and resources to deliver the behind-the-scene tour is also a problem for the Conservation Office to engage with a broader audience.

Strength and Weakness

Conservation is professional actions and measures aimed at safeguarding tangible cultural heritage concerning its significance and physical properties to ensure its accessibility for presentation to future generations (ICOM-CC, 2008). It also tremendously impacts the display of museum collections in front of the public. During the tour, visitors are allowed a closer look at the museum objects undergoing conservation treatment (Conservation Office, 2021). Conservators interact with the public to explain their decision-making processes and the materials, equipment, and tools used to investigate and conserve museum objects. The ongoing preservation of the museum collections can construct authority, signal hierarchies of value, and bring inclusion and meaning to individuals, groups, or cultures (Saunders 2014). Live conservation interaction for demonstration or conversation on conservation stories and science behind the decision process offers a new experience and engagement with the museum visitors.

There are some constraints in the delivery of the tour. Conservators have daily duties where staff time and resources are insufficient for regular delivery of the laboratory tours. Museum objects are sensitive to the environment and have security restrictions in access and handling. Conservation laboratories are in controlled environments with security that protects museum objects from thieves and risk of deterioration. Chemicals used in the laboratories for conservation treatment may put visitors at risk (Drago, 2011). As health and safety are a concern to the visitors, visitors below the age of 16 are restricted access to the conservation laboratories (Science Museum, 2019). Accessibility in laboratories, possibly with a crate and large museum objects, is an issue that may hinder disabled visitors. In addition, the complexity level of conservation is also a concern that may not be able to communicate in a brief conversation or lay terms. (Drago, 2011). How to convey this to the general public is a challenge.

Opportunities and Threats

Public awareness of conservation activities in public museums has been raised recently through thematic exhibitions such as "Unlocking the Secrets: The Science of Conservation at the Palace Museum" exhibition held at the Science Museum (2019) or "Heritage Over a Century: Tung Wah Museum and Heritage Conservation" exhibition held at Hong Kong Heritage Discovery Centre (Antiquities and Monuments Office, 2020). Conservation is the exhibition theme, and related work by the Conservation Office is depicted on text panels and multimedia. Yet, the exhibits in this kind of exhibition remain the central focus where conservation treatments, considering educational information for enhancing the museum experience (van Saaze, 2011). Interest in learning more about the conservation work at public museums increases during these exhibitions.

In Hong Kong, other cultural heritage sites and museums offer similar behind-the-scene information to the public. For example, Tai Kwun (2021), the center for heritage and arts, provides behind-the-scene tours video online. The duration of the videos was around 6 to 8 minutes, and the conservation specialist explained the work involved for the audience to discover the conservation effort in revitalizing the heritage site. M+ is a visual art museum that opened recently. M+ does not offer live behind-the-scene tours due to the social distancing measure of the COVID-19 outbreak (Meng et al., 2021; Yu, Lam, & Chiu, 2022; Huang et al., 2021; 2022). Behind-the-scene articles supplement with videos of the conservation work behind the museum opening, where conservators work at the gallery, store, and laboratories to install or treat a specific kind of museum collection (M plus, 2021). They are potential competitors for the Conservation Office that does not post related conservation articles or videos on its website.

LITERATURE REVIEW

Museum's conservation practices as the backstage activities are concealed from the public. Conservation activities may lead to heated debates, devaluation of the monetary value of artifacts or artworks, and even loss of reputation responsible for supposed mistakes (van Saaze, 2011). Positive examples show conservation work can arouse public curiosity and attract visitors' attention which is increasingly made more visible and transparent to a diverse audience (van Saaze, 2011). Extension activities in conservation work support the public museums' aim by positively impacting the inclusion of non-professionals (Saunders, 2014). Creating a connection with visitors through conservation activities and museum collections contributes to the current museum's participatory goals (Henderson, 2020).

Previous studies on live conservation showing at galleries during the Conservation Focus exhibition at the British Museum were to bring a message on the roles of conservators are: (1) reveal information about objects; (2) ensure the preservation of museum objects; (3) enable a better understanding of objects; and (4) collaborate for research in the museum (Drago, 2011). Visitors felt privileged and were more engaged with live interaction, using texts and images as supporting delivery mechanisms (Drago, 2011). The most frequently asked questions by the visitors on their approaches to treating the museum objects, followed by how long it takes to treat the object and what the conservators are doing (Drago, 2011). Thus, implementing similar activities as the ongoing program is recommended.

Porter's (1985) value chain offers a systematic way of examining all the activities a firm performs and how they interact, aiming to analyze the sources of competitive advantages. The value chain analysis was proposed as a strategy paradigm to identify and further add value to the product and services for developing the competitive advantages of an organization (Porter, 1985). The value chain helps identify primary activities (including inbound logistics, operations, outbound logistics, marketing and sales, and services) and distinguishes them from support activities (including infrastructure, procurement, technological development, and human resource management) (Porter, 1985). This framework suggests adopting non-profit sectors such as cultural institutions and cultural heritage management to improve the quality of museum services, breaking them down into production processes in museum institutions that make up the values through digital technologies (Simone et al., 2021; Yu, Chiu, & Chan, 2022; Cheung et al., 2021; Li et al., 2023).

Research Gap

The development of new visitor engagement activities with the use of museum collection to engage with the community for more understanding of museum work and fulfill museum educational purposes (Antomarchi et al., 2021; Hide & Pemberton, 2021; Barron & Leask, 2017). A few small scope studies on the behind-the-scene tours as open collection storage to visitors with positive responses as feeling more personally engaged with the museum collection in the questionnaires and interviews (Gallimore &

Wilkinson, 2019). Scant studies focus on conservation practices at museums through visiting conservation laboratories to learn more about the day-to-day conservation work as an education and extension activity in the museum studies and conservation literature. Also, there are limited studies on the value created in non-profit organizations such as museums in the new digital information environment (Simone et al., 2021). Further, this study also benefits the general public by understanding the operation behind the museums and the effort in the long-term preservation of museum collections.

VALUE CHAIN ANALYSIS

Figure 1. Value chain in delivering behind-the-scene tours

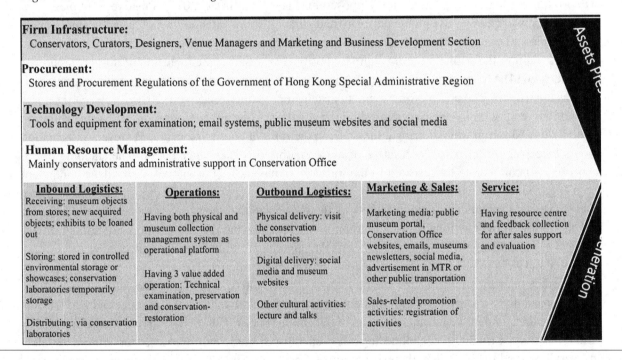

Figure 1 shows the five primary activities of the value chain analysis (Porter, 1985; Cheung et al., 2021; Li et al., 2023) with related support activities, from museum conservation practices to delivery as an education and extension activity for visitor engagement using museum objects and interactions with experts in delivering the behind-the-scene tour by the Conservation Office. In general, the staffing in public museums includes curators in art, history, and science streams, conservators in Conservation Office, designers, venue managers, and support from administrative staff and the marketing and business development section. The conservators are responsible for delivering the activities. The source of funding in public museums can be government-funded or sponsorships. Since they are under the management of LCSD, all procurement should follow the Stores and Procurement Regulation of the Government of Hong Kong Special Administrative Region (Financial Services and the Treasury Bureau, 2021).

The value chain analysis focuses on the dynamics of information from supplier to the value-added in the production and processing activities to consumer consumption in the supply chain (Claudine et al.,

2012). Museum collections seem autonomous and exist in galleries or stores. Their continued existence needs a lot of conservation work. In the case of public museums in Hong Kong, conservation is the production process, including identification, safeguarding, and conserving museum cultural heritage for public appreciation (Simone et al., 2021).

Inbound logistics

Inbound logistics involve receiving, storing, and disseminating the raw materials to the products (Tuomula, 2014; Porter, 1985). Museum objects consider information, and the museum is an information system with care, storage, and accessibility to information seekers (Latham, 2011). Museum objects associated with object information are accessible by the public on the centralized collection management system (LCSD, 2017).

Receiving raw materials. Museum objects retrieved from collection stores should usually be on display or loaned to other institutions. New acquisition objects from purchase or donation require documenting their conditions before entering collection stores.

Storing raw materials. Museum objects are stored in controlled environment storage or exhibition showcases to prevent the risk of deterioration. Conservation laboratories are securer places that temporarily store the museum object for conservation works.

Distributing raw materials. The museum objects are retrieved by curators responsible for collection management or acquired through subject curators and sent to the laboratories with the help of conservators or transportation, packing, and handling services from art handlers.

Table 1. Primary and supportive activities related to inbound logistics

Primary activities related to inbound logistics	Details of the primary activities related to inbound logistics
Receiving	● Retrieval of objects from stores ● New acquired objects by purchase or by donation ● Exhibits to be loaned out
Storing	● In a controlled environment, storage, or exhibition showcases ● Conservation laboratories as temporarily store
Distributing	● Objects are distributed to conservation laboratories
Supportive activities	**Details of the support activities**
Firm infrastructure	● Curators ● Conservators
Procurement	● Museum collections are mainly donated or acquired through authentic sources ● Conservation materials are procured from special suppliers
Technology development	● 24/7 environmentally controlled room
Human resource management	● Curators responsible for receiving the objects and Conservators involved in the whole inbound logistics process

Operations

The operation process transforms the raw material into final products (Grant, 2010; Porter, 1985). Conservation adds value by prolonging the life of museum objects and clarifying artistic or historical messages without losing authenticity (Simone et al., 2021). Also, the better technical condition of a museum object increases its monetary value (Parowicz, 2019).

A conservator is a trained professional who has the knowledge, skills, and experience to act ethically on the long-term preservation of cultural heritage (Parowicz, 2019). The activities of the conservator consist of technical examination, preservation, and conservation-restoration of museum objects (ICOM-CC, 1984). The technical examination involves sophisticated technology aids to identify the material's composition and ultraviolet or infrared viewing systems and stereomicroscopes to examine object details. Preservation is the protective intervention measure in maintaining the object's current condition (Parowicz, 2019). Preventive conservation mitigates the risk of deterioration through environmental control and protection care in handling, storage, and maintenance (ICOM-CC, 2008). Conservation-restoration is also remedial conservation and restoration treatment to reinforce the damages or stabilize the fragile condition of the objects (ICOM-CC, 2008).

Conservators diagnose the museum object and document its condition or related testing and input on the Centralised Collection Management System. The condition report is available for museum staff to research material culture and decisions for future collection management work. The physical operation platform is the laboratories with specific needs for different types of objects such as paper, paintings, objects of metal, etc. These laboratories are equipped with specialized scientific equipment and inspection tables with artificial lighting to examine museum objects.

Table 2. Primary and supportive activities related to operations

Primary activities related to operations	Details of the primary activities related to operations
Value-added operation	● Technical examination ● Preservation ● Conservation-restoration
Operation platform	● Physical: three laboratories located at the Museum of Art, Museum of History, and Museum of Heritage ● Museum objects condition reports on Centralized Collection Management System
Supportive activities	**Details of the support activities**
Firm infrastructure	● Conservators in Conservation Office
Procurement	● Consumable conservation materials and chemicals ● Conservation equipment and tools
Technology development	● Labs with sophisticated technological aids for the identification of the materials composition of objects ● Ultraviolet and infrared viewing systems and stereomicroscopes for object examination
Human resource management	● Conservators specialized in different fields involved in the whole process ● Conservators/Curators/IT Contractors assist in the development and maintenance of the Centralized Collection Management System

Outbound Logistics

Outbound logistics is the scheduling and dissemination of the final products or services to the consumer (Libertore and Millerm, 2016; Porter, 1985). The conservation process may take a long-time to produce the final products, which are the museum objects that have undergone conservation services. The objects' owners or caretakers approached a conservator as the consumer of the conservation services (Parowicz, 2019). Curators responsible for collection management or exhibitions are the internal consumers. Since it is public-funded, Hong Kong citizens and those who visit museums are the final consumers to appreciate the museum objects displayed at galleries.

Since it is a behind-the-scene tour, not all museum objects are ready for display, loan-out, or return to the store. During the tour, conservators present their works on museum objects with technology aids explaining the artistic or historical techniques, rationale, and usage of conservation materials applied to objects. Interaction with the conservators is also an important service to engage the visitors (Jiang et al., 2022; Deng & Chiu, 2022; Wang et al., 2022). Currently, only an onsite visit to the conservation laboratories is available, limiting the promotion of conservation awareness. Similar to the tour, a few case studies are shared on the Conservation Office and Hong Kong Museum of Art websites. Brief text and before-and-after conservation treatment photos are regularly shared on the social media run by the Conservation Office. Conservators sometimes hold special lectures or talks on topics on conservation work in special projects related to the exhibition in detail.

Table 3. Primary activities and related supportive activities of outbound logistics

Primary activities related to outbound logistics	Details of the primary activities related to outbound logistics
Physical delivery	● Visiting the conservation laboratories
Digital delivery	● Share photographs and some brief text on social media ● Conservation Office website ● Museum website e.g. Hong Kong Museum of Art
Other cultural activities	● Lecture and talks
Supportive activities	**Details of the supportive activities**
Firm infrastructure	● Conservators
Technology development	● Aids such as display tablets showing before-and-after conservation treatment photos, microphones ● Museum websites and social media platforms
Human resource management	● Conservators are involved in the whole outbound logistics process ● Venue managers assist in escort or crowd management

Marketing and Sales

Marketing and sales are promotion-related activities. It includes advertising and dissemination channels that aim to distribute the final product to the consumer (Li, 2014). Museums significantly shift from collection focus to visitor engagement to build a diverse audience (Barron & Leask, 2017; Lo et al., 2019; Deng et al., 2022). Conservation practices such as museum behind-the-scene activities have an attractive power to non-specialist visitors as consumers (Jiang et al., 2022). Most visitors appreciate historical objects and are eager to learn as they know relatively little about conservation practices (Parowicz, 2019). Internet and social media are compelling sites to promote awareness and demystify museum conservation practices. Conservators posted their works regularly on the Conservation Office Instagram to raise public interest and share simple conservation knowledge with young audiences (Lam et al., 2022; Chan et al., 2020).

The tour is part of the museum's annual festival targeting new museum visitors and teenagers. The disseminated channels are the festival web pages, public museum portals, and related publicity on media or advertisements (Wang et al., 2022). Event information is published in the newsletters of three LCSD major museums where the laboratories are located, targeted at usual museum visitors. The Conservation Office maintains participants' email lists for sending any future extension and education activities, and the event information is also posted on the Office's website. For the sales-related distribution activities, all the tours are free, and the Office takes care of all the registration.

Table 4. Primary and supportive activities for marketing and sales

Primary activities related to marketing and sales	Details of the primary activities related to marketing and sales
Marketing	● Targeted at regular museum visitors and also new audiences through Museum Festival and International Museum Day ● Marketing media: public museum portal and Conservation Office websites, emails, museums newsletters, social media, and advertisements in the subway or other public transportation ● Promotion interviews with curators and conservators for the whole festival
Sales	● Registration of the activities
Supportive activities	**Details of the supportive activities**
Firm infrastructure	● Conservators ● Marketing and Business Development Section in LCSD
Procurement	● Advertisement
Technology development	● Email to previous participants who wish to receive promotional information ● Social media platform ● Museum websites
Human resource management	● The staff in the marketing and business development section assist in the marketing and sale process, and the conservators assist in providing the materials for marketing on social media, promotional email, and registration

Services

Services support the customer to obtain a better product experience and consider performance maintenance activities after production (Grant, 2010; Chiu, 2013; Leung et al., 2010). Live interaction or the conservator directly communicating with the customer is the best service that visitors feel privileged in the behind-the-scene tour (Drago, 2011). More information about conservation as background knowledge should be provided to the general public for better services (Parowicz, 2019). The resource center with conservation books, publications, and journals is open freely for public access as a reference library in the museum during office hours. The public can also send inquiries to the Office. After the activities, feedback collection improves future offerings (Wang et al., 2022), including (1) attendance or application of the activity, (2) evaluation form, (3) oral feedback, and (4) LCSD customer appreciation card in both paper and e-form.

Table 5. Primary and supportive activities related to services

Primary activities related to services	Details of the primary activities related to services
Resource center	● Open freely for public access as a reference library ● Email inquiry
Feedback collection	● Attendance or application of the activity ● Evaluation form of the activity ● Oral feedback ● LCSD customer appreciation card in paper and e-form
Supportive activities	**Details of the supportive activities**
Firm infrastructure	● Conservators
Technology development	● Not much technology was used in the evaluation and feedback collection ● E-form LCSD customer appreciation card
Human resource management	● Conservators and venue management involve in coordinating the statistical figure on attendance and feedback to the Marketing and Business Development Section ● Conservators involve in the evaluation of the behind-the-scene tours

DISCUSSION AND RECOMMENDATIONS

Value chain analysis is a tool for examination of the current production chain of the behind-the-scene tour and identifying room for improvement in the future (Dent, 2012). Since the core of activities shows the conservation measures geared toward the long-term preservation of museum objects (Depcinski, 2020), operation activities are essential and add value to the whole process. There is room for improvement in the outbound logistics, marketing and sales, and services areas based on the findings of the value chain analysis.

Digitalization

Since the museum objects that require conservation services depend on the curators' research needs, exhibitions, etc., the inbound materials may be limited when preparing the tour. Unable to choose the inbounding raw materials, using digital technology helps (Sun et al., 2022; Tse et al., 2022). Video clips recording previous conservation cases to launch online streaming channels offer informative behind-the-scenes experiences (Hsu et al., 2018; Jiang et al., 2022). Companying with free online digitized images of archival documents or artworks for navigation on the museums' websites is much better than merely texts or limited photo images without details in creating digital connectivity with the audience (Valeonti et al., 2019). The use of new digital technology and interactive tools makes museum walls permeable (Deng et al., 2022). Virtual reality creates an immersive experience for the public to view the conservation laboratories online (Lo et al., 2019; Suen et al., 2020). Everyone can access the conservation laboratories to learn more about the works of conservators online (Art Institute of Chicago, 2021). Users can click on the tools and equipment in the virtual conservation laboratories tour with short descriptions of their use in conservation (George Eastman Museum, 2019). The digital content can also be uploaded into kiosks or tablets beside each laboratory for educating visitors (Deng et al., 2022). Thus, the activities can shift from providing physical space with constraints in the delivery to online for a wider audience.

Since conservation is also part of the collection management process, bringing conservation practices to the public can be integrated into the museum collection system. In adopting open access to cultural heritage, the condition information and before-and-after conservation treatment photos without copyright restrictions can be released for public access via the Centralised Museum Collection System (Chung & Chiu, 2016). The use of machine-generated keywords helps users browse and search the museum collection based on the topic and related conservation documentation (Villaespeasa & Crider, 2021), revealing digital stories on museum objects. Social network and blogging technologies also enable users to discuss museum collections online and promote their interest in conservation (Lam et al., 2022; Deng & Chiu, 2022).

Sensorial Experience

Museums are transforming to use emerging technologies to create new sensorial experiences innovatively to provide a personalized experience based on visitors' needs (Capece & Chivaran, 2020; Lo et al., 2019). Also, finding interesting content to work for the majority of visitors and keeping the extension and education activities fresh were considered the main challenges of experience design (Easson & Leask, 2020). The use of social media to create and support conversational spaces for social interactions (Mutibwa et al., 2020; Wang et al., 2022) with the conservators. A live camera lets visitors closely watch ongoing conservation activities done by the conservators at the laboratories (Terhune, 2006), and participants can ask questions promptly on social media. This kind of live interaction with participants can be flexibly organized, while conservators can work ad-hoc anytime without the constraint on venue and museum objects from the inbound logistic. Digital outreach has become a priority, emphasizing user interactions from both sides rather than generic communication of information from one direction from websites and blogs (Reeve, 2021; Cheng et al., 2020; Fong et al., 2020).

Conservation practices can be every part of public museum permanent exhibitions. They show visitors the techniques applied to museum objects and knowledge related to conservation, such as how light, temperature, and pests affect museum objects. Augmented reality can be adopted in galleries to stimulate simple conservation tasks (Suen et al., 2021), such as performing examinations removing dust, measuring light intensity, relative humidity, etc., for visitors to experience virtually. QR codes can be used to link the related work depicted in basic conservation practices on websites to provide an in-depth understanding of conservation (Science Museum, 2019).

Social Media Marketing

While the above suggestions are tailored for outbound logistics, social media marketing in conservators' daily practice is an out-of-the-box marketing strategy widely practiced in cultural organizations (Parowicz, 2019; Deng & Chiu, 2022; Lam et al., 2022; Cheng et al., 2020; Cheung et al., 2022). Marketing campaigns should focus on conservation work inside the museum as a central theme to introduce social and cultural awareness (Jiang et al., 2022). For example, the Masters in the Forbidden City documentary series depicts the craft inheritance between masters and apprentices to explore the craftsmanship in restoring and conserving museum objects at the Palace Museum. This kind of documentary was later published as a good conservation marketing book by raising public awareness and increasing the sustainability of cultural heritage preservation on the local and national levels (Parowicz, 2019).

Another marketing strategy is to motivate people towards voluntary engagement in sustaining social good by preserving cultural heritage (Parowicz, 2019). So far, LCSD has a museum volunteer scheme, and the Conservation Office has a Facebook group for its volunteer. This kind of marketing is a good way for more people to involve and by word-of-mouth to promote conservation awareness (Jiang et al., 2022). Besides, internships should be arranged to educate interested students and offload simpler museum conservation tasks (Li & Chiu, 2021; Ng et al., 2021; Cheung et al., 2021).

Enhancing Services with other Contemporary Technologies

To provide after-production services to visitors who had participated in the conservation laboratories tour, the resource center and the Conservation website are the sources for interested parties. The website needs to revamp to provide more general conservation knowledge and case studies, with updates posted on social media (Cheng et al., 2020). The resource center of the Conservation Office supplied with conservation books, journals, and publications as a reference library may not be user-friendly for the general public. Using three-dimensional (3D) scanning and multi-view 3D reconstruction to document museum objects can produce high-resolution 3D models (Pan et al., 2019) for adopting the makerspace concept and demonstrating 3D printing of museum object restoration. Conservation tools can also be displayed aside from videos to arouse interest in conservation. Learning kits with simple broken museum objects and conservation materials can be used in education programs for school students (Deng et al., 2022; Chan et al., 2023).

CONCLUSION

This study has applied the value chain analysis to the current practices in the delivery of visit conservation laboratories as a behind-the-scenes tour in the Hong Kong public museums. Recommendations on adopting the emerging technology in onsite and virtual tours are thus provided to help the museums to reach border audiences and younger generations. The use of contemporary technologies facilitates better management in service delivery, reducing the risks of deterioration of collections through frequent access and close visitor examinations without the protection of display cases. Such technologies also help provide higher-quality educational content and sensory engagement. Overall, these kinds of activities are well-received by the public and should be arranged more frequently with more interactions with the conservators or museum objects using high-resolution imaging and printing techniques instead of the original museum object.

Since this study references information on websites and researcher observation, there are no visitor feedback data such as surveys or focus groups for analyzing the effectiveness and engagement of the behind-the-scenes tours and conservation practices. These further studies are on our research agenda. Further investigation of the visitors' feedback helps evaluate the convey of complex conservation knowledge and proposed educationally and entertaining approaches in delivery. Moreover, we plan to apply emerging technology as an educational aid to engage with border audiences and social media marketing to promote conservation awareness in Hong Kong public museums.

REFERENCES

American Institute for Conservation. (2021). *AIC Wiki. Collection Care.* Retrieved from https://www.conservation-wiki.com/wiki/Collection_Care

Amsel-Arieli, M. (2012). Cabinets of Curiosity (Wunderkammers). *Histoire Magazine.* http://www.micheleleigh.net/wp-content/uploads/2019/01/Cabinets-of-Curiosity.pdf

Antiquities and Monuments Office. (2020). *"Heritage Over a Century: Tung Wah Museum and Heritage Conservation" Exhibition.* Retrieved from https://www.amo.gov.hk/en/whatsnew_20200511.php

Art Institute of Chicago. (2021). *360 Tours: Taking a Spin through Conservation.* Retrieved from https://www.artic.edu/articles/935/360-tours-taking-a-spin-through-conservation

Barron, P., & Leask, A. (2017). Visitor engagement at museums: Generation Y and 'Lates' events at the National Museum of Scotland. *Museum Management and Curatorship*, *32*(5), 473–490. doi:10.1080/09647775.2017.1367259

Bennett, T. (1995). *The Birth of the Museum (Culture: policies and politics).* Routledge.

Capece, S., & Chivăran, C. (2020). The Sensorial Dimension of the Contemporary Museum between Design and Emerging Technologies. *IOP Conference Series. Materials Science and Engineering*, *949*(1), 12067. doi:10.1088/1757-899X/949/1/012067

Ch'ng, E., Cai, S., Leow, F., & Zhang, T. (2019). Adoption and use of emerging cultural technologies in China's museums. *Journal of Cultural Heritage*, *37*, 170–180. doi:10.1016/j.culher.2018.11.016

Chan, T. T. W., Lam, A. H. C., & Chiu, D. K. W. (2020). From Facebook to Instagram: Exploring user engagement in an academic library. *Journal of Academic Librarianship*, *46*(6), 102229. doi:10.1016/j.acalib.2020.102229 PMID:34173399

Chan, V. H. Y., & Chiu, D. K. W. (2023). Integrating the 6C's Motivation into Reading Promotion Curriculum for Disadvantaged Communities with Technology Tools: A Case Study of Reading Dreams Foundation in Rural China. In A. Etim & J. Etim (Eds.), *Adoption and Use of Technology Tools and Services by Economically Disadvantaged Communities: Implications for Growth and Sustainability.* IGI Global.

Chen, Y., Chiu, D. K. W., & Ho, K. K. W. (2018). Facilitating the learning of the art of Chinese painting and calligraphy at Chao Shao-an Gallery. *Micronesian Educators*, *26*, 45–58.

Cheng, W. W. H., Lam, E. T. H., & Chiu, D. K. W. (2020). Social media as a platform in academic library marketing: A comparative study. *Journal of Academic Librarianship*, *46*(5), 102188. doi:10.1016/j.acalib.2020.102188

Cheung, T. Y., Ye, Z., & Chiu, D. K. W. (2021). Value chain analysis of information services for the visually impaired: A case study of contemporary technological solutions. *Library Hi Tech*, *39*(2), 625–642. doi:10.1108/LHT-08-2020-0185

Cheung, V. S. Y., Lo, J. C. Y., Chiu, D. K. W., & Ho, K. K. W. (2022). Predicting Facebook's influence on travel products marketing based on the AIDA model. *Information Discovery and Delivery*. Advance online publication. doi:10.1108/IDD-10-2021-0117

Chiu, D. K. W. (Ed.). (2012). *Mobile and Web Innovations in Systems and Service-oriented Engineering*. IGI Global.

Chung, A. C. W., & Chiu, D. K. (2016). OPAC Usability Problems of Archives: A Case Study of the Hong Kong Film Archive. *International Journal of Systems and Service-Oriented Engineering*, *6*(1), 54–70. doi:10.4018/IJSSOE.2016010104

Conservation Office. (2016). *2015-16 Annual Report*. Retrieved from https://www.lcsd.gov.hk/CE/Museum/Conservation/documents/101 18435/10118860/2015-16_Annual_Report_EN_final_RE.pdf

Conservation Office. (2021a). *Conservation "Behind-the-Scenes" Tour*. Retrieved from https://www.lcsd.gov.hk/CE/Museum/Conservation/en_US/web/co/education_and_extension_programs_imd.html

Conservation Office. (2021b). *Education and extension programs*. Retrieved from https://www.lcsd.gov.hk/CE/Museum/Conservation/en_US/web/co/education_and_extension_programs_muse_2018.html

Conservation Office. (2021c). *School Culture Day Scheme*. Retrieved from https://www.lcsd.gov.hk/CE/Museum/Conservation/en_US/web/co/school_culture_day_scheme.html

Conservation Office. (2021d). *Vision, Mission and Values*. Retrieved from https://www.lcsd.gov.hk/CE/Museum/Conservation/en_US/web/co/vision_and_values.html

Deng, S., & Chiu, D. K. W. (2022). Analyzing Hong Kong Philharmonic Orchestra's Facebook Community Engagement with the Honeycomb Model. In M. Dennis & J. Halbert (Eds.), *Community Engagement in the Online Space*. IGI Global.

Deng, W., Chin, G. Y.-l., Chiu, D. K. W., & Ho, K. K. W. (2022). Contribution of Literature Thematic Exhibition to Cultural Education: A Case Study of Jin Yong's Gallery. *Micronesian Educators*, *32*, 14–26.

Depcinski, M. C. (2020). Conservation in Museums. In C. Smith (Ed.), *Encyclopedia of Global Archaeology*. Springer. doi:10.1007/978-3-030-30018-0_796

Drago, A. (2011). 'I feel included': The Conservation in Focus exhibition at the British Museum. *Journal of Insect Conservation*, *34*(1), 28–38. doi:10.1080/19455224.2011.566473

Easson, H., & Leask, A. (2020). After-hours events at the National Museum of Scotland: A product for attracting, engaging and retaining new museum audiences? *Current Issues in Tourism*, *23*(11), 1343–1356. doi:10.1080/13683500.2019.1625875

Financial Services and the Treasury Bureau. (2021). *Guide to Procurement.* Retrieved from https://www.fstb.gov.hk/en/treasury/gov_procurement/guide-to -procurement.htm

Fong, K. C. H., Au, C. H., Lam, E. T. H., & Chiu, D. K. W. (2020). Social network services for academic libraries: A study based on social capital and social proof. *Journal of Academic Librarianship, 46*(1), 102091. doi:10.1016/j.acalib.2019.102091

Gallimore, E., & Wilkinson, C. (2019). Understanding the Effects of 'Behind-the-Scenes' Tours on Visitor Understanding of Collections and Research. *Curator (New York, N.Y.), 62*(2), 105–115. doi:10.1111/cura.12307

George Eastman Museum. (2019). *Virtual Tour.* Retrieved from https://www.eastman.org/360-conservation-lab-tour

Grant, R. M. (2010). *Contemporary Strategy Analysis* (7th ed.). John Wiley & Sons.

Hide, L., & Pemberton, D. (2021). Mobilising Collections Storage to Deliver Wide-Ranging Strategic Objectives at the Sedgwick Museum. *Museum International, 73*(1-2), 110–119. doi:10.1080/13500775.2021.1956753

Huang, P. S., Paulino, Y. C., So, S., Chiu, D. K. W., & Ho, K. K. W. (2021). Editorial. *Library Hi Tech, 39*(3), 693–695. doi:10.1108/LHT-09-2021-324

Huang, P.-S., Paulino, Y. C., So, S., Chiu, D. K. W., & Ho, K. K. W. (2022). Guest editorial: COVID-19 Pandemic and Health Informatics Part 2. *Library Hi Tech, 40*(2), 281–285. doi:10.1108/LHT-04-2022-447

ICOM-CC. (1984). *Definition of the profession.* Retrieved from https://www.icom-cc.org/en/definition-of-the-profession-1984

ICOM-CC. (2008). *Terminology for conservation.* Retrieved from https://www.icom-cc.org/en/terminology-for-conservation

Jiang, X., Chiu, D. K. W., & Chan, C. T. (2022). Application of the AIDA model in social media promotion and community engagement for small cultural organizations: A case study of the Choi Chang Sau Qin Society. In M. Dennis & J. Halbert (Eds.), *Community Engagement in the Online Space.* IGI Global.

Lam, A. H. C., Ho, K. K. W., & Chiu, D. K. W. (2022). (in press). Instagram for student learning and library promotions? A quantitative study using the 5E Instructional Model. *Aslib Journal of Information Management.* Advance online publication. doi:10.1108/AJIM-12-2021-0389

Legislative Council Panel on Home Affairs. (n.d.). *Progress Report on Enhancement of Programming, Audience Building and Collection Management of Public Museums.* Retrieved from https://www.legco.gov.hk/yr17-18/english/panels/ha/papers/ha20171221cb2-553-3-e.pdf

Leisure and Cultural Services Department. (2017). *LCSD Museums Collection Management System.* Retrieved from https://mcms.lcsd.gov.hk/Search/search/enquire?timestamp=1639643297962

Leisure and Cultural Services Department. (2020b). *LCSD Customer Appreciation Card.* Retrieved from https://www.lcsd.gov.hk/en/aboutlcsd/forms/appr_card.html

Leisure and Cultural Services Department. (2021). *About Us.* Retrieved from https://www.museums.gov.hk/en_US/web/portal/about-us.html#

Leisure and Cultural Services Department. (2020a). *LCSD Annual Report 2019-2020.* Retrieved from https://www.lcsd.gov.hk/dept/annualrpt/2019-20/en/cultural-services/museums

Leung, H. F., Chiu, D. K. W., & Hung, P. C. (Eds.). (2010). *Service Intelligence and Service Science: Evolutionary Technologies and Challenges: Evolutionary Technologies and Challenges.* IGI Global.

Li, K. K., & Chiu, D. K. W. (2021). A Worldwide Quantitative Review of the iSchools' Archival Education. *Library Hi Tech.* Advance online publication. doi:10.1108/LHT-09-2021-0311

Li, L. (2014). Supply Chain Management and Strategy. *Managing Supply Chain and Logistics*, 3-36.

Li, S. M., Lam, A. H. C., & Chiu, D. K. W. (2023). Digital transformation of ticketing services: A value chain analysis of POPTICKET in Hong Kong. In R. Pettinger, B. B. Gupta, A. Roja, & D. Cozmiuc (Eds.), *Handbook of Research on the Digital Transformation Digitalization Solutions for Social and Economic Needs.* IGI Global.

Liberatore, M. J., & Miller, T. (2016). Outbound logistics performance and profitability: Taxonomy of manufacturing and service organisation. *Business and Economics Journal*, *7*(2), 1000221.

Lo, P., Chan, H. H. Y., Tang, A. W. M., Chiu, D. K. W., Cho, A., Ho, K. K. W., See-To, E., He, J., Kenderdine, S., & Shaw, J. (2019). Visualising and Revitalising Traditional Chinese Martial Arts – Visitors' Engagement and Learning Experience at the 300 Years of Hakka KungFu. *Library Hi Tech, 37*(2), 273–292. doi:10.1108/LHT-05-2018-0071

McLee, L., Luke, J., Ong, A., & University of Washington. Museology. (2018). *The Collections Connection: Understanding the Attitudes of Participants in Behind-the-Scenes Museum Tours.* Museology Master of Arts Theses, University of Washington.

Meng, Y., Chu, M. Y., & Chiu, D. K. W. (2022). (in press). The impact of COVID-19 on museums in the digital era: Practices and Challenges in Hong Kong. *Library Hi Tech.* Advance online publication. doi:10.1108/LHT-05-2022-0273

Mutibwa, D., Hess, A., & Jackson, T. (2020). Strokes of serendipity: Community co-curation and engagement with digital heritage. *Convergence (London, England), 26*(1), 157–177. doi:10.1177/1354856518772030

Ng, T. C. W., Chiu, D. K. W., & Li, K. K. (2021). Motivations of choosing archival studies as major in the i-School: Viewpoint between two universities across the Pacific Ocean. *Library Hi Tech.* Advance online publication. doi:10.1108/LHT-07-2021-0230

Pan, R., Tang, Z., & Da, W. (2019). Digital stone rubbing from 3D models. *Journal of Cultural Heritage*, *37*, 192–198. doi:10.1016/j.culher.2018.11.013

Parowicz, I. (2018). *Cultural Heritage Marketing: A Relationship Marketing Approach to Conservation Services.* Palgrave.

Porter, M. (1985). Competitive advantage: Creating and sustaining superior performance. Collier Macmillan.

Reeve, J. (2021). The Museum as Changemaker. *A Pathmaking Arts Quarterly*, *72*(3), 64-71.

Saunders, J. (2014). Conservation in Museums and Inclusion of the Non-Professional. *Journal of Conservation & Museum Studies*, *12*(1), 6. doi:10.5334/jcms.1021215

Science Museum. (2019). *Unlocking the Secrets: The Science of Conservation at the Palace Museum*. Retrieved from https://hk.science.museum/ms/con2019/activities-EN.html

Science Museum. (2020). *Conservation Laboratory – Unlocking the Secrets of Artefact Conservation*. Retrieved from https://hk.science.museum/web/scm/se/cl.html

Simone, C, Cerquetti, M., & La Sala, A. (2021). Museums in the Infosphere: Reshaping value creation. *Museum Management and Curatorship, 36*(4), 322-341.

Suen, R. L. T., Tang, J., & Chiu, D. K. W. (2020). Virtual reality services in academic libraries: Deployment experience in Hong Kong. *The Electronic Library*, *38*(4), 843–858. doi:10.1108/EL-05-2020-0116

Sun, X., Chiu, D. K. W., & Chan, C. T. (2022). Recent Digitalization Development of Buddhist Libraries: A Comparative Case Study. In S. Papadakis & A. Kapaniaris (Eds.), *The Digital Folklore of Cyberculture and Digital Humanities* (pp. 251–266). IGI Global. doi:10.4018/978-1-6684-4461-0.ch014

Terhune, L. (2006). *Smithsonian Museums Make Art Conservation Part of the Show: Lunder Conservation Center allows visitors to see conservators at work*. Federal Information & News Dispatch, LLC. Retrieved from Research Library http://eproxy.lib.hku.hk/login?url=https://www.proquest.com/reports/smithsonian-museums-make-art-conservation-part/docview/190012923/se-2?accountid=14548

Tse, H. L., Chiu, D. K., & Lam, A. H. (2022). From Reading Promotion to Digital Literacy: An Analysis of Digitalizing Mobile Library Services With the 5E Instructional Model. In A. Almeida & S. Esteves (Eds.), *Modern Reading Practices and Collaboration Between Schools, Family, and Community* (pp. 239–256). IGI Global. doi:10.4018/978-1-7998-9750-7.ch011

van Saaze, V. (2011) *Going Public: Conservation of Contemporary Artworks. Between Backstage and Frontstage in Contemporary Art Museums*. Retreived from http://hdl.handle.net/10362/16705

Wang, J., Deng, S., Chiu, D. K. W., & Chan, C. T. (2022). Social Network Customer Relationship Management for Orchestras: A Case Study on Hong Kong Philharmonic Orchestra. In N. B. Ammari (Ed.), *Social Customer Relationship Management (Social-CRM) in the Era of Web 4.0*. IGI Global. doi:10.4018/978-1-7998-9553-4.ch012

Yu, H. H. K., Chiu, D. K. W., & Chan, C. T. (2022). Resilience of symphony orchestras to challenges in the COVID-19 era: Analyzing the Hong Kong Philharmonic Orchestra with Porter's five force model. In W. Aloulou (Ed.), *Handbook of Research on Entrepreneurship and Organizational Resilience During Unprecedented Times* (pp. 586–601). IGI Global. doi:10.4018/978-1-6684-4605-8.ch026

Yu, P. Y., Lam, E. T. H., & Chiu, D. K. W. (2022). Operation management of academic libraries in Hong Kong under COVID-19. *Library Hi Tech*. Advance online publication. doi:10.1108/LHT-10-2021-0342

Chapter 11
Readiness for Implementing an E–Voting System in Ethiopia:
A Gap Analysis From the Supply Side

Lemma Lessa
https://orcid.org/0000-0002-2890-9721
Addis Ababa University, Ethiopia

Mekuria Hailu
Addis Ababa University, Ethiopia

ABSTRACT

Extant literature revealed that elections conducted in traditional ways mostly result in conflicts. Intending to address such challenges, e-voting technology is being used in some countries to conduct a transparent election. However, the application of this new system encountered different challenges due to a lack of readiness to exploit its value, especially in developing countries contexts. To that end, the readiness of government, citizens, and political parties needs to be assessed before using e-voting as an electoral system. The main purpose of this study is to assess the gaps in the readiness of Ethiopia for e-voting system implementation. A qualitative research method is employed, and a thematic analysis was used to analyze the data. The finding revealed that Ethiopia is not ready in terms of information communication technology (ICT) infrastructure, human resources, and legality measures for e-voting technology. Finally, recommendations are forwarded for policymakers and practitioners for action.

BACKGROUND

Democratic voting is very important for a nation because it provides an opportunity for people to vote for their opinion and vote for what they believe in. It holds elected officials accountable for their behavior while in office and prevents minority from dictating the policies of the majority. Among the different types of voting, paper-based voting is the most dominating in many countries of the world. Currently, technological advancement provides an opportunity for countries of the world to use ICTs in the gov-

DOI: 10.4018/978-1-6684-4102-2.ch011

ernment's democratic processes. Among the process electronic voting is the one that uses technological devices such as stand-alone or movable voting machines to cast votes.

International Democracy Election Assistance (IDEA) (2014) defines the e-voting system as a voting system which uses information and communication technologies to record, cast, and count votes in political elections and referendums. Electronic voting can be thought of as a better form of voting for eliminating the drawbacks of the paper-voting system. Even though e-voting technology plays an important role in the reduction of the problems of paper-based voting, its implementation is affected by several factors such as human resources, technology, and legal structure. Lubis (2018) said that the social structure, human resources, and ICT infrastructure of the country should be considered before adopting the system. There are also nine significant substances related to the previous adoption from other countries including internet vulnerabilities, democracy drawbacks, e-voting unconstitutionality, privacy, and confidentiality confusion, technology insecurity, fraud proneness, adverse experience, technical preparation, and hacker ability (Ardiyanti, 2016). According to a statistic from the National Election Board of Ethiopia (NEBE), the first nationwide election was held under the provision of the current constitution in June 1994 to elect members of local government. A general election has since been held in 1995, 2000, 2005, 2010, and 2015, yet the public reaction to those six election results was rigged which result in the death of many citizens in opposition to some polls. These concerns arise from the fact that voters do not have trust in the election result of the paper-based voting.

According to the Amnesty International Report (2015) among the problems of the Ethiopian election that resulted in the paper-based voting system was the lack of trust of the conservative parties that was mainly due to the delay of the official election results being announced. Also, some of the ruling party's leaders would notice the stealing of the votes of opposition parties by intimidating observers and opposition party members at some polling stations. Also, the lack of opportunity for disabled citizens and citizens living outside the country to participate in the voting is a problem. These concerns are the main initiator for assessing the readiness of Ethiopia to implement the e-voting system. Rubin (2004) stated that elections require the citizen to choose the people they consider fit to serve. Naturally, the honesty of the political decision process is fundamental to the trustworthiness of democracy itself. He further claimed that any system designed for election must be a system capable of withstanding any attack. It must also be a system that the electorate can embrace and the election results should be acceptable by different candidates without any dispute. However, election results are most frequently manipulated to influence their outcome.

The electronic election provides several benefits for Ethiopia like reducing voting fraud that was raised in the previous elections, accessibility to eligible citizens across the country, quick election result announcement, reduce costs and secure trustfulness. In this research, the researchers sought to assess the readiness of the supply side for the implementation of the e-voting system in Ethiopia, identify factors that affect the readiness of e-voting systems such as ICT infrastructure, the legal framework and human resources.

RELATED WORKS

The researcher strongly finds related literature that helps to understand the concept of e-voting systems using different search parameters. The issues discussed here are: an overview of e-voting systems, a requirement of e-voting system implementation and factors affecting e-voting system implementation.

Overview of E-Voting System

E-voting is one of the platforms of e-government that has various approaches and models that contribute to the theory and practice of e-democracy. Collord (2013) describes e-voting as a fully-electronic means of capturing and counting ballots for an election. The term "electronic voting" depicts the use of some electronic means or machinery that is more or less computer-supported in voting and ensures better security, reliability, and transparency (Hossain, Shakur, Ahmed, and Paul, 2012). International Democracy Election Assistance (IDEA) (2011) classified the e-voting system as Poll-site electronic voting systems (PEVS) and remote electronic voting systems (REVS). In a poll-site electronic voting system, the voting process can be performed in a pre-defined and controlled environment by the electoral commission. The remote electronic voting systems do not require the presence of the voter in pre-defined voting places: the vote can be cast from anywhere using the internet as a medium of communication between the system and the voter. The direct record electronic voting system, optical scanning system, and punch card are under the PEVS, while internet voting (i-voting) is in the REVS. An E-voting system is crucial since it provides a solution for problems of traditional voting system like voter-verifiable audit trail, multiple voting, over-voting, security and confidentiality. Vries and Bokslag (2016) state the benefits of the e-voting system as fast counting, less labor-intensive, cheaper, and accessibility. E-voting will also increase the security and reliability of elections, reduce and simplify the work of authorities significantly and will lead to cost-saving through the reduction of electoral officers and personnel in polling stations.

Requirements for E-Voting System

Council of Europe (2005) states the e-voting technology requirement as a legal requirement, technical requirements, and procedural requirement. These three categories of requirements all include provisions concerning all stages of elections and referendums (i.e. the pre-voting stage, the actual casting of votes, and the post-voting stage). The legal requirement relates to the legal context in which the e-voting system is permitted. While the technical requirements related to the construction and operation of electronic voting equipment and software. The Procedural requirement relates to how e-voting hardware and software should be operated and maintained.

Factors Affecting E-Voting System Implementation

The Social structure, human resources, and ICT infrastructure are important factors that should be considered before the implementation of e-voting technology. Achieng and Ruhode (2013) describe the factors that affect the implementation of e-voting systems as ICT resources and infrastructures, the usefulness of the technology, ease of use, trust in the technology, and environment.

RESEARCH DESIGN AND METHODS

This study aims to analyze gaps in Ethiopia's readiness for e-voting system implementation from the supply side, identifying the gaps in ICT resources and infrastructure, human resources, government willingness, and legal framework. To answer the research questions and to accomplish the objectives of the research qualitative data collection and analysis techniques have been used. Through semi-structured and unstructured interviews, the researcher collects qualitative data about the ICT resource and infrastructure, legal framework, and human resource which are considered a pillar to implement the e-voting system. The targeted population of the study is four Directorate of the National Election Board of Ethiopia and two Directorate of the Ministry of Innovation and Technology. Purposive sampling is used to select two interviewees from one hundred and seven of the Human Resource Directorate, two interviewees from eight of the IT Directorate, two interviewees from three of the Legal Service Directorate, and two interviewees from thirty Election Operation and Logistic Directorate of the National Election Board of Ethiopia. The reason why the researcher purposively selects is that they are familiar with the challenges of e-voting technology implementation and the factor that affects e-voting technology readiness such as human resources, government willingness and the legal structures. Also, the target population of the study includes the Government Electronic Service Application Development and Administrative Directorate, the Government ICT Network Development and Administrative Directorate of Ministry of Innovation and Technology (MInT). With this in mind, one interviewee from each directorate would be selected from a total of thirty to obtain detailed information about the ICT infrastructure readiness of the country.

Thematic analysis was used to analyze the data which was gathered from the in-depth interviews and document reviews because it is a more powerful tool when combined with research methods such as interviews, observation, and the use of archival records. Thematic analysis is a highly flexible method of research that has been widely used in Library and Information Science (LIS) studies with varying research goals and objectives.

FINDINGS AND DISCUSSION

Summary of Findings

The finding of the study revealed that the adoption of e-voting technology is difficult in light of the current state of Ethiopia. Participants of this study from the Electoral Board indicated much needs to be done to adopt e-voting technology, such as expanding ICT resources and infrastructure, train adequate human resources, and extract legal structure. The willingness of the government to use technology for conducting a free and fair election and experience of e-voting technology from different nations is also the groundwork for the adoption of e-voting technology. The study also revealed a potential gap in ICT infrastructures, human resources, and the legal framework for e-voting technology implementation. According to the findings of the study, Ethiopia's current ICT resources and infrastructure are not mature enough to implement a nationwide e-voting system. Successful implementation of e-voting technology requires a secure data center, accessible telecommunication networks, Internet connectivity and voting equipment. Ethiopia's current telecommunications infrastructures do not encompass the entire parts of the country. There is no sufficient network and internet in most rural areas of the country.

The participant of this study from MInT indicated that the mechanisms used by the government of Ethiopia to protect ICT infrastructures and resources are inadequate. As a result, the country's existing ICT resources and infrastructure are venerable to theft and destruction. There is also a gap in the IT infrastructure of the Ethiopian electoral board. The electoral board did not have a secure data center and internet connectivity to count and store the election data. The study found that there were no qualified staffs' supporting the implementation of the e-voting system in the country. Participants of this study from the Electoral Board indicated that the unavailability of qualified human resources inhibits the implementation of e-voting technology. No expert assesses the environment of the e-voting system in the electoral board. Before implementing e-voting technology continuous training should be given for staff involved in elections. Participants of this study in the Electoral Board revealed that there is a gap in providing technology-related training for employees. The study also found that there is no legal structure for e-voting technology in the country. The study participant from the Electoral Board indicates that there is no legal framework for e-voting technology in Ethiopia. This is because no election was conducted through e-voting. However, in the electoral Proclamation article 51 number 8l, there is an idea about technology which says the Board may deploy technology to assist with the voting and following the vote-counting processes.

Discussion

The purpose of this study was to identify Ethiopia's readiness for the e-voting system implementation from the supply side. The study aimed to address the following research questions to identify the gaps in the country's readiness for e-voting system adoption.

How ready is Ethiopia to adopt the e-voting system from the supply side?

E-voting system can reduce the problem of the traditional ballot voting system of Ethiopia. The paper-based voting systems, which are practiced by Ethiopia, encounter many problems. They are often tedious, insecure, expensive, non-inclusive and delay elections (Achieng and Ruhode, 2013). The interviewees support this, where the Team Leader and Election Expert of Election Operation and Logistic Directorate of NEBE mentioned the problems of paper-based voting systems of Ethiopia as vulnerable to theft and fraud of vote, delay in voting result, inaccessible to disabled voters, expensive, over-voting and problems in the vote-counting process. The e-voting system is a solution to update the electoral process of the country, to reduce the problem of the traditional voting system and to gain trust among the community and opposition parties. Achieng and Ruhode (2013) also said nations like Namibia, Brazil, and Australia adopt electronic voting systems to address numerous challenges related to costs of the physical ballot paper and other overheads, electoral delays, distribution of electoral materials, and general lack of confidence in the electoral process.

E-voting technology provides several benefits. IDEA (2011) stated the benefits of e-voting technology as faster in voting and tabulation, avoiding human error and laborious counting procedures, improving presentation of complicated ballot papers, increasing participation, preventing fraud at polling stations and during the transmission and tabulation of results, increasing accessibility with internet voting for disabled voters and voters from abroad. Also, it offers long-term cost savings by reducing poll worker time and by reducing costs for the manufacturing and distributing of ballot papers. In E-voting system, there are no shipment costs, delays in sending out material and receiving back.

A finding from the study of Krimmer and Schuster (2008) identified the determinant that affects the readiness of the e-voting system as IT infrastructures and resources, legal structures, human resources, and social structures. E-voting readiness also requires the readiness of the Electoral Board to use technology for the election process within its internal structures. The interviewees of the Election Operation and Logistic Directorate agreed that the country is not ready to adopt the e-voting technology. Eight years ago, the Ethiopian Electoral Board and MInT jointly began a trial to deploy e-voting technology. However, the trial was not completed, due to inadequate IT infrastructures, lack of qualified human resources, lack of budget, and continual reforms. The election expert stated that much needs to be done by NEBE and other stakeholder to be ready for e-voting technology adoptions such as expanding IT infrastructure and resources of the country and the electoral board, training adequate human resources and extracting legal framework. The Election Expert also said that extensive research settings are required for adopting e-voting technology. The study revealed that there was no readiness in ICT infrastructures and resources of the country as well as the Electoral Board for e-voting technology adoption. There were also no knowledgeable, experienced, and qualified human resources in the country that support e-voting technology adoption. The study also found that there was no legislative framework for e-voting technology in the country. As a result, the country is not ready to adopt e-voting technology from the supply side.

What are the possible gaps for successful implementation of e-voting systems?

A finding from the study of Kunle Ajayi (2013) identified several factors that affect the implementation of the e-electoral system in Africa. This includes leadership ineptitude and lack of political will, regional insensitivity, ICT infrastructure and resources, qualified human resources, legal framework, corruption, mass poverty, secrecy, and lack of open governance, economic hardship, illiteracy, poor quality of leadership and leadership failure.

One of the several factors that influence the implementation of e-voting technology is ICT resources and infrastructures. The ICT infrastructure team leader of MInT supports this. She said inadequate ICT resources and infrastructure affect the implementation of the e-voting system. Ethiopia's existing ICT resources are not matured enough to implement technological innovation, specifically the e-voting system. The lack of ICT resources and infrastructures within the county makes the implementation of technological innovations difficult.

The finding of the study of Agbesi Samuel (2013) identified ICT infrastructural requirements for the successful nationwide implementation of an e-voting system. This includes fiber links, microwave links and voting devices. The Project Manager of MInT support this; he said the implementation of an e-voting system requires an accessible telecommunications network, Internet connectivity, computer hardware, and software, voting devices, a secure data transmission system, and a secure data center. ICT resource and infrastructure serve as a keystone for any types of e-voting technology. E-voting technology uses ICT resources and infrastructures in pre-voting, while-voting, and post-voting processes. The process includes voter registrations, candidate display, vote casts, vote counting, result transmission, and result notification. To transmit the election results from polling stations to central consolidation centers telecommunication networks and internet connectivity are required. The finding of this study shows country's existing telecommunications infrastructure does not encompass the entire parts. In most rural areas of the country, there is no sufficient network and internet coverage. The mobile network converges of Ethiopia do not reach the entire parts, even if there is no absolute 2G mobile network coverage. This indicates a gap in the telecommunications network requirements for the successful e-voting system implementation.

In Ethiopia, Ethio Telecom is the only government organization that provides telecommunications and internet services for about 114,963,588 peoples, which indicates a gap in the investment of ICT infrastructures and resources. The existing ICT resource and infrastructure of the country are exposed to theft and destruction. The finding of this study shows that there is a gap in coordination across the country to protect ICT infrastructures and resources. Some government and private organizations damage the underground IT infrastructures and resources when they carry out their separate task. Recently Ethio Telecom loses 100 million birrs due to cable theft and vandalism. The infrastructure robbery and damage highly disrupt the internet service and affect the quality of the network. The government needs to conduct several tasks to protect the country's ICT infrastructures and resources by raising awareness of various governmental and non-governmental organizations as well as communities and by building fences for cellular network towers.

From the different types of e-voting systems, the i-voting system is the one that uses the internet for the voting process. Through i-voting voters are not expected to go to the polling station, rather they vote from anywhere. I-voting enables Ethiopians residing around the world to participate in the voting. The government of Australia used I-voting in 2007 G.C national election, which was limited to members of the Australian Defense Force who were serving in various locations, including Iraq and Afghanistan. The implementation of I-voting requires a stable, secure, and fast internet connection. The participants of the interviewees at the MInT have assured that the internet interruption and instability of the country are barriers for implementing web-based applications like I-voting. As stated by the Project Manager of Government Electronic Service Application Development and Administrative Directorate of MInT, increasing internet speed and avoiding interruption are the first task to be done before implementing I-voting and other web-based applications.

The implementation of e-voting technology also requires a well-organized internal structure of the organization regarding technology. Organizational readiness refers to the degree to which an organization is ready to implement IT innovation (in this case e-voting) in line with its internal structural characteristics. The finding of this research indicated that the national election board of Ethiopia does not have a secure, and well-organized IT infrastructure. The IT Director and the Network Administrator of NEBE stated that the existing ICT infrastructures of NEBE could not support the implementation of the e-voting system. There is a lack of IT resources and infrastructure in the organization; this implies the election board is not fully ready for e-voting system implementation. As stated by the IT Director of the NEBE, inadequate IT infrastructure was one of the several reasons for the failure of the 2012 G.C e-voting system trail. E-voting technology requires secure data center of the Electoral Board for storing and counting vote data. The study revealed a gap in the IT infrastructure security mechanism of the election board. The IT Director of the NEBE said that to protect the IT facilities of the election board some physical security mechanisms are applied such as a lock of the data center room and switch boxes. The election board uses Local Area Network (LAN) connectivity to transmit the data at a very fast rate. Also, many users of the organization have data access through the organization LAN. The finding of this study shows that the election board has planned to facilitate IT resources and infrastructure. According to the data obtain from interviewees, there is a plan to connect the nine regions branches with each other and with headquarter via Virtual Private Network (VPN) to increase the speed of communication with branches and head office. The other plan of NEBE related to ICT facilities is changing existing IT devices of branch offices and headquarter into the same brand. The NEBE will need additional ICT infrastructures and resources before e-voting system implementation will be possible.

Adequate and skillful human resources are also other factors that affect the implementation of e-voting technology. Participants of the interviews support this idea. The Human Resource Director and Higher Personnel Expert of the NEBE stated that the implementation of the e-voting system requires qualified and knowledgeable human resources. There are no sufficient human resources in the electoral board for e-voting technology implementation. Before implementing the e-voting technology continuous training should be given for staff involved in elections by the concerned body to reduce technical constraints (Oguejiofor, 2018). In the NEBE, there is a gap in providing e-voting technology-related training to employees. As a result, it is difficult to say that all office staffs are aware of e-voting technology. However, some employees of the election board went to Belgium, India, and Namibia to take experience in e-voting technologies. There is also a gap in leadership support. NEBE Human Resource Director said that the top executives of the election board have an interest in the e-voting system, but they do not support its adoption because of the on-going reform that has taken place in the election board. A finding from the study of Thao & Hwang (2011) identified the main factor-affecting employee's performances in the organizations. This includes the leadership style of the organization, motivation, and training. Reward, recognition, promotion, job enrichment, and payment system of an organization are included in the motivation factors of employees' performance. The finding of this research indicates that the reward, recognition, and payment system of NEBE affects the employees' performance. There is no tradition of a reward and a recognition system in the NEBE. This is because of the organization's structure. Some directorates work throughout the year, while some work only at the time of the election making it difficult to measure the performance of employees. There is a high turnout in the election board due to low salaries and unfulfilling benefits. The Higher Personnel Expert of the NEBE said that in recent times, the board has launched an actual plan to meet employees' benefits and increase wages. The management staff of the electoral board does not influence employees without motivating them. Organizational factors also affect the performance of employees. Some offices of the NEBE have a lack of facilities such as water supply. There is a strategic problem in the headquarters of the election board, i.e. not all directorates are located in the same place making it difficult to communicate and exchange information with each other.

The government of Ethiopia has a gap in the provision of human resources who are knowledgeable in e-voting system. The Human Resource Director of the NEBE stated that if the government and election board decide to conduct elections through e-voting, it is possible to train human resources by communicating in cooperation with the educational institutions. A democracy-oriented legal framework and constitutional requirements are necessary for the implementation of the e-voting system (Mitrou, et al, 2009). The Legal Service Director of the NEBE supports this and has stated that following international law there must be regulation and direction that shows how the elections are conducted. Such regulation and direction shall be included in the legal framework. The legality of e-voting technology must be considered before the election is conducted through the e-voting system. The legal framework of the e-voting system helps to protect the basic democratic rights of the peoples in the election process.

The finding of the study of Bishop and Hoeffler (2016) shows that the legal framework determines the law of the game in the election by ensuring the right to vote and run for office, and that election is held at regular intervals. The NEBE's Legal Service Director explained the benefits of establishing a legal framework for the e-voting system, which include making e-voting technology compatible with existing laws and principles, making the e-voting system lawful and regulated, and listing the e-voting system's transparency and security mechanisms. E-voting technologies are protected under the law that appears in the legal framework. There is no legal framework for e-voting technology in Ethiopia because no election was conducted through e-voting. In Ethiopian, Political Parties Registration, and Election's Code of Conduct Proclamation noting about the e-voting technology, but in Article 51 number 8l there is an idea about technology which says the Board might deploy technology to assist with the voting and, following the vote-counting processes. This shall be done in consultation with contesting political parties and the particulars shall be determined by a directive to be issued by the Board. Formulating a legal framework for e-voting technology requires the willingness of the legislative body of the country. The way of extracting legal framework for e-voting technology stated by the Higher Legal Expert of NEBE as to when the election board and government want to use e-voting technology as the electoral system, the legislative body of the country extract legal framework for e-voting system and present it to House of People Representative (HPR) for approval. Based on the grounded framework the election board establishes specific principles and regulations that help to execute the legislative structure.

CONCLUSION

The traditional voting system used by the Government of Ethiopia is prone to several problems. E-voting technology is a solution to reduce the problems of the traditional voting system, to update the electoral system of the country, and to make the election results reliable for the public and the opposition parties. Along with the successful implementation of the e-voting system, the assessment of readiness is crucial. The following conclusion was drawn from the above findings and discussion.

E-voting technology uses ICT resources and infrastructures to record, cast, and count votes. Ethiopia's existing ICT resources and infrastructure are inadequate to implement e-voting technology. There were no secure and expandable ICT resources and infrastructures across the country which is sufficient for nationwide e-voting technology implementation. There is a gap in coordination across the country to protect ICT infrastructures and resources.

Concerning qualified human resources for the successful implementation of the e-voting system, it was found that there were no qualified, knowledgeable, and experienced staff supporting the implementation of the e-voting system. There are no e-voting experts in Ethiopia and the existing election experts has no depth idea about e-voting technology. Some of the staff of the election board did not hear about e-voting technology which indicates gaps in providing technology-related training to employees. No written documents were available in the country on the legal structure of e-voting technology. Furthermore, there are no standards, rules, and guidance for e-voting technology.

RECOMMENDATIONS

Based on the findings of this research, the researchers recommend the following measures be taken to improve Ethiopia's readiness for the e-voting system implementation.

- The National Election Board of Ethiopia should be ready to communicate with the government and other stakeholders to identify the challenges of the adoption of the e-voting system. The management of the electoral board needs to identify the gaps in e-voting readiness and make efforts to address these gaps.
- A great deal of effort is needed to make the country's current ICT infrastructure and resources adequate for the e-voting technology implementation. The government of Ethiopia needs to engage the private sector in the telecom investment to extend the existing ICT resource and infrastructure.
- Ethio-Telecom and the Ministry of Innovation and Technology should work together to protect the ICT infrastructures and resources of the country. Guidelines should be issued to ensure that governments and public institutions do not interfere with the telecommunications infrastructure when they carry out their tasks.

- The NEBE should be ready to use ICT infrastructure for implementing the e-voting system. The board should conduct various activities to make the existing IT resources enough for e-voting system implementation.
- In terms of qualified human resources, a thoughtful effort is required to implement the e-voting system. Skillful and experienced human resources are needed to address technical problems that may arise on e-voting hardware and software before, during, and after the election. The election board should assign experts who are responsible for assessing the e-voting system environment.
- Awareness about e-voting technology should also be given to employees of the election board. The board shall deploy several tasks for fulfilling the benefits and wages of employees. The top management staff of the election board should provide continual support and guidance for the employees.
- Formulating a legal framework for e-voting technology needs the willingness of the legislative body of the country. Preparing a legal framework for the e-voting system is one of the indicators of e-voting readiness. The election board and legislative body of the country should prepare a legal framework for the e-voting system.
- The legislative body of the country must incorporate the e-voting system into the electoral proclamation. The election board shall also prepare a list of directions and rules that help to execute the e-voting system proclamation.

LIMITATION OF THE STUDY

Referencing or citing earlier research studies provides a theoretical foundation for the research context. The lack of previous research studies in e-voting system readiness put a limitation on the relevant literature review of the research. Although the researchers intended to interview more election experts, legal staff, and IT personnel from the Ethiopian Electoral Board, the election season approaches, time and resource constraints put confinement on the number and selection of interviewees. Another limitation of the study was the lack of information on the country's IT infrastructure and resources. Also, the refusal of Ethio-Telecom limited the amount of information on IT infrastructure and resources.

FUTURE WORK

This research primarily focuses on identifying the gaps in the readiness of Ethiopia for e-voting system implementation from the supply side. This research is a preliminary research in relation to Ethiopia's readiness for e-voting system implementation, and it will serve as the keystone for future research in the area. However, the researchers recommend future studies in the following area: Factors that affect the implementation of e-voting technology; the readiness of the eligible citizens and political parties to conduct elections through the e-voting system; and better types of e-voting systems to engage citizens who living aboard in the election.

ACKNOWLEDGMENT

This research received no specific grant from any funding agency in the public, commercial, or not-for-profit sectors. The researchers would like to express their gratitude to study participants for their assistance in responding to the interview questions.

REFERENCES

Achieng, M., & Ruhode, E. (2013). The Adoption and Challenges of Electronic Voting Technologies Within the South African Context. *International Journal of Managing Information Technology, 5.*

Amnesty International. (2015). *Ethiopia's Human Rights Violations Report 2015.* https://www.amnesty.org/en/countries/africa/ethiopia/report-ethiopia

Bishop, S., & Hoeffler, A. (2016). Free and fair elections: A new database. *Journal of Peace Research, 53*(4), 608–616. doi:10.1177/0022343316642508

Bokslag, W., & Vries, M. (2016). *Evaluating e-voting theory and practice.* Department of Information Security Technology, Technical University Eindhoven.

Bowen, G. A. (2019). Document Analysis as a Qualitative Research Method. *Qualitative Research Journal, 9*(2), 27–40. doi:10.3316/QRJ0902027

Council of Europe. (2005). *Legal, Operational and Technical Standards for E-Voting.* http://www.com.int/t/dgap/democracy/activities/keytexts/recommendat ions/Eng_Evoting_and_Expl_Memo_en.pdf.'

Creswell, J. (2013). *Research design: Qualitative, quantitative, and mixed methods approach* (2nd ed.). SAGE Publications.

Crossman, A. (2019). *Understanding Purposive Sampling. An overview of the method and its application.* https://www.thoughtco.com/purposive-sampling-3026727

Data-Monitor. (2008). *The benefit of e-voting system Feature analysis.* Data-Monitor.

Frankland, R., & Volkamer, M. (2011). *The readiness of various e-Voting systems for complex elections.* Technical Report. https://www.researchgate.net/publication/262933684

Goldsmith, B., & Ruthrauff, H. (2013). *Case Study Report on the Philippines 2010 Elections.* Academic Press.

Gritzalis, D. (2003). *Secure Electronic Voting systems.* Dept. of Informatics Athens University of Economics & Business &Data Protection Commission of Greece. doi:10.1007/978-1-4615-0239-5

Grönlund, Å., & Horan, T. (2004). Introducing e-Gov: History, Definitions, and Issues. *Communications of the Association for Information Systems, 15.*

Habibu, T., Sharif, K., & Nichola, S. (2017). Design and Implementation of Electronic Voting System. *International Journal of Computer & Organization Trends*, *45*(1), 7.

International Institute for Democracy and Electoral Assistance. (2014). *A Brief Assessment Report on Electronic Voting and the 2014 Namibian General Elections*. Author.

International Telecommunication Union. (2018). *Measuring the Information Society Report 2018* (vol. 1). ITU. https://www.itu.int/en/ITU- Retrieved from D/Statistics/Documents/publications/misr2018/MISR-2018-Vol-1-E

jayi, K. (2014). The ICT culture and transformation of Electoral Governance and politics in Africa: the challenges and prospects. *5th European Conference on African Studies*.

Johnson, R. B., & Christensen, L. B. (2004). *Educational research: Quantitative, qualitative, and mixed approaches*. Allyn and Bacon.

Joppe, M. (2000). *The Research Process*. http://www.ryerson.ca/~mjoppe/rp.htm

Krimmer, R., & Schuster, R. (2008). *The E-Voting Readiness Index*. Competence Centre for Electronic Voting and Participation.

Kumar, S., & Walia, E. (2011). Analysis of Electronic Voting System in Various Countries. *International Journal on Computer Science and Engineering*, *3*(5).

Lambrinoudakis, C., & Kokolakis, S. (2002). Functional Requirements for a Secure Electronic Voting System. *Conference Paper*.

Lubis, M., Kartiwi, M., & Durachman, Y. (2017). *Assessing Privacy and Readiness of Electronic Voting System in Indonesia*. International Islamic University Malaysia and Syarif Hidayatullah State Islamic University. doi:10.1109/CITSM.2017.8089242

Makarava, Y. (2011). *Critical Assessment of the Relationship between E-governance and Democracy*. Mid Sweden University.

Mitrou, L., Gritzalis, D., Katsikas, S., & Quirchmayr, G. (2009). *E-Voting: Constitutional and Legal Requirements and Their Technical Implications*. Academic Press.

Mokodir, P. E. (2011). *E-voting readiness in Kenya: A case study of Nairobi County*. University of Nairobi School of Computing and Informatics.

National Election Board of Ethiopia. (n.d.). In *Wikipedia, the Free Encyclopaedia*. https://en.wikipedia.org/w/index.php?title=National_Election_Board_of_Ethiopia&oldid=791863032

Ntulo, G., & Otike, J. (2013). *E-Government: Its Role, Importance, and Challenges*. School of Information Sciences Moi University Eldoret.

O'Meara, M. (2013). *Survey and Analysis of E-Voting Solutions*. Master in Computer Science Trinity College Dublin.

Oguejiofor, O. O. (2018). Advancing Electronic Voting Systems in Nigeria's Electoral Process. *Afe Babalola University: J. of Sust. Dev. Law & Policy, 9*(2).

Onwe, S., Nwogbaga, D., & Ogbu, M. (2015). *Effects of Electoral Fraud and Violence on Nigeria Democracy: Lessons from 2011 Presidential Election.* Academic Press.

Rogers, M. E. (2009). *Diffusion of Innovations.* The Free Press.

Rubin, D. A. (2003). *Analysis of an Electronic Voting System.* Information Security Institute.

Samuel, A. (2013). *Investigating the feasibility of implementing the E-Voting system in Ghana.* Department of Computer Science, Kwame Nkrumah University of Science and Technology.

Thao, H., & Hwang, J. (2010). *Factors affecting employee performance. Evidence from Petro Vietnam engineering consultancy J.S.C.* Academic Press.

Yildiz, M. (2007). Decision making in e-government projects. The case of Turkey. In Handbook of Decision-making. Marcel Dekker Publication.

Yin, R. K. A. (2003). *Case study research; design and methods (3rd ed.).* Sage Publications Inc.

Chapter 12
The Digitalization of Health Behaviors:
A Bibliometric Analysis

Ece Özer Çizer
iD https://orcid.org/0000-0002-8597-2073
Yıldız Technical University, Turkey

ABSTRACT

The number of studies aiming to change consumers' health behaviors by digitizing is increasing day by day. Health behaviors have become the focus of different sciences and disciplines, both socio-cultural and technological. Especially after the COVID-19 pandemic, investments and research in this field have increased considerably. In the literature, there is a rapid increase in the number of studies on digital health behaviors. This study aims to systematically examine the studies in the literature on digitalized health behaviors. For this purpose, bibliometric analysis was applied to 357 studies in the Web of Science database using the R Studio program in order to determine the most frequently studied areas and possible gaps in the literature.

INTRODUCTION

Today, digitalization in every aspect of life has become a necessity, not an option (Gupta et al., 2020). Business processes and ways of doing business have changed, transformed by digitalization (Galyarski & Mironova, 2021). In the digital world, digital platforms have increased (Cozmiuc and Pettinger, 2021). When it comes to digital platforms, e-commerce platforms come to mind first. However, it would not be right to limit the scope of these platforms only to e-commerce. Many sectors, especially education, health, and service, have transformed and taken their place on digital platforms (Dwivedi et. al., 2020). Digitizing industries have created their digital ecosystem (Subramaniam, 2020). In this digital ecosystem, customers, payment systems, products, services, and more have become digital (Gruia et al., 2020). Especially the Covid-19 pandemic has accelerated the digitalization process for every sector (Nanda et al., 2021).

DOI: 10.4018/978-1-6684-4102-2.ch012

Covid-19, which emerged in the city of Wuhan, China in December 2019, has taken the whole world under its influence in a short time (Sedik et al., 2022). Covid-19 was declared a pandemic in a short time by the World Health Organization (Arora and Gray, 2020). With the declaration of the pandemic, countries have also started to apply quarantine and take measures to protect against the virus. Along with the quarantine application and the measures taken, the Covid-19 pandemic has suddenly changed people's lives, ways of doing business and behaviors (Aydın et al, 2021). During the pandemic period, both states and individuals have started to take various measures to protect themselves from the virus. As a result of both the pandemic and the measures taken, the Covid-19 pandemic has caused radical changes in human behavior (Jetten et al, 2021). Online shopping has increased to reduce physical contact, and individuals have become more attentive to their health status (Donthu & Gustafsson, 2020). One of the measures taken during the pandemic period is curfews. With the curfews, people have started to spend more time at home (Andronico et al, 2021). The increase in time spent at home has led people to practice sports at home, prepare healthier meals at home instead of ordering food from outside, and to use food supplements to maintain body resistance. In addition to these, individuals who cannot leave the house and cannot socialize with other people have started to benefit from various meditation and online psychological support practices to avoid the negative psychological effects caused by social isolation.

BACKGROUND

In this section, while examining the digitalized health habits of consumers, first of all, the Covid-19 pandemic period, which accelerates the transfer of health behaviors to the digital environment, will be discussed. Then, the most used theories in the literature to explain health behaviors will be given.

DIGITALIZED CONSUMER BEHAVIOR IN THE COVID-19 PANDEMIC

The Covid-19 pandemic has facilitated the rapid digitalization of consumers' health behaviors. Many studies have been conducted to examine the health habits and changes in health habits of consumers during and after the pandemic period (Adams et al., 2020; Arora and Grey, 2020; Bahl et al., 2020; Castañeda-Babarro et al., 2020). These studies were discussed by Özer Çizer (2022) under the headings of nutrition, physical activity, mental health and sleep patterns, telemedicine services, and hygiene behaviors. In the study, it is seen that the changing health habits of consumers are also digitalized. For example, in the studies carried out during the pandemic period, it has been revealed that individuals who cannot go out focus on sports activities at home. As a result of this focus, the number of online fitness apps and wearable device users has also increased day by day (Castañeda-Babarro et al., 2020). Individuals who do sports at home have also benefited from digital content such as online fitness training, yoga, and pilates classes (Mutz et al., 2021). In addition, individuals who want to do sports within a certain program have increased their physical activity at home by making use of the applications of various sports brands such as Nike Training Club, Adidas Training, and MAC +. In the post-pandemic normalization period, users of fitness applications continue to use these applications, which are also adapted for external use, in gyms, while jogging or walking. In the Google Play Store, App Store, Windows Store, and Amazon Appstore, physical activity applications that encourage sports at home, which have gained even more popularity during the pandemic period, are still in the first place under the health and fitness category

(Al-Abbadey et al., 2021). In addition to health and meditation applications, smartwatches and wristbands have increased their sales even more in this period as tools that provide both sleep tracking and physical activity measurement. Individuals who want to monitor their exercise and sleep can easily access the recorded data with the applications of the watches and wristbands they download to their phones. Smartwatches and wristbands also provide information such as sitting alert, calories burned, the number of steps taken, sleep quality, and time spent in sleep, as well as heart rate information and the amount of oxygen in the blood. Fitbit, Apple Watch, Samsung Galaxy Watch, Xiaomi Mi Band, and Huawei Band are among the most preferred smartwatch and wristband brands in the world (Peckham, 2021).

The rapid spread of Covid-19 has increased awareness, especially about health and well-being. Studies on public health during the pandemic period have revealed that people are most concerned about their health and the health of their relatives (Accenture, 2020). Individuals emphasized that they started to adopt a healthier lifestyle to keep their worries under control. During the pandemic process, the use of practices such as online rehabilitation services and therapies, well-being, and yoga has increased among people who want to protect both themselves and their relatives by adopting a healthier lifestyle (WHO, 2020). In the post-pandemic period, meditation and sleep regulation applications such as Meditopia and Calm, and yoga applications such as Asana Rebel and Yoga Studio are still very popular applications and they continue to be used (Özer Çizer, 2022).

Telemedicine services are used as a method of remote patient treatment through communication technologies. These services ensure the transfer of medical information and data from the patient to the specialist through advanced technologies. Since no physical environment is required during this transfer, telemedicine applications have taken their place among the most popular applications of the pandemic period (Jnr, 2020). While these applications help to maintain social distance, they also allow people to easily access health services from their homes (Bahl et al., 2020).

When the usage status of telemedicine applications in the world is researched, the usage levels of such applications are associated with the availability of suitable infrastructure conditions. The USA, China, some European countries, and Australia are given as examples of the countries with developed infrastructure that use telemedicine technologies the most (Kalhori et al., 2021). During the pandemic process, the sensitivity to the delivery of necessary health services, especially to children with chronic diseases has led doctors to telemedicine applications. Utilizing telemedicine applications to avoid both the virus and the time and transportation costs is a good alternative for many patients with chronic diseases (Vaishya et al., 2020).

THEORIES OF HEALTH BEHAVIOR

Behavioral theories provide a conceptual framework that uncovers key structures for creating a new behavior or changing an existing behavior (Rhodes et al., 2019). Behavioral theories that are currently used are also used to explain the health behaviors of consumers (Fortuna et al., 2019). While explaining the health behaviors of consumers who have changed by digitalization, the Social Cognitive Theory, Health Belief Model, and Transtheoretical Model are frequently used, especially the Theory of Planned Behavior (Medlock and Wyatt, 2019).

Theory of Planned Behavior: Theory of Planned Behavior explains the role of behavioral intention as a precursor to performing a behavior (Erul et al., 2020). According to this theory, a person's behavioral intention is directly affected by their attitudes, subjective norms, and social norms (Liu et al., 2018).

Planned Behavior Theory is frequently used in studies in the literature on changing health behaviors, dietary patterns, quitting smoking habits, and e-health literacy (Raman et al., 2013; Hou et al., 2014; Levin-Zamir and Bertsch, 2019; Athamneh et al., 2017).

Social Cognitive Theory: Social Cognitive Theory proposes that behavior change occurs in the context of interactions between personal factors, behavioral factors, and environmental factors (Kursan Milaković, 2021). The theory emphasizes that there are several key determinants of behavior that enable it to emerge. According to the theory; Outcome expectations are an individual's beliefs about the outcome that may result from performing a behavior and the perceived value of that outcome (Schunk and DiBenedetto, 2020). Social outcome expectations take into account an individual's beliefs about how other people will evaluate them if they engage in a behavior and how other people value the consequences of that behavior. Self-evaluation results expectations refer to an individual's beliefs about how they will feel about themselves if they decide to perform a particular behavior (Naslund et al., 2017). Self-efficacy, which is the most widely known construct of Social Cognitive Theory, expresses the confidence in an individual's ability to perform a behavior (Lin et al., 2020). Self-efficacy is influenced by individuals' past experiences, perceptions of the environment and social context, and intellectual and physical abilities (Kamkari, 2022). Studies using Social Cognitive Theory in the field of health behavior change have mainly focused on increasing self-efficacy. Social Cognitive Theory is among the theories frequently used in the field of health behavior in the literature on diabetes management, dietary habits change, skill training patients, and lifestyle behavioral changes (Ghoreishi et al., 2019; Javed and Charles, 2018; Li et al., 2019; Beauchamp et al., 2019).

Health Belief Model: The Health Belief Model is a model that explains behavior based on individuals' beliefs and attitudes (Shmueli, 2021). Self-efficacy in this model refers to one's confidence in one's ability to act. Self-efficacy is an important predictor of initiating or adopting behavior (Shiau et al., 2020). In the Health Belief Model, it is argued that individuals' health behaviors will be affected by their beliefs, values and attitudes. If the beliefs and attitudes that are seen as a problem are determined, the health education to be given or the treatment methods to be applied will be determined more suitable for that person (Naslund et al., 2017). The Health Belief Model is a model frequently used in health-related studies in the literature. The Health Belief Model is used in determining the obstacles to the treatment of patients, especially in studies on mental illnesses where privacy is at the forefront (O'connor et al., 2014; Zhou et al., 2021; Marashi et al., 2021).

Transtheoretical Model: The Transtheoretical Model proposes that the decision-making process is a cyclical process and includes six stages that show how the decision-making process takes place over time (Naslund et al., 2017). According to the model, an individual goes through six stages while changing behavior.

-Precontemplation: At this stage, the individual has no intention of changing his behavior in the near future. Individuals at this stage have a very low awareness of the harm caused by their behavior. They may resist change or be unmotivated because of the failures they experienced when they wanted to change their behavior before (Inman et al., 2022).

-Contemplation: At this stage, the individual has an intention to change the behavior in the near future. Individuals at this stage are aware of the harms of their behavior, change. They believe it will be helpful. They explore the pros and cons of behavior change (Rai and Arokiasamy, 2021).

-Preparation: At this stage, the individual does not want to postpone changing his behavior, he wants to experience change immediately. Individuals at this stage may have started to engage in activities such as getting health education, talking to a doctor, and buying health-related books in order to change their behavior (Mansuroğlu and Kutlu, 2022).

-Action: At this stage, the individual has already changed his behavior. Individuals at this stage have changed their problematic behaviors or adopted healthier behaviors. These individuals also intend to continue to make further behavioral changes going forward (Inman et al., 2022).

-Maintenance: At this stage, the individual continues his/her new behavior uninterruptedly without returning to the previously changed behavior. Individuals in this stage become attached to their new behavior to maintain behavior changes and avoid reverting to earlier stages in the long run (Clarkson et al., 2022).

-Termination: Individuals in the final stage are completely attached to their new behavior and no longer think of returning to their old behavior (Akdaş and Cismaru, 2021).

The Transtheoretical Model shows how health behaviors are adopted and maintained over time. In the literature, the Transtheoretical Model is frequently used in studies on smoking cessation, receiving psychological support, healthy lifestyle interventions, and adopting physical activities (Tseng et al., 2022; Evans et al., 2018; Mansuroğlu and Kutlu, 2022; Jiménez-Zazo et al., 2020).

METHODOLOGY

In this study, it is aimed to examine the reflections of the digitalized health behaviors of consumers in the digitalized world in the literature. For this purpose, 357 articles between 1998 and 2022 containing the keywords "digital*", and "health behavior" will be selected from the Web of Science database and bibliometric analysis will be applied with the R Studio analysis program. As a result, the analysis, it is aimed to create a comprehensive framework for the digitalized health behaviors of consumers. The digitalized health behaviors of consumers will be discussed under the headings within the framework of the main themes that emerged as a result of the bibliometric analysis.

Figure 1. Word Cloud

In Figure 1, the 50 most repetitive words in the data set were examined. Among these words, it is seen that words such as "mhealth", "digital health", "behavior change", "physical activity" and "ehealth" are common words in articles.

Figure 2. Word Tree

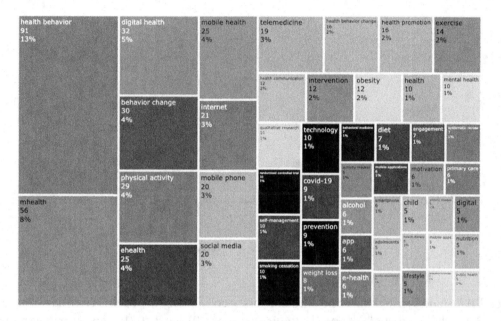

In the word tree shown in Figure 2, the distribution of the frequencies of the words is given in order. In addition to the leading words in the word cloud "mobile health", "internet", "mobile phone", "social media" and "telemedicine" are frequently discussed in the literature.

Figure 3. Trend Topics

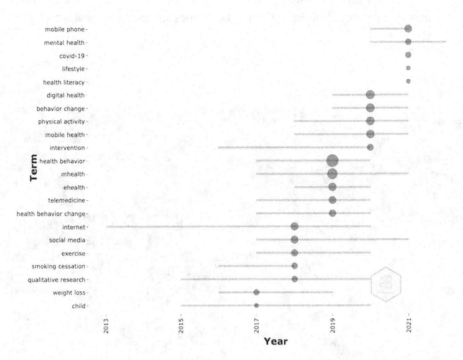

Since 2013, studies on digitalized health behaviors of consumers have been carried out in the literature. In Figure 3, the distribution of the trending topics between 2013 and 2021 is given. In the studies in the literature, starting from 2013 until the beginning of the 2020s, the main theme studied was "internet". Between 2015-2020 it is seen that the concepts of "qualitative research" and "child" have come to the fore. There has been an increase in the number of studies on "weight loss" in 2017 and "smoking cessation" in 2018. Between 2017-2020, publications on "health behavior", "mhealth", "telemedicine", "health behavior change", "social media" and "exercise" topics were produced. Between 2018 and 2020, it is seen that "physical activity", "mobile health", and "ehealth". studies accelerated. In 2020 "digital health" and "behavior change" topics are studied. After 2020, "mobile phone", "mental health", "covid-19", "lifestyle", and "health literacy" topics gain popularity.

Figure 4. Thematic Map

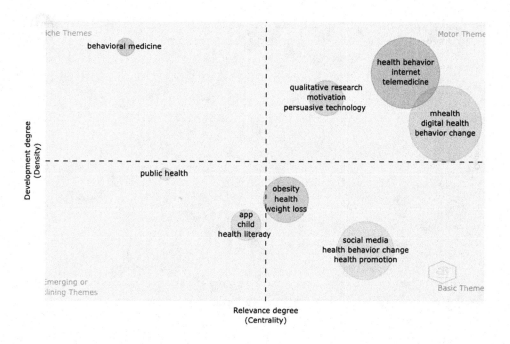

In the study, a thematic map divided into four topological regions based on density and centrality was created and is shown in Figure 4. The upper right region offers "motor" themes indicated with high intensity and centrality; These topics are gathered around three clusters. Topics in each cluster refer to topics that are frequently studied together in the literature. The first cluster includes "health behavior", "internet" and "telemedicine"; the second cluster includes "mhealth", "digital health" and "behavior change"; the last cluster includes "qualitative research", "motivation" and "persuasive technology". These topics in the motor themes need to be further developed given their importance for future research.

In the upper left region, "behavioral medicine" topic is high density but low centrality. These topics in this region illustrate specific and underrepresented issues that are areas of rapid development.

Themes in the lower left region are shown with low centrality and density. Two clusters are shown in this region. The first cluster of this region includes "app", "child" and "health literacy" topics and the second cluster of this region include only "public" topic. These topics in this area of the thematic map now include topics with a downward trend.

Finally, the lower right region contains the core themes shown with high centrality but low intensity. This field includes two clusters. The first and biggest cluster of this region includes "social media", "health behavior change" and "health motivation" other cluster includes "obesity", "health" and "weight loss". These subjects in this region refer to the general importance of the studies to be carried out.

Figure 5. Most Cited Countries

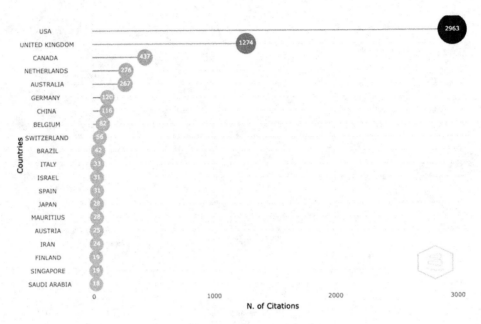

According to Figure 5, it is seen that the most cited publications in studies on digitalized health behaviors of consumers are presented by authors from USA, United Kingdom, Canada, Netherlands and Australia, respectively.

Figure 6. Annual Scientific Production

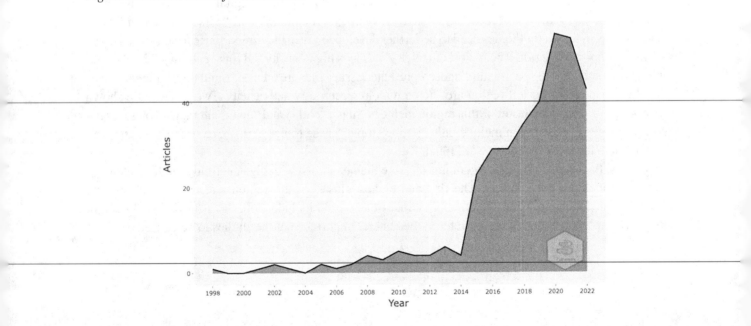

In Figure 6, it is seen that the first studies on digitalized health behaviors of consumers were put forward in 1998. In addition, the number of studies increased rapidly starting from 2014 and reaching the highest level in 2020. The reason for this obvious increase in 2020 can be shown as the digitalization of health behaviors as well as the digitalization of all processes with the emergence of the Covid-19 epidemic.

Figure 7. Corresponding Author's Country

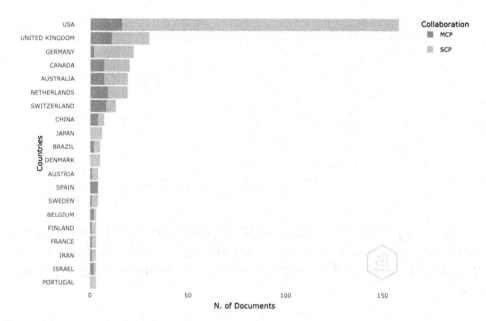

The countries of the responsible authors for the publications were examined and shown in Figure 7. According to Figure 7, it is seen that the country with the most publications is the United States of America. United States of America is followed by United Kingdom, Germany, and Canada respectively. While the green colors in the table show that the broadcasts are made in a single country, the orange-colored parts indicate the number of broadcasts from multiple countries.

Figure 8. Most Relevant Sources

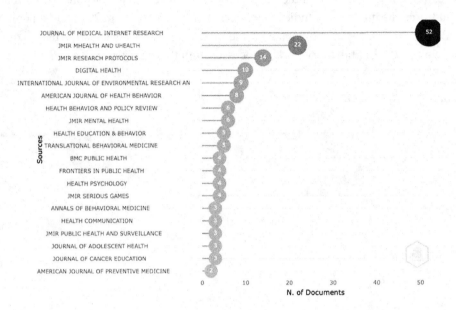

The most preferred journals were searched to publish studies on digitalized health behaviors of consumers. According to the most relevant sources in Figure 8, "Journal of Medical Internet Research", "JMIR MHealth and UHealth", and "JMIR Research Protocols" are the most popular scientific journals, respectively.

Figure 9. Co-occurrence Network

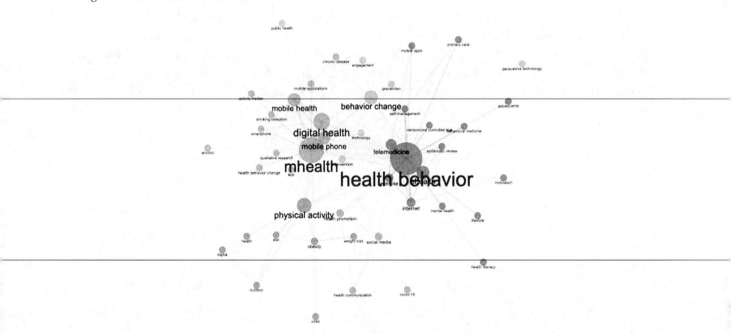

Figure 9 shows the interactivity of the keywords used in the studies in the literature with each other. Keywords are divided into five separate clusters. Each subset is colored with a different color. The red cluster shows the most frequently used word network and the words in this network are concentrated under the keyword "health behavior", while in the blue cluster it is observed that the keywords are distributed around "mhealth". The words in the purple cluster are concentrated around the keyword "physical activity" and the words in the green cluster are concentrated around the keyword "behavior change". The last cluster is shown orange color. Keywords in the orange clusters show a dispersed distribution without centralization.

DISCUSSION

This study is a comprehensive review that examines digital health behaviors as a result of the bibliometric analysis performed within the scope of the keywords "digital" and "health behavior". In this review, the articles in the globally published Web of Science database were examined, and the current state of the literature, which topics were studied in the past studies, and what were the touchstones for future studies were revealed.

FUTURE RESEARCH DIRECTIONS

Based on the findings of the study, a number of recommendations will be made for future research in this area. In the study, studies on digital health behaviors in the literature were examined. The theoretical background, trends, most studied topics, and gaps in the literature were investigated. As a result of the literature review and bibliometric analysis carried out in the study, it has been determined that there are theoretical inadequacies in understanding the health behavior of consumers that have changed and become digital. While the health behaviors of consumers are digitalized in the studies in the literature, this digitalization has been considered as a change and tried to be explained with the existing behavioral theories in the literature. Future studies should be handled within the scope of innovative theories that will be put forward within the scope of digital-health-behavior by making use of existing behavioral theories. An ingenious combination of theories used in the digital field and those used in the behavioral field will be more effective in explaining the digitalized health behaviors of consumers. According to the results of the study, it has been observed that there has been an increase in the number of studies on "mental health", "life style" and "health literacy" in recent years. Innovative studies on these topics, which have become trendy with the Covid-19 pandemic, should contribute to the literature. The thematic map presented in the study shows the general view of the studies in the literature. By making use of this map, studies can be carried out on "app", "child" "health literacy" and "public health" subjects, which are recommended to be developed within the scope of frequently studied subject clusters.

CONCLUSION

The purpose of this systematic review is to reveal the current view in the literature by examining the change and development of consumers' digitalized health behaviors. In the study, trend topics, the most used theories, and the pioneers that trigger the digitalization of consumers' health behaviors are revealed. Based on the findings of the study, a road map was drawn for future studies and suggestions were made.

REFERENCES

Accenture. (2020). *Data-driven insights into consumer behavior How is COVID-19 changing the retail consumer?* https://www.accenture.com/_acnmedia/PDF-130/Accenture-Retail-Research-POV-Wave-Seven.pdf

Adams, K. K., Baker, W. L., & Sobieraj, D. M. (2020). Myth Busters: Dietary Supplements and COVID-19. *The Annals of Pharmacotherapy, 54*(8), 820–826. doi:10.1177/1060028020928052 PMID:32396382

Akdaş, O., & Cismaru, M. (2021). Promoting mental health during the COVID-19 pandemic: The transtheoretical model of change and social marketing approach. *International Review on Public and Nonprofit Marketing*, 1–28.

Al-Abbadey, M., Fong, M. M., Wilde, L. J., Ingham, R., & Ghio, D. (2021). Mobile health apps: An exploration of user-generated reviews in Google Play Store on a physical activity application. *Digital Health, 7*, 20552076211014988. doi:10.1177/20552076211014988 PMID:34017609

Andronico, A., Tran Kiem, C., Paireau, J., Succo, T., Bosetti, P., Lefrancq, N., Nacher, M., Djossou, F., Sanna, A., Flamand, C., Salje, H., Rousseau, C., & Cauchemez, S. (2021). Evaluating the impact of curfews and other measures on SARS-CoV-2 transmission in French Guiana. *Nature Communications, 12*(1), 1–8. doi:10.103841467-021-21944-4 PMID:33712596

Arora, T., & Grey, I. (2020). Health behaviour changes during COVID-19 and the potential consequences: A mini-review. *Journal of Health Psychology, 25*(9), 1155–1163. doi:10.1177/1359105320937053 PMID:32551944

Athamneh, L., Essien, E. J., Sansgiry, S. S., & Abughosh, S. (2017). Intention to quit water pipe smoking among Arab Americans: Application of the theory of planned behavior. *Journal of Ethnicity in Substance Abuse, 16*(1), 80–90. doi:10.1080/15332640.2015.1088423 PMID:26720395

Aydemir, S., Ocak, S., Saygılı, S., Hopurcuog, D., Has, F., ˘rul Kıykım, E., … Canpolat, N. (n.d.). *Telemedicine Applications in a Tertiary Pediatric Hospital in Turkey During COVID-19 Pandemic.* Academic Press.

Aydin, K., Çizer, E. Ö., & Köse, Ş. G. (2021). Analyzing attitude towards COVID-19 vaccine in the context of the health industry: The role of country of origin image. *Duzce Medical Journal, 23*(1), 122–130.

Bahl, S., Singh, R. P., Javaid, M., Khan, I. H., & Vaishya, R., & Suman, R. (2020). Telemedicine Technologies for Confronting COVID-19 Pandemic. *RE:view, 5*(4), 547–561.

Beauchamp, M. R., Crawford, K. L., & Jackson, B. (2019). Social cognitive theory and physical activity: Mechanisms of behavior change, critique, and legacy. *Psychology of Sport and Exercise, 42*, 110–117. doi:10.1016/j.psychsport.2018.11.009

Castañeda-Babarro, A., Coca, A., Arbillaga-Etxarri, A. ve Gutiérrez-Santamaría, B. (2020). Physical Activity Change during COVID-19 Confinement. *International Journal of Environmental Research and Public Health, 17*(18), 6878.

Clarkson, P., Stephenson, A., Grimmett, C., Cook, K., Clark, C., Muckelt, P. E., O'Gorman, P., Saynor, Z., Adams, J., Stokes, M., & McDonough, S. (2022). Digital tools to support the maintenance of physical activity in people with long-term conditions: A scoping review. *Digital Health, 8*, 20552076221089778. doi:10.1177/20552076221089778 PMID:35433017

Cozmiuc, D. C., & Pettinger, R. (2021). Consultants' Tools to Manage Digital Transformation: The Case of PWC, Siemens, and Oracle. *Journal of Cases on Information Technology, 23*(4), 1–29. doi:10.4018/JCIT.20211001.oa7

Donthu, N., & Gustafsson, A. (2020). Effects of COVID-19 on business and research. *Journal of Business Research, 117*, 284–289. doi:10.1016/j.jbusres.2020.06.008 PMID:32536736

Dwivedi, Y. K., Hughes, D. L., Coombs, C., Constantiou, I., Duan, Y., Edwards, J. S., Gupta, B., Lal, B., Misra, S., Prashant, P., Raman, R., Rana, N. P., Sharma, S. K., & Upadhyay, N. (2020). Impact of COVID-19 pandemic on information management research and practice: Transforming education, work and life. *International Journal of Information Management, 55*, 102211. doi:10.1016/j.ijinfomgt.2020.102211

Erul, E., Woosnam, K. M., & McIntosh, W. A. (2020). Considering emotional solidarity and the theory of planned behavior in explaining behavioral intentions to support tourism development. *Journal of Sustainable Tourism, 28*(8), 1158–1173. doi:10.1080/09669582.2020.1726935

Evans, M., Malpass, A., Agnew-Davies, R., & Feder, G. (2018). Women's experiences of a randomised controlled trial of a specialist psychological advocacy intervention following domestic violence: A nested qualitative study. *PLoS One, 13*(11), e0193077. doi:10.1371/journal.pone.0193077 PMID:30481185

Fortuna, K. L., Brooks, J. M., Umucu, E., Walker, R., & Chow, P. I. (2019). Peer support: A human factor to enhance engagement in digital health behavior change interventions. *Journal of Technology in Behavioral Science, 4*(2), 152–161. doi:10.100741347-019-00105-x PMID:34337145

Galyarski, E., & Mironova, N. (2021). Digitalization and its impact on business processes. *Economics and Management, 18*(1), 81–89.

Ghoreishi, M. S., Vahedian-Shahroodi, M., Jafari, A., & Tehranid, H. (2019). Self-care behaviors in patients with type 2 diabetes: Education intervention base on social cognitive theory. *Diabetes & Metabolic Syndrome, 13*(3), 2049–2056. doi:10.1016/j.dsx.2019.04.045 PMID:31235135

Gruia, L. A., Bibu, N., Roja, A., Danaiață, D., & Năstase, M. (2020). Digital Transformation of Businesses in Times of Global Crisis. In *Griffiths School of Management and IT Annual Conference on Business, Entrepreneurship and Ethics* (pp. 43-62). Springer.

Gupta, B. B., Chaudhary, P., & Gupta, S. (2020). Designing a XSS defensive framework for web servers deployed in the existing smart city infrastructure. *Journal of Organizational and End User Computing, 32*(4), 85–111. doi:10.4018/JOEUC.2020100105

Hou, S. I., Charlery, S. A. R., & Roberson, K. (2014). Systematic literature review of Internet interventions across health behaviors. *Health Psychology and Behavioral Medicine*, *2*(1), 455–481. doi:10.108 0/21642850.2014.895368 PMID:25750795

Inman, R. A., Moreira, P. A., Faria, S., Araújo, M., Cunha, D., Pedras, S., & Correia Lopes, J. (2022). An application of the transtheoretical model to climate change prevention: Validation of the climate change stages of change questionnaire in middle school students and their schoolteachers. *Environmental Education Research*, *28*(7), 1003–1022. doi:10.1080/13504622.2021.1998382

Javed, A., & Charles, A. (2018). The importance of social cognition in improving functional outcomes in schizophrenia. *Frontiers in Psychiatry*, *9*, 157. doi:10.3389/fpsyt.2018.00157 PMID:29740360

Jetten, J., Bentley, S. V., Crimston, C. R., Selvanathan, H. P., & Haslam, S. A. (2021). COVID-19 and social psychological research: A silver lining. *Asian Journal of Social Psychology*, *24*(1), 34–36. doi:10.1111/ajsp.12465 PMID:33821140

Jiménez-Zazo, F., Romero-Blanco, C., Castro-Lemus, N., Dorado-Suárez, A., & Aznar, S. (2020). Transtheoretical model for physical activity in older adults: Systematic review. *International Journal of Environmental Research and Public Health*, *17*(24), 9262. doi:10.3390/ijerph17249262 PMID:33322327

Jnr, B. A. (2020). Use of telemedicine and virtual care for remote treatment in response to COVID-19 pandemic. *Journal of Medical Systems*, *44*(7), 1–9. PMID:32542571

Kalhori, S. R. N., Bahaadinbeigy, K., Deldar, K., Gholamzadeh, M., Hajesmaeel-Gohari, S., & Ayyoubzadeh, S. M. (2021). Digital health solutions to control the COVID-19 pandemic in countries with high disease prevalence: Literature review. *Journal of Medical Internet Research*, *23*(3), e19473. doi:10.2196/19473 PMID:33600344

Kamkari, K. (2022). Communication model of self-concept and Exercise self-efficacy on mental health of physical education staff in Iranian universities of medical sciences of Tehran. *Majallah-i Ulum-i Pizishki-i Razi*, *28*(11), 14–23.

Kursan Milaković, I. (2021). Purchase experience during the COVID-19 pandemic and social cognitive theory: The relevance of consumer vulnerability, resilience, and adaptability for purchase satisfaction and repurchase. *International Journal of Consumer Studies*, *45*(6), 1425–1442. doi:10.1111/ijcs.12672 PMID:33821146

Levin-Zamir, D., & Bertschi, I. (2019). Media health literacy, eHealth literacy and health behaviour across the lifespan: Current progress and future challenges. International handbook of health literacy, 275.

Li, A. S. W., Figg, G., & Schüz, B. (2019). Socioeconomic status and the prediction of health promoting dietary behaviours: A systematic review and meta-analysis based on the theory of planned behaviour. *Applied Psychology. Health and Well-Being*, *11*(3), 382–406. doi:10.1111/aphw.12154 PMID:30884154

Lin, C. Y., Imani, V., Majd, N. R., Ghasemi, Z., Griffiths, M. D., Hamilton, K., Hagger, M. S., & Pakpour, A. H. (2020). Using an integrated social cognition model to predict COVID-19 preventive behaviours. *British Journal of Health Psychology*, *25*(4), 981–1005. doi:10.1111/bjhp.12465 PMID:32780891

Liu, S., Chiang, Y. T., Tseng, C. C., Ng, E., Yeh, G. L., & Fang, W. T. (2018). The theory of planned behavior to predict protective behavioral intentions against PM2. 5 in parents of young children from urban and rural Beijing, China. *International Journal of Environmental Research and Public Health, 15*(10), 2215. doi:10.3390/ijerph15102215 PMID:30309043

Mansuroğlu, S., & Kutlu, F. Y. (2022). The Transtheoretical Model based psychoeducation's effect on healthy lifestyle behaviours in schizophrenia: A randomized controlled trial. *Archives of Psychiatric Nursing, 41*, 51–61. doi:10.1016/j.apnu.2022.07.018 PMID:36428075

Marashi, M. Y., Nicholson, E., Ogrodnik, M., Fenesi, B., & Heisz, J. J. (2021). A mental health paradox: Mental health was both a motivator and barrier to physical activity during the COVID-19 pandemic. *PLoS One, 16*(4), e0239244. doi:10.1371/journal.pone.0239244 PMID:33793550

Medlock, S., & Wyatt, J. C. (2019). Health behaviour theory in health informatics: Support for positive change. *Studies in Health Technology and Informatics, 263*, 146–158. PMID:31411160

Mutz, M., Müller, J., & Reimers, A. K. (2021). Use of digital media for home-based sports activities during the COVID-19 pandemic: Results from the German SPOVID survey. *International Journal of Environmental Research and Public Health, 18*(9), 4409. doi:10.3390/ijerph18094409 PMID:33919180

Nanda, A., Xu, Y., & Zhang, F. (2021). How would the COVID-19 pandemic reshape retail real estate and high streets through acceleration of E-commerce and digitalization? *Journal of Urban Management, 10*(2), 110–124. doi:10.1016/j.jum.2021.04.001

Naslund, J. A., Aschbrenner, K. A., Kim, S. J., McHugo, G. J., Unützer, J., Bartels, S. J., & Marsch, L. A. (2017). Health behavior models for informing digital technology interventions for individuals with mental illness. *Psychiatric Rehabilitation Journal, 40*(3), 325–335. doi:10.1037/prj0000246 PMID:28182469

O'connor, P. J., Martin, B., Weeks, C. S., & Ong, L. (2014). Factors that influence young people's mental health help-seeking behaviour: A study based on the Health Belief Model. *Journal of Advanced Nursing, 70*(11), 2577–2587. doi:10.1111/jan.12423 PMID:24720449

Peckham, J. (2021). *Best smartwatch 2021: The top wearables you should buy today.* TechRadar. https://www.techradar.com/news/wearables/best-smart-watches-what-s-the-best-wearable-tech-for-you-1154074

Rai, B., & Arokiasamy, P. (2021). Identifying stages of smoke and smokeless tobacco cessation among adults in India: An application of transtheoretical model. *Journal of Substance Use, 26*(4), 343–350. doi:10.1080/14659891.2020.1807634

Raman, J., Smith, E., & Hay, P. (2013). The clinical obesity maintenance model: An integration of psychological constructs including mood, emotional regulation, disordered overeating, habitual cluster behaviours, health literacy and cognitive function. *Journal of Obesity, 2013*, 2013. doi:10.1155/2013/240128 PMID:23710346

Rhodes, R. E., McEwan, D., & Rebar, A. L. (2019). Theories of physical activity behaviour change: A history and synthesis of approaches. *Psychology of Sport and Exercise, 42*, 100–109. doi:10.1016/j.psychsport.2018.11.010

Schunk, D. H., & DiBenedetto, M. K. (2020). Motivation and social cognitive theory. *Contemporary Educational Psychology*, *60*, 101832. doi:10.1016/j.cedpsych.2019.101832

Sedik, A., Hammad, M., El-Samie, A., Fathi, E., Gupta, B. B., El-Latif, A., & Ahmed, A. (2022). Efficient deep learning approach for augmented detection of Coronavirus disease. *Neural Computing & Applications*, *34*(14), 11423–11440. doi:10.100700521-020-05410-8 PMID:33487885

Shiau, W. L., Yuan, Y., Pu, X., Ray, S., & Chen, C. C. (2020). Understanding fintech continuance: Perspectives from self-efficacy and ECT-IS theories. *Industrial Management & Data Systems*, *120*(9), 1659–1689. doi:10.1108/IMDS-02-2020-0069

Shmueli, L. (2021). Predicting intention to receive COVID-19 vaccine among the general population using the health belief model and the theory of planned behavior model. *BMC Public Health*, *21*(1), 1–13. doi:10.118612889-021-10816-7 PMID:33902501

Subramaniam, M. (2020). Digital ecosystems and their implications for competitive strategy. *Journal of Organization Design*, *9*(1), 1–10. doi:10.118641469-020-00073-0

Tseng, M. F., Huang, C. C., Tsai, S. C. S., Tsay, M. D., Chang, Y. K., Juan, C. L., ... Wong, R. H. (2022). Promotion of Smoking Cessation Using the Transtheoretical Model: Short-Term and Long-Term Effectiveness for Workers in Coastal Central Taiwan. *Tobacco Use Insights, 15*.

Vaishya, R., Bahl, S., & Singh, R. P. (2020). Letter to the editor in response to: telemedicine for diabetes care in India during COVID19 pandemic and national lockdown period: guidelines for physicians. *Diabetes & Metabolic Syndrome*, *14*(4), 687–688. doi:10.1016/j.dsx.2020.05.027 PMID:32442918

WHO. (2021). *WHO Coronavirus (COVID-19) Dashboard | WHO Coronavirus (COVID-19) Dashboard With Vaccination Data*. https://covid19.who.int/

Zhou, C., Yue, X. D., Zhang, X., Shangguan, F., & Zhang, X. Y. (2021). Self-efficacy and mental health problems during COVID-19 pandemic: A multiple mediation model based on the Health Belief Model. *Personality and Individual Differences*, *179*, 110893. doi:10.1016/j.paid.2021.110893 PMID:36540084

KEY TERMS AND DEFINITIONS

Digital Health: Digital health is an umbrella term and refers to the use of information and communication technologies in health to prevent diseases and health risks and improve health. Digital health includes wearable devices, mobile health, telehealth, health information technology and the use of telemedicine.

Digitalization: In its simplest form, it is the process of transitioning to a digital business. It is also defined as the use of digital technologies to change a business model and provide new revenue and value-generating opportunities.

E-Literacy: E-literacy refers to the degree of skill required to effectively use electronic materials, tools and resources while navigating an online learning environment.

Health Behavior: It is the whole of the behaviors that an individual believes and does in order to protect, improve, maintain and protect their health.

mHealth: mHealth (mobile health) refers to the use of mobile phones and other wireless technologies in medical care. The most common applications associated with mHealth are the use of mobile devices to educate consumers about preventive health care.

Telemedicine: Telemedicine is the practice of medicine whose term includes the use of technology to provide remote care to individuals. This application uses a telecommunication infrastructure to provide doctor care to a remote patient.

Chapter 13
Zara Joins the Trends:
Analysis of Its Content Creation Within the Framework of Its Business Strategy

Belén Ramírez Barredo
https://orcid.org/0000-0001-9403-8613
CEU San Pablo, Spain

Beatriz Guerrero González-Valerio
https://orcid.org/0000-0002-7552-9553
CEU San Pablo, Spain

Davinia Martín Critikián
https://orcid.org/0000-0002-6921-4707
CEU San Pablo, Spain

ABSTRACT

Brands are the universal symbols of the modern era, comprising a set of values with which the consumer identifies. To activate a brand is to make its promise tangible through different marketing and advertising tools. The fashion film is established in the retail sector as a form of brand communication, which can encompass fashion, photography, film, advertising, music, and art, becoming heir to different and varied artistic disciplines: a format that allows interaction with customers by breaking traditional boundaries through transmediality. The aim of this chapter is to analyse the series of three fashion films, "Zara Scenes," launched in 2019, in the context of the communication of changes in the brand's business strategy. With this purpose, the different chapters are analysed from the fashion film perspective as social media content and a brand communication tool.

DOI: 10.4018/978-1-6684-4102-2.ch013

ZARA'S BACKGROUND

To study Zara is to delve into the growth of a small company that, in a quarter of a century, has become one of the largest fashion distribution companies in the world, thanks to the business vision of its founder, Amancio Ortega. In twenty-five years, the company has experienced significant growth and international expansion, which we summarise below, taking key moments in its history as a reference.

The business began in 1963 in a modest workshop - Confecciones GOA - in A Coruña, Spain. In this Galician city the firm opened its first shop, Zara, in 1975. In 1977 they established their headquarters in Arteixo with the construction of two factories: GOA and Samlor. 1983 was a year of expansion for the company nationwide. In 1984 it opened its first logistics centre with a heavy investment in state-of-the-art facilities and a surface area of 10,000 square metres. In 1985 Inditex was founded: the Group's holding company brought together under its umbrella the different brands that make up the Group. This laid the foundations for a distribution system in line with the demands of the market and with the capacity to adapt to a very rapid growth rate. In 1988, with the opening of the first shop in Oporto, Zara began its international expansion, arriving on Lexington Avenue (Manhattan, New York) in 1989. In 1990 it set up in France, on Rue Halévy in Paris. During the years that followed, Zara began to operate, for the first time, in numerous countries, oppening its shops in the most emblematic areas of major cities. In 2000 they opened new offices in Arteixo. In 2001 the brand was listed on the stock exchange for the first time. In 2003 Zara Home was born, focused on the sale of household goods. In 2005, Pablo Isla was appointed Deputy Chairman and CEO of Inditex. In 2006, the Group presented its Strategic Environmental Plan, with the aim of to ensure that all its operations are environmentally sustainable. The first chain of the holding company to sell online is Zara Home. This was introduced in e-commerce in 2007, while Zara did not take the step until 2010. In 2011, Pablo Isla was appointed chairman of Inditex, by which time all the Group's brands were already operating online. In the course of 2013, the holding's brands implemented their new image in their flagship shops. Larger than the previous ones, they were also located in premium shopping areas. 2014 saw the opening of a new logistics center in Cabanillas, Guadalajara, Spain. 2016 brings Inditex international recognition in terms of sustainability. 2019 brings the boost of the digital and sustainable transformation of the company, a strategic challenge that had been set during the previous years, which entailed the almost complete renovation of the commercial area, and the incorporation of advanced technology in all of its aspects. The digital transformation strategy launched in 2012 through the integrated platform of shops and online will experience strong growth in 2020, a key year in the transformation of the company, during which the global environment is marked by the pandemic (Inditex, n.d.).

Amancio Ortega envisioned fashion as an industry whose *raison d'être* is novelty, turning the "just in time" formula into a hallmark of business identity and a new business model (O'Shea, 2008). The JIT strategy is based on three concepts: the first is product rotation: products do not remain on the website or in the physical shops for long, so that a new collection is launched every 10-15 days to ensure that there is always something new. In this way, the articles sell out quickly and generate trends. The second concept is to offer the customer a unique experience through its physical shops; Zara's stores are among the best in each city, giving the consumer immediate satisfaction. Finally, there is technological innovation and its incorporation in the development of new products - such as personalised items - making production faster and more efficient.

Ever since 2011, Zara has been among the five most valuable Spanish brands (Pérez, 2020; de Lemus, 2021), occupying the forty-fifth position worldwide in 2021 (Interbrand, 2021). Its model of integration and flexibility have been essential for its expansion in the international market, as the key components behind the success achieved by this brand are its insourced manufacturing process, its vertical integration, the JIT approach, a diversification of supply based on the market's features, and its precise knowledge of consumer preferences (Martínez, 2008; O'Shea, 2008).

In terms of its graphic identity, the brand has changed its logo three times in 44 years. The original logo dates back to 1975, when it first appeared on the market. In 2011, coinciding with the premiere of its first sporadic documentaries on YouTube, the first variation was produced, which was aimed at promoting simplicity and minimalism (García, 2019). The 2019 redesign, with lines closer to the editorial sphere and the work of French art director Fabien Baron, conveys the brand's desire to consolidate its position as a market leader in the "high fast fashion" category (Salido, 2019). It coincides its transformation with the start of the aesthetic and tactical renewal of the online shop and social networks (Gràffica, 2019), communicating an image of Zara as the leader in the mix of high street and high fashion.

Zara's marketing strategies have focused on market research, technological investment, social networks (digital marketing), window dressing and strategic location of its shops since its inception (Seghelmeble, 2020). For this firm, shops have always been its best communication support, as these are perceived as physical spaces in which consumers can become emotionally involved with the brand (Velilla, n.d.). This channel has traditionally played the lead in the field of the brand's advertising and marketing. The bonding strategy that the firm has opted for translates into a non-intrusive communication, reinforcing its positioning in the market through the shopping experience. Far from using paid advertising, Zara relies on its products and on the customer's own role as brand ambassador (Ortega & Mera, 2019; Strauss, 2020). Other tactical actions of the firm include collaborating with celebrities to present limited edition garments.

Zara's priorities are to offer attractive and responsible fashion, and to constantly improve customer experience. One example is the global implementation of the Radio-Frequency Identification System (RFID), which uses state-of-the-art technology to locate garments in the stores and in its warehouses, in order to improve customer service (Inditex, n.d.). Zara's business policy can be summarized as giving customers what they want and getting it to them faster than anyone else. This commercial strategy is mainly based on two sources of competitiveness: the speed of trends and the speed to market. The former has to do with the firm's understanding of culture; the latter with its flexible, data-driven supply chain (Andjelic, 2021).

SETTING THE STAGE

Zara has been evolving and adapting over the last twenty-five years to new markets and emerging social networks, to the style of customers, to new generations and to technological innovation (online sales). As mentioned in previous lines, Zara's incursion into e-commerce did not take place until 2007 (Regueira, 2019) with the opening of a website for Zara Home. Slowly and belatedly, the brand entered the online space later than its competitors. However, the digital revolution has been the brand's big bet in recent years. The strategy of connecting all sales channels, the use of augmented reality to improve the overall perception of the customer in the purchasing process, and the commitment to the online sphere have been key in its development, expansion and consolidation (ABC, 2019).

Inditex allocates a high percentage of its investment to the promotion and constant technological updating of the linked model of shops. This sale on different platforms, both physical and online, synthesizes a "flexible fashion", integrated, omnichannel and sustainable model (Lara and Mas, 2018). Omni-channel sales entail an evolution and overcoming of the multi-channel approach and the cross-channel model, for the sake of a sales strategy focused on the customer. This requires the company to be able to interact 360° with the consumer, "who obtains an absolutely integrated, homogeneous and seamless shopping experience, regardless of the channel in which it takes place" (Mirosa de Villalobos, 2021).

The omnichannel strategy has been implemented in the brand through different actions carried out since 2012, as a result of the investment made in technology. These include, among others, the following:

· In 2014 Zara launched its app, where customers can buy products from the collection. With the rise in popularity of Instagram, the brand began to publish content on this social network, under the name @zara_worlwide.
· In 2016 it achieved the integration of online and offline commerce strategies, through the opening of national online shops in different countries, the search for alliances whilst maintaining maximum caretaking of its physical shops.
· In 2018, it launched an augmented reality app, allowing customers to see the brand's models in real time, focusing the camera on the storefront and/or on some spots located inside the shops, hence displaying the option of buying directly, either through the app or in the physical shop (Baldé, 2018).
· In 2019, a plan to break down international barriers and deliver its orders anywhere in the world is put in place (Calviño, 2020).
· In 2020, it updated its app, making it possible to pay virtually without queuing, reserve a supplier from a mobile phone or use the shop mode to geolocate an item in the store itself. It also generates QR codes with which customers can remove alarms and pay (Malagón, 2020). The arrival of the coronavirus was the opportunity to increase sales in the digital environment by creating a new project: "Inditex Open Platform", a proprietary technological base with which all the company's operations work.

If we go into more detail regarding social media aspects - an optimal stage for the development of branded content - we find the following data:

· Zara entered the world of networks with Facebook; in 2016 it had 23.4 million followers; today it has 29.7 million. It posts almost daily on its social media networks, thanks to the creation of new ones such as Instagram - a more innovative platform that offers a wider range of possibilities for the brand. Its account, which has 47.6 million followers, is interactive and offers a wealth of information for users, making it its star social media platform. It has recently taken on TikTok, where its videos have gone viral on multiple occasions. Despite having 780,400 followers, its videos reach millions of views (Rodríguez, 2021).
· It is the Spanish brand with the most followers on Instagram - currently 47.6 million. The number of comments on its posts ranges between 300 and 400. Users give their opinion on whether they like or dislike the garments or mention friends so that they can see the post. The number of "likes" varies between 40,000 and 50,000. Zara does not usually respond to these comments. The number of weekly feed-posts is between five and twelve (with hardly any activity on weekends) and almost daily on stories. The content is the same on all of its networks, with the exception of Youtube, where it uses branded content. It is characterized by a certain monotony, by focusing on videos or images of its collections, and by the absence of interaction with the user, maintaining a formal tone. Its posts generate an engagement rate of 0.15%, while its interactions per post amount to 61,740.

· The brand has 87,600 followers on YouTube. This platform has been used primarily to publish content about its collections. Since 2018, it has opted for the branded content format. Its publications are usually monthly. While user comments are quite rare (between 1 or 3), and usually show their opinion about the content, the number of likes varies between 10 and 20 per publication and the average number of views is 1,000 (https://zara.com).

· On Twitter it has 1.3 million followers. The number of likes per post ranges from 50 to 100. As for the comments, between 10 to 20 per post, which usually express doubts about the designs, questions about product references or opinions about the content. The number of retweets is between 7 and 20 per post. The brand posts tweets daily on this platform, with the exception of weekends. Its content, which is rather informative, is very similar to that of the other platforms: videos or photos of the designs of its new collections, mostly accompanied by brief phrases, keywords, hashtags or links to its website, although not very creative (Inditex Annual Report, 2020).

· With respect to Facebook, it currently has more than 29 million followers. The frequency of publication is usually weekly, although in the last month it has not published any content. The number of likes ranges between 900 and 1,000, while the comments, with an average of 40 per post, are both positive and negative and show, above all, the problems that customers have had with the brand. In this case, Zara usually responds to users' comments, to solve their doubts or any potential issues. The content of the publications is very similar to that of Instagram, showing the latest news from the collections, through images or videos, using a formal tone and with brief messages. Its posts generate an engagement rate of 0.01%, while number of interactions per publication is 2,512.

CASE DESCRIPTION

"The showcase that the digital environment and social networks represent for individuals means that brands find in this space a new potential connection point where they can exploit new audiovisual formats that are attractive to the audience" (Valderrama, Mallo & Crespo, 2020, p.240). In e-commerce, as in offline sales, people are not only looking to purchase a product; they also value the emotions and experiences they perceive at all stages of the purchasing process (Nambisan & Watt, 2011; Zhang et al., 2014; Ruiz, Riano & Aguado, 2019). The vehicles of expression that the branded content tool can acquire are many and varied. The content can be of a diverse nature: editorial, musical, audiovisual, etc. Some formulas are directly related to the presence of the brand itself; others focus on producing or creating content that represents and makes the brand's values tangible, while connecting with current and potential consumers. In any case, the crux lies in the existence of a logical relationship between the brand and the way in which the content is expressed (Ramírez, 2016), so that the firm "will only capitalize on value when it manages to create quality entertainment content that responds both to the interests of its target audience and to the values with which it wants to associate its brand" (Martorell, 2009, p.10). The ultimate goal is to achieve brand loyalty, highlighting its brand image, engagement, notoriety and interaction with the consumer (Del Pino-Romero & Castelló-Martínez, 2015, p. 112). It should not be forgotten that "the engagement or identification that the consumer manages to feel is what drives them to make a decision or take an action" (Arbaiza & Huertas, 2018, p. 11). Social networks have become an essential means for brands to make their image and products known. The Internet is increasingly used as a shopping channel. The proportion of people who combine online and offline shopping continues to grow, standing at 58%. In addition, the majority of users make research on social networks before consuming products and almost

50% of them state that social networks have influenced their final purchase. As for the products that are most purchased online, fashion is one of the four main categories, according to the E-commerce Spain 2021 Annual Study. The most used social networks by companies in Spain in terms of e-commerce are: Facebook and Instagram, followed by Whatsapp, Youtube and Twitter.

Zara released its first documentaries on social networks in 2011, the year in which it inaugurated its YouTube channel, showing videos of its product-centric collections. Their appearances were sporadic and were usually limited to sharing an aesthetic, or showing their new collections. In 2018, they launched a series entitled Behind Your Click, where they showed the company's work with images. However, it did not achieve the expected success. That same year, in order to expand its presence in social media, it opened its Twitter account (Regueira, 2019). In 2019, Zara Scenes was born on its YouTube channel, a section in which the brand, using the fashion film format, launched a new series based on a storytelling common to all its chapters, defined by themselves as: "a new way to tell stories". In 2021, the brand made a foray into the virtual world by collaborating with the South Korean brand, Ader Error, to create "AZ Collection", the first collection in the Zara metaverse, aimed at generation Z, based on the identity and uniqueness of each individual. Through communication strategies more focused on transmitting the values of its identity than on the characteristics of the products, the brand connects with the mind and emotions of the consumer, establishing an emotional bond with them. The flagship of the Inditex group made its debut in the world of big fashion films last December with the project "O Night Divine", where it unveiled its evening collection proposals through an emotional story, thus promoting its "Zara Atelier line, which seeks to raise the value of craftsmanship and moves closer to the realm of affordable luxury by betting on innovation" (El Publicista, 2021). Prior to the "O Night Divine" project, the company presented on its website the Zara Origins line, a project focused on the construction of a contemporary wardrobe that offers essential and timeless garments, made with high quality materials. The aesthetics and styling used to generate this audiovisual piece of branded content reflects the values of Origins: simplicity, sophistication, modernity and timelessness.

Technology Concerns

Since 2012, with Pablo Isla at the helm of Inditex, Zara has promoted a digital transformation strategy. This project has materialized in a gradual process of almost complete renovation of the retail space and the incorporation of advanced technology in all the axes of the model.

· In 2012 Zara opened its eco-efficient Oxford Street shop in London (UK). This same year they inaugurated an innovative distribution centre in Tordera (Barcelona). In Spain, it is the first of its kind to have received LEED gold environmental certification.
· In 2014, the new logistics centre at Cabanillas in Guadalajara began operations - also obtaining LEED gold environmental certification. The brand continues to expand the integrated shop model (online and physical), with access to new markets, new flagship shops and with the refurbishment of existing shops. The new Technology Centre in La Coruña, at that time the only one of its kind in the world, begins operating.
· In 2016 Inditex is recognised as a sector leader in the Dow Jones Sustainability Index and tops Greenpeace's Detox Catwalk ranking for its commitment to zero dumping of hazardous chemicals. It strengthens its agreement with the international trade union federation IndustriALL and launches

the 2016-2020 Strategic Environmental Plan. Physical shops are opened for the first time in five new markets and online operations are launched in eight new markets.

· In 2017, in line with the development of the integrated shop and online model, the number of markets with online sales was expanded with the opening of zara.com in India, Malaysia, Singapore, Thailand and Vietnam.

· The evolution of the integrated shop and online model accelerated the online expansion of the Group's brands. In 2018, Zara launched its global shop zara.com/ww in 106 markets where the retailer did not have a physical shop. Inditex expands its headquarters in Arteixo (La Coruña, Spain), which incorporates new workspaces for different departments and new services for employees.

· In 2019, the year in which Zara Scenes is launched, Inditex's sales reach 202 markets, with a physical presence in 96 of them and its own online platforms integrated in 66. (Inditex, n.d.)

Technology Components

Understanding Zara's social media tactics requires starting from its omnichannel strategy. An over time analysis of the branded content proposals made by the brand allows us to appreciate its evolution towards a hybrid business model, which combines luxury and fast retail strategy, with customer experience at its core. Fast fashion responds to trends at the pace of popular culture and novelty. Luxury is deployed through limited edition capsule collections, the development of a universe of sub-brands made up of key singular pieces, with limited distribution, and an aesthetic akin to high fashion retail.

The premiere of the fashion film series "Zara Scenes" is part of the company's drive for digital and sustainable transformation, which in 2019 was visualised through the almost complete renovation of the retail space, and the incorporation of advanced technology in all the axes of the model (Inditex, n.d.). Thanks to the investment made in technology, the company became capable of supplying its products to any point on the planet. Inditex's sales reached 202 markets at that time, with a physical presence in 96 of them and its own online platforms integrated in 66 of those.

Management and Organizational Concerns

The use of branded content as a support for the development of the company's omnichannel strategy comprises a series of key moments that allow a global understanding of its purpose for the brand and its evolution over time.

1. In 2011, the brand sporadically released a series of audiovisuals with the apparent aim of sharing or inspiring its very own aesthetic (Regueira, 2019). One example is the documentary Lucy Chadwick. A Selby film for Zara.
2. Progressively, the technological innovations introduced by the company, such as the augmented reality app, in-store product localisation, etc., were aimed at improving the shopping experience. During this time, the branded content strategy was focused on its YouTube channel, where it limited itself to premiering its collections on video.
3. In 2018 it premiered the series entitled "Behind the click", focused on showing the work of the organisation behind an online order (Regueira, 2019).
4. From 2019 onwards, the firm took a step forward: in order to revalue the brand, it established collaborations with relevant figures in the luxury and fashion sector, simultaneously focusing the

content on building and enhancing the aspirational dimension of the brand in the customer, overcoming the concept of fast and affordable fashion that had prevailed until then.

The Zara Scenes series premiered, as noted above, in 2019 on the brand's YouTube channel. In the words of the brand it was "a new form of storytelling"; "a mix of dreams, realities, music, dance, casual jokes and other writing inventions to present our latest women's collection in a more experimental and relatable way" (Regueira, 2019).

Through short narrations, starring the Argentinean model Mica Argañaraz, the firm suggested to the viewer the acquisition of the garments that made up the model's "total look" during these short films, a format to which it would later resort on its website. The fashion films in the series were an original method through which the brand launched its new season, differentiating itself from other "low-cost" brands while seeking to connect with the public in a more intimate way. The garments of the collection would be discovered little by little, in each chapter, all united by a common storytelling that gave continuity to the story (Llanos, 2019).

Shot in California, Zara Scenes becomes the spearhead of the brand to present a new way of shopping and telling stories that will remain. Thus, by the hand of Fabien Baron in the series Zara Spring Summer 2021, Zara Man Studio Fall '21 Collection, Zara Woman Studio Fall '21 Collection, renowned models (Marisa Berenson, Sasha Pivovarova, Chiharu Okunugi or Yumi Nu) pose for Steven Meisel's lens. In line with the business strategy, these fashion films incorporate the visual change of the brand and its evolution on an artistic level. The renewal of a company's logo or visual identity often goes hand in hand with a profound change at a strategic or business level, incorporating a more complex message than a simple change of appearance.

The analysis of Zara Scenes is approached firstly from the perspective of branded content aimed at social networks; and secondly, it is analysed from the point of view of fashion films.

Analysis of Zara Scenes. Focus on Branded Content in Social Networks.

The first chapter entitled "The wake up call" re-creates the deserts of California, which will be travelled through by the Argentinean model Mika Argañaraz in a convertible. (Llanos, 2019) The weekly chapters will develop the mysterious story and discover the new designs of this brand. In addition to being a new form of communication, Zara Scenes becomes a new online shopping experience. Now, in addition to being able to buy the products in the physical shops or on the website, they can also be purchased through the videos. (Vázquez, 2019)

a) Key dates: September 2019 saw the launch of Zara Scenes, a new way of selling online, showing its seasonal looks in different videos as a series. The first chapter, "The wake up call" is uploaded on 9 September 2019; the next, "Sounds around you", two days later; and a final episode, "How to get the keys", on 13 September. (Vázquez, 2019). During 2020, Zara did not publish any content in this section. However, in 2021, in its spring-summer season, it launched a new series again, also consisting of three episodes, each one specialising in one of its target audiences: Zara Kids, Zara Women and Zara Man. Finally, in November 2021, the brand presents a new collection through three stories that will focus on the world of sport, entitled: "Athleticz training" (Bonillo, 2021).

b) Social Networks: Zara Scenes was initially published on its YouTube channel. It generated a great impact on the audience, as it was a more subtle and narrative way of showing the new collection, without explicitly referring to the sale of the designs. Later a new section was added on the website, from which this content could also be accessed. (Llanos, 2019)

c) Target audience: Zara Scenes is aimed at a target audience between 18 and 35 years old. The brand's philosophy is to be accessible to everyone, imitating the models of luxury catwalks. With this new communication model - more dynamic and less aggressive - Zara manages to capture a young audience, through a story that is inspired by the brand's own universe and defends a spirit of freedom that anyone in their youth would like to have. (Inditex Annual Report, 2020)

d) Content: The three chapters of Zara Scenes, in English, are created in order to show its new collection in a different way; the economic intention of the brand is relegated to the background. The first chapter, "The wake up call", is set in the desert; it combines the colour black, with eighties-style garments such as leather jackets, ankle boots, biker trousers... (Llanos, 2019). In "Sounds around you", the black and eighties looks are still present, but with a different setting. The episode takes place in a house, where the model prepares a milkshake while listening to the sounds around her. In this case, the common thread of the story is the sounds of the blender, the music of the ra-dio, the clatter of the spoons... (Vázquez, 2019). Finally, in "How to get the keys", Mika goes to the Roosevelt Hotel in Los Angeles. On the way, her chauffeur informs her that she has the terrace with the best views of the city, but that she must pass some tests to get to the key. Mika follows the instructions and reaches the rooftop. In this episode, in addition to the black and eighties theme, the model wears party looks. (facebook.com/Zara)

e) Impact: The launch of the first episode of Zara Scenes generated, at first, a strong impact and curiosity among the public, not only because it was a new way of showing its latest collection, but also because it presented a new online shopping model through videos. In the subsequent episodes, the audience started to decrease.

Currently, "The wake up call" has 600,004 views, but only 738 likes and no comments. "Sounds around you" has 32,942 views, with 548 likes, and no comments. In the last episode, the number of views increases - although it remains well below the first one - with 36,098 views, yet the number of likes drops to 287.

In the following fashion films produced by Zara, the reproductions of each episode fall to 20,000, in the case of the Campaign collection. This being said, the last episodes published do not exceed 2,000 views. Despite the notable drop, the third episode of Campaign unexpectedly reaches 1.2 million visits (Zara.com).

Analysis From the Perspective of Fashion Films

As Del Pino-Romero and Castelló-Martínez point out, "good content in all the social platforms where a company is present is key: there is no room for improvery" (2015, p. 110). Likewise, "low-cost fashion brands, whose fundamental value lies in knowing how to respond to the demands of the present, are inscribed in this position with videos that praise youth, take a stand on social issues or portray the latest trends" (Díaz Soloaga & García Guerrero, 2013, p. 368).

· In the case of the three Zara Scene videos, the aim is to show the latest trends, presenting the Fall Winter 2019 collection, as it appears at the beginning of each one of them. All three are shot in Los Angeles (California) and star the model Mica Argañaraz, seeking greater involvement and immersion on the part of the viewers. Likewise, the dialogues in the first and third films (there are none in the second) are in English. These aspects highlight how the aesthetics respond to an international taste, in tune with the reality of Zara.

· The length is quite similar in all three - around 2 minutes - the second one being slightly shorter. In all of them, reality has been fictionalised in an curious mixture with dreams. In the first one, it is not until the end that the viewer can distinguish which part represents reality and which part is supposedly dreamt by the protagonist. In the second one, the effect of cutlery, books and chairs moving to the rhythm of the music is played with. And, in the third one, the viewer discovers that the driver was joking with Mica - the main character - at the end of the story.

· In the three videos we can observe several of the characteristics of fashion films: the use of storytelling, or, in other words, narration through stories. The advantage of this method is that the viewer gets emotionally carried away. In all of these, fashion is represented as a symbol of aspiration, as well as values such as freedom, transgression, sophistication and minimalism. These are seductive stories, in which great importance has been given to visual aesthetics and glamour (a convertible, a horse in the desert in the first one; an idyllic house in the second one; a chauffeur, a hotel with the best views of Los Angeles in the third one). In all three we witness a perfect integration of storytelling (narrative structure), brand image and artistry. Moreover, the use of a series format allows for the displaying of more garments from the collection and, in addition, generating expectations from the viewer.

· In addition to their virality, these stories have been designed in the context of a transmedia narrative, as they were first created for YouTube and were then broadcasted on the brand's website.

· These fashion films have four objectives: to entertain, to generate en-gagement, to communicate the brand's identity and, finally, to promote the products of the 2019 winter collection. All the afore-mentioned actions are part of a brand strategy and form part of a comprehensive communication plan for the Inditex group, as nowadays multi-disciplinary and transmedia communication is a must. The company understood non-conventional forms of communication as an alternative, resorting more to experiential and relational marketing (interactivity, connectivity and a lot of creativity). Likewise, all communication and advertising actions have been carried out this way.

· With all these tactics, the Inditex group manages to create visibility for the brand; credibility – as it is positioned as a connoisseur through the contents; and establishes a link through storytelling - gener-ating engagement. On top of this, it also generates traffic on search engines and on its own website.

Therefore, the content of fashion films, through the aesthetics of the images and the rhythm of the plot, has become an experience for the user. We can speak of a long-term strategy, which seeks to approach digital natives, since many of them are already consumers or could become ones.

CURRENT CHALLENGES FACING THE ORGANISATION

There is a relationship between Zara's branded content strategies and the evolution experienced at a corporate level with the appointment of Pablo Isla as chairman of Inditex in 2011. His presidency is marked by the expansion and growth of the brand thanks, to a large extent, to the impulse of the digital transformation process.

The branded content actions carried out by the brand after the premiere of "Zara Scenes" are focused on presenting and promoting the new lines opened by Zara. These fashion films constitute the pre-sentation, to a certain extent, of the technological process developed since 2012 by the Holding Company. The actions taken by the brand in this regard respond to the challenge that companies are currently facing to evolve and rethink the current way of presenting the online environment, which is focused on product cataloguing. "If a website wants to be a true flagship and create a rich and exciting shopping experience, it will have to face a huge creative challenge. (...) It won't be enough to create appealing products and try to sell them, it will have to entertain the potential consumer who, at this moment, regards fashion as an escape route and needs constant sources of entertainment" (Baron, as cited in Ferrero, 2020).

1. With the Zara Beauty collection, the brand enters the cosmetics sector in 22 shops around the world through its own physical space, and through its online channel, which recreates, virtually, the same architectural characteristics. A line conceived to create an "immersive" experience (Cinco Días, 2021).
2. The Zara Athleticz Training series presents a series of proposals to stay fit, launching a site with healthy lifestyle trending content.
3. On the website we find short films such as Chloë Sevigny in a bubble bath or Charlotte Gainsbourg promoting her line with Zara, while Baron presents Peter Lindbergh's photography in Zaratribute, with matching products. The collection that reinvents the basics, Zara Origins, recovers the idea of water as the origin of life in a timeless yet contemporary setting, in resonance with the line's identity.
4. O Night Divine (2021), the brand's first feature-length branded content film, reinforces the Zara Atelier line. It is a precious audiovisual project directed by Luca Guadagnino, which shows a com-municative level closer to luxury and premium than to fast fashion (Pérez, 2021).

Since 2019, we can observe that the branded content actions of the brand have a triple purpose:

· To transform its online shop into a flagship, within the colligated model of shops. Every action is a reflection and consolidation of its own identity. The pattern of product rotation, characteristic of the firm, translates into the permanent renewal of content and visual appearance of the website. It adapts to seasonal movements, such as the sales period, Christmas, or the launch of new summer and winter collections. In the more exclusive sub-brands, unique and coveted pieces are introduced (Zaratribute), as well as limited edition, current and timeless items (Zara Origins, Studio Collec-tion, and collaborations such as the one with Charlotte Gainsbourg). Likewise, the fashion films or other branded content that it releases on its website have an expiry date and are subsequently available on its YouTube channel.
· To present new lines of business, oriented towards a brand approach that goes beyond the concept of fast fashion. In this sense, the new brand extensions, sub-brands, products and experiences play an equivalent role in their role of bringing the brand to life as gateways to the brand universe. There is no hierarchy between them, they are all images of company values, where price/product coher-ence, creative production and aesthetics play a key role.
· At the beginning, the fashion sector focused its communication efforts on specialized magazines. With the arrival of social networks, companies have lost control of information, although they have gained other advantages. In this context, Zara is committed to incorporating an editorial line to its

online platform, with a similar aesthetic to traditional fashion magazines such as Harpers Bazaar, offering an image of knowledge of the sector, innovation and leadership.

SOLUTIONS AND RECOMMENDATIONS

In line with its marketing strategy focused on the in-store consumer experience, Zara is pursuing the consolidation of its web platform as the flagship of online retailing. The development of e-commerce demands a different way of communicating in order to maintain the brand experience, through the shop. The brand uses branded content as a tool for expressing its values and lifestyle, presenting new lines and sub-brands, and business model. It is in its online shop where Zara deploys the creativity and quality work that it distributes on its social media networks, based on their characteristics and content.

With an omnichannel sales strategy, the brand's identity and values are reflected in both its sub-brands and its own lines, emphasizing a sustainable, responsible and committed business model. The website is the space where the customer can discover the initiatives that the brand is promoting in this sense, such as the join life line to promote sustainability; the annual report of activities - where it emphasizes the search for a sustainable business model in a sustainable way; the annual report of activities, where it emphasizes the search for a sustainable business model through investments in the technological field, scientific collaborations on fabrics, and its own model without stock thanks to knowledge of the environment and the customer.

The communication model that the firm has opted for, with a modern visual language, an aesthetic similar to that of luxury brands, and collaboration with renowned figures from the cultural sphere of fashion (art directors, photographers, models...) is aimed at reinforcing a hybrid business model between luxury strategy and fast retail. However, Zara seeks to differentiate itself from its competitors in many ways: the redesign of its logo distances itself from the minimalism towards which luxury brands are evolving; the prices of its new, higher-quality product lines do not involve renouncing affordable prices and the principle of the democratization of fashion; all of this in line with the concept of integration and universality. "Zara is building a hybrid business model, where it combines fast fashion with cues of luxury (curated capsules, art direction, photography, model selection, styling). Zara is still in the business of the mass consumption of standardized products, but now it is also in the business of considered consumption of signaling products (Peter Lindbergh merch, Charlotte Gainsbourg collection, Studio Collection)" (Andjelic, 2021).

The fashion film, as a new format with its own entity, which has emerged with great impetus in the fashion sector, forms part of the integral communication strategy of Inditex. It is a clear example of how the term "advertising has evolved towards another term, which, without being new, is more complete: communication" (Del Pino-Romero & Castelló-Martínez, 2015, p. 118).

However, the brand goes beyond fashion films, incorporating multiple other branded content formats. In a way, Zara's website is branded content. The art direction, photography, styling and props of the recently incorporated content - marked by the collaboration of Fabien Baron, Steven Meisel and legendary models (Marisa Berenson, Sasha Pivovarova, Chiharu Okunugi or Yumi Nu) - are transformed into branded content.

REFERENCES

ABC. (2019). Zara cambia el concepto de «moda rápida» por el de «flexible». *Abc. es*. Recuperado de https://www.abc.es/economia/abci-zara-cambia-concepto-moda-r apida-flexible-201901110217_noticia.html?ref=https%3A%2F%2Fw ww.abc.es%2Feconomia%2Fabci-zara-cambia-concepto-moda-rapida -flexible-201901110217_noticia.html%3Fref%3Dhttps%3A%2F%2Fwww.google.com%2F.

Andjelic, A. (2021). *Deconstructing Zara's strategy*. https://andjelicaaa.medium.com/deconstructing-zaras-strategy -8daf218af9e8

Arbaiza, F., & Huertas, S. (2018). Comunicación publicitaria en la industria de la moda: Branded content, el caso de los fashion films. *Revista de Comunicación, 17*(1), 9–33. doi:10.26441/RC17.1-2018-A1

Baldé, G. (2018). Zara presenta su primera app de realidad aumentada. *It Publicidad*. https://www.itfashion.com/moda/zara-presenta-su-primera-app- de-realidad-aumentada/

Blackett, T. (1989). The role of brand valuation in marketing strategy. *Marketing and Research Today*, (17), 245–248.

Blancart García, N. (2015). *Importancia de la monitorización de las redes sociales*. https://zaguan.unizar.es/record/32588/files/TAZ-TFG- 2015-34 65.pdf

Bonillo, M. (2021). *Amancio Ortega nos quiere en forma*. https://www.65ymas.com/sociedad/zara-athleticz-recetas-ejerc icios-moda- deportiva_32894_102.html

Calviño Lorenzo, S. (2020). *Zara: Desmontando el mito de usar publicidad tradicional para llegar al éxito*. https://uvadoc.uva.es/handle/10324/42311

Castelló, A., & Del Pino-Romero, C. (2014). La comunicación publicitaria se pone de moda: branded content y fashion films. *Revista Mediterránea de Comunicación*. https://rua.ua.es/dspace/bitstream/10045/44253/1/ReMedCom_06 _01_07.pdf

Cinco Días. (2021). *Así es el interior de Zara Beauty, los espacios de cosmética de Inditex*. Recuperado de https://cincodias.elpais.com/cincodias/2021/05/12/companias/ 1620824085_266540.html

Cristófol Rodríguez, C., Martínez Sala, A., & Segarra Saavedra, M. (2018). Estrategia de comunicación digital en el sector franquicias de moda. El caso de Zara en Facebook. *Revista de comunicación audiovisual y publicitaria*. https://revistas.ucm.es/index.php/ARAB/article/view/60999

De Lemus, T. (2021). *Brand Finance España 100 2021. Las marcas más valiosas de España pierden valor por tercer año consecutivo, según Brand Finance*. Recuperado de https://brandirectory.com/rank ings/spain/

Del Pino-Romero, C., & Castelló-Martínez, A. (2015). La comunicación publicitaria se pone de moda: Branded content y fashion films. *Mediterránea de Comunicación*, 6(1), 105–128. doi:10.14198/MED-COM2015.6.1.07

Díaz Soloaga, P., & García Guerrero, L. (2013). Los fashion films como estrategia de construcción de marca a través de la seducción. Persuasión Audiovisual: formas, soportes y nuevas estrategias, 349-371

El Publicista. (2021). *Así es el fashion film de Zara que cautiva a millones de audiencia a nivel mundial en 24 horas.* Recuperado de https://www.elpublicista.es/anunciantes/asi-fashion-film-zara-cautiva-millones-audiencia-nivel-mundial

Ferrero, C. (2020). *Fabien Baron, diseñador del logo de Zara: «La moda se está enfrentando a las consecuencias de haber sido elitista y blanca».* Recuperado de https://smoda.elpais.com/moda/fabien-baron-disenador-del-logo-de-zara-la-moda-se-esta-enfrentando-a-las-consecuencias-de-haber-sido-elitista-y-blanca/

García, M. (2019). *Zara presenta un nuevo logo por segunda vez en 44 años.* Recuperado de https://brandemia.org/zara-presenta-un-nuevo-logo-por-segunda-vez-en-44-anos

González Fernández, A. (2021). *Comunicación comercial a través de redes sociales de empresas de moda y belleza.* https://iabspain.es/wp-content/uploads/2017/12/estudio-content-native- advertising-2017-vcorta.pdf

Gràffica. (2019). *Así es el nuevo logo Zara.* Recuperado de https://graffica.info/asi-es-el-nuevo-logo-de-zara/

Inditex. (n.d.). *Nuestra historia.* https://www.inditex.com/es/quienes-somos/nuestra-historia

Interbrand. (2021). *Best Global Brands 2021. The Decade of Possibility.* Interbrand Best Global Brands 2021Report.

Lara, L., & Mas, J. (2018). *Por qué unas tiendas venden y otras no en la era digital: Claves del éxito del New Retail.* Libros de Cabecera.

Launchmetrics Content Team. (2019). *La evolución en marketing digital de las marcas de moda y lujo.* Launchmetrics. https://www.launchmetrics.com/es/recursos/blog/la-evolucion-digital-de-las-marcas-de-lujo

Llanos, P. (2019). Zara transforma tienda online y cambia la manera de vender sus colecciones. *Revista ELLE.* https://www.elle.com/es/moda/noticias/a28962635/zara-tienda-online-nueva/

Lusch & Harvey. (1994). The case for an off -balance- sheet controller. *Sloan Management Review*, (2), 101–105.

Malagón, P. (2020). Así es la nueva App de Zara. *Libertad Digital.* https://www.libremercado.com/2020-09-16/app-nueva-zara-inditex-6660510/

Martínez Barreiro, A. (2008). Hacia un nuevo sistema de la moda. El modelo Zara. *Revista Internacional de Sociologia*, *66*(51), 105–122. doi:10.3989/ris.2008.i51.111

Martorell, C. (2009). Y ahora pasamos a publicidad...si usted quiere. El advertainment como alternativa al modelo de comunicación basado en la interrupción. In *Actas del I Congreso Internacional Brand Trends*. Dpto. de Comunicación Audiovisual, Publicidad y Tecnología de la Información. CEU Universidad Cardenal Herrera, Alfara del Patriarca. Recuperado de https://goo.gl/xibOeQ

Memoria anual inditex 2020 y Estado de Información no financiera del Grupo Inditex. (2020). https://www.inditex.com

Méndiz Noguero, A., Regadera González, E., & Pasillas Salas, G. (2018). Valores y storytelling en los fashion film. El caso de Tender Stories (2014-2017), de Tous. *Revista de Comunicación*, *17*(2), 316–334. doi:10.26441/RC17.2-2018-A14

Mirosa de Villalobos, R. (2021). Bienvenidos a la era de la venta omnicanal. *Harvard Business Review, 308*. Recuperado de https://www.harvard-deusto.com/bienvenidos-a-la-era-de-la-venta-omnicanal

Nambisan, P., & Watt, J. H. (2011). Managing customer experiences in online product communities. *Journal of Business Research*, *64*(8), 889–895. doi:10.1016/j.jbusres.2010.09.006

O'Shea, C. (2008). *Así es Amancio Ortega, el hombre que creó Zara*. La Esfera de los Libros.

Ortega, A., & Mera, R. (2019). Plan de comunicación y marketing de Zara - Case Study de The Apartment. *The Apartment*. Retrieved https://www.theapartment.es/blog/plan-de-comunicacion-y-marketing-de-zara/

Ortegón Cortázar, L. (2014). *Gestión de Marca. Conceptualización, diseño, registro, construcción y evaluación*. Editorial Politécnico Grancolombiano. Recuperado de https://alejandria.poligran.edu.co/handle/10823/798

Parra, R., Nieves, G. D., & Sánchez, H. (2017). 8 Tipos de contenido que funcionan en Instagram. *Postedin*. https://www.postedin.com/blog/8-tipos-de-contenido-que-funcionan-en-instagram/

Pérez, M. J. (2021). *Zara y la película navideña que sí apetece ver*. Recuperado de https://www.elmundo.es/yodona/moda/2021/12/14/61b89826e4d4d82b338b45af.html

Pérez, R. (2020). *LaLiga, DIA y Desigual entran en el ranking de las 30 marcas españolas más valiosas de 2020*. Recuperado de https://www.kantar.com/es/inspiracion/marcas/laliga-dia-y-desigual-entran-en-el-ranking-de-las-30-marcas-espanolas-mas-valiosas-de-2020

Ramírez Barredo, B. (2016). *Los títulos de crédito, marca de las películas* [Tesis doctoral]. Universidad Complutense, Madrid.

Regueira, J. (2019). *El nuevo contenido de Zara: ¿apuesta por el Branded Content? No Content No Brand*. Recuperado de https://www.javierregueira.com/contenido-de-zara/

Rodríguez, R. (2021). Zara tiene más seguidores en Instagram que en Facebook, Twitter y Tik Tok juntos. *Galicia*. Economía. https://www.economiadigital.es/galicia/empresas/zara-tiene-m a s - s e g u i d o r e s - e n - i n s t a g r a m - q u e - e n - f a c e b o o k - t w i t t e r - y - t i k - t o k-juntos .html

Romero Coves, A., Carratalá Martínez, D., & Segarra Saavedra, J. (2020). Influencers y moda en redes sociales. Análisis de las principales modelos españolas en Instagram. *Revista de Marketing Aplicado*. https://revistas.udc.es/index.php/REDMARKA/article/view/redm a.2020.24.2.7053

Ruiz Vega, A. V., Riano Gil, C., & Aguado Gonzalez, A. (2019). ¿Cómo se forma el estado de flow entre los internautas compradores de moda? un estudio comparativo de los sitios web de Zara y H&M. *International Journal of Information Systems and Software Engineering for Big Companies*, 6(1), 79–95. www.ijisebc.com

Seghelmeble, M. (2020). *Análisis de las estrategias de marketing empleadas por Zara que contribuyeron a su desarrollo y renombre internacional en el mercado de la moda del siglo XXI* [Trabajo de investigación]. Monografía, Centro Educativo Particular San Agustín.

Strauss, L. (2020). 10 estrategias que utiliza Zara para vender más y que puedes llevar a cabo en tu empresa. *Muy Negocios & Economía*. https://www.muynegociosyeconomia.es/negocios/articulo/10-est ra tegias-que-utiliza-zara-para-vender-mas-y-que-puedes-llevar-a-cabo-en-tu-empresa-481584704915

Tauber, E. M. (1988). Brand Leverage-Strategy for Growth in a Cost-Control World. *Journal of Advertising Research*, 28(4), 26–30.

Valderrama Santomé, M., Mallo Méndez, S., & Crespo Pereira, V. (2020). Nuevas narrativas en el marketing de moda: estudio de caso Fashion Dramas de Vogue, Redmarka. Revista de Marketing Aplicado, 24(2), 238-250. doi:10.17979/redma.2020.24.2.7161

Vázquez, S. (2019). ¿Qué es Zara Scenes? *Revista Woman*. h t t p s : / / w o m a n . e l p e r i o d i c o . c o m / m o d a / s h o p p i n g / z a r a - s c e n e s - n u e v a-coleccion-cambios-web-compras-tienda-online

Velilla, J. (n.d.). *Estrategias corporativas: el caso de Zara*. Recuperado de https://comuniza.com/blog/ estrategias-corporativas-zara

Zara. (2011). *Lucy Chadwick: A Selby film for ZARA*. YouTube. https://www.youtube.com/ watch?v=s3gDnnTfE2E&t=136s

Zara. (2019). *Zara scenes | the wake up call*. YouTube. https://www.youtube.com/watch?v=2UqbRq_ JMMY

Zara. (n.d.). h t t p s : / / w w w . i n d i t e x . c o m / e s / q u i e n e s - s o m o s / n u e s t r a s - m a r c a s / z a r a

Zhang, H., Lu, Y., & Guta, S., & Zhao, L. (2014). What motivates customers to participate in social commerce? The impact of techological environments and virtual customer experiences. *Information & Management*, 52(4), 496–505.

Chapter 14
Post–Truth Politics as a Threat to Democracy

Kingsley Mbamara Sabastine

https://orcid.org/0000-0002-4046-7858

The John Paul II Catholic University of Lublin, Poland

ABSTRACT

Politics has long been associated with mendacity, disinformation, manipulation, and at odd with the truth. In recent times, the term post-truth has been used to further characterise politics, which implies a fresh phenomenon in the conflict between truth and politics. The chapter examines the concept of post-truth and post-truth politics. The chapter argues that the application of post-truth rhetoric in politics implies a novelty in politics and in the relationship between truth and politics, which undermines democracy. It is arguable that post-truth negatively impacts individual ability to discriminate between what is true or false, taking into consideration the volume of disinformation on the one hand and on the other hand the need to make informed decisions and choices without having to consult experts at the critical time that the stakes involved in such decisions and choices are urgent and crucial.

INTRODUCTION

The term post-truth does not mean or imply that the idea and concern for truth is a thing of the past. Instead, post-truth describes and denounces the trending manner of communication that reflects a disregard for truth and deflects from reliable means of knowing what is true. A person's readiness to discover and accept the truth correlates with his/her overall development: the ability to stand up for what is right, the capacity for social critique, and the capacity to stand up to power. The post-truth condition is counterproductive because misinformation and manipulation negatively affect a person's decision-making process at a time when the stakes involved in these decisions and choices are becoming increasingly high (McIntyre 2018, Block 2019, and Kalpokas 2019).

In 2016 the term "post-truth" was declared the word of the year by the Oxford English Dictionary (OED) following a hick in its usage (cf. McIntyre 2018, 1). The declaration further popularised it, leading to "a rise in its use in a growing number of domains as its original meaning becomes stretched and

DOI: 10.4018/978-1-6684-4102-2.ch014

mangled" (Block 2019, 3). As a result, it has caught the interest of the Media and intellectuals. This is evident in the growing number of literature and media coverage on the phenomenon of post-truth. This also has resulted in a more precise conceptualisation and description of its manifestations, consequences, origin, and relations to various domains such as philosophy, religion, psychology, economy, science, and technology, particularly the Internet, media and specifically the social media.

Earlier characterisation and conceptualisation of the idea of post-truth appeared in the year 2004 by Ralph Keyes in a book he titled *The Post-Truth Era: Dishonesty and Deception in Contemporary Life.* Keyes in this publication offers a broad critique of the lack of sincerity and the prevalence of deceit in contemporary society. A year later, Harry G. Frankfurt, in his essay *On bullshit* (2005), offers a similar critique of modern society. He says that "one of the most salient features of our culture is that there is so much bullshit. Everyone knows this. Each of us contributes his share" (2005, 1). However, the term "post-truth" has been used increasingly to describe the contemporary era (McIntyre 2018, Lockie 2017, Block 2019, and Kalpokas 2019). The close association of post-truth with populist politics and new communication technologies coupled with the understanding of it as manipulative and relying on misrepresentation gives post-truth the status of being both familiar and strange. However, this does not undermine the fact that "there are still differences between old-style lies and conspiracies, and post-truth manipulation" (Yilmaz 2019, 240).

The paper analyses the contemporary association of politics with post-truth and defends the thesis that this association not only denigrates politics but undermines the value and trust in democracy (cf. Suiter 2016, 17-25). The argument is developed in three steps. The first section explores the concept of post-truth and post-truth politics to argue that the phenomenon of post-truth politics poses a significant threat to democratic politics. These threats are illustrated by explicating the consequences and implications of post-truth rhetoric in the recent political campaigns as manifestations of post-truth politics. The second section situates the challenges of post-truth and post-truth politics in the broader context of the crisis of truth to argue that post-truth politics poses a threat to democracy in general by undermining the value of truthfulness in democratic politics. The last section concludes with a proposal on countering the challenge of post-truth and post-truth politics.

THE CONCEPT OF POST-TRUTH

According to McIntyre (2019, 123 -125), post-truth has its remote origin within the academic discussions concerning the "standard of evidence, critical thinking, scepticism, cognitive bias, and so on" but in connection with postmodernists' approach that questioned everything. This approach ended up in perspectivism that denied the possibility of objective truth and indirectly attacked evidence-based reasoning. However, the term "post-truth" was meant to describe the kind of political rhetoric known today as post-truth politics. Hopkin and Rosamond (2017, 3) described the contemporary context in which post-truth emerged as word of the year as follows: "The rise of populist and anti-elite movements and the rejection of basic principles of reason and veracity characteristic of much of their political discourse." This was evident in the political discourse that characterised the Brexit campaign, the presidential campaign of Donald Trump in 2016. We may place the change campaign in Nigeria in 2015 in the same category. Post-truth politics is not limited to the above-mentioned countries. We can also name countries like Venezuela and Kenya as seen in the Nicolas Maduro's and Uhuru Kenyatta's campaigns, respectively. The political rhetoric that accompanied these political campaigns raised perennial concern about the relationship between truth and politics to new levels. The campaign rhetoric was primarily based not on evidence/facts, truth, and viable policies (cf. Hopkin and Rosamond 2017, 5). Narratives were coloured by sentiments, fabricated facts, and misleading data used to create political rhetoric that had no direct bearing on the truth or the concrete situation of things. Hence, in the words of Rose (2017, 555-558), "[t]hese electoral outcomes not only represented serious challenges to the established political norms, but they also exposed serious fault lines between different groups of citizens." Within this context, the term "post-truth" emerged as an expression of that concern. It highlights a new form of politics and relationship between politics and truth that is the combination of all that was traditionally known as political lies, mendacity, political spin, bullshit with manipulation and exploitation of passions through the power of rhetoric to win arguments and consequently electoral votes and endorsement rather than the use of logic/reason and evidence. According to Suiter (2016, 17-25): "currently, what seems to matter most is not the truth of any given interpretation on this history but the ability of a nativist or populist leader to appeal to instincts and nostalgic emotions of his group." This is what post-truth is about. Taking Donald Trump as an example, Zerilli (2020, 4) states:

With Trumpism, however, we seem to be moving from the register of the deliberate lie into another register [...]. In this new register, the lie is not so much put forward and taken for truth. Rather, the very distinction between true and false cease to exist with consequences far more corrosive of democratic politics than anything cooked up by inveterate liars such as Nixon.

According to Hopkin and Rosamond (2017, 2) post-truth, "rhetorical utterances can be understood as typical of a particular mode of reasoning [...] or indeed as a form of politics that more or less manifests itself *in toto* as a distinctive mode of communication [...]" that appeals to emotions instead of facts or evidence. The current hype in the use of the term is a phenomenon peculiar to contemporary time. This point is sufficiently made in the 2016 OED definition of post-truth. The term was defined as "Relating to or denoting circumstances in which objective facts are less influential in shaping public opinion than appeals to emotion and personal belief" (OED 2016 as cited in McIntyre 2019, 1). Closely following that definition, McIntyre (2019, 3-9) asserts that post-truth means more than the prevalence of falsehood, lying, deceit, manipulation, misinterpretation and falsification of facts, self-deception, and delusion. McIntyre concludes that:

Yet all seem sufficiently hostile to the truth to qualify as post-truth. [...] As presented in the current debate, the word post-truth is irreducibly normative. It is an expression of concern by those who care about the concept of truth and feel that it is under attack. [...] In its purest form, post-truth is when one thinks that the crowd's reaction does change the facts about a lie.

Similarly, Kalpokas (2019, 12-13) asserts that:

Any claims that post-truth consists of 'misrepresentations at best, and at worst, lies,' even including a routinisation of blatant lies ... are somewhat simplistic, since the idea of a 'lie' is itself anachronistic in the post-truth environment. [...] Hence, the prefix 'post-' does not indicate that we have moved to 'beyond' or 'after' truth as such but that we have entered an era where the distinction between truth and lie is no longer important; hence, we had also moved beyond an era when a consensus about the content of truth was possible.

Furthermore, post-truth entails a conscious and deliberate manner of speaking and acting or relating with the truth that disagrees with the facts of the situation or disregard of evidence but not without an interior motive. In that sense, Kalpokas (2019, 2) states that "post-truth does not have to involve discarding truth and embracing lies; it refers, instead, to the blurring of the distinction between the two." In the typical political lie, the distinction between true and false still holds, but in post-truth, the difference between true and false is erased. In a sense, it is a cognitive manipulation; hence, post-truth actors like their populists' counterparts "cannot be accused of lying any longer since it has succeeded in abrogating the very idea of truth" (Jay 2010, 149). Post-truth political actors exploit the resulting atmosphere of scepticism and confusion as Arendt (1973 in Bendall and Robertson 2018, 6) explained: "Left vulnerable by socio-economic upheavals, a muddled mass could be susceptible to Goebbelian big lies."

However, there is still more about post-truth that these definitions do not capture and therefore do not translate seamlessly to our current predicament concerning the relationship between truth, politics, and the public that is genuinely new in our situation. The public accepts the lies and cooperates with the liars to further their interests. If the lies do not immediately and negatively affect their well-being, nobody cares. The pursuit of truth for its sake does not matter and could be ignored if it does not bring any material benefit. Zerilli (2020, 3-4) aptly describes this new situation because of post-truth:

On this view of the problem of so-called post-truth democracies, people know they are being lied to, but they refuse to acknowledge it. They refuse to accord the lie any public significance because buying into the lie pays, so to speak. Accordingly, material interests outweigh fidelity to truth, but truth itself remains in principle knowable. It assumes that citizens are poised to recognise what is right before their eyes if only their material interests could be properly aligned with what is real. It is a view of mystification and deception familiar to anyone who has worked on the classic question of ideology, where how things appear is a distortion of what really is, but a distortion in which subjects are invested because it aligns with what they take their interests to be. Understood as ideological mystification, this account of post-truth suggests that reality is there to be seen by all those who have an interest in seeing it and are conscious of what that interest is.

The quotation above further explains the reality of post-truth condition. Truth does not matter as much as emotions/feelings and the existing situation in which one cannot objectively say what is true and what is untrue (cf. McIntyre 2018, 116). We are not hallowing or romanticising any period or exempting any side of the political divide, party, and political actor as free of post-truth but exploring the idea of post-truth as a mode of expression, which has become a defining characteristic of contemporary politics in some countries. This point is sufficiently made by McIntyre (2019, 172), when he writes that "whether we are liberals or conservatives, we are all prone to the sorts of cognitive biases that can lead to post-truth. One should not assume that post-truth arises only from others, or that its results are somebody else's problem." Hence according to Keane (2018, 2), "post-truth is not simply the opposite of truth, however, that is defined; it is more complicated. It is better described as an omnibus term, a word for communication comprising a salmagundi or assemblage of different but interconnected phenomena [...]. Post-truth has recombinant qualities."

THE NATURE OF POST-TRUTH POLITICS

When applied to politics, the concept of post-truth has a deeper meaning beyond dishonesty and deceit and must be distinguished from a long tradition of political lies, propaganda, fake news, and mendacity. McIntyre (2019, 13) says post-truth politics

amounts to a form of ideological supremacy, whereby its practitioners are trying to compel someone to believe in something whether there is good evidence for it or not. And this is a recipe for political domination. [...] If one looks at the Oxford definition, and how all of this has played out in the recent public debate, one gets the sense that post-truth is not so much a claim that truth does not exist but that facts are subordinate to our political point of view.

This is because post-truth and post-truth politics appeal to non-cognitive faculties such as feelings, emotions, and guts are dominant and form the basis for decisions and choices. Consequently, facts are ignored, objectivity in judgment is abandoned, and priority is given to appearances and subjective assessment that results in holding a distinct perspective so that "there is no such thing as objective truth. [...] any profession of truth is nothing more than a reflection of the political ideology of the person who is making it" (McIntyre 2019, 126). This sort of situation as Suiter (2016, 25-27) notes produced swathes of expressive voters moved by dangerous rhetoric, nativism, irritation, and anger to vote in a certain direction without regard to facts or for evidence. But the crucial question that may be asked is how was that possible?

Advancement in science and technology have created new and effective forms of communication. This development makes it easy and through social media, for post-truth political actors to tape in and create online communities that cut across borders and class distinctions with shared interests, desires, aspirations, and a common sensitivity. This is possible through mining and harvesting of personal data of individuals and communities on a larger scale delegated to algorithms, statistical and computing programs (Kalpokas 2019, 22-28). Once created these communities become crucial hubs for the dissemination of post-truth fiction. Davis (in Kalpokas 2019, 29) states that "post-truth involves political actors openly tailoring a pitch to a selected segment of the population by entertaining its members with fantasies and myths that have a particular appeal to them."

Furthermore, Kalpokas (2019, 2) notes that post-truth should be "seen as deeply embedded in everyday practices and developments (most notably, mediatisation) and innermost human drives (primarily, the striving for pleasure as a means of persevering in existence). Hence, what matters is how we experience and emotionally connect with information." In this case, cognitive bias becomes a factor as people tend to follow only information consonant with their sentiments which are made readily available or preselected for them through algorism online. Thus, according to Yilmaz (2019, 237-238),

While considering post-truth and its application by politicians, one needs to consider the reception side, namely the public's perceptions and acceptance of post-truth statements. As stressed, public opinion is inclined towards arguments with misinformation, fake and/or falsified stories, untruths and half-truths, and assertions without factual basis if the arguments appeal emotionally and if they are closer to one's own belief system.

This scenario is not possible without the complacency and cooperation of the wider society with post-truth political actors. Hence post-truth politics is not merely a manipulation but a co-created fiction and in fact, a collusion between the post-truth political actors and the wider society. The reason says Kalpokas (2019, 29) is that;

Data is created by users themselves, which is a permanent process in the current era of ubiquitous connectivity: messaging records, social media posts, browsing and search history etc as well as data generated by various connected smart devices and appliances that gather and transmit data by default is collected, collated, and analysed, sparing data users the need to specifically collect what is necessary for them, ultimately allowing for complete quantification and datafication of the subject, from their walking patterns to meals ordered and friends met. The more convenience, user-specific tailoring, and proactivity there is in the services one uses, the more data is ultimately being collected.

However, the fact that people's data online are collected and used without their knowledge and approval for such political manipulations and intent, make them unconscious or unwilling co-operators (Kalpokas 2019, 29-31) which rules out the possibility of meaningful consent and responsible political action. This threatens the heart of democracy, for democracy is not merely about a system of voting. Democracy is an expression of political freedom which is akin to human dignity. It is about a choice that has a moral value that a candidate or political party represents. Or it is a choice of a value that a party represents, and therefore it is a moral choice. It is a violation of conscience to persuade and manipulate peoples' emotions or sentiments to make the wrong choice or come to the wrong conclusion. The significance of this cannot be overemphasised in connection to democracy because it creates forms of participating politically that are separated from the satisfaction of desire or the endorsement and promotion of what is in vogue over that which is good. Kalpokas (2019, 30) explains that,

The Cambridge Analytica scandal is illustrative here: while the harvesting of user data has allowed for campaign planning in the most rational-qua-efficiency-maximising sense, it may not have led to the most rational outcome as far as electoral choices of the affected societies are concerned.

This raises many concerns. First, regarding the moral and legal permissibility of taping into the private lives of citizens and using such data or information for political purposes. Second, concerning the legitimacy of the advantage and the results due to such an advantage over an opponent. Third, it raises legitimate moral and legal concern over the use of such indirect means for political campaigns and the extent to which the public have been manipulated into giving their consent notwithstanding whether it is for their good or not. Fourth, the lack of shame and guilt on the part of post-truth political actors for manipulating the public further raises concern over the intent of gaining political power (Chen et al., 2019, 57-72). Political rhetoric as employed by post-truth politicians is primarily concerned not about truth but power and the sustenance of that power. To sustain that power, it is important, that the public is bewildered and confused by rhetoric. This kind of rhetoric as noted earlier blurs the line between truth and falsity so that the people's capacity for veracity is lost, especially if they are at the same time offered narratives that they would like to believe as true (Pinter 2012, 10). Therefore, in the words of Zerilli (2020, 6),

[T]he real danger of what we call post-truth politics is not so much ideological fervour or political provocation but the erosion of a common world in which things can be judged to be true or false. ... This distinction is not eroded overnight of course but emerges through, among other things, continual lying: The result of a consistent and total substitution of lies for factual truth is not that the lies will now be accepted as truth, and the truth be defamed as lies, but that the sense by which we take our bearings in the real world—and the category of truth vs. falsehood is among the mental means to this end—is being destroyed. That was the condition that characterised 20th Century totalitarianism and that is the condition into which we seem to be moving today.

Therefore, post-truth politics is a well-calculated strategy in some advanced democracies to entrench neoliberalism with its emphasis on individual political action that places cognitive responsibility on citizens to shape and make sense of their political world but would restrict the extent of political actions and decisions that democratic and social institutions are allowed to pursue (cf. Hopkin and Rosamond 2017, 2).

Post-truth politics may represent a modern form of grasping power achieved through (the force of) datafication to claim to some form of democratic endorsement in elections. According to Kalpokas (2019, 30) "datafication is a key term here, referring to a process whereby any online action is turned into exploitable data, and that data, in turn, becomes the epicentre of business models, either as a tradable commodity or as a key input into business planning." Legitimate questions could be raised about the use of datafication as a campaign strategy or mechanism for mass persuasion and profiling, particularly the extent to which online personal data are used to create political/campaign content that deliberately distracts the public from core political issues and into making bad decisions and choices. Political rhetoric produced through datafication is tailored and can be understood as a distinct type of reasoning and communication that is manipulative by intentionally anchoring political utterances on emotions/feelings instead of basing them on verifiable facts. This form of politics is consequential as it is employed to keep rhetoric at bay with the truth and can be illustrated in contemporary campaigns in which we have seen the apparent disconnection in the relationship between political rhetoric and truth. For instance, research by Blyth (as cited by Hopkin and Rosamond 2017, 2-3) shows that the austerity policies widely adopted by many countries as the result of the world economic meltdown were contrary to expert opinions of academicians and the austerity policies were sustained despite this opposition being borne out by the expected poor results. Hence, the popularity of ideas that prove to be unviable in the political sphere were evidence of their operation as bullshit but should be understood as post-truth politics because the decision was not based on existing true knowledge and evidence. According to Forough, Gabriel, and Fotaki (2019, 18), "Human progress is not assured, and the environment in which post-truth narratives have taken hold poses many threats." The continuous rise of post-truth politics despite these threats is rooted in the erosion of the culture of good democratic politics and campaign culture (cf. Hopkin and Rosamond (2017, 3).

POST-TRUTH AS ANTI-DEMOCRACY

Politics and truth have for long been at odds (cf. Arendt 2006, 128). Rose (2017, 555) asserts that "the belief among voters that politicians lie is near-ubiquitous in contemporary political systems, and politicians, in general, are routinely placed at or towards the bottom of indices of truth." It is arguable whether this has any significant effect during democratic elections involving such politicians. Likewise, post-truth rhetoric is known to be untruths or half-truths and manipulations. Nevertheless, it happens that such knowledge or discovery has little or no impact on the legitimacy of post-truth political actors where sentiments or emotions outweigh the judgement of reason. This situation, eventually, could lead to indifference and denial of established truths and values. Kinna (in Yilmaz 2019, 240-241) stresses the fact that lies are

no longer told with the pretence of upholding a public good, the lies that characterise post-truth are not designed to pass undetected, as Machiavellian lie was indented to do. A prince discovered to be dishonest and to tell untruths was hardly well equipped to rule a virtuous republic.' In other words, today's […] politicians employing a post-truth strategy do not even pretend they are not misinforming the public, nor that the 'facts' they are creating or rewriting are for the public good. Interestingly, despite their detection, politicians employing post-truth strategies have so far not suffered major repercussions or delegitimization.

In political and democratic cultures where liars are not reprimanded or punished for mendacity, they are at liberty to invent non-existent realities that match their aspirations/ambitions (cf. Yilmaz 2019, 241). This scenario is cementing the belief that politics is a dirty game, with no clear-cut moral norms for good political conduct. Living ethically and actively involved in politics does no longer resonate with what is acceptable in a political and democratic community. Notably according to Shapin (2019, 8),

English High Court judges, asked to consider a summons against Boris Johnson for demonstrable false-hoods in the referendum on EU membership, ruled against the petitioners, saying that everyone knew that lying was part of politics.

Similarly, we know about post-truth and the post-truth politics and its combination with modern technologies, social media, fake news, and populism to manipulate, polarise and "with the specific intention of shaping voter opinion and exciting emotions through inciting fear and hatred of the "other"" (Yilmaz 2019, 237-238). Yet, there seems to be a quiet resignation from the public notwithstanding that fact-checking tools are accessible, there seems not much willingness and need in that area. This is happening at a time when according to Yilmaz, (2019, 240) the multiplication of means of communication such as social media due to advancement in media technology has propelled post-truth politics to new heights and at the same has dwindled the impact and influence of traditional journalism and traditional media.

There is no doubt about the effectiveness of social media networks to easily spread unverified information, misinformation, as well as fake news to a large population across borders. Political actors can easily attract a large group of followers across borders, thereby making social media not only a platform to circulate fake news and post-truth utterances that misinform the public on purpose. Media technology has given politicians direct access to communicate with the public which again has the impact of decreasing the influence of mass rallies and party influence while empowering and popularising individual politicians and at the same time increasing individual voting autonomy. "In the end, 'the value or credibility of the media has somewhat faded in comparison to personal opinions. The facts themselves take second place, while "how" a story is told takes precedence over what. It is, therefore [. . .] about listening, seeing, and reading the version of facts which more closely fits with each person's ideology" (Yilmaz 2019, 240). At a time when political lies and fake news are shaking the very foundations of democracy and the world as we know it, the increasing use of post-truth politics assisted by social media as a campaign strategy and weapon by political actors further exacerbates such concerns and creates new challenges.

Post-truth situation significantly undermines the ability of individuals to differentiate what is false or true and to know what is accurate or inaccurate. In a society full of misinformation vis-à-vis the challenge of making informed decisions/choices without the requisite expertise when the stakes involved are high post-truth makes it more dangerous. Post-truth has left ordinary citizens without access to the truth by misusing the technological power of the internet and Media and as a result, undermines human potential and orientation to truth. We thought that with access to the internet it could no longer be possible to isolate people from the truth and manipulate them. However, with the same internet we are doomed switching from one source of information to the other in search of objectivity and truth without any hope of reaching certainty.

Consequently, trust in the credibility of experts and authorities is low in many areas or aspects of public interests, for instance; the media/press, economics/finance, policies, medicine and so on. This is fuelled by truth denial and the dissemination of which constitute partly what is called fake news. For instance; the recent connection in some quarters of the spread of coronavirus pandemic and 5G telecommunication technology and the connection between coronavirus vaccine, the implants of chips and the biblical mark of the beats called 666 and so on. It is important to note in passing how this sort of misinformation or fake news touch on critical sensibilities of the public beginning from science, technology, religion, and culture (cf. COSMOS). It is a crisis because the line separating truth and false, facts and mere opinions are blurred by all shades of emotions/sentiments and reasoning that fact-checking does immediately tell you where the truth lies. This is the era and situation of post-truth society where truth is not just relative but it is no longer a truth as *adequatio intellectus et rei*. It should be noted that because there is a concept of truth as consistency within a system, now we have post-truth, which says adequacy to emotions or consistency with feelings is a criterion of what is true and false. This means that to influence the choice I must influence not reason with facts but emotions. In this critical circumstance, a lot is at stake and there is no reason for the ignorance that some of our natural virtues such as trust, reasonableness and sincerity are undermined or compromised. There arises immediately the need to re-examine or diagnose some aspects of our current culture which are in a critical state and isolate them and determine, what the solutions are. Firstly, our task is to break the wall of lies. It is a difficult challenge. We need to mobilise our technological and human potentials to win the battle for truth. Our emphasis is on the human potentials and orientation to truth. In which case, the individual through personation has a significant role to play. This is moral duty that we owe to each other.

Secondly, there is the problem with some contemporary philosophical accounts of truth, which do not only misunderstand but ignore the moral nature of truth. Hence, academic discussion of truth is far removed from real life, its metaphysical relation and, restricted to inference/logic and placeholder of reality. This is only a clue about how narrowly truth is understood or conceived; so that what is missed or not stated or understated is significant enough not to be ignored and can be against the very truth which is defended. This is the case with half-truths or when truth becomes a matter of personal convenience, that is, truth as it suits one's situation and needs. Hence, the objectivity and universality of truth are denied and rejected in favour of a subjective and individualistic approach to and affirmation of truth. Consequently, it is difficult if not impossible to see the current problem of truth as a crisis. The crisis might be missed or go unnoticed because the demand for truth and its usage is hardly made in daily living. The question of truth will normally arise in situations that entail a search, a questioning, and an inquiry about the state of an affair or what is/was the case. A proposition about the state of an affair may describe a situation or state the truth about a particular case, for instance, in a testimonial or attestation in a law court. The demand for truth is made in efforts to delineate it from falsehood and when it is considered that the integrity and trust have been tempered or compromised, for instance, when the owner of Facebook, Mark Zuckerberg was questioned about the privacy and safety of personal information of users and how they are used (cf. Quartiroli 2011, 7). Another example is the demand to know the truth about the safety and morality of Artificial Intelligence such as when robots perform some duties like driving and health diagnoses on human beings.

Discussion about truth is heard and the truth is sought at critical and crucial circumstances. Truth matters when public officials, experts, and commissions/committees make statements or declarations of intent and purpose. However, truth matters not less but more in intellectual pursuit and is akin to such disciplines as philosophy, religion, and science. What is truth or what is the nature of truth and what is the truth of the case, feature more in the abovementioned disciplines but especially in philosophy where different and competing theories of truth have been propounded and debated without a clear winner. This is an indication that we commonly know that there exists such a thing as truth. We recognise that truth is important but as to what it is and how it should be construed or understood, philosophers do not agree.

Such disagreements do not exist when it comes to the recognition of truth or absence of truth and the important need to disentangle truth from falsehood. Failure in this respect gives the problem of truth a moral charge. We are strictly careful besides human error and ignorance that truth is not made to look like a lie and a lie is not made to look like truth by any form of rhetoric and description. A lie is not just the opposite of truth but a statement that has no match in the actual world and neither realisable in any world or circumstance; for instance, when religious persons or scientists make so-called future predictions; or when politicians make promises that cannot be fulfilled, or when scientists in valedictory speech say there are out in search of the reality. Further instances can be made from other areas such as the economy, finance, national security, social security, and public safety concerning consumer goods and services. The least consideration we can make of any such statement or claim is that it is nonsense. But it is fraudulent and constitutes mendacity, that is, intentional deception of people. It is depressing that such fears, concerns, and questions are raised in every aspect of life. Shapin (2019, 10) notes that

credibility and legitimacy problems attending a number of cultural institutions are not new. We have never needed critical analysis to support the belief that governments lie. Machiavelli recommended deceit as sound policy, and, in a famous 17th-century formulation, the English diplomat Henry Wotton defined an ambassador as "an honest gentleman sent to lie abroad for the good of his country.

Therefore, it is evident we are dealing with a crisis which is merely the lack or absence of truth but it is more of the situation in which people appeal and respond more to emotions than to truth or facts. This is typical of post-truth politics. Hence, the supposition or the claim that we are living in a Post-Truth Culture. The consequences are both psychological and material as the result of the widespread culture of lying, covering up the truth, and fraud. It weakens trust, spread suspicion, and increases the burden and scope of verification and vigilance. We want other persons to trust and believe what we say and do, in the same manner, and for the same reasons that we want to trust and believe what they say and do. Thus, where there are persons, authority, or experts, we take for granted their responsibilities and our expectations of mutual reliability, integrity, and trust. Public awareness and involvement in the detailed technicalities and intricacies of government, politics, economy, health policies and issues, finances, investments manufacturing, security, and so on is limited for lack of knowledge or ignorance. To take advantage of this ignorance to lie and deceive ask questions about the essence of representative democracy.

Besides, to know that some institutions and their representatives lack credibility and still leave them as normal without challenging the status quo as unacceptable is partly the expression of the Post-Truth Culture. The practice of feeding the electorates with lies and false campaign promises in the name of politics is the abnormal culture or practice that has become normal. Stamping out bad politics and entrenching a healthy political and democratic culture necessarily means a zero-tolerance to lies and liars. Morality is absent where truth is lacking and both morality and truth are needed for the rise of any social group and the advancement of human well-being more than the appeal to emotions.

THE RESPONSE TO POST-TRUTH IN POLITICS

This way of understanding and interpreting the post-truth phenomenon demonstrates that countering post-truth demands more than fact-checking skills, truth-telling or standing up for a notion of truth or even the idea of allowing the facts to speak for themselves. These may not be sufficient to counter post-truth narratives. We may quickly recall that Kalpokas (2019) describes post-truth as collusion between politicians and the public. Besides that, Zerilli (2020) has exposed post-truth as an interest-based phenomenon both from the end of politicians and the public. McIntyre (2019) and likewise Kalpokas (2019) link post-truth with manipulation and exploitation but in connection to our cognitive biases, affectivity, and science through modern technologies such as social media to effectively shape public opinion and perception of reality. These accounts of post-truth suggest that it is a deliberate act that has a premeditated outcome. Thus, the post-truth problem is not due to ignorance or because of error; nor is it a problem of understanding. Post-truth politics knowingly reject truth/facts for political ends. Thus, it is not a cognitive issue. That means post-truth politicians know the facts and fact-checking does little to change the narrative or sway public opinion, particularly where and when the facts in question are not commonly accepted and upheld as objective truths (cf. Zerilli 2020, 6). There is often more than one perspective about reality supported by different sets of facts. From this perspective fact-checking is not enough. Zerilli asserts (2020, 5),

what if the obsessive fact-checking that has become second-nature to those who would contest the delusional reality show (Dietz) of Trumpism actually worked against the public acceptance of the facts that are checked? What if fact-checking undermines the truth of opinion [...] crucial to caring about factual truths at all? If the problem of post-truth is the loss of our allegiance to a fact-based reality, we need to understand wherein that allegiance consists, that is, the worldly atmosphere in which facts, once checked, can be received as true in a politically significant way.

Similarly, McIntyre basing his arguments on research work and experimental results in psychology has noted the limitations of fact-checking as a strategy in dealing with post-truth. He acknowledges that the repetition of facts does eventually have an effect (McIntyre, 2019, 158). But he further argues that merely exposing people to facts or truths is often not enough to make them change their beliefs. This is because McIntyre (2019, 34) notes, "the selective use of facts that prop up one's position, and the complete rejection of facts that do not, seem part and parcel of creating the new post-truth reality."

Furthermore, fact-checking offers little in countering post-truth amidst growing "media fragmentation, information bias, the decline of objectivity, the threat of not just to knowing the truth but to the idea of truth itself" (McIntyre 2019, 153). Hence, McIntyre's proposal in dealing with post-truth is to stand up for a notion of truth and fight back. He states that "in the era of post-truth we must challenge each and every attempt to obfuscate a factual matter and challenge falsehoods before they are allowed to fester" (McIntyre 2019, 157). It is a moral duty not to accord any form of tolerance to lies and structures or organisations that support them. Consequently, McIntyre (2019, 157 - 158) has challenged and rejected the idea of a media balancing act, which means telling both sides of the story. The rule of media balancing act accords the same time to opposing views at the split-screen debates hosted by television media houses. The practice accords a lie the status of a competing opinion while being in fact a falsehood that should have been outrightly rejected. In effect, all facts become mere opinions in the subjective sense of it. Zerilli (2020, 9) notes that "this transformation of fact into mere opinion destroys the common world about which to exchange opinions and form judgment." For that reason, Zerilli's solution to post-truth is centred on her idea of the truth-teller as a fact-checker. It is in substance the same with McIntyre's proposal, particularly where she argues that "whatever resilience factual truths have depended on the continual testimony of human beings. In our ordinary speech and action, it is we who affirm a world held in common, reality as shared" (Zerilli 2020, 8).

There is no doubt to the merits of McIntyre's, and Zerilli's solutions and others of like minds to the problem of post-truth. However, their positions and solutions suffer the same weakness for which fact-checking is judged to be insufficient in countering post-truth. Those solutions are reactionary responses to the problem. In other words, the solutions arose only at those instances that truth is known or on suspicion to have been violated and necessitate a response, otherwise, there would be none. However, post-truth like a political lie and manipulation are all designed not to be dictated by their victims which is often the case. Thus, Williams (2002, 207) argues that "if it is a good thing, other things being equal, for the people to be truthful, it is a good thing for people in government to be truthful." Where and when post-truth (political lie or bullshit) succeeds, it takes advantage of public trust as one of the primary virtue and condition of democracy in liberal societies, which is the basis of the consent to govern. That trust is not and should not be dependent on the ability of citizens to be able to know when those saddled with the responsibility to govern are dishonest. Where and when that ability is lacking "the search for truth as the telos of inquiry and action" (Jay 2010, 23) would hardly arise. It is on this basis that it is argued here that fact-checking, truth-telling, and standing up for a notion of truth come short of solving the problem of post-truth. These solutions confine the responsibility and pursuit of truth and truthfulness to only those who know. Meanwhile, a vast number of citizens are ignorant not only in many matters of politics but of life in general and are not less concerned about truth. Due to human weakness, our knowledge is limited in many areas of interest/need, in which others are experts. This necessitates our mutual dependence, trust, and truthfulness. Liberal democratic societies are made of interdependent relationships based on mutual trust and truthfulness that democratic leadership cannot be dispensed from, for the continued sustenance and maintenance of the consent to govern and consequently, for the growth and survival of democracy and society (cf. MacIntyre 2014, 1-13).

Walzer (1997, 9) asserts that "coexistence requires a politically stable and morally legitimate arrangement and this too is an object of value." From the foregoing, the fight against post-truth should be a commitment to living, acting, and speaking truthfully/frankly that take cognisance of our human interdependence and mutual relationship as agents. "Truthfulness implies a respect for the truth" (Jay 2010, 11). And truthfulness is akin to the virtue of accuracy, sincerity, honesty, and commitment to acquire true beliefs and knowledge, so that what one says and does reveal what he/she knows and believes. Authority should be rooted first in truthfulness before competence so that it is careful not to lie and ruin trust. What must be emphasised, however, is that the basic cultivation of truthfulness in relation to everyday truth is only the beginning, not the whole story" (Jay 2010, 12). This recognition makes truth an independent value acceptable as a natural virtue not measured exclusively in terms of its effects. Williams (2002, 210) makes the same argument in a slightly different manner.

Liberal societies are democracies, and it may look as though it is the democratic element in the liberal complex that has a particular connection with a demand for government truthfulness. The people are the source of the government's authority (under various substantial restrictions) even of its policies. Government is in some sense a trust; there is a special relationship between government and people, and it is a violation of this conception for secrecy or falsehood to come between trustee and people. It is a feature of democracy, obviously that the citizens are supposed to be able to trust the government. … One relationship that by its nature excludes deceit is agency, in the sense of an agent's doing things on behalf of the or in place of a principal, things that the principal is poorly placed, for instance, or too occupied to do for himself.

Agency is an existential human condition that engenders trusting others on the things and areas of life that we do not know or have the capacity to understand. In this sense, trust and sincerity are natural virtues and the exercise of which are duties we owe each other without exceptions. "The general rule is, that truth should never be violated, because it is of utmost importance to the comfort of life, that we should have a full security by mutual faith…" (McIntyre 2006, 102). From this perspective, the bridge of trust or compromise of sincerity has moral and legal implications. If everyone lies or conceals the truth to everyone in our common and personal search for meaning and truth, common actions/goals will be impossible and our long-time survival and progress will be frustrated. Human beings are by nature social and political animals that embody those relationships of giving and receiving through which our individual and common goods can be achieved (MacIntyre 2014, 129). Thus, we have roles and responsibilities arising from belonging to the political community. This shared human survival depends on the collective recognition of the virtues of truth, trust, and sincerity as the foundation of the liberal democratic society. According to MacIntyre (2006, 114) "What is needed is the identification of some mode of institutionalised social practice within which generally established norms and reflective habits of judgment and action could sustain a coherent and rationally justifiable allegiance to a rule concerning truth and lying in a way and to a degree very different from the present dominant culture."

Politics is a genuine human activity that engenders responsibility which often comes in conflict with morals or truth. In such a situation of conflict, truth is a virtue and a value when removed or neglected or even concealed is a deliberate choice with implications. It is a failure not to live up to the moral truth and even to recognise certain higher standards or values. More accurately, it is a failure not to recognise that truth does not merely regulate a practice or activity; truth defines the kind of practice or activity called politics and at the same time legitimises it. We give away or concede too much to politics if politics is allowed to determine truth. Truth should determine politics.

Disclosure Statement: No potential conflict of interest was reported by the author.

REFERENCES

Arendt, H. (1973). *Origin of Totalitarianism*. Harvest Books.

Bendall, M. J., & Robertson, C. (2018). Crisis of democratic culture. *International Journal of Media and Cultural Politics*, *14*(3), 383–391. doi:10.1386/macp.14.3.383_7

Block, D. (2019). *Post-Truth and Political Discourse*. Springer Nature Publishers. doi:10.1007/978-3-030-00497-2

Cerovac, I. (2016). 'The Role of Experts in a Democratic Society'. *Journal of Education Culture and Society*, *2*(2), 75–88. doi:10.15503/jecs20162.75.88

Chen, M. (2019). Modelling Public Mood and Emotion: Blog and News Sentiment and Politico-economic Phenomena. In A. Visvizi & M. D. Lytras Effat (Eds.), *Politics And Technology In The Post-Truth Era* (pp. 57–72). Emerald Publishing Limited.

Collaboratarium for Social Media and Online Behavioural Studies (COSMOS). (2020). https://www.cosmos.ualr.edu/misinformation

Elkins, J. (2012). Concerning Practices of Truth. In J. Elkins & A. Norris (Eds.), *Truth and Democracy*. University of Pennsylvania Press. doi:10.9783/9780812206227.19

Ellington, J. W. (Ed.). (1994). *Immanuel Kant Ethical Philosophy*. Hackett Publishing Company.

Forough, H., Gabriel, Y., & Fotaki, M. (2019). Leadership in a Post-Truth Era: A new Narrative Disorder? The Sage. doi:10.1177/1742715019835369

Frankfurt, H. (2005). *On Bullshit*. Princeton University Press. doi:10.1515/9781400826537

Gaita, R. (2016). *When Politicians Lie: Reflections On Truth, Politics And Patriotism*. https://griffithreview.com/contributors/raimond-gaita/

Greenberg, D. (2015). *Truth in Politics Now*. https://prospect.org/article/truthpolitics-now

Hopkin, J., & Rosamond, B. (2017). Post-truth Politics, Bullshit and Bad Ideas. 'Deficit Fetishism' in the UK. *New Political Economy*. Advance online publication. doi:10.1080/13563467.2017.1373757

Human Rights Watch. (2019). *Nigeria: Widespread Violence Ushers in President's New Term*. Human Rights Watch. https://www.hrw.org/news/2019/06/10/nigeria-widespread-violence-ushers-presidents-new-term

Jay, M. (2010). *The Virtues of Mendacity*. University of Virginia Press.

Kalpokas, I. (2019). *A Political Theory of Post-Truth*. Springer Nature Publishers. doi:10.1007/978-3-319-97713-3

Keane, J. (2018). *Post-Truth Politics and why the Antidote isn't simply 'Fact-Checking' and Truth*. https://theconversation.com/post-truth-politics-and-why-the-antidoteisnt-simply-fact-checking-and-truth-87364

Keyes, R. (2004). *The Post-Truth Era*. St Martin's Press.

Lee, M. Y. H. (2015, July 8). Donald Trump's False Comments connecting Mexican immigrant and Crime. *The Washington Post.* https://www.washingtonpost.com/news/fact-checker/wp/2015/07/08/donald-trumps-false-comments-connecting-mexican-immigrants-and-crime/

Lockie, S. (2017). Post-truth Politics and the Social Sciences. *Environmental Sociology, 3*(1), 1–5. doi:10.1080/23251042.2016.1273444

Mair, P. (2013). *Ruling the Void. The Hollowing of Western Democracy.* Verso.

Marty, E. M., & Moore, J. (2000). *Politics Religion and the Common Good.* Jesse-Bass Publisher.

Maryniarczyk, A. (2016). Rationality and Finality of the World of Persons and Things. *Polskie Towarzyztwo Tomasza z Akwinu.*

McInery, R. (Ed.). (1998). *Aquinas, T. Selected Writings.* Penguin Books.

McIntyre, A. (2006). *Ethics and Politics: Selected Essays* (Vol. 2). Cambridge University Press. doi:10.1017/CBO9780511606670

McIntyre, A. (2007). *After Virtue* (3rd ed.). University of Notre Dame Press.

McIntyre, L. (2019). *Post-Truth.* The MIT Press.

Mearsheimer, J. J. (2011). *Why Leaders Lie: The Truth About Lying in International Politics.* Oxford Press.

Pashkova, V. (2016). *Arendt's political thought: The relationship between truth and politics.* http://hdl.handle.net/1959.7/uws:37675

Pinter, H. (2012). Art, Truth and Politics. In J. Elkins & A. Norris (Eds.), *Truth and Democracy.* University of Pennsylvania Press. doi:10.9783/9780812206227.9

Quartiroli, I. (2011). *The Digitally Divided Self: Relinquishing Our Awareness to the Internet.* Silens Publishers.

Rose, J. (2017). Brexit, Trump, and Post-Truth Politics. *Public Integrity, 19*(6), 555–558. doi:10.1080/10999922.2017.1285540

Shapin, S. (2019). Is there a Crisis of Truth. *Los Angeles Review of Books.* https://lareviewofbooks.org/article/is-there-a-crisis-of-truth/

Suiter, J. (2016). Post-Truth Politics. Political Insight, 7(3), 25-27. doi:10.1177/2041905816680417

Vogler, C. (2016). Anscombe on Two Jesuits and Lying. In L. Gormally, D. A. Jones, & R. Teichmann (Eds.), *The Moral Philosophy of Elizabeth Anscombe.* Imprint Academic.

Williams, B. (2002). *Truth and Truthfulness: An Essay in Genealogy.* Princeton University Press.

Yilmaz, G. (2019). Post-truth politics in the 2017 Euro-Turkish crisis. *Journal of Contemporary European Studies, 27*(2), 237–246. doi:10.1080/14782804.2019.1587390

Zerilli, L. M. G. (2020). Fact-Checking and Truth-Telling in an Age of Alternative Facts. *Le foucaldien, 6*(1), 1–22. . doi:10.16995/lefou.68

Chapter 15

New Skills for Sustainability in the Socially Challenging Era of Digital Transformation:
Opportunities and Challenges

Shweta Patel

Amity University, Raipur, India

Pratiksha Mishra

iD https://orcid.org/0000-0002-1802-1841

Amity University, Raipur, India

ABSTRACT

There are both opportunities and challenges because of technological advancements. The world as we know it has undergone a significant transformation in the digital age. And, to keep up with the changes, organizations have placed a greater emphasis on digital transformation. We are presently in the midst of a technological revolution with the intention to basically affect people, work, and the mode of interplay with one another. While the people will want to work with technology, there may additionally be a developing want of people to increase specialized skills for the way they interact with one another. To keep pace with rapid disruption, organizations should also emphasize the need to invest and reskill their people to stay competitive in the fast advancement of technology, which is altering the nature of skills and abilities required in the workplace, which also needs to make a mental shift among the employees who can make that technology useful.

DOI: 10.4018/978-1-6684-4102-2.ch015

INTRODUCTION

The times we live in is one that is filled with both opportunities and challenges because of technological advancements. The world as we know it today has undergone a significant transformation into the digital age. And, to keep up with the changes, organizations have placed a greater emphasis on digital transformation. We are currently during a technological revolution that will fundamentally impact people, work, and the mode of interaction with one another.

Transformation is not something new; it has been going on since the dawn of time right from First Industrial revolution which took place during the 18th centuries which focused on mechanization, the Second revolution focused on Electrification focusing on the skills of division of labour and mass production. The Third industrial revolution was all about automation and globalization through electronics and information technology and the Fourth Industrial revolution which was coined by Klaus Schwab, the founder of the World Economic Forum, at the WEF meeting in Davos in 2016, brought the era of Digitalization through robotics, artificial intelligence, augmented reality, and virtual reality. Eventually Marc Benioff the founder of Salesforce advised the World Economic Forum that One cannot do commercial enterprise in the Fourth Industrial Revolution without the believe of your personnel and your clients and partners." Hence the Fifth revolution might be all approximately Personalization, Innovation, purpose, and inclusivity which calls for deep, multi-stage cooperation among human beings and machines. It is a common misconception that to succeed in this era of transformative technologies, there will be need to learn highly technical or scientific skills. However, the fact cannot be denied that new skills will be needed for the new jobs forcing people to adapt, incorporate and respond to new requirements in the Digital Age. While the people will need to work with technology, there is also a growing need for people to develop specialized skills for how they interact with one another. People are more important than technology in the digital transformation. While the future is more ambiguous and uncertain than ever, organizations should also focus on bridging the digital divide and work together in reskilling and upskilling employees so that they are better equipped to adapt to change. To keep pace with rapid disruption organizations should also emphasize on the need to invest and reskill their people to stay competitive in the fast advancement of technology which is altering the nature of skills and abilities required in the workplace which also needs to make a mental shift among the employees who can make that technology useful. Hence to reveal its transformative power, digital possibilities must be combined with skilled employees. As a result, digital transformation necessitates the use of both technology and people.

Modern technologies are boosting productivity, improving people's lives, and reshaping our world. Adapting to this digital age requires organizations also to adapt themselves with new capabilities. The overall objective of this chapter is to discuss several challenges faced when pursuing sustainable change owing to technology and the changing skills and competencies needed for the new professionals in the Digital Age in pursuing sustainable technological change.

The authors also identify some avenues for future research. The discussions are based on five challenges: (a) Digital Literacy (b) adaptability and resilience (c) Critical Thinking for conceptualizing, synthesizing, and analysing data (d) emotional intelligence and (e) Communication & collaboration, dealing with challenges encountered when pursuing sustainable technological change. The article argues that sustainable technological change will require a re-assessment of the skills and that future research should be towards addressing the challenges required to be adapted in the era of digital transformation.

Figure 1. Author's contribution

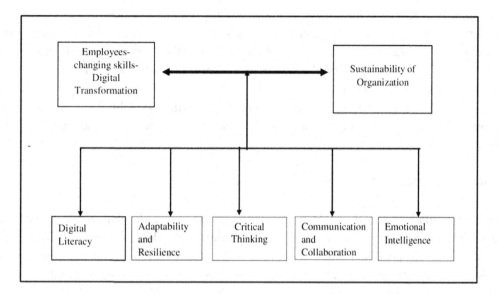

1. Digital Literacy

Much of the current academic and professional interest in digital transformation and corporate enterprise structures has focused on the generation or the organization's external forces, leaving the internal factors, particularly personnel, out. The purpose of this chapter is to identify employee digital literacy as an organizational affordance for capturing contextual variables where digital technology is found and used. (Kutanoglu, D. C., & Abedin, B. (2020)

In the late 1990s, Gilster (1997) defined 'digital literacy' in academic terms, recognizing the internet's essential however progressive strong point and figuring out the digitally literate scholar as owning a selected set of facts skills (e.g., evaluation, searching) carried out to textual content and multimedia facts observed at the internet and located in a formal, school-primarily based totally mastering context. Digital literacy has evolved, changed, and prolonged on the grounds that Glister's authentic concept, turning into an increasing number of crucial to cultural, civic, and monetary participation (Aabo 2005). Gilster, P. (1997). A lack of digital literacy is increasingly threatening one's ability to be a capable student, an empowered employee, or an active citizen. Digital literacy is generally thought of as a school-based skill, although it is taught and cultivated in a variety of informal learning settings, including libraries, museums, social organisations, online affinity spaces, and even the home.

Managers who understand their organization's degree of digital literacy may improve it and better prepare their employees for future challenges area of digital transformations. They may be able to win the "emotions and reasoning" of their employees in this way, allowing them to surpass competitors that treat their people like machines (Westerman, 2016)

The major problems for firms, according to Murawski and Bick (2017), are adapting their culture, mind-set, and competencies to the new, digital style of working rather than technology trends, disruptive technologies, or changing consumer habits. The move toward culture, mindset, and competencies necessitates a focus on the workforce. It may be difficult for them to endure digital transformations if they are not digitally literate. A digital savvy individual understands when and how to use digital resources to address an information need -a knowledge or understanding gap that demands further research.

Depending according to whether managers or decision makers actively or inadvertently introduce modern technology or make changes to current technologies, the change in norms may be direct or indirect. Employees can then opt to conform with new standards and tailor them to their jobs, or reject them, based on their "active interpretations."

As a result, better understanding the role of employees in digital technology usage through their digital literacy may help us better grasp how to align the changing interaction between employees and technology.

2. Adaptability and Resilience

It is critical, particularly during digital transformation processes, to address the associated challenges in the workplace and to be able to adapt to the changes. Modern technologies provide businesses with new chances and options to boost their productivity and competitiveness. Humans in the organisation play a crucial role in realising the promise of digitization. As a result, it's critical that staff take a constructive attitude toward the accompanying transformation procedures. (Peschl, A., & Schüth, N.J.,2022). As we speed our transition to a new, digital economy, it is critical that we foster adaptation and resilience in our workforce. According to Mona Cheriyan, President and Group Head–HR, Thomas Cook India, organisations must invest in preparing people with new skills and competences to deal with the "new normal."

Various forms of dangers that come from the external environment are frequently encountered by enterprises. As a result, in today's increasingly uncertain world, resilience has become essential for service businesses (Zeya He et al,2021). They suggested three alternative approaches to interpret Organizational Resilience: Reacting, Adapting, and Changing. The reactive approach is based on crisis management research, which defines OR as an organization's ability to revert to its prior state, or normality. The adaptive perspective sees OR as the ability to not only survive but also recover from a crisis through a sequence of adaptive interventions, while the transformative perspective sees OR as a conscious endeavour to improve one's ability to cope with surprise.

(Trenerry, B., et.al, 2021) through a multi-level approach, gave concepts that bridged the significant gaps by identifying and combining key aspects important for an organization's overarching digital transformation. They identified five broad variables for successful digital transformation among employees at the individual level: technology adoption, perceptions and attitudes toward technological change, skills and training, workplace resilience and adaptation, and work-related wellbeing. They identified three key aspects for digital transformation at the group level: team communication and collaboration, workplace relationships and team identification, and team flexibility and resilience. Finally, they recommended three components for digital transformation at the organisational level: leadership, human resources, and organisational culture/climate.

3. Critical Thinking

Increased adoption of modern technology is coupled by growing labour market skills shortages, making reskilling and upskilling personnel one of the most pressing concerns that businesses and governments confront. According to industry surveys, most businesses will face growing skills gaps in the next years, with employers looking for individuals with a diverse set of talents, including critical thinking, analytic, and problem-solving abilities, as well as self-management, adaptability, and resilience (World Economic Forum, 2020; McKinsey, 2021). According to a recent poll by McKinsey (2021), most organisations throughout the world (89 percent) have or will have a skills gap in the next few years. Employer's value critical thinking, analytic and critical thinking skills, self-management, and other soft skills in addition to highly specialised talents (Chuang and Graham, 2018). Employers also value critical thinking, analytic, and problem-solving abilities, as well as self-management, adaptability, and resilience, as essential qualities in today's employment (World Economic Forum, 2020). Individuals' capacity to learn new skills and receptivity to instruction are thus becoming a more relevant focus of research as the pace of digital transformation accelerates.

According to the World Economic Forum's (WEF) Future of Jobs 2020 report, critical thinking and analysis, complex problem-solving, and self-management skills such as active learning, resilience, stress tolerance, and flexibility are among the top skills and skill groups that global employers view as important in the run-up to 2025.

4. Communication & Collaboration

It is simple to combine online and offline communication while keeping employees engaged on their own devices. This enables "anywhere, anytime" access to workplace papers and resources, which has become increasingly ubiquitous in our daily lives. Furthermore, digital links frequently assist in bridging generational divides and bringing personnel of all ages together. Collaboration in departments and throughout the enterprise is facilitated by new digital communication. At the end of the day, the digital transformation should feel less like "technology" and more like natural ways to execute activities as mentioned (Daniel Newman,2017). The accrued virtuality of the work team facilitates task-oriented and relative communication among team members, leading to positive cooperative work outcomes like economical knowledge sharing and knowledge flow, fast and precise task coordination, and increased transparency of labour processes (Alshawi and Ingirige, 2003; Grudin, 2006; ellison et al., 2014; Anders, 2016).

In each geographically proximate and distributed work teams, the extent of team virtuality is enhanced as organisations endure digital transformation by implementing a spread of advanced and innovative collaboration technologies (Leonardi et al., 2013; Ellison et al., 2014; Anders, 2016).

Guinan et al. (2019) emphasized the importance of flexible and diverse team collaboration in achieving digital transformation goals, based on findings from a multilevel study.

According to studies, it is miles vital to permit worker affinity for the usage of digital technology for collaboration and to become aware of inner digital experts (Berghaus and Back, 2016). On the contrary, old group collaboration and verbal exchange behaviors are likely to stymie digital transformation processes and outcomes. Inertia over pre-current on-web website online collaboration and face-to-face verbal exchange routines often results in inadaptability or resistance to the shift to digitalized paintings and verbal exchange processes (Alshawi and Ingirige, 2003; Hur et al., 2019). In the face of digital revolution, enhancing group verbal exchange and collaboration via social and technological scaffolds is vital.

4. Emotional Intelligence

People who can connect with others, show empathy and understanding, and understand emotions are needed in our workplace. Emotional intelligence is now more than ever a "nice to have," but a "fundamental capability for the future." As quoted by Pip Russell, Strategy, innovation, and commercial operations vice-president, Schneider Electric signifies the importance of Emotional Intelligence in workplace.

Automation and AI have a significant impact on the workforce, causing uncertainty and change and necessitating the development of new skills to deal with the changes. Emotional intelligence will provide people and businesses with the skills they require, and as a result, it is quickly gaining traction. Organizations will soon require a baseline level of emotional intelligence as a required qualification, even for nonsupervisory roles, according to our research. Employees and companies benefit from EI in the form of increased productivity, job satisfaction, and lower attrition, among other things.

According to Capgemini Research Institute, Emotional Intelligence Research, Executive Survey, August–September 2019, N=750 executives concluded with the following analysis.

EI will become a "must-have" ability, according to 74 percent of executives and 58 percent of non-supervisory employees. It will become such in the next one to five years, according to 61% of executives and 41% of non-supervisory staff. Displacement of routine tasks, evolving job roles, and the inability to automate certain tasks will be key reasons for an increase in demand for EI skills.

Although automation and AI will impact all career levels, organizations currently focus more on building EI skills at senior levels than at non-supervisory levels. Organizations do not conduct enough training in building EI skills for employees across grades. Enhanced productivity, high employee happiness, and improved revenue are among the top benefits for companies. Reduced attrition and increased market share

Employee benefits include increased happiness, a lower risk of losing their work, and a higher willingness to adapt to new situations, the preservation of human jobs in the face of technological advancements.

Investing in EI skills might result in returns that are up to four times higher.

CONCLUSION

As digital technologies disrupt industry after industry, many businesses are adopting large-scale change initiatives to reap the benefits of these trends or merely to stay competitive.

Smart technology, artificial intelligence (AI) and automation, robotics, cloud computing, and the Internet of Things (IoT) are all rapidly advancing digital technologies that are profoundly changing the nature of work and raising concerns about the future of jobs and businesses. To be competitive in the face of fast change, businesses must update and transform their business models. Meanwhile, advances in technology are altering the types of skills and competences required in the workplace, necessitating a mental shift among individuals, teams, and organisations. The commencement of the Fourth Industrial Revolution has been signalled by a series of ground-breaking, developing technologies throughout the last decade. Many companies in the private sector have reoriented their strategic direction to take advantage of the opportunities presented by these technologies. By 2025, machine and algorithm capabilities will be more widely used than in previous years, and the number of hours spent working by machines will equal the number of hours spent working by humans. Workers' employment prospects will be disrupted across a wide range of industries and geographies because of work augmentation. According to new data from the Future of Jobs Survey, 15 percent of a company's staff is at danger of disruption in the horizon up to 2025, and on average, 15 percent of a company's personnel is at risk of disruption in the horizon up to 2025. (World Economic Forum, 2020) Employees can help with the change by acting proactively and promoting company (digital) strategies, as digital transformation affects businesses beyond organisational boundaries. As a result, intrapreneurial skills contribute significantly to digitalization and digital transformation. (Blanka, C.,2022)

During a digital transition, change occurs at all levels, especially when it comes to talent and competencies. One of the keys to transformation success is the addition of such a leader. The engagement of transformation-specific roles, such as leaders of individual initiatives and program-management or transformation office executives who are dedicated full time to the change effort, is also important. Leadership dedication is another crucial factor in achieving success.

Adapting to new jobs amid a digital disruption can feel like transitioning to work and living in a different culture. When the tasks to be completed change and become more important, employee positions in the organisation change as well. While learning new duties is vital in transitions — for example, transitioning to working during the epidemic necessitated learning to use videoconferencing capabilities - it is not the only item to consider.

REFERENCES

Aabø, S. (2005). The role and value of public libraries in the age of digital technologies. *Journal of Librarianship and Information Science*, *37*(4), 205–211. doi:10.1177/0961000605057855

Alshawi, M., & Ingirige, B. (2003). Web-enabled project management: An emerging paradigm in construction. *Automation in Construction*, *12*(4), 349–364. doi:10.1016/S0926-5805(03)00003-7

Berghaus, S., & Back, A. (2016, September). Stages in digital business transformation: results of an empirical maturity study. In MCIS (p. 22). Academic Press.

Blanka, C., Krumay, B., & Rueckel, D. (2022). The interplay of digital transformation and employee competency: A design science approach. *Technological Forecasting and Social Change*, *178*, 121575. doi:10.1016/j.techfore.2022.121575

Ellison, N. B., Vitak, J., Gray, R., & Lampe, C. (2014). Cultivating social resources on social network sites: Facebook relationship maintenance behaviors and their role in social capital processes. *Journal of Computer-Mediated Communication*, *19*(4), 855–870. doi:10.1111/jcc4.12078

Gilster, P. (1997). *Digital literacy*. John Wiley & Sons, Inc.

Grudin, J. (2006, January). Enterprise knowledge management and emerging technologies. In *Proceedings of the 39th Annual Hawaii International Conference on System Sciences (HICSS'06)* (Vol. 3, pp. 57a-57a). IEEE. 10.1109/HICSS.2006.156

Guinan, P. J., Parise, S., & Langowitz, N. (2019). Creating an innovative digital project team: Levers to enable digital transformation. *Business Horizons*, *62*(6), 717–727. doi:10.1016/j.bushor.2019.07.005

He, Z., Huang, H., Choi, H., & Bilgihan, A. (2022). Building organizational resilience with digital transformation. *Journal of Service Management*. https://fowmedia.com/helping-your-employees-adapt-to-digital-transformation/ https://www.capgemini.com/wp-content/uploads/2019/10/Digital-Report-%E2%80%93-Emotional-Intelligence.pdf

Kozanoglu, D. C., & Abedin, B. (2020). Understanding the role of employees in digital transformation: Conceptualization of digital literacy of employees as a multi-dimensional organizational affordance. *Journal of Enterprise Information Management.*

Leonardi, P. M., Huysman, M., & Steinfield, C. (2013). Enterprise social media: Definition, history, and prospects for the study of social technologies in organizations. *Journal of Computer-Mediated Communication, 19*(1), 1–19. doi:10.1111/jcc4.12029

McKinsey. (2021). *Five Fifty: The skillful corporation.* Available online at https://www.mckinsey.com/business-functions/mckinsey-accelerate/our-insights/five-fifty-the-skillful-corporation

Meyers, E. M., Erickson, I., & Small, R. V. (2013). Digital literacy and informal learning environments: An introduction. *Learning, Media and Technology, 38*(4), 355–367. doi:10.1080/17439884.2013.783597

Murawski, M., & Bick, M. (2017). Digital competences of the workforce–a research topic? *Business Process Management Journal, 23*(3), 721–734. doi:10.1108/BPMJ-06-2016-0126

Peschl, A., & Schüth, N. J. (2022). Facing digital transformation with resilience–operational measures to strengthen the openness towards change. *Procedia Computer Science, 200*, 1237–1243. doi:10.1016/j.procs.2022.01.324

SouleD. L.PuramA.WestermanG. F.BonnetD. (2016). Becoming a digital organization: The journey to digital dexterity. Available at SSRN 2697688.

Trenerry, B., Chng, S., Wang, Y., Suhaila, Z. S., Lim, S. S., Lu, H. Y., & Oh, P. H. (2021). Preparing workplaces for digital transformation: An integrative review and framework of multi-level factors. *Frontiers in Psychology, 12*, 822. doi:10.3389/fpsyg.2021.620766 PMID:33833714

World Economic Forum. (2020). *The Future of Jobs Report.* Available online at: https://www3.weforum.org/docs/WEF_Future_of_Jobs_2020.pdf

Chapter 16
Impact of Digitalization on Economic and Social Aspects

Subhanil Banerjee
https://orcid.org/0000-0001-7485-9967
PES University, Bengaluru, India

Souren Koner
Royal School of Business (RSB), The Assam Royal Global University, Guwahati, India

Arakhita Behera
School of Humanities and Social Science, K.R. Mangalam University, Gurgaon, India

Suhanee Gupta
School of Humanities, K.R. Mangalam University, Gurugram, India

ABSTRACT

The chapter aims to understand the impact of mobile and broadband subscriptions on the ease of doing business and per 100 people trade volume that might be critical in facing a pandemic of recent type in the future. On the one hand, the ease of doing business has been gaining popularity over the years as an indicator of where to set business. On the other hand, the spread of mobile subscriptions and broadband and their conjugation has significantly impacted modern-day business. Moreover, trade has already proven itself as the engine of growth. With this background, the chapter intends to determine the impact of the digital duo on ease of doing business and per 100 trade volume. The study is holistic in nature. It considered 128 countries in the world regarding the variables ease of doing business ranking (dependent), per 100 capita trade volume (dependent), and per 100 broad brand penetration (independent variable) along with per 100 mobile subscriptions (independent variable) for the years 2019 and 2020, and then opted for a pooled regression analysis with robust standard errors.

DOI: 10.4018/978-1-6684-4102-2.ch016

INTRODUCTION

It is no exaggeration that we live in a digital era. Machines have replaced manual activities, and the entire human genome has almost become a cyborg. Digital intervention in human life was there before; there were dishwashers, jukeboxes, and many other home appliances easily called substitutes for human activities. But the surge that came in the 1990s and is still following us promoted the digital world to a new echelon. Such changes became more prominent after the internet and smartphone conjugation, which offered a time and spaceless environment and penetrated our personal lives, especially social interaction.

The telephone itself was a path-breaking invention, but it was geographically restricted, whereas mobile phones keep a person in touch with others while moving. Social websites like Facebook, Twitter, and Snap Chat have spanned the human arena and kept them in constant touch and interaction. Various government agencies in different countries have digitized their mechanisms. So that people can approach them more frequently and easily.

So far, it is apparent that digitalization has made social and economic interaction easier. But can it be a concluding remark? Internet forgery has increased, and increased interaction among people has given birth to further friction.

In this background, the present chapter delves into this vital aspect: how digitalization may have impacted the economic and social aspects, the associated risks, and what lies in the future.

The production regime that followed the Fordist era was unique in its availability of information at a lower cost, irrespective of the geographical boundaries around the world. A conjugation of microelectronic revolution and independence from the regulatory regimes following Uruguay and GATT negotiations around the world made this possible (Garnham, 1990; Goddard & Gillespie, 2017; Hepworth, 1990). Harvey (2010, 2020) coined it as the convergence of time and space, and Poster (1990) portrayed the same as an electronic uprising that transformed knowledge's content and the process of attaining it. This particular production mode (Castells, 1989) was backed by a series of collective calamities that were random and consistent in tandem. The collapse of the Bretton-Wood method, resorting to floating from a fixed exchange rate, and the OPEC crisis during the second half of the 1970s unconfined an enormous amount of Petro-dollars, which tossed the West into an ambiance constructed by stagnation and inflation, popularly known as stagflation, are some of them. The developing countries' debt escalated to a towering height, and a swap between equity debt and debt became usual (Corbridge, 1984).

Moreover, Japan has poised itself as the emerging financial epicentre of the world. The explosion of the Euro market, proceeded by the demur of time-tested industrial nations like the United Kingdom and the United States of America, paved mainly by the Thatcher and Regan governments, respectively, created an emptiness in the global economic ambiance and Japan from Asia, along with Germany from Europe, were more than apt and willing to fill that space (Vogel, 1985; Walter, 1988).

Furthermore, the intelligent mode of production that opted for real-time inventory systems through automation and innovation by multinational corporations who believed in spreading their business irrespective of time and space eventually replaced the state monopoly and embraced globalization. Finally, the much-awaited fragmentation of Soviet Russia led to the realization that the world ambiance from now on would be more dynamic than static (Graham et al., 1988).

The amalgamation of the financial systems around the world was made possible by the aforementioned alteration of telecommunication (Marshall, 1987). Simply, a country's comparative advantage over others becomes subject to the trio of data processing, data transformation, and improved telecommunication (Gillespie & Williams, 2016).

The transformation of data from analogue to digital led to the amalgamation of computer services and telecommunication. Here, the importance of successive privatizations of the telecommunication system that made the sector competitive and innovative cannot be overlooked. Bearing in mind the significance of telecommunication, the United Nations (UN) established The Integrated Service Digital Network to balance the dynamism of the technological changes linked to telecommunications around the world among its member countries. The global spread of optical fibre, mostly owing to the pro-activeness of the United States of America, brought equality and symmetry in information all over the world (Warf, 2016).

The landline telephone has been a proven success in business and personal communication. However, it confined communication within the periphery of its chord. In a dynamic world, this is a significant paucity; still, its contribution cannot be demeaned (Brewster, 1994). The introduction of mobile phones has overcome the extremely local nature of this communication, and this particular evolution from a static to a dynamic form of communication has been coined the digital divide. Mobile phones might sound new, but its prototype was built in 1973 before it became available in the market from 1983 (Tiong et al., 2012). However, the transition from market to mass took some more time, and the new millennium is nothing less than the millennium for mobile phones (Regazzi, 2018).

In this modified world, all the aspects of human life and activities come with direct interaction with information. The intra-personal, interpersonal, intra-institution and inter-institute communication has become dynamic and is no longer bounded by time and space. It is undoubtedly significant to state that such a changeover provided no less than three benefits to global trade, cutting accounting and opportunity costs through RFID. For example, that particular technology reduced the cost of transit for German automobile giant BMW by 11% to 14% (Manyika et al., 2014). The same is true for ebay.com "Conducting cross-border business on a digital platform like eBay can reduce transaction costs by 64 percent" (Lund & Manyika, 2016, p.5). All these were possible as a drive towards a 'Pareto Optimal' ambiance through more competition and simplification of the entire trade process has been initiated by the digital divide.

These elements led to the emergence of digital entrepreneurs and created employment opportunities for the ever-increasing global population. Financial transactions, one of the most important prerequisites, also became more holistic regarding geographical coverage and less time-consuming following the digital duo.

Also, this new ambiance significantly reduced remoteness and the associated issues regarding trade participation. Previously insignificant countries on the international trade map have started participating spontaneously in global trade. In this particular aspect, explaining the virtual market becomes very important. In a capitalist world, time is money, and life is becoming overly busy. Physically visiting a market or mall is becoming tiresome and often suffocating. The advent of the virtual market in this background is a path-breaking and pro-human development. It is quite apparent that people have become tech-savvy, and with the digital duo, they are opting for online marketplaces similar to Amazon, Flipkart, and others. Different-fitting mobile apps presented by these virtual entities through visual appeal often influence impulse purchases. Consumers can now compare products while sitting within the comfort of their own four walls. Again, the choice of products is no longer a slave to the geographical periphery. Any consumer who has internet access can purchase a foreign product in this digitally augmented globalized market. Such international transactions will increase the trade volume, as rightly observed in the current article "The digitalization of trade is a reality" (Janow & Mavroidis, 2019, p.1).

Following the aforementioned discussion, the concerned chapter opts for a pooled regression of data from 128 countries worldwide to examine and corroborate the association between global trade and cellular subscriptions. It is essential to mention that smartphones are usually cheaper and handier than desktops or laptops. Furthermore, smartphones also ensure connectivity between traders, producers, and consumers while travelling.

The present chapter begins with an introduction, followed by a thorough review of existing literature, opts for a brief highlight of the present status of digital trade before moving into econometric analysis, and ends with a conclusion.

LITERATURE REVIEW

From historical times, global trade has drawn the attention of economists. Different schools of thought gave varied explanations regarding the conventional determinants of the same. Obviously, in a dynamic world, as depicted in the introduction, some of these determinants are augmented, and some new determinants are introduced. The current literature disregards the literature that delves into the usual causal factors leading to trade and emphasizes the literature that discusses the interaction of digital duo with trade.

Digital Conjugation and Global Trade: Existing, literature

Soobramanien and Worrall (2017) mentioned that the export of conventionally imported goods and services and the growth of e-commerce are positively correlated in the Caribbean islands. Quéré et al. (2015) highlighted the bi-directional positive variation between trade and digitalization. Through a quantitative approach, Freund and Weinhold (2004) illustrated that a 10% rise in internet access would result in a 0.2% rise in exports. Marcella and Davies (2004) emphasized the three kinds of marketing forms constructed through the internet and their comparative relevance in explaining modern-day commerce and business; they also highlighted their peer influence on consumer behaviour. Bhavnani et al. (2008) expressed their hope that trade through mobile devices might eventually become a potent armour against poverty. "The survey of 300 sardine fishing units was conducted every Tuesday, from September 3, 1996, to May 29, 2001. Data on the amount of fish caught, costs of operation, sale conditions (market, price, quantity, time, etc.), weather conditions, and whether they used a mobile phone were obtained. The survey found that the largest boats are prone to phones, owing to their higher purchasing power and eventual arbitrage gain. This study concluded that the use of mobile phones (a) increased consumer surplus (by an average of 6%), (b) increased the fishermen's profits (by an average of 8%), (c) reduced price dispersion (by a decline of 4%), and reduced waste (which was averaging 5-8% of daily catch, before the use of mobile phones)." (Bhavnani et al. 2008, p.16). Meijers (2013) indicated a high degree of direct alliance between global trade and the internet base.

Shaikh and Karjaluoto (2015) emphasized the astronomical impact of e-commerce on global commerce and business. Ling & Typhina, 2017 mentioned, "A high mobility and high context app would allow someone to play a locative game of hide-and-seek with their friend, while a low mobility and low context app would consist of someone checking the hours of a local business before leaving home." (Ling & Typhina, 2016). Abeliansky and Hilbert (2017) took into account the differential impact of augmented technology on trade in a two-by-two framework. They have concluded that the quality and quantity of telecommunication are essential for developing and developed countries, respectively. González and Jouanjean (2017) depicted digitalization as a harbinger of change in global trade and commerce. Maiti and Kayal (2017) opined, considering India, that the revolution in telecommunications has boosted the service sector and, thereby, international trade for the concerned country.

The aforementioned illustration reveals that the internet and mobile phones have led to the digital trade. Moreover, the extension of the definition with the possibility of becoming broader in the near future is now passing different types of trade as digital trade.

Smart Phones and Internet

Osibanjo & Nnorom (2008) decorated the extraordinary expansion of mobile and quick demur in landline subscriptions in Nigeria: "The country's tele density increased from a mere 0.4 in 1999 to 10 in 2005. More than 25 million new digital mobile lines had been connected by June 2006... Consequently, the contribution of fixed lines decreased from about 95% in 2000 to less than 10% in March 2005 " (Osibanjo & Nnorom, 2008, pp. 199-200). Meltzer (2016) added that in developing countries, mobile phones and the internet are complementary (Meltzer, 2016, p. 198).

The paucity of holistic attempts to measure the effect of mobile and fixed broadband subscriptions on world trade is quite clear from the above literature. However, it is also clear that digitalization and mobile mutiny are the latest driving factors of international trade.

DIGITAL DUO AND TRADE: STATISTICAL EVIDENCE

Even back in 2013, the volume of digital international trade was no less than $105 billion (PayPal, 2013). This increase is frequently linked with the rise of the bourgeois. According to Akrimi and Khemakhem (2022), "In 2017, ecommerce was responsible for around $2.3 trillion in sales and is expected to hit $4.5 trillion in 2021." (Akrimi & Khemakhem, 2022, p.27). Again, "In the US alone, ecommerce represents almost 10% of retail sales and that number is expected to grow by nearly 15% each year!" (Akrimi & Khemakhem, 2022, p.27). Also, the internet aligned minor businesses, such as SMBs, to world commerce. Over 95% of the USA's SMBs opt to use eBay to showcase their excellence. Considering their sustainability, it is worth mentioning that 74% were still operational in 2015 (eBay, 2015). As per the United Nations Conference on Trade and Development (UNCTAD), business-to-business commerce nine years ago was 95% of total e-commerce (UNCTAD, 2015). Ever mounting influence and importance of the digital duo may be understood from the statement that almost seven years ago, one billion viewers were devoted to YouTube, and half a billion of them belonged to the United States of America (Dadaczynski et al., 2021).

Regarding business, developed and developing countries are equal shareholders through the digital platform. As per an old estimate by the World Bank, just three years after 2022, developing countries may pose a business worth USD 96 billion (World Bank, 2013).

Banerjee et al. (2022) marked the Internet-generated percentage of B2B transactions as 57 percent for manufacturing, 26.5 percent for merchant wholesalers, and only 3.5 percent for services. Particularly for USA-based firms, digital businesses accumulated almost one trillion USD in goods and services online in 2012, which is equivalent to around 6.3 percent of the US gross domestic product (GDP). Almost 25% of the same accounts for exports (Banerjee et al., 2022). Moreover, the authors remarked that "US firms in digitally intensive industries purchased US $471.4 billion in products and services online in 2012, including US$106.2 billion in imports, implying a net surplus on digital trade." (Banerjee et al., 2022, p.6). Again, they focused on that in 2015, global mobile broadband penetration reached 47 percent, with some 3.5 billion mobile broadband subscriptions and 800 million fixed-line subscriptions (Banerjee et al., 2022). The authors also commented that 1.4 billion users for Facebook, 83 percent of which are outside the US and Canada (but blocked in China); more than 1 billion users for Google search; and more than 1 billion YouTube users in 61 languages (Banerjee at al. 2022). This considerable tech-savvy population has many yet to be revealed horizons.

Lund and Manyika (2016) commented that between 2002 and 2012, cross-border Internet traffic grew by 60 percent a year. "This reflects an increase in Internet users worldwide and a sevenfold increase in cross-border Internet usage. By 2025, on conservative assumptions, we estimate cross-border Internet traffic could grow another eightfold." (Lund & Manyika, 2016, p.1). They have also mentioned that the "surplus from the US and Europe alone is close to €250 billion (USD 266.4 billion) each year." (Lund & Manyika, 2016, p.4). In an international scenario, "The result has been an explosion in e-commerce." Global e-commerce sales reached over US\$ 1.3 trillion in 2014 — nearly 2 percent of the global GDP. While most e-commerce sales are within a country, a growing share is cross-border. About 40 percent of Amazon's net sales in 2014 came from sales outside of North America. Alibaba, the leading e-commerce platform in China that includes marketplaces for business-to-business (B2B), business-to-consumer (B2C), and peer-to-peer (P2P) e-commerce, posted a gross merchandise value of US\$370 billion in 2014, larger than Amazon and eBay combined." (Lund & Manyika, 2016, p.5).

The performance of South-East and South Asia has been impeccable as far as digital trade is concerned. "In 2016 and 2017, retail sales are estimated to have grown at 64.3 percent and 55.3 percent, respectively, in Indonesia, 21 percent and 22.6 percent respectively in Malaysia, 27.3 percent and 25 percent, respectively in the Philippines, 15.2 and 14 percent, respectively in Singapore, 20.7 percent and 19.5 percent respectively in Thailand, and 24.3 percent and 22 percent respectively in Vietnam…100 percent of Thai online sellers export, on an average, to 46 markets and in Indonesia, 100 percent of eBay sellers export to 34 markets" (Suominen, 2017, p.4). Again, the growth rate of these sellers are way ahead of that of the concerned country "The average compound annual growth rate of these sellers in 2011 to 2015 was 6 percent in Thailand and 11 percent in Indonesia—well above these countries' economic growth rates (3 and 5.5 percent, respectively) during the period" (Suominen, 2017, p.4). More strikingly, the sustainable chance of digital sellers beyond the 1st year of survival is 80% greater than 33% of physical sellers (Suominen, 2017, p.5).

Suominen (2017) also opined that parcel transaction over the four years, starting from 2011 to 2015, has amplified by 73%, around seven times quicker than global trade. During the same time, Intraregional parcel transaction in the Association of South East Asian Nations has improved by 140% (Suominen, 2017). Above all, the service sector and digitalization are almost synonymous "Developing countries increased their exports of commercial services from 25 percent to 32 percent of global services exports between 2006 to 2015" (Suominen, 2017, p.5). A company, Fresh Desk, originated in India and started with two employees in 2010 and multi-folded to 900 in the next seven years. The concerned company now stands beside more than 0.1 million companies worldwide committed to improved consumer satisfaction (Suominen, 2017).

It is noticeable that global trade is going through an optimistic alteration through the digital duo. Moreover, the performance of ASEAN and the Asia-pacific region is impeccable. However, there is a sharp contrast between South and South-East Asia "South Asia contributes to only 2.26% of the total global trade volume. Worse, the intra-regional trade is in an even sorrier state – the intra-regional trade volume as a percentage of the region's trade with the rest of the world stands at a low 1.46%. In sharp contrast, ASEAN contributes 6.5% of the total global goods trade volume, which is much higher than the contribution to global trade volume by the SAARC nations. Moreover, the intraregional trade volume of the ASEAN constitutes 12.33% of its trade volume with the rest of the world. A close consideration of the overseas investment behaviour of the biggest nation of BIMSTEC, India, reveals the urge of the concerned country to spread out and interact with the ASEAN nations. Whereas Indian overseas investment in BIMSTEC nations for the month of October 2017 was only 1.29%, the same for ASEAN nations was almost ten times higher" (Chatterjee & Banerjee, 2018).

METHODOLOGY

The analysis is poised to approximate two separate regression equations as précised below through pooled data analysis. The first regression analysis considers the ease of business ranking (edbi) as the dependent variable taken from https://data.worldbank.org/indicator/IC.BUS.EASE.XQ?locations=LK and per 100 people mobile subscriptions (pmssit), along with per 100 people fixed broadband connections (pcbit) as independent variables for 128 countries over 2019 and 2020 collected from the World Development Indicator interactive database (databank.worldbank.org). The second regression analysis considers data from 128 countries over 2019 and 2020 on per 100 people trade volume as the dependent variable while the independent variables are kept the same as the first regression analysis. Data on Trade Volume was considered from WTO Trade Database (wto.org) covering both merchandise trade and trade in commercial services (BPM6); for both the regression analysis, the robust standard errors have been opted irrespective of the homo/heteroskedastic nature of the error terms (Angrist & Pischke, 2009; Banerjee et al., 2020; Imbens & Kolesár, 2016). However, first, the multicollinearity has been tested using the variance inflation factor.

The entire analysis considers two econometric analyses.

Business and Digitalization an econometric analysis

Data Analysis

The regression equations to be estimated are–

$edb_{it} = a_1 + b_1 \cdot pcb_{it} + c_1 \cdot pmss_{it} + v_{it}$, where; i =1, 2, 3…128 and t = 2019 and 2020……1

$pctv_{it} = a + b \cdot pmss_{it} + c \cdot pcb_{it} + u_{it}$, where; i =1, 2, 3….128 and t=2019 and 2020…….2

For the above equations, 'a' and '$_{a1}$'is the constant term, b, b_1, c and c_1, are the coefficients of pcb_{it}, and $pmss_{it}$, respectively. u_{it} and v_{it} is the stochastic error term. Here both regression analysis is considered simultaneously. Considering the nature of the data presence of multicollinearity has been tested.

Table 1. Results

edb$_{it}$(Dependent variable)	pcb$_{it}$, pmss$_{it}$(Independent variables)	
Econometric tests	**Results**	
Multicollinearity: Variance inflation factors (vif)	Stat	Prob
	1.23<2	NA

Source: Computed

Table 2. Econometric test results

pctv$_{it}$ (Regressand)	pcb$_{it}$, pmss$_{it}$ (Regressors)	
Econometric tests	**Results**	
Multicollinearity: Variance inflation factors (vif)	Stat	Prob
	1.23<2	NA

Table 3. Regression Results (Regressand edbit)

Regressors	Coefficient	t statistics	Probability	R2	Adjusted R2	F Statistics (2, 253)	Probability
pmss$_{it}$	-0.31	-4.58	0.000				
pbc$_{it}$	-2.19	-15.27	0.000	0.5459	0.5423	158.39	0.000
Constant	153.29	19.14	0.006				

Source: Computed

As illustrated in Tables 1 and 2, there is no multicollinearity.

Table 3 depicts that ease of doing business rank is negatively proportionate with the chosen independent variables, since low is better regarding ease of doing business, hence digital penetration has positive impact over ease of doing business The high F value justifies the model and the independent variables are statistically significant and negative.

Table 4. Regression Results (Regressand $_{pctvit}$)

Regressors	Coefficient	t statistics	Probability	R2	Adjusted R2	F Statistics (2, 253)	Probability
pmss$_{it}$	0.022	2.38	0.018				
Pbc$_{it}$	-0.129	6.31	0.000	0.2574	0.2514	28.10	0.000
Constant	-2.66	-2.80	0.006				

Source: Computed

It is clear from Table 4, that ^{pbc}it and ^{pmss}it have a positive and statistically significant impact on ^{pctv}it. The F statistics indicate that the model is significant. It implies that digitalization has a positive impact on business facilitation. As with the increase in both the independent variables, namely mobile subscription per hundred people and fixed broadband connection per hundred people, the per hundred population trade volume will increase, which will boost global trade.

Nowadays, digital business is very usual. Two digital tycoons Amazon and Flipkart are subject to an exponential growth rate. Pay if the satisfied concept has turned customers optimistic for digital markets. The demanding corporate life has also squeezed consumers and led them to the virtual market. An integrated digital community fuelled by WhatsApp and Facebook created a consumption spiral boosting the digital market like never before.

Social Interaction on a Digital Platform

Digital interaction has reduced the spatial barrier among people now it is possible to be in touch with everyone both virtually and audibly even when they are in motion. Social platform like Facebook and WhatsApp has given the liberty to engage in important and recesses equally. Twitter has become a platform for the expression of thought. Instagram may cater with photos that you may like. YouTube itself has become a big phenomenon in terms of talent searching. There are bloggers of different genres. The pros of this digital angel cannot be completed within a paper.

However, this angel has a devil side too. Identity theft and internet-based forgery are the two broad heads of this evil. A person can adopt another person's identity through Facebook, Twitter can become a place to talk nonsense. A vengeful boy or girlfriend might use Instagram for vengeance. Above all, there is cyber porn and especially child porn. Another typical problem is spreading rumours and bluffing people financially through the internet. As mentioned, the pros and cons of the digital duo cannot be summarized within a paper so it is better to move to the conclusion now.

CONCLUSION

The emphasis of this chapter is on digitalization and its chemistry with business as perceived by 128 countries of the world. The pooled analysis illustrates that fixed broadband and mobile subscription facilitates business and improves per capita trade volume.

The popularity of digital platforms regarding trade and commerce owes a lot to improved refund policies, ease of payment, and bonus bonanzas. Digital platforms also significantly curbed the need for added advertisement costs. In a capitalist world time is a scarce resource, hence consumers might turn into digital customers operating through the digital duo as far as purchasing is concerned. The digital platform also comes with an international flavor that attracts consumers in the era of liberalization, globalization, and privatization. Again, through digitalization, the market remains open 24 X 7 hours a week.

On behalf of the producer in this digital ambiance owning a shop is no longer necessary, and the number of intermediaries also has significantly reduced through globalization. This particular change has eventually increased the consumer and producer surplus. Digitalization contacted the shoreline of the Globe. Augmented revenue succeeding digitalization, added consumer contentment owing to less time-consuming yet more varied options of products and increased competition eventually decreased the price of the products. Producers on the other hand are blessed with a wider consumer base. So, whether it is ease of doing business or per capita trade volume digital duo might be considered as a pro-commerce policy for the entire world. It might unravel many untapped potentials, explore new unknown horizons, and eventually gear towards a better future for all.

REFERENCES

Abeliansky, A. L., & Hilbert, M. (2017). Digital technology and international trade: Is it the quantity of subscriptions or the quality of data speed that matters? *Telecommunications Policy*, *41*(1), 35–48. doi:10.1016/j.telpol.2016.11.001

Akrimi, Y., & Khemakhem, R. (2022). Website Usability, Website Interactivity, and Website Personality as Drivers of Online Purchase. In Social Customer Relationship Management (Social-CRM) in the Era of Web 4.0 (pp. 26-47). IGI Global.

Angrist, J. D., & Pischke, J.-S. (2009). *Mostly harmless econometrics: An empiricist's companion*. Princeton University Press. doi:10.1515/9781400829828

Banerjee, S., Bose, P., & Siddiqui, I. N. (2022). Digital Dynamics and International Trade: Experiences of South and South-East Asia. *International Journal of Asian Business and Information Management*, *13*(1), 1–16. doi:10.4018/IJABIM.297849

Banerjee, S., Sar, A. K., & Pandey, S. (2020). *Improved yet Unsafe: An Aquatic Perspective of Indian Infant Mortality*. doi:10.1177/0972063420908379

Bhavnani, A., Chiu, R. W. W., Janakiram, S., Silarszky, P., & Bhatia, D. (2008). *The role of mobile phones in sustainable rural poverty reduction*. documents1.worldbank.org/curated/fr/644271468315541419/pdf/446780WP0Box321bile1Phones01PUBLIC1.pdf

Brewster, R. L. (Ed.). (1994). *Data Communications and Networks 3*. The Institute of Electrical Engineers. doi:10.1049/PBTE031E

Castells, M. (1989). *The informational city: Information technology, economic restructuring, and the urban-regional process*. Oxford. http://dk.fdv.uni-lj.si/dr/dr10Gantar1.PDF

Chatterjee, B., & Banerjee, S. (2018). Trade and Economic Integration. *Seminar*, *703*, 2–5.

Corbridge, S. (1984). Political geography of contemporary events V: Crisis, what crisis? Monetarism, Brandt II and the geopolitics of debt. *Political Geography Quarterly*, *3*(4), 331–345. doi:10.1016/0260-9827(84)90017-X

Dadaczynski, K., Okan, O., Messer, M., Leung, A. Y. M., Rosário, R., Darlington, E., & Rathmann, K. (2021). Digital Health Literacy and Web-Based Information-Seeking Behaviors of University Students in Germany During the COVID-19 Pandemic: Cross-sectional Survey Study. *Journal of Medical Internet Research*, *23*(1), 1–17. doi:10.2196/24097 PMID:33395396

eBay (2015). US Small Business Global Growth Report. Available at https,//www.ebaymainstreet.com/sites/default/files/2015-us-small-biz-global-growth-report_0.pdf, Accessed on 7th March 2022

Freund, C. L., & Weinhold, D. (2004). The effect of the Internet on international trade. *Journal of International Economics*, *62*(1), 171–189. https://doi.org/10.1016/S0022-1996(03)00059-X

Garnham, N. (1990). *Capitalism and communication: Global culture and the economics of information*. Sage Publications. https://scholar.google.com/scholar?hl=en&as_sdt=0%2C5&q=Capitalism+and+communication%3A+Global+culture+and+the+economics+of+information&btnG=

Gillespie, A., & Williams, H. (2016). Telecommunications and the reconstruction of regional comparative advantage. *Environment and Planning A: Economy and Space, 20*(10), 1311–1321. doi:10.1068/A201311

Goddard, J., & Gillespie, A. (2017). Advanced telecommunications and regional economic development. In *In managing the city* (pp. 84–109). Routledge. https://www.taylorfrancis.com/chapters/edit/10.4324/9781315178264-6/advanced-telecommunications-regional-economic-development-goddard-gillespie

González, L., & Jouanjean, J. (2017). Digital Trade: Developing a Framework for Analysis. In *Trade Policy Papers, No 205*. OECD Publishing. https://www.oecd-ilibrary.org/trade/digital-trade_524c8c83-en

Harvey, D. (2010). Between Space and Time: Reflections on the Geographical Imagination1. *Annals of the Association of American Geographers*, *80*(3), 418–434. https://doi.org/10.1111/J.1467-8306.1990.TB00305.X

Harvey, D. (2020). The condition of postmodernity. In Knowledge and Postmodernism in Historical Perspective (pp. 494–507). Taylor and Francis. https://doi.org/10.4324/9781003060963-38/CONDITION-POSTMODERNITY-DAVID-HARVEY.

Hepworth, M. E. (1990). Geography of the information economy. *NETCOM : Réseaux, Communication et Territoires. NETCOM*, *4*(1), 266–267. https://doi.org/10.3406/NETCO.1990.1055

Imbens, G. W., & Kolesár, M. (2016). Robust Standard Errors in Small Samples: Some Practical Advice. *The Review of Economics and Statistics*, *98*(4), 701–712. https://doi.org/10.1162/REST_A_00552

Janow, M. E., & Mavroidis, P. C. (2019). Digital trade, e-commerce, the WTO and regional frameworks. *World Trade Review*, *18*(S1), S1–S7.

Ling, R., & Typhina, E. (2016). Mobile communication. In A. de S. e Silva (Ed.), *Dialouges on Mobile Communication* (p. 222). Routledge. https://www.routledge.com/Dialogues-on-Mobile-Communication/Silva/p/book/9781138691582

Lund, S., & Manyika, J. (2016). *How digital trade is transforming globalisation*. International Centre for Trade and Sustainable Development (ICTSD). http://e15initiative.org/wp-content/uploads/2015/09/E15-Digital-Lund-and-Manyika.pdf

Maiti, M., & Kayal, P. (2017). Digitization: Its Impact on Economic Development & Trade. *Asian Economic and Financial Review, 7*(6), 541–549. doi:10.18488/JOURNAL.AEFR.2017.76.541.549

Manyika, J., Bughin, J., Lund, S., Nottebohm, O., Poulter, D., Jauch, S., & Ramaswamy, S. (2014). *Global flows in a digital age: How trade, finance, people, and data connect the world economy.* McKinsey Global Institute. https://scholar.google.com/scholar?cluster=31932466010355353 85&hl=en&oi=scholarr

Marcella, R., & Davies, S. (2004). The use of customer language in international marketing communication in the Scottish food and drink industry. *European Journal of Marketing, 38*(11/12), 1382–1395. https://doi.org/10.1108/03090560410560155

Meijers, H. (2013). Does the internet generate economic growth, international trade, or both? *International Economics and Economic Policy, 11*(1), 137–163. https://doi.org/10.1007/S10368-013-0251-X

Meltzer, J. P. (2016). Maximizing the Opportunities of the Internet for International Trade. *ICTSD and World Economic Forum, 35.* doi:10.1109/APEC.2015.7104315

Osibanjo, O., & Nnorom, I. C. (2008). Material flows of mobile phones and accessories in Nigeria: Environmental implications and sound end-of-life management options. *Environmental Impact Assessment Review, 28*(2-3), 198–213. https://doi.org/10.1016/j.eiar.2007.06.002

PayPal. (2013). *Modern Spice Routes The Cultural Impact and Economic Opportunity of Cross-Border Shopping.* Scientific Research - An Academic Publisher. https://www.scirp.org/(S(i43dyn45teexjx455qlt3d2q))/reference/ReferencesPapers.aspx?ReferenceID=1379307

Pecchioli, R. M. (1983). *The internationalisation of banking: The policy issues.* Organisation for Economic Co-operation and Development.

Porges, A., & Enders, A. (2016). *Data Moving Across Borders: The Future of Digital Trade Policy.* International Centre for Trade and Sustainable Development (ICTSD). Available at e15initiative.org/wp-content/uploads/2015/09/E15-Digital-Economy-Porges-and-Enders-Final.pdf

Poster, M. (1990). Words without Things: The Mode of Information. *October, 53,* 63–67. doi:10.2307/778915

Quéré, B. P., Nouyrigat, G., & Baker, C. R. (2015). A Bi-Directional Examination of the Relationship Between Corporate Social Responsibility Ratings and Company Financial Performance in the European Context. *Journal of Business Ethics, 148*(3), 527–544. https://doi.org/10.1007/S10551-015-2998-1

Regazzi, J. J. (2018). The Shifting Sands of the Information Industry. In *Information Retrieval and Management: Concepts, Methodologies, Tools, and Applications* (Vol. 4, pp. 1–23). IGI Global. doi:10.4018/978-1-5225-5191-1.CH001

Shaikh, A. A., & Karjaluoto, H. (2015). Mobile banking adoption: A literature review. *Telematics and Informatics, 32*(1), 129–142. https://doi.org/10.1016/J.TELE.2014.05.003

Soobramanien, T. Y., & Worrall, L. (Eds.). (2017). *Emerging Trade Issues for Small Developing Countries: Scrutinising the Horizon.* Commonwealth Secretariat. https://books.google.co.in/books?hl=en&lr=&id=M2VnDwAAQBAJ&oi=fnd&pg=PA111&dq=the+export+of+conventionally+imported+goods+and+services+and+the+growth+of+e-commerce+are+positively+correlated+in+the+Caribbean+islands&ots=zeeBEYxdBU&sig=MeegIUQEdDA8m7OWfzxyRmvR11k&redir_esc=y#v=onepage&q&f=false

Suominen, K. (2017). *Fuelling Trade in the Digital Era: Policy Roadmap for Developing Countries.* Geneva: International Centre for Trade and Sustainable Development (ICTSD). Available at https,//www.ictsd.org/sites/default/files/research/suominen_fuelling_trade_in_the_digital_era_0.pdf

Tiong, T. K., Tianyi, G., Sopra, R., & Sharma, R. (2012). The Issue of Fragmentation on Mobile Games Platforms. In R. S. Sharma, M. Tan, & F. Pereira (Eds.), *Understanding the Interactive Digital Media Marketplace* (pp. 89–96). IGI Global. doi:10.4018/978-1-61350-147-4.CH009

UNCTAD. (2015). *Information economy report 2015-unlocking the potential of e-commerce for developing countries.* United Nations. unctad.org/en/PublicationsLibrary/ier2015_en.pdf

Vogel, E. (1985). Pax Nipponica. *Foreign Affairs, 64,* 752. https://heinonline.org/hol-cgi-bin/get_pdf.cgi?handle=hein.journals/fora64&section=52

Walter, I. (1988). *Global Competition in Financial Services: Market Structure, Protection and Trade Liberazation.* Ballinger Publishing Company. https://scholar.google.com/scholar?hl=en&as_sdt=0%2C5&q=Global+Competition+in+Financial+Services%2C+Market+Structure%2C+Protection%2C+and+Trade+Liberalization+by+walter&btnG=

Warf, B. (2016). Telecommunications and the changing geographies of knowledge transmission in the late 20th century. *Urban Studies (Edinburgh, Scotland), 32*(2), 361–378. https://doi.org/10.1080/00420989550013130

World Bank. (2013). *Crowdfunding's Potential for the Developing World.* Info Dev, Finance and Private Sector Development Department, World Bank. Available at https,//www.infodev.org/infodev-files/wb_crowdfundingreport-v12.pdf

Compilation of References

Aabø, S. (2005). The role and value of public libraries in the age of digital technologies. *Journal of Librarianship and Information Science, 37*(4), 205–211. doi:10.1177/0961000605057855

Aalst, W. (2016). *Process Mining: Data Science in Action.* Academic Press.

Aalst, W. (1998). The Application of Petri Nets to Workflow Management. *Journal of Circuits, Systems, and Computers, 8*(01), 21–66. doi:10.1142/S0218126698000043

Aalst, W., Hee, K., Ter, A., & Sidorova, N. (2011). Soundness of workflow nets: Classification, decidability, and analysis. *Formal Aspects of Computing, 23*(3), 333–363. doi:10.100700165-010-0161-4

Aalst, W., & Medeiros, A. K. A. (2005). Process Mining and Security: Detecting Anomalous Process Executions and Checking Process Conformance. *Electronic Notes in Theoretical Computer Science, 121*, 3–21. doi:10.1016/j.entcs.2004.10.013

Aarikka-Stenroos, L., & Ritala, P. (2017). Network management in the era of ecosystems: Systematic review and management framework. *Industrial Marketing Management, 67*, 23–36. doi:10.1016/j.indmarman.2017.08.010

Abbu, H., Mugge, P., Gudergan, G., & Kwiatkowski, A. (2020). Digital leadership-Character and competency differentiates digitally mature organizations. In *2020 IEEE International Conference on Engineering, Technology and Innovation (ICE/ITMC)* (pp. 1-9). IEEE. 10.1109/ICE/ITMC49519.2020.9198576

ABC. (2019). Zara cambia el concepto de «moda rápida» por el de «flexible». *Abc. es.* Recuperado de https://www.abc.es/economia/abci-zara-cambia-concepto-moda-rapida-flexible-201901110217_noticia.html?ref=https%3A%2F%2Fwww.abc.es%2Feconomia%2Fabci-zara-cambia-concepto-moda-rapida-flexible-201901110217_noticia.html%3Fref%3Dhttps%3A%2F%2Fwww.google.com%2F.

Abeliansky, A. L., & Hilbert, M. (2017). Digital technology and international trade: Is it the quantity of subscriptions or the quality of data speed that matters? *Telecommunications Policy, 41*(1), 35–48. doi:10.1016/j.telpol.2016.11.001

Acatech. (2017). *Industrie 4.0 Maturity Index.* Managing the Digital Transformation of Companies. Retrieved April 12th, 2020, from https://www.acatech.de/wp-content/uploads/2018/03/acatech_STUDIE_Maturity_Index_eng_WEB.pdf

Accenture. (2017). *Digital Transformation of Industries. Demystifying Digital and Securing $100 Trillion for Society and Industry by 2025.* Retrieved on June 30th, 2021, from: https://www.accenture.com/t00010101t000000z__w__/ru-ru/_acnmedia/accenture/conversion-assets/dotcom/documents/local/ru-ru/pdf/accenture-digital-transformation.pdf

Accenture. (2019). *An AI roadmap to maximize the value of AI.* Retrieved on June 30th, 2021, from: https://www.accenture.com/us-en/insights/artificial-intelligence/ai-roadmap

Accenture. (2019). *Rethink, reinvent, realize.* Retrieved April 12th, 2020, from https://www.accenture.com/_acnmedia/thought-leadership-assets/pdf/accenture-ixo-industry-insights-hightech.pdf

Accenture. (2020). *Data-driven insights into consumer behavior How is COVID-19 changing the retail consumer?* https://www.accenture.com/_acnmedia/PDF-130/Accenture-Retail-Research-POV-Wave-Seven.pdf

Accenture. (2020). *Industry X.0.* Retrieved April 12th, 2020, from https://www.accenture.com/us-en/services/industryx0-index

Acharya, V., Kumar, S., Sunand, S., & Gupta, K. (2018). Analyzing the factors in industrial automation using analytic hierarchy process. *Computers & Electrical Engineering*, *71*, 877–886. doi:10.1016/j.compeleceng.2017.08.015

Achieng, M., & Ruhode, E. (2013). The Adoption and Challenges of Electronic Voting Technologies Within the South African Context. *International Journal of Managing Information Technology, 5.*

Adam, N., Atluri, V., & Huang, W. (1998). Modeling and Analysis of Workflows Using Petri Nets. *Journal of Intelligent Information Systems*, *10*(2), 131–158. doi:10.1023/A:1008656726700

Adams, K. K., Baker, W. L., & Sobieraj, D. M. (2020). Myth Busters: Dietary Supplements and COVID-19. *The Annals of Pharmacotherapy*, *54*(8), 820–826. doi:10.1177/1060028020928052 PMID:32396382

Adner, R., & Kapoor, R. (2010). Value Creation In Innovation Ecosystems: How the Structure of Technological Interdependence Affects Firm Performance in New Technology Generations. *Strategic Management Journal*, *31*(3), 306–333. doi:10.1002mj.821

Agentia pentru Dezvoltare Regionala Vest. (2022). *1.1.A Entitati de inovare...* Retrieved on October 26th, 2022, from https://adrvest.ro/ghidul-specific-pi-1-1-a/

Agrawal, A. (2020). COVID-19 – Driving Digitization, Digitalisation & Digital Transformation in Healthcare. *Science Reporter*, 20–21.

Akdaş, O., & Cismaru, M. (2021). Promoting mental health during the COVID-19 pandemic: The transtheoretical model of change and social marketing approach. *International Review on Public and Nonprofit Marketing*, 1–28.

Akdil, K. Y., Ustundag, A., & Cevikcan, E. (2018). *Maturity and Readiness Model for Industry 4.0 Strategy. In Industry 4.0: Managing the Digital Transformation.* Springer.

Akgül, H., Akgül, B., & Ayer, Z. (2018). Sanayi 4.0 sürecinde gazetecilik sektöründe çalışacak personelin mesleki yetenek ve yeterliliğine yönelik değerlendirme ve öngörüler. *Avrasya Sosyal Ve Ekonomi Araştırmaları Dergisi*, *5*(8), 198–205.

Akrimi, Y., & Khemakhem, R. (2022). Website Usability, Website Interactivity, and Website Personality as Drivers of Online Purchase. In Social Customer Relationship Management (Social-CRM) in the Era of Web 4.0 (pp. 26-47). IGI Global.

Al-Abbadey, M., Fong, M. M., Wilde, L. J., Ingham, R., & Ghio, D. (2021). Mobile health apps: An exploration of user-generated reviews in Google Play Store on a physical activity application. *Digital Health*, *7*, 20552076211014988. doi:10.1177/20552076211014988 PMID:34017609

Albukhitan, S. (2020). Developing Digital Transformation Strategy for Manufacturing. *Procedia Computer Science*, *170*, 672–679. doi:10.1016/j.procs.2020.03.173

Alcayaga, A., Wiener, M., & Hansen, E. G. (2019). Towards a Framework of Smart Circular Systems: An Integrative Literature Review. *Journal of Cleaner Production*, *221*, 622–634. doi:10.1016/j.jclepro.2019.02.085

Alexander, C. (1977). *A Pattern Language*. Oxford University Press.

Alexander, C. (1979). *The Timeless Way of Building*. Oxford University Press.

Alexander, C., Silverstein, M., Angel, Sh., Ishikawa, S., & Abrams, D. (1975). *The Oregon Experiment*. Oxford University Press.

Alrahili, R. (2021). *Towards Employing Process Mining for Role Based Access Control Analysis: A Systematic Literature Review*. Academic Press.

Alshawi, M., & Ingirige, B. (2003). Web-enabled project management: An emerging paradigm in construction. *Automation in Construction*, *12*(4), 349–364. doi:10.1016/S0926-5805(03)00003-7

Altman, R. (2014). *Hype Cycle for Application Architecture*. Gartner.

American Institute for Conservation. (2021). *AIC Wiki. Collection Care*. Retrieved from https://www.conservation-wiki.com/wiki/Collection_Care

Amnesty International. (2015). *Ethiopia's Human Rights Violations Report 2015*. https://www.amnesty.org/en/countries/africa/ethiopia/report-ethiopia

Amsel-Arieli, M. (2012). Cabinets of Curiosity (Wunderkammers). *Histoire Magazine*. http://www.micheleleigh.net/wp-content/uploads/2019/01/Cabinets-of-Curiosity.pdf

Anand, N., & Daft, R. L. (2007). What is the right organization design? *Organizational Dynamics*, *36*(4), 329–344. doi:10.1016/j.orgdyn.2007.06.001

Anderl, R., Picard, A., Wang, Y., Fleischer, J., Dosch, S., Klee, B., & Bauer, J. (2015). *Guideline Industrie 4.0 – Guiding Principles for the Implementation of Industrie 4.0 in Small and Medium Sized Businesses*. In VDMA Forum Industrie, Frankfurt, Germany.

Andjelic, A. (2021). *Deconstructing Zara's strategy*. https://andjelicaaa.medium.com/deconstructing-zaras-strategy-8daf218af9e8

Andronico, A., Tran Kiem, C., Paireau, J., Succo, T., Bosetti, P., Lefrancq, N., Nacher, M., Djossou, F., Sanna, A., Flamand, C., Salje, H., Rousseau, C., & Cauchemez, S. (2021). Evaluating the impact of curfews and other measures on SARS-CoV-2 transmission in French Guiana. *Nature Communications*, *12*(1), 1–8. doi:10.103841467-021-21944-4 PMID:33712596

Angrist, J. D., & Pischke, J.-S. (2009). *Mostly harmless econometrics: An empiricist's companion*. Princeton University Press. doi:10.1515/9781400829828

Aniche, M., & Gerosa, M. (2010). *How Test-Driven Development Influences Class Design: A Practitioner's Point of View. Department of Computer Science*. University of Sao Paulo. USP.

Antiquities and Monuments Office. (2020). *"Heritage Over a Century: Tung Wah Museum and Heritage Conservation" Exhibition*. Retrieved from https://www.amo.gov.hk/en/whatsnew_20200511.php

Apache Ignite. (2022). *Digital Integration Hub With Apache Ignite.* The Apache Software Foundation. https://ignite.apache.org/use-cases/digital-integration-hub.html

Apitz, S. E., Elliott, M., Fountain, M., & Galloway, T. S. (2006). European Environmental Management: Moving to an Ecosystem Approach. *Integrated Environmental Management: Moving to an Ecosystem Approach, 2*(1), 80–85. doi:10.1002/ieam.5630020114 PMID:16640322

Arbaiza, F., & Huertas, S. (2018). Comunicación publicitaria en la industria de la moda: Branded content, el caso de los fashion films. *Revista de Comunicación, 17*(1), 9–33. doi:10.26441/RC17.1-2018-A1

Arcitura (2020). *SOA Patterns.* Arcitura. https://patterns.arcitura.com/soa-patterns

Arcitura. (2020). *SOA Patterns.* Arcitura. https://patterns.arcitura.com/soa-patterns

Arendt, H. (1973). *Origin of Totalitarianism.* Harvest Books.

Aron, R., Clemons, E. K., & Reddi, S. (2005). Just right outsourcing: Understanding and managing strategic risk. *Journal of Management Information Systems, 22*(2), 37–35. doi:10.1080/07421222.2005.11045852

Arora, T., & Grey, I. (2020). Health behaviour changes during COVID-19 and the potential consequences: A mini-review. *Journal of Health Psychology, 25*(9), 1155–1163. doi:10.1177/1359105320937053 PMID:32551944

Art Institute of Chicago. (2021). *360 Tours: Taking a Spin through Conservation.* Retrieved from https://www.artic.edu/articles/935/360-tours-taking-a-spin-through-conservation

Athamneh, L., Essien, E. J., Sansgiry, S. S., & Abughosh, S. (2017). Intention to quit water pipe smoking among Arab Americans: Application of the theory of planned behavior. *Journal of Ethnicity in Substance Abuse, 16*(1), 80–90. doi:10.1080/15332640.2015.1088423 PMID:26720395

Augusto, A., Deitz, T., Faux, N., Manski-Nankervis, J.-A., & Capurro, D. (2022). Process mining-driven analysis of COVID-19's impact on vaccination patterns. *Journal of Biomedical Informatics, 130,* 104081. doi:10.1016/j.jbi.2022.104081 PMID:35525400

Awatef, Gueni, Fhima, Cairns, & David. (2015). *Process Mining in the Education Domain.* Academic Press.

Aydemir, S., Ocak, S., Saygılı, S., Hopurcuog, D., Has, F., ˙rul Kıykım, E., … Canpolat, N. (n.d.). *Telemedicine Applications in a Tertiary Pediatric Hospital in Turkey During COVID-19 Pandemic.* Academic Press.

Aydin, K., Çizer, E. Ö., & Köse, Ş. G. (2021). Analyzing attitude towards COVID-19 vaccine in the context of the health industry: The role of country of origin image. *Duzce Medical Journal, 23*(1), 122–130.

Azizan, S., Ismail, R., Baharum, A., & Hidayah Mat Zain, N. (2021). Exploring The Factors That Influence The Success Of Digitalization In An Organization's IT Department. *2021 6th IEEE International Conference on Recent Advances and Innovations in Engineering (ICRAIE),* 1–6. 10.1109/ICRAIE52900.2021.9704018

Backlund, F., Chroneer, D., & Sundqvist, E. (2014). Project management maturity models – A critical review: A case study wish Swedish engineering and construction organizations. *Procedia: Social and Behavioral Sciences, 119,* 837–846.

Bahl, S., Singh, R. P., Javaid, M., Khan, I. H., & Vaishya, R., & Suman, R. (2020). Telemedicine Technologies for Confronting COVID-19 Pandemic. *RE:view, 5*(4), 547–561.

Bai, C., Quayson, M., & Sarkis, J. (2021). COVID-19 pandemic digitization lessons for sustainable development of micro-and small- enterprises. *Sustainable Production and Consumption, 27,* 1989–2001. doi:10.1016/j.spc.2021.04.035 PMID:34722843

Baldé, G. (2018). Zara presenta su primera app de realidad aumentada. *It Publicidad.* https://www.itfashion.com/moda/zara-presenta-su-primera-app-de-realidad-aumentada/

Baldellon, Fabre, & Roy. (2011). Modeling distributed realtime systems using adaptive Petri nets. *Actes de la 1re journ´ee 3SL,* 10.

Baldwin, C. Y., & Clark, K. B. (1997). Managing in an age of modularity. *Harvard Business Review, 75*(5), 84–93. PMID:10170333

Baldwin, C. Y., & Clark, K. B. (2000). *Design rules: The power of modularity.* The MIT Press. doi:10.7551/mitpress/2366.001.0001

Banerjee, S., Sar, A. K., & Pandey, S. (2020). *Improved yet Unsafe: An Aquatic Perspective of Indian Infant Mortality.* doi:10.1177/0972063420908379

Banerjee, S., Bose, P., & Siddiqui, I. N. (2022). Digital Dynamics and International Trade: Experiences of South and South-East Asia. *International Journal of Asian Business and Information Management, 13*(1), 1–16. doi:10.4018/IJABIM.297849

Barata, J., & da Cunha. (2017). Climbing the Maturity Ladder in Industry 4.0: A Framework for Diagnosis and Action that Combines National and Sectorial Strategies. *Twenty-third Americas Conference on Information Systems,* Boston, MA.

Barboza, R. J. G., Michalke, S., & Siemon, D. (2022). Towards the Design of a Digital Business Ecosystem Maturity Model for Personal Service Firms: A Pre-Evaluation Strategy. *PACIS 2022 Proceedings,* 86. https://aisel.aisnet.org/pacis2022/86

Barro, R. J. (2001). Education and economic growth. *The contribution of human and social capital to sustained economic growth and well-being, 79,* 13–41.

Barron, P., & Leask, A. (2017). Visitor engagement at museums: Generation Y and 'Lates' events at the National Museum of Scotland. *Museum Management and Curatorship, 32*(5), 473–490. doi:10.1080/09647775.2017.1367259

Bause & Kritzinger. (2013). *Stochastic Petri Nets -An Introduction to the Theory.* Academic Press.

BCG. (2015). *Industry 4.0: The Future of Productivity and Growth in Manufacturing Industries.* Retrieved April 14th, 2020, from https://www.bcgperspectives.com/content/articles/engineered_products_project_business_industry_40_future_productivity_growth_manufacturing_industries/?chapter=2

BCG. (2019). *What Is Your Business Ecosystem Strategy?* Retrieved on August 20th, 2022, from https://www.bcg.com/publications/2022/what-is-your-business-ecosystem-strategy

BCG. (2020a). *Digital Acceleration Index.* Retrieved April 14th, 2020, from https://www.bcg.com/capabilities/technology-digital/digital-acceleration-index.aspx

BCG. (2020b). *Digital Transformation*. Retrieved April 10th, 2020, from https://www.bcg.com/digital-bcg/digital-transformation/overv iew.aspx

Beauchamp, M. R., Crawford, K. L., & Jackson, B. (2019). Social cognitive theory and physical activity: Mechanisms of behavior change, critique, and legacy. *Psychology of Sport and Exercise, 42*, 110–117. doi:10.1016/j.psychsport.2018.11.009

Beauvoir, P., & Sarrodie, J.-B. (2018). *Archi-The Free Archimate Modelling Tool*. User Guide. The Open Group.

Bebensee, B., & Hacks, S. (2019). Applying Dynamic Bayesian Networks for Automated Modeling in ArchiMate-A Realization Study. In *2019 IEEE 23rd International Enterprise Distributed Object Computing Workshop (EDOCW)*. IEEE. DOI 10.1109/EDOCW.2019.0001

Ben Letaifa, S. (2014). *The Uneasy Transition from Supply Chains to Ecosystems*. Academic Press.

Benazeer, S. (2018). *On the feasibility of applying the concept of modularity in the context of IS/IT outsourcing* [Doctoral dissertation, University of Antwerp]. University of Antwerp Research Repository. https://repository.uantwerpen.be/docman/irua/199d4e/153326.p df

Benazeer, S., De Bruyn, P., & Verelst, J. (2017). Applying the concept of modularity to IT outsourcing: A financial services case. In R. Pergl, R. Lock, E. Babkin, & M. Molhanec (Eds.), *Enterprise and organizational modeling and simulation* (pp. 68–82). Springer. doi:10.1007/978-3-319-68185-6_5

Benazeer, S., De Bruyn, P., & Verelst, J. (2020). The concept of modularity in the context of IS/IT project outsourcing: An empirical case study of a Belgian technology services company. *International Journal of Information System Modeling and Design, 11*(4), 1–17. doi:10.4018/IJISMD.2020100101

Benazeer, S., Huysmans, P., De Bruyn, P., & Verelst, J. (2018). The concept of modularity and normalized systems theory in the context of IS outsourcing. In M. Khosrow-Pour (Ed.), *Encyclopedia of information science and technology* (pp. 5317–5326). IGI Publications.

Bendall, M. J., & Robertson, C. (2018). Crisis of democratic culture. *International Journal of Media and Cultural Politics, 14*(3), 383–391. doi:10.1386/macp.14.3.383_7

Benedict, N., Smithburger, P., Donihi, A. C., Empey, P., Kobulinsky, L., Seybert, A., Waters, T., Drab, S., Lutz, J., Farkas, D., & Meyer, S. (2017). Blended Simulation Progress Testing for Assessment of Practice Readiness. *American Journal of Pharmaceutical Education, 81*(1), 14.

Benedict, N., Smithburger, P., Donihi, A. C., Empey, P., Kobulinsky, L., Seybert, A., Waters, T., Drab, S., Lutz, J., Farkas, D., & Meyer, S. (2017). Blended Simulation Progress Testing for Assessment of Practice Readiness. *American Journal of Pharmaceutical Education, 81*(1), 14. doi:10.5696/ajpe81114

Bennett, T. (1995). *The Birth of the Museum (Culture: policies and politics)*. Routledge.

Berghaus, S., & Back, A. (2016, September). Stages in digital business transformation: results of an empirical maturity study. In MCIS (p. 22). Academic Press.

Bergsjo, D. (2009). *Product Lifecycle Management: Architectural and Organizational Perspectives* [Doctoral Dissertation]. Chalmers University of Technology, Goteborg, Sweden.

Bernard, G̈., & Andritsos, P. (2019). Accurate and transparent path prediction using process mining. In *European Conference on Advances in Databases and Information Systems* (pp. 235–250). Springer. 10.1007/978-3-030-28730-6_15

Bernard, G., Boillat, T., Legner, C., & Andritsos, P. (2016). When sales meet process mining: A scientific approach to sales process and performance management. *Proceedings of the 37th International Conference on Information Systems (ICIS 2016).*

Berti, van Zelst, & Aalst. (2019). *Process Mining for Python (PM4Py): Bridging the Gap Between Process- and Data Science.* Academic Press.

Bertolini, M., Esposito, G., Neroni, M., & Romagnoli, G. (2019). Maturity models in industrial internet: A review. *Procedia Manufacturing, 39.* Advance online publication. doi:10.1016/j.promfg.2020.01.253

Beulen, E. J. J., & Ribbers, P. M. A. (2003). International examples of large-scale systems - theory and practice: A case study of managing IT outsourcing partnerships in Asia. *Communications of the AIS, 11*, 357–376. doi:10.17705/1CAIS.01121

Bhavnani, A., Chiu, R. W. W., Janakiram, S., Silarszky, P., & Bhatia, D. (2008). *The role of mobile phones in sustainable rural poverty reduction.* documents1.worldbank.org/curated/fr/644271468315541419/pdf/446780WP0Box321bile1Phones01PUBLIC1.pdf

Bibby, L., & Dehe, B. (2018). *Defining and Assessing Industry 4.0 Maturity levels - Case of the Defence sector.* Production Planning and Control. Retrieved April 14th, 2020, from https://pure.hud.ac.uk/ws/files/14152037/0_final_accepted_manuscript_plain_text_with_T_F_pure.pdf

Binder, J. (2007). *Global Project Management*, Communication, Collaboration and Management Across Borders. Routledge Taylor and Francis Group.

Bingham, Ch., Eisenhardt, K., & Furr, N. (2007). What makes a process a capability? Heuristics, strategy, and effective capture of opportunities. Strategic Entrepreneurship Journal. doi:10.1002ej.1

Bishop, S., & Hoeffler, A. (2016). Free and fair elections: A new database. *Journal of Peace Research, 53*(4), 608–616. doi:10.1177/0022343316642508

BizzDesign. (2022). *Modeling a SWOT analysis.* BizzDesign. https://support.bizzdesign.com/display/knowledge/Modeling+a+SWOT+analysis

Blackett, T. (1989). The role of brand valuation in marketing strategy. *Marketing and Research Today*, (17), 245–248.

Blair, M. M., O'Connor, E. O., & Kirchhoefer, G. (2011). Outsourcing, modularity, and the theory of the firm. *BYU Law Review, 2011*(2), 262–314.

Blancart García, N. (2015). *Importancia de la monitorización de las redes sociales.* https://zaguan.unizar.es/record/32588/files/TAZ-TFG-2015-3465.pdf

Blank, S. (2009). *Customer Development at Startup2Startup.* Retrieved on August 20th, 2022, from https://www.slideshare.net/sblank/customer-development-at-startup2startup/28-Thanks_Startup_Lessons_

Blank, S. (2014). *What Founders Need to Know: You Were Funded for a Liquidity Event – Start Looking.* Retrieved on August 20th, 2022, from https://steveblank.com/category/venture-capital/

Blanka, C., Krumay, B., & Rueckel, D. (2022). The interplay of digital transformation and employee competency: A design science approach. *Technological Forecasting and Social Change, 178*, 121575. doi:10.1016/j.techfore.2022.121575

Blank, S. (2013). Why the Lean Start-Up Changes Everything. *Harvard Business Review.*

Block, D. (2019). *Post-Truth and Political Discourse*. Springer Nature Publishers. doi:10.1007/978-3-030-00497-2

Block, Z., & MacMillan, I. C. (1985). Milestones for Successful Venture Planning. *Harvard Business Review, 63*(5), 84–90.

Bloomberg. (2018). *Digitization, Digitalization, And Digital Transformation: Confuse Them At Your Peril*. https://www.forbes.com/sites/jasonbloomberg/2018/04/29/digitization-digitalization-and-digital-transformation-confuse-them-at-your-peril/?sh=6f9ca6142f2c

Bloomberg. (2018, August 28). *Digitization, Digitalization, and Digital Transformation: Confuse Them at Your Peril*. Author.

Bocanett, W. (2022). *Break the monolith: Chunking strategy and the Strangler pattern-Build decoupled microservices to strangle your monolithic application*. IBM.

Bokslag, W., & Vries, M. (2016). *Evaluating e-voting theory and practice*. Department of Information Security Technology, Technical University Eindhoven.

Bonchek, M., & Choudary, S. P. (2013). Three Elements of a Successful Platform Strategy. *Harvard Business Review*.

Bonillo, M. (2021). *Amancio Ortega nos quiere en forma*. https://www.65ymas.com/sociedad/zara-athleticz-recetas-ejercicios-moda-deportiva_32894_102.html

Bonnet, D., & Westerman, G. (2021). The New Elements of Digital Transformation. *MIT Sloan Management Review, 62*(2), 83–89. https://sloanreview.mit.edu/article/the-new-elements-of-digital-transformation/

Bose, van der Aalst, Zliobaite, & Pechenizkiy. (2011). Handling concept drift in process mining. In *International Conference on Advanced Information Systems Engineering* (pp. 391–405). Springer.

Boston Consulting Group. (2012). *The Most Innovative Companies 2012*. Retrieved on August 20th, 2022, from https://www.bcgperspectives.com/Images/The_Most_Innovative_Companies_2012_Dec_2012_tcm80-125210.pdf

Boston Consulting Group. (2013). *The Most Innovative Companies 2013*. Retrieved on August 20th, 2022, from https://www.bcgperspectives.com/Images/Most-Innovative-Companies-2013_tcm80-186913.pdf

Boston Consulting Group. (2014). *The Most Innovative Companies 2014*. Retrieved on August 20th, 2022, from https://www.bcgperspectives.com/Images/Most_Innovative_Companies_2014_Oct_2014_tcm80-174313.pdf

Boston Consulting Group. (2015). *The Most Innovative Companies 2015*. Retrieved on August 20th, 2022, from https://www.bcgperspectives.com/Images/BCG-Most-Innovative-Companies-2015-Dec-2015_tcm80-203388.pdf

Boston Consulting Group. (2016). *The Most Innovative Companies 2016*. Retrieved on August 20th, 2022, from https://media-publications.bcg.com/MIC/BCG-The-Most-Innovative-Companies-2016-Jan-2017.pdf

Bowen, G. A. (2019). Document Analysis as a Qualitative Research Method. *Qualitative Research Journal, 9*(2), 27–40. doi:10.3316/QRJ0902027

Bradley, A. (2014). *The Connected Enterprise Maturity Model*. Rockwell Automation.

Brennen, J. S., & Kreiss, D. (2016). Digitalization. In K. B. Jensen, E. W. Rothenbuhler, J. D. Pooley, & R. T. Craig (Eds.), *The International Encyclopedia of Communication Theory and 239 International Journal for Modern Trends in Science and Technology Philosophy* (pp. 556–566). Wiley-Blackwell.

Brettel, M., Friederichsen, N., Keller, M., & Rosenberg, M. (2014). How Virtualization, Decentralization And Network Building Change The Manufacturing Landscape: An Industry 4.0 Perspective. *International Journal of Information and Communication Engineering, 8*(1), 37–44.

Brewster, R. L. (Ed.). (1994). *Data Communications and Networks 3*. The Institute of Electrical Engineers. doi:10.1049/PBTE031E

Brusakova, I. A. (2022). Comparative Analysis of Models for Assessing the Digital Maturity of the Transformation Infrastructure. *2022 XXV International Conference on Soft Computing and Measurements (SCM)*, 209–211. 10.1109/SCM55405.2022.9794829

BSkyB v. EDS, High Court of Justice, UK (2010).

Buijs, J., Dongen, B., & Aalst, W. (2012). On the Role of Fitness, Precision. *Generalization and Simplicity in Process Discovery., 7565*(09), 305–322.

Burattin, A. (2015). *Process Mining Techniques in Business Environments* (Vol. 207). doi:10.1007/978-3-319-17482-2

Burbeck, S. (n.d.). *Application Programming in Smalltalk-80: How to use Model-View-Controller (MVC)*. University of Illinois in Urbana-Champaign (UIUC) Smalltalk Archive.

Buschmann, F., Meunier, R., Rohnert, H., Sommerlad, P., & Stal, M. (1996). *Pattern-Oriented Software Architecture: A System of Patterns*. Wiley.

Calviño Lorenzo, S. (2020). *Zara: Desmontando el mito de usar publicidad tradicional para llegar al éxito*. https://uvadoc.uva.es/handle/10324/42311

Camarinha-Matos, L. M. (2012). *Scientific research methodologies and techniques- Unit 2: Scientific method. Unit 2: Scientific methodology. PhD program in electrical and computer engineering*. Uninova.

Cambridge Dictionary. (2022). *Cambridge Dictionary: Maturity*. https://dictionary.cambridge.org/dictionary/english/maturity

Campagnolo, D., & Camuffo, A. (2010). The concept of modularity in management studies: A literature review. *International Journal of Management Reviews, 12*(3), 259–283.

Camuffo, A. (2004). Rolling out a world car: Globalization, outsourcing, and modularity in the auto industry. *Korean Journal of Political Economy, 2*(1), 183–224.

Cao, Y., Guo, Y., She, Q., Zhu, J., & Li, B. (2021). Prediction of medical expenses for gastric cancer based on process mining. *Concurrency and Computation, 33*(15), e5694. doi:10.1002/cpe.5694

Capece, S., & Chivăran, C. (2020). The Sensorial Dimension of the Contemporary Museum between Design and Emerging Technologies. *IOP Conference Series. Materials Science and Engineering, 949*(1), 12067. doi:10.1088/1757-899X/949/1/012067

CapGemeni. (2011). *Digital transformation: a roadmap for billion-dollar organizations*. Retrieved April 12th, 2020, from https://www.capgemini.com/wp-content/uploads/2017/07/Digital_Transformation__A_Road-Map_for_Billion-Dollar_Organizations.pdf

CapGemeni. (2014). *Digital Transformation Review, Crafting a Compelling Customer Experience.* Retrieved on August 20th, 2022, from https://www.capgemini-consulting.com/digital-transformation-review-6

CapGemeni. (2014). *Digital transformation review. Crafting a Compelling Customer Experience.* Retrieved April 12th, 2020, from https://www.capgemini.com/wp-content/uploads/2017/07/digital-transformation-review-6_3.pdf

CapGemeni. (2016a). *The Digital Strategy Imperative: Steady Long-Term Vision, Nimble Execution.* Retrieved on August 20th, 2022, from https://www.capgemini-consulting.com/dti/digital-strategy-review-9

CapGemeni. (2016b). *The New Innovation Paradigm for the Digital Age: Faster, Cheaper and Open.* Retrieved on August 20th, 2022, from https://www.capgemini-consulting.com/digital-transformation-review-8

CapGemeni. (2016c). *Strategies for the Age of Digital Disruption.* Retrieved on August 20th, 2022, from https://www.capgemini-consulting.com/resource-file-access/resource/pdf/digital_transformation_review_7_1.pdf

CapGemeni. (2016d). *Gearing Up for Digital Operations.* Retrieved on August 20th, 2022, from https://www.capgemini-consulting.com/digital-transformation-review-5

CapGemeni. (2017). *The Digital Advantage: How digital leaders outperform their peers in every industry.* Retrieved April 15th, 2020, from https://www.capgemini.com/wp-content/uploads/2017/07/The_Digital_Advantage__How_Digital_Leaders_Outperform_their_Peers_in_Every_Industry.pdf

CapGemeni. (2018). *Industry 4.0 Maturity Model – Mirroring today to sprint into the future.* Retrieved April 15th, 2020, from https://www.capgemini.com/fi-en/2018/09/industry-4-0-maturity-model-mirroring-today-to-sprint-into-the-future/

Capgemini. (2007). *Trends in Business transformation - Survey of European Executives.* Capgemini Consulting and The Economist Intelligence Unit.

Capgemini. (2009). *Business transformation: From crisis response to radical changes that will create tomorrow's business. A Capgemini Consulting survey.* Capgemini.

Caplin, M. (2012). *MVC Techniques with jQuery, JSON, Knockout, and C#.* www.codeproject.com

Casadesus-Masanell, R., & Ricart, J. E. (2011). How to Design a Winning Business Model. *Harvard Business Review.*

Castañeda-Babarro, A., Coca, A., Arbillaga-Etxarri, A. ve Gutiérrez-Santamaría, B. (2020). Physical Activity Change during COVID-19 Confinement. *International Journal of Environmental Research and Public Health, 17*(18), 6878.

Castelló, A., & Del Pino-Romero, C. (2014). La comunicación publicitaria se pone de moda: branded content y fashion films. *Revista Mediterránea de Comunicación.* https://rua.ua.es/dspace/bitstream/10045/44253/1/ReMedCom_06_01_07.pdf

Castells, M. (1989). *The informational city: Information technology, economic restructuring, and the urban-regional process.* Oxford. http://dk.fdv.uni-lj.si/dr/dr10Gantar1.PDF

Cennamo, C., Dagnino, G. B., Di Minin, A., & Lanzolla, G. (2020). Managing Digital Transformation: Scope of Transformation and Modalities of Value Co-Generation and Delivery. *California Management Review*, *62*(4), 5–16. doi:10.1177/0008125620942136

CenturyLink. (2019). *Application Lifecycle Management (ALM)*. CenturyLink.

Cerovac, I. (2016). 'The Role of Experts in a Democratic Society'. *Journal of Education Culture and Society*, *2*(2), 75–88. doi:10.15503/jecs20162.75.88

Ch'ng, E., Cai, S., Leow, F., & Zhang, T. (2019). Adoption and use of emerging cultural technologies in China's museums. *Journal of Cultural Heritage*, *37*, 170–180. doi:10.1016/j.culher.2018.11.016

Chan, T. T. W., Lam, A. H. C., & Chiu, D. K. W. (2020). From Facebook to Instagram: Exploring user engagement in an academic library. *Journal of Academic Librarianship*, *46*(6), 102229. doi:10.1016/j.acalib.2020.102229 PMID:34173399

Chan, V. H. Y., & Chiu, D. K. W. (2023). Integrating the 6C's Motivation into Reading Promotion Curriculum for Disadvantaged Communities with Technology Tools: A Case Study of Reading Dreams Foundation in Rural China. In A. Etim & J. Etim (Eds.), *Adoption and Use of Technology Tools and Services by Economically Disadvantaged Communities: Implications for Growth and Sustainability*. IGI Global.

Chatterjee, B., & Banerjee, S. (2018). Trade and Economic Integration. *Seminar*, *703*, 2–5.

Cheng, W. W. H., Lam, E. T. H., & Chiu, D. K. W. (2020). Social media as a platform in academic library marketing: A comparative study. *Journal of Academic Librarianship*, *46*(5), 102188. doi:10.1016/j.acalib.2020.102188

Chen, M. (2019). Modelling Public Mood and Emotion: Blog and News Sentiment and Politico-economic Phenomena. In A. Visvizi & M. D. Lytras Effat (Eds.), *Politics And Technology In The Post-Truth Era* (pp. 57–72). Emerald Publishing Limited.

Chen, Y., Chiu, D. K. W., & Ho, K. K. W. (2018). Facilitating the learning of the art of Chinese painting and calligraphy at Chao Shao-an Gallery. *Micronesian Educators*, *26*, 45–58.

Chesbrough, H. (2012). *Open Innovation. Where We've Been and Where We're Going*. Research Technology Management.

Chesbrough, H. (2002). Making Sense of Corporate Venture Capital. *Harvard Business Review*. PMID:11894386

Cheung, T. Y., Ye, Z., & Chiu, D. K. W. (2021). Value chain analysis of information services for the visually impaired: A case study of contemporary technological solutions. *Library Hi Tech*, *39*(2), 625–642. doi:10.1108/LHT-08-2020-0185

Cheung, V. S. Y., Lo, J. C. Y., Chiu, D. K. W., & Ho, K. K. W. (2022). Predicting Facebook's influence on travel products marketing based on the AIDA model. *Information Discovery and Delivery*. Advance online publication. doi:10.1108/IDD-10-2021-0117

Chidamber, S. R., & Kemerer, C. F. (1994). A metrics suite for object-oriented design. *IEEE Transactions on Software Engineering*, *20*(6), 476–493. doi:10.1109/32.295895

Chiu, D. K. W. (Ed.). (2012). *Mobile and Web Innovations in Systems and Service-oriented Engineering*. IGI Global.

Chopra, M., Singh, D. S. K., Gupta, A., Aggarwal, K., Gupta, B. B., & Colace, F. (2022). Analysis & prognosis of sustainable development goals using big data-based approach during COVID-19 pandemic. *Sustainable Technology and Entrepreneurship*, *1*(2), 100012. doi:10.1016/j.stae.2022.100012

Choueiri, A. C., Denise, M. V. S., Scalabrin, E. E., & Eduardo, A. P. S. (2020). An extended model for remaining time prediction in manufacturing systems using process mining. *Journal of Manufacturing Systems*, *56*, 188–201. doi:10.1016/j.jmsy.2020.06.003

Christensen, C., Kaufman, S., & Shih, W. (2008). Innovation Killers: How Financial Tools Destroy Your Capacity to Do New Things. *Harvard Business Review*, *86*(1), 98–105, 137. PMID:18271321

Chung, A. C. W., & Chiu, D. K. (2016). OPAC Usability Problems of Archives: A Case Study of the Hong Kong Film Archive. *International Journal of Systems and Service-Oriented Engineering*, *6*(1), 54–70. doi:10.4018/IJSSOE.2016010104

CIMdata. (2016). *Free Resources*. Retrieved June 2022, from: https://www.cimdata.com/en/resources/about-plm

CIMdata. (2022). *CIMdata Publishes PLM Market and Solution Provider Report*. Retrieved June 2022, from: https://www.cimdata.com/en/news/item/6459-cimdata-publishes-plm-market-and-solution-provider-report

Cinco Días. (2021). *Así es el interior de Zara Beauty, los espacios de cosmética de Inditex*. Recuperado de https://cincodias.elpais.com/cincodias/2021/05/12/companias/1620824085_266540.html

Ciric, Z., & Rakovic, L. (2010). Change management in information system development and implementation projects. *Management Information Systems*, *5*(2), 23–28.

Clark, D. (2002). *Enterprise Security: The Manager's Defense Guide*. Addison-Wesley Professional.

Clarkson, P., Stephenson, A., Grimmett, C., Cook, K., Clark, C., Muckelt, P. E., O'Gorman, P., Saynor, Z., Adams, J., Stokes, M., & McDonough, S. (2022). Digital tools to support the maintenance of physical activity in people with long-term conditions: A scoping review. *Digital Health*, *8*, 20552076221089778. doi:10.1177/20552076221089778 PMID:35433017

CMMI Product Team. (2007). *CMMI for Acquisition*. Technical Report. Retrieved June 2022, from: ftp://ftp.sei.cmu.edu/pub/documents/07.reports/07tr017.pdf

Cohen, L., & Young, A. (2006). *Multisourcing: Moving beyond outsourcing to achieve growth and agility*. Harvard Business School Press.

Collaboratarium for Social Media and Online Behavioural Studies (COSMOS). (2020). https://www.cosmos.ualr.edu/misinformation

Colli, M., Bergerb, U., Bockholta, M., Madsena, O., Møllera, C., & Vejrum Wæhrens, B. (2019). A maturity assessment approach for conceiving context-specific roadmaps in the Industry 4.0 era. *Annual Reviews in Control*, *48*, 165–177. doi:10.1016/j.arcontrol.2019.06.001

Colli, M., Madsen, O., Berger, U., Møller, C., Wæhrens, B. V., & Bockholt, M. (2018). Contextualizing the outcome of a maturity assessment for Industry 4.0. *IFAC-PapersOnLine*, *51*(11), 1347–1352. doi:10.1016/j.ifacol.2018.08.343

Conservation Office. (2016). *2015-16 Annual Report*. Retrieved from https://www.lcsd.gov.hk/CE/Museum/Conservation/documents/101 18435/10118860/2015-16_Annual_Report_EN_final_RE.pdf

Conservation Office. (2021a). *Conservation "Behind-the-Scenes" Tour*. Retrieved from https://www.lcsd.gov.hk/CE/Museum/Conservation/en_US/web/co/education_and_extension_programs_imd.html

Conservation Office. (2021b). *Education and extension programs*. Retrieved from https://www.lcsd.gov.hk/CE/Museum/Conservation/en_US/web/co/education_and_extension_programs_muse_2018.html

Conservation Office. (2021c). *School Culture Day Scheme*. Retrieved from https://www.lcsd.gov.hk/CE/Museum/Conservation/en_US/web/co/school_culture_day_scheme.html

Conservation Office. (2021d). *Vision, Mission and Values*. Retrieved from https://www.lcsd.gov.hk/CE/Museum/Conservation/en_US/web/co/vision_and_values.html

Cooper, R. G. (2008). Perspective: The Stage-Gate®Idea-to-Launch Process-Update, What's New, and Nex Gen Systems. *Journal of Product Innovation Management, 25*, 213–232.

Corallo, A., Errico, F., Latina, M. E., & Menegoli, M. (2019). Dynamic Business Models: A Proposed Framework to Overcome the Death Valley, Springer. *Journal of the Knowledge Economy, 10*(1). Advance online publication. doi:10.100713132-018-0529-x

Corbridge, S. (1984). Political geography of contemporary events V: Crisis, what crisis? Monetarism, Brandt II and the geopolitics of debt. *Political Geography Quarterly, 3*(4), 331–345. doi:10.1016/0260-9827(84)90017-X

Corradini, F., Marcantoni, F., Morichetta, A., Polini, A., Re, B., & Sampaolo, M. (2019). Enabling auditing of smart contracts through process mining. In *From Software Engineering to Formal Methods and Tools, and Back* (pp. 467–480). Springer. doi:10.1007/978-3-030-30985-5_27

Correani, A., Massis, A. D., Frattini, F., Petruzzelli, A. M., & Natalicchio, A. (2020). Implementing a Digital Strategy: Learning from the Experience of Three Digital Transformation Projects. *California Management Review, 62*(4), 37–56. doi:10.1177/0008125620934864

Council of Europe. (2005). *Legal, Operational and Technical Standards for E-Voting*. http://www.com.int/t/dgap/democracy/activities/keytexts/recommendations/Eng_Evoting_and_Expl_Memo_en.pdf.'

Cozmiuc, D. C., & Pettinger, R. (2021). Consultants' Tools to Manage Digital Transformation: The Case of PWC, Siemens, and Oracle. *Journal of Cases on Information Technology, 23*(4), 1–29. doi:10.4018/JCIT.20211001.oa7

Cozzolino, A., Corbo, L., & Aversa, P. (2021). Digital platform-based ecosystems: The evolution of collaboration and competition between incumbent producers and entrant platforms. *Journal of Business Research, 126*, 385–400. doi:10.1016/j.jbusres.2020.12.058

Creswell, J. (2013). *Research design: Qualitative, quantitative, and mixed methods approach* (2nd ed.). SAGE Publications.

Cristófol Rodríguez, C., Martínez Sala, A., & Segarra Saavedra, M. (2018). Estrategia de comunicación digital en el sector franquicias de moda. El caso de Zara en Facebook. *Revista de comunicación audiovisual y publicitaria*. https://revistas.ucm.es/index.php/ARAB/article/view/60999

Crossman, A. (2019). *Understanding Purposive Sampling. An overview of the method and its application*. https://www.thoughtco.com/purposive-sampling-3026727

Crosswell, A. (2014). *Bricks and the TOGAF TRM*. National Institute of Health.

Cunningham, R. (2014). *Information environmentalism: A governance framework for intellectual property rights*. Edward Elgar Publishing. doi:10.4337/9780857938442

Czerwinski, Lasota, Lazi´c, Leroux, & Mazowiecki. (2020). The Reachability Problem for Petri Nets Is Not Elementary. *Journal of the ACM, 68*, 1–28.

Dadaczynski, K., Okan, O., Messer, M., Leung, A. Y. M., Rosário, R., Darlington, E., & Rathmann, K. (2021). Digital Health Literacy and Web-Based Information-Seeking Behaviors of University Students in Germany During the COVID-19 Pandemic: Cross-sectional Survey Study. *Journal of Medical Internet Research*, *23*(1), 1–17. doi:10.2196/24097 PMID:33395396

Daft, R. L., & Lewin, A. Y. (1993). Where are the theories for the new organizational forms? *Organization Science*, *4*(4), i–vi.

Dakic, Stefanovic, Cosic, Lolic, & Medojevic. (2018). Business process mining application: A literature review. *Annals of DAAAM & Proceedings*, 29.

Data-Monitor. (2008). *The benefit of e-voting system Feature analysis*. Data-Monitor.

Daum, J. H. (2003). *Intangible Assets and Value Creation*. Wiley.

De Carolis, A., Macchi, M., Negri, E., & Terzi, S. (2017), Guiding manufacturing companies towards digitalization a methodology for supporting manufacturing companies in defining their digitalization roadmap. *IEEExplore International Conference on Engineering, Technology and Innovation*. https://doi.org/10.1145/3477911.3477924

De Carolis, A., Macchi, M., Negri, E., & Terzi, S. (2017). A maturity model for assessing the digital readiness of manufacturing companies. In *IFIP International Conference on Advances in Production Management Systems*. Springer.

De Carolis, A., Macchi, M., Negri, E., & Terzi, S. (2017). A Maturity Model for Assessing the Digital Readiness of Manufacturing Companies. In H. Lödding, R. Riedel, K. D. Thoben, G. von Cieminski, & D. Kiritsis (Eds.), *Advances in Production Management Systems. The Path to Intelligent, Collaborative and Sustainable Manufacturing. APMS 2017. IFIP Advances in Information and Communication Technology, 513*. Springer. doi:10.1007/978-3-319-66923-6_2

De Lemus, T. (2021). *Brand Finance España 100 2021. Las marcas más valiosas de España pierden valor por tercer año consecutivo, según Brand Finance*. Recuperado de https://brandirectory.com/rankings/spain/

de Leoni, M. (2022). Foundations of Process Enhancement. *Springer International Publishing.*, *8*. Advance online publication. doi:10.1007/978-3-031-08848-3

De Santos, J. A., & Francisco, E. de R. (2021). Digital Maturity Level of a B2B Company: Case Study of a Brazilian Complex Manufacturing Company. *International Conference on Information Resources Management (CONF-IRM)*, 12. https://aisel.aisnet.org/confirm2021/22/

Dedrick, Carmel, & Kraemer (2017). A dynamic model of offshore software development. In L. P. Wilcocks, M. C. Lacity, & C. Sauer (Eds), *Outsourcing and offshoring business services* (pp. 281-320). Springer.

Del Pino-Romero, C., & Castelló-Martínez, A. (2015). La comunicación publicitaria se pone de moda: Branded content y fashion films. *Mediterránea de Comunicación*, *6*(1), 105–128. doi:10.14198/MEDCOM2015.6.1.07

Delens, G. P. A. J., Peters, R. J., Verhoef, C., & van Vlijmen, S. F. M. (2016). Lessons from Dutch IT-outsourcing success and failure. *Science of Computer Programming*, *130*(32), 37–68. doi:10.1016/j.scico.2016.04.001

Deloitte. (2018a). *Digital Maturity Model. Achieving digital maturity to drive growth*. Retrieved April 12th, 2020, from https://www2.deloitte.com/content/dam/Deloitte/global/Documents/Technology-Media-Telecommunications/deloitte-digital-maturity-model.pdf

Deloitte. (2018b). *Digital Transformation with new SAP Technologies*. Retrieved April 13th, 2020, from: https://www2.deloitte.com/content/dam/Deloitte/lu/Documents/technology/lu-digital-transformation-sap.pdf

Deloitte. (2019). *Pivoting to digital maturity. Seven capabilities central to digital transformation.* Retrieved April 12th, 2020, from https://www2.deloitte.com/us/en/insights/focus/digital-maturity/digital-maturity-pivot-model.html

Deloitte. (2020). *Uncovering the connection between digital maturity and financial performance. How digital transformation can lead to sustainable high performance.* Retrieved April 12th, 2020, from https://www2.deloitte.com/us/en/insights/topics/digital-transformation/digital-transformation-survey.html

Deng, S., & Chiu, D. K. W. (2022). Analyzing Hong Kong Philharmonic Orchestra's Facebook Community Engagement with the Honeycomb Model. In M. Dennis & J. Halbert (Eds.), *Community Engagement in the Online Space*. IGI Global.

Deng, W., Chin, G. Y.-l., Chiu, D. K. W., & Ho, K. K. W. (2022). Contribution of Literature Thematic Exhibition to Cultural Education: A Case Study of Jin Yong's Gallery. *Micronesian Educators*, *32*, 14–26.

Depcinski, M. C. (2020). Conservation in Museums. In C. Smith (Ed.), *Encyclopedia of Global Archaeology*. Springer. doi:10.1007/978-3-030-30018-0_796

Desfray, Ph. (2011). *Using OMG Standards with TOGAF*. SOFTEAM – Modeliosoft. www.modeliosoft.com

Diamantini, C., Genga, L., Potena, D., & Aalst, W. (2016). Building Instance Graphs for Highly Variable Processes. *Expert Systems with Applications*, *59*, 59. doi:10.1016/j.eswa.2016.04.021

Díaz Soloaga, P., & García Guerrero, L. (2013). Los fashion films como estrategia de construcción de marca a través de la seducción. Persuasión Audiovisual: formas, soportes y nuevas estrategias, 349-371

Dietz, J. L. G. (2006). *Enterprise ontology: Theory and methodology*. Springer. doi:10.1007/3-540-33149-2

Dijkstra, E. W. (1974). *On the role of scientific thought. E.W. Dijkstra Archive (EWD447), Center for American History*. The University of Texas at Austin.

Dogan, O., Fernandez-Llatas, C., & Oztaysi, B. (2019). Process mining application for analysis of customer's different visits in a shopping mall. In *International Conference on Intelligent and Fuzzy Systems* (pp. 151–159). Springer.

Dombrowski, U., Krenkel, P., Falkner, A., Placzek, F., & Hoffmann, T. (2018). Potenzialanalyse von Industrie 4.0-Technologien: Zielorientiertes Auswahlverfahren. ZWF - Zeitschrift für wirtschaftlichen Fabrikbetrieb, 107–111.

Dombrowski, U., Richter, T., & Ebentreich, D. (2015). *Auf dem Weg in die vierte industrielle Revolution: Ganzheitliche Produktionssysteme zur Gestaltung der Industrie-4.0*. Architektur. zfo Zeitschrift Führung +Organisation.

Dominguez, C., Garcia-Izquierdo, F. J., Jaime, A., Perez, B., Rubio, A. L., & Zapata, M. A. (2021). Using Process Mining to Analyze Time Distribution of Self-Assessment and Formative Assessment Exercises on an Online Learning Tool. *IEEE Transactions on Learning Technologies*, *14*(5), 709–722. doi:10.1109/TLT.2021.3119224

Dong, J., & Yang, Sh. (2003). *Visualizing Design Patterns With A UML Profile. Department of Computer Science*. University of Texas.

Donthu, N., & Gustafsson, A. (2020). Effects of COVID-19 on business and research. *Journal of Business Research*, *117*, 284–289. doi:10.1016/j.jbusres.2020.06.008 PMID:32536736

Downey, K. (2003). *Architectural Design Patterns for XML Documents*. https://www.xml.com/pub/a/2003/03/26/patterns.html

Draghici, G., & Draghici, A. (2009). Collaborative Multisite PLM Platform. *CENTER-IS 2009, Conference on Enterprise Information Systems, Conference: CENTERIS 2009, Conference on Enterprise Information Systems*. https://www.researchgate.net/publication/235673601_Collaborative_Multisite_PLM_Platform

Drago, A. (2011). 'I feel included': The Conservation in Focus exhibition at the British Museum. *Journal of Insect Conservation, 34*(1), 28–38. doi:10.1080/19455224.2011.566473

Drucker. (1994). What Is a Business Model? *Harvard Business Review*.

Duan, H., Li, J., Fan, S., Lin, Z., Wu, X., & Cai, W. (2021). Metaverse for social good: A university campus prototype. *Proceedings of the 29th ACM International Conference on Multimedia*, 153-161. 10.1145/3474085.3479238

Duffy, J. (2001). Maturity models: Blueprints for evolution. *Strategy and Leadership, 29*(6), 19–26.

Dupuis, A., Cam'elia, D., & Agard, B. (2022). Predicting crop rotations using process mining techniques and Markov principals. *Computers and Electronics in Agriculture, 194*, 106686. doi:10.1016/j.compag.2022.106686

Duvander, A. (2021). *API Design Patterns for API. Stoplight*. https://blog.stoplight.io/api-design-patterns-for-rest-web-services

Dwivedi, Y. K., Hughes, D. L., Coombs, C., Constantiou, I., Duan, Y., Edwards, J. S., Gupta, B., Lal, B., Misra, S., Prashant, P., Raman, R., Rana, N. P., Sharma, S. K., & Upadhyay, N. (2020). Impact of COVID-19 pandemic on information management research and practice: Transforming education, work and life. *International Journal of Information Management, 55*, 102211. doi:10.1016/j.ijinfomgt.2020.102211

Easson, H., & Leask, A. (2020). After-hours events at the National Museum of Scotland: A product for attracting, engaging and retaining new museum audiences? *Current Issues in Tourism, 23*(11), 1343–1356. doi:10.1080/13683500.2019.1625875

Easterbrook, S., Singer, J., Storey, M., & Damian, D. (2008). *Guide to Advanced Empirical Software Engineering-Selecting Empirical Methods for Software Engineering Research* (F. Shull, Ed.). Springer.

eBay (2015). US Small Business Global Growth Report. Available at https,//www.ebaymainstreet.com/sites/default/files/2015-us-small-biz-global-growth-report_0.pdf, Accessed on 7th March 2022

Ebert, C., Gallardo, G., Hernantes, J., & Serrano, N. (2016). *DevOps4AI*. IEEE.

El Publicista. (2021). *Así es el fashion film de Zara que cautiva a millones de audiencia a nivel mundial en 24 horas*. Recuperado de https://www.elpublicista.es/anunciantes/asi-fashion-film-zara-cautiva-millones-audiencia-nivel-mundial

Elkins, J. (2012). Concerning Practices of Truth. In J. Elkins & A. Norris (Eds.), *Truth and Democracy*. University of Pennsylvania Press. doi:10.9783/9780812206227.19

Eller, R., Alford, P., Kallmünzer, A., & Peters, M. (2020). Antecedents, consequences, and challenges of small and medium-sized enterprise digitalization. *Journal of Business Research, 112*, 119–127. doi:10.1016/j.jbusres.2020.03.004

Ellington, J. W. (Ed.). (1994). *Immanuel Kant Ethical Philosophy*. Hackett Publishing Company.

Ellison, N. B., Vitak, J., Gray, R., & Lampe, C. (2014). Cultivating social resources on social network sites: Facebook relationship maintenance behaviors and their role in social capital processes. *Journal of Computer-Mediated Communication, 19*(4), 855–870. doi:10.1111/jcc4.12078

Enterprise Architecture Solutions. (2021). *Essential Meta Model.* Enterprise Architecture Solutions. https://enterprise-architecture.org/docs/introduction/essential_meta_model/

Eremina, Y., Lace, N., & Bistrova, J. (2019). Digital Maturity and Corporate Performance: The Case of the Baltic States. *Journal of Open Innovation, 5*(3), 54. doi:10.3390/joitmc5030054

Erul, E., Woosnam, K. M., & McIntosh, W. A. (2020). Considering emotional solidarity and the theory of planned behavior in explaining behavioral intentions to support tourism development. *Journal of Sustainable Tourism, 28*(8), 1158–1173. doi:10.1080/09669582.2020.1726935

Escalona, J., & Koch, N. (2004). *Requirements engineering for web applications – A comparative Study.* University of Seville.

Essig, M., Glas, A. H., Selviaridis, K., & Roehrich, J. K. (2016). Performance-based contracting in business markets. *Industrial Marketing Management, 59*, 5–11. doi:10.1016/j.indmarman.2016.10.007

Ethiraj, S. K., Levinthal, D., & Roy, R. R. (2008). The dual role of modularity: Innovation and imitation. *Management Science, 54*(5), 939–955. doi:10.1287/mnsc.1070.0775

European Commission. (2018). *EU businesses go digital: Opportunities, outcomes and uptake.* Retrieved on August 12th, 2020, from https://ec.europa.eu/growth/tools-databases/dem/monitor/sites/default/files/Digital%20Transformation%20Scoreboard%202018_0.pdf

European Commission. (2020a). *Parteneriat în exploatarea Tehnologiilor Generice Esențiale (TGE), utilizând o PLATformă de interacțiune cu întreprinderile competitive TGE-PLAT.* Accesat în iunie 2022: https://www.imt.ro/TGE-PLAT/e-news/Concept%20adus%20la%20TRL%207-8.pdf

European Commission. (2020b). *The European Commission's science and knowledge service.* Joint Research Centre, Accesat în iunie 2022: https://s3platform-legacy.jrc.ec.europa.eu/documents/20182/443954/M.+Ranga+-+Day+3+TRLs+and+tech+transfer.pdf/f9214fc2-7514-4b67-a18f-962d45cdd5ab

European Commission. (2022a). *Documents download module.* Retrieved April 13th, 2020, from: https://ec.europa.eu/research/participants/documents/downloadPublic?documentIds=080166e5cce7b21b&appId=PPGMS

European Union. (2022a). *H2020l Model for ERC Proof of Concept Grants 2 (H2020 ERC PoC — Multi).* Retrieved April 13th, 2020, from: file:///D:/Desktop/h2020-mga-erc-poc-multi_v2.0_en.pdf

European Union. (2022b). *Access 2EIC National Contact Points for Innovation.* Retrieved April 13th, 2020, from: https://access2eic.eu/wp-content/uploads/2020/04/Access2EIC_How-to-apply-successfully_template.pdf

Evans, M., Malpass, A., Agnew-Davies, R., & Feder, G. (2018). Women's experiences of a randomised controlled trial of a specialist psychological advocacy intervention following domestic violence: A nested qualitative study. *PLoS One, 13*(11), e0193077. doi:10.1371/journal.pone.0193077 PMID:30481185

EY. (2018). *Digital Readiness Assessment. Does your business strategy work in a digital world?* Retrieved April 14th, 2020, from https://digitalreadiness.ey.com/

EY. (2020a). *Digital Strategy and Transformation.* Retrieved on June 30th, 2021, from: https://www.ey.com/en_gl/digital/transformation

EY. (2020b). *Strategic Roadmap.* Retrieved on June 30th, 2021, from: https://www.ey.com/en_se/advisory/strategic-roadmap

Fan, X. Y., Shan, X. S., Day, S., & Shou, Y. Y. (2022). Toward Green Innovation Ecosystems: Past Research on Green Innovation and Future Opportunities from an Ecosystem Perspective. *Industrial Management & Data Systems, 122*(9), 2012–2044. Advance online publication. doi:10.1108/IMDS-12-2021-0798

Farhoomand, A. (2004). *Managing (e)business transformation.* Palgrave Macmillan. doi:10.1007/978-1-137-08380-7

Fasterholdt, I., Lee, A., Kidholm, K., Yderstræde, K., Møller, B., & Pedersen, K. (2018). A qualitative exploration of early assessment of innovative medical technologies. *BMC Health Services Research, 18*(1). Advance online publication. doi:10.118612913-018-3647-z

Ferrero, C. (2020). *Fabien Baron, diseñador del logo de Zara: «La moda se está enfrentando a las consecuencias de haber sido elitista y blanca».* Recuperado de https://smoda.elpais.com/moda/fabien-baron-disenador-del-logo-de-zara-la-moda-se-esta-enfrentando-a-las-consecuencias-de-haber-sido-elitista-y-blanca/

Financial Services and the Treasury Bureau. (2021). *Guide to Procurement.* Retrieved from https://www.fstb.gov.hk/en/treasury/gov_procurement/guide-to-procurement.htm

Fixson, S. K., Ro, Y., & Liker, J. (2005). Modularity and outsourcing: Who drives whom? A study of generational sequences in the U.S. automotive cockpit industry. *International Journal of Automotive Technology and Management, 5*(2), 166–183. doi:10.1504/IJATM.2005.007181

Fletcher, G., & Griffiths, M. (2020). Digital Transformation During a Lockdown. *International Journal of Information Management, 55,* 102185. doi:10.1016/j.ijinfomgt.2020.102185 PMID:32836642

Flyvbjerg, B. (2006). Five misunderstandings about case study research. *Qualitative Inquiry, 12*(2), 219–245. doi:10.1177/1077800405284363

Fong, K. C. H., Au, C. H., Lam, E. T. H., & Chiu, D. K. W. (2020). Social network services for academic libraries: A study based on social capital and social proof. *Journal of Academic Librarianship, 46*(1), 102091. doi:10.1016/j.acalib.2019.102091

Forough, H., Gabriel, Y., & Fotaki, M. (2019). Leadership in a Post-Truth Era: A new Narrative Disorder? The Sage. doi:10.1177/1742715019835369

Fortuna, K. L., Brooks, J. M., Umucu, E., Walker, R., & Chow, P. I. (2019). Peer support: A human factor to enhance engagement in digital health behavior change interventions. *Journal of Technology in Behavioral Science, 4*(2), 152–161. doi:10.100741347-019-00105-x PMID:34337145

Fowler, M. (2003). *Catalog of Patterns of Enterprise Application Architecture.* https://martinfowler.com/eaaCatalog

Fowler, M. (2014). *Microservices.* https://martinfowler.com/articles/microservices.html. USA.

Fowler, M. (1996). *Analysis Patterns: Reusable Object Models.* Addison-Wesley.

Fowler, M., Rice, D., Foemmel, M., Hieatt, E., Mee, R., & Stafford, R. (2002). *Patterns of Enterprise Application Architecture.* Addison Wesley.

Frankfurt, H. (2005). *On Bullshit.* Princeton University Press. doi:10.1515/9781400826537

Frankland, R., & Volkamer, M. (2011). *The readiness of various e-Voting systems for complex elections*. Technical Report. https://www.researchgate.net/publication/262933684

Freund, C. L., & Weinhold, D. (2004). The effect of the Internet on international trade. *Journal of International Economics, 62*(1), 171–189. https://doi.org/10.1016/S0022-1996(03)00059-X

Fuego. (2006). *BPM Process Patterns-Repeatable Designs for BPM Process Models*. Fuego.

Gaita, R. (2016). *When Politicians Lie: Reflections On Truth, Politics And Patriotism*. https://griffithreview.com/contributors/raimond-gaita/

Gallimore, E., & Wilkinson, C. (2019). Understanding the Effects of 'Behind-the-Scenes' Tours on Visitor Understanding of Collections and Research. *Curator (New York, N.Y.), 62*(2), 105–115. doi:10.1111/cura.12307

Gallo, A. (2017). A Refresher on Discovery Driven Planning. *Harvard Business Review*.

Galyarski, E., & Mironova, N. (2021). Digitalization and its impact on business processes. *Economics and Management, 18*(1), 81–89.

Gamma, E., Vlissides, J., Helm, R., & Johnson, R. (1994). *Design Patterns: Elements of Re-usable Object-Oriented Software*. Addison-Wesley.

García, M. (2019). *Zara presenta un nuevo logo por segunda vez en 44 años*. Recuperado de https://brandemia.org/zara-presenta-un-nuevo-logo-por-segunda-vez-en-44-anos

Garnham, N. (1990). *Capitalism and communication: Global culture and the economics of information*. Sage Publications. https://scholar.google.com/scholar?hl=en&as_sdt=0%2C5&q=Capitalism+and+communication%3A+Global+culture+and+the+economics+of+information&btnG=

Gartner. (2005). *External Service Providers' service oriented architecture Frameworks and Offerings: Capgemini*. Gartner.

Gartner. (2013a). *Scenario Toolkit: Using EA to Support Business Transformation*. Gartner Inc.

Gartner. (2013b). *Hype Cycle for Business Process Management*. Gartner.

Gartner. (2014a). *What the Business Process Director Needs to Know About Enterprise Architecture*. Gartner.

Gartner. (2016). Gartner's 2016 Hype Cycle for ICT in India Reveals the Technologies that are Most Relevant to Digital Business in India Analysts to Explore Key Technologies and Trends at Gartner Symposium/ITxpo 2016, 15-18 November, in Goa, India. Retrieved April 3, 2018, from https://www.gartner.com/newsroom/id/3503417

Gartner. (2022). *Gartner Information Technology Glossary: Digitization*. https://www.gartner.com/en/information-technology/glossary/digitization

Gavrila Gavrila, S., & De Lucas Ancillo, A. (2022). Entrepreneurship, innovation, digitization and digital transformation toward a sustainable growth within the pandemic environment. *International Journal of Entrepreneurial Behaviour & Research, 28*(1), 45–66. doi:10.1108/IJEBR-05-2021-0395

Gay, C. L., & Essinger, J. (2000). *Inside outsourcing: The insider's guide to managing strategic sourcing*. Brealey.

George Eastman Museum. (2019). *Virtual Tour*. Retrieved from https://www.eastman.org/360-conservation-lab-tour

Gewald, H., & Dibbern, J. (2005). *The influential role of perceived risks versus perceived benefits in the acceptance of business process outsourcing: Empirical evidence from the German banking industry* (Working Paper Nr. 2005-9). J. W. Goethe University, Germany: E-Finance Lab.

Ghoreishi, M. S., Vahedian-Shahroodi, M., Jafari, A., & Tehranid, H. (2019). Self-care behaviors in patients with type 2 diabetes: Education intervention base on social cognitive theory. *Diabetes & Metabolic Syndrome, 13*(3), 2049–2056. doi:10.1016/j.dsx.2019.04.045 PMID:31235135

Gillespie, A., & Williams, H. (2016). Telecommunications and the reconstruction of regional comparative advantage. *Environment and Planning A: Economy and Space, 20*(10), 1311–1321. doi:10.1068/A201311

Gill, M., & VanBoskirk, S. (2016). *The Digital Maturity Model 4.0 Benchmarks: Digital Business Transformation Playbook.* Forrester. https://www.forrester.com/report/The-Digital-Maturity-Model-40/RES130881

Gilster, P. (1997). *Digital literacy.* John Wiley & Sons, Inc.

Gioratra, K., & Netessine, S. (2014). Four Paths to Business Model Innovation. *Harvard Business Review.*

GitBook. (2021). *ArchiMate Guide-ArchiMate and TOGAF Layers.* GitBook. https://archimatetool.gitbook.io/project/archimate-and-togaf-layers

Global Center for Digital Business Transformation. (2015). *Digital Vortex. How Digital Disruption Is Redefining Industries.* Retrieved April 13th, 2020, from https://www.cisco.com/c/dam/en/us/solutions/collateral/industry-solutions/digital-vortex-report.pdf

Global I. P. Center. (2022). *US Chamber of Commerce.* https://www.theglobalipcenter.com/

Gobble, M. M. (2018). Digitalization, Digitization, and Innovation. *Research Technology Management, 61*(4), 56–59. doi:10.1080/08956308.2018.1471280

Goddard, J., & Gillespie, A. (2017). Advanced telecommunications and regional economic development. In *In managing the city* (pp. 84–109). Routledge. https://www.taylorfrancis.com/chapters/edit/10.4324/9781315178264-6/advanced-telecommunications-regional-economic-development-goddard-gillespie

Godin. (2005). *The Linear Model of Innovation: The Historical Construction of an Analytical Framework.* Retrieved on August 20th, 2022, from http://www.csiic.ca/PDF/Godin_30.pdf

Goebl, W., Guenther, M., Klyver, A., & Papegaaij, B. (2020). Enterprise design patterns-35 ways to radically increase your impact on the enterprise. Intersection Group.

Goikoetxea, A. (2004). A mathematical framework for enterprise architecture representation and design. *International Journal of Information Technology & Decision Making, 3*(1), 5–32.

GoldbeckG.SimperlerA. (2019). *Business Models and Sustainability for Materials Modelling Software, Projects.* European Materials Modelling Council, EMMC - European Materials Modelling Council. Doi:10.5281/zenodo.2541722

Goldsmith, B., & Ruthrauff, H. (2013). *Case Study Report on the Philippines 2010 Elections.* Academic Press.

Gomes, L. A. D., Facin, A. L. F., Salerno, M. S., & Ikenami, R. K. (2018). Unpacking the Innovation Ecosystem Construct: Evolution, Gaps and Trends. *Technological Forecasting and Social Change, 136*, 30–48. doi:10.1016/j.techfore.2016.11.009

Gong, C., & Ribiere, V. (2021). Developing a unified definition of digital transformation. *Technovation, 102*, 102–117. doi:10.1016/j.technovation.2020.102217

González Fernández, A. (2021). *Comunicación comercial a través de redes sociales de empresas de moda y belleza.* https://iabspain.es/wp-content/uploads/2017/12/estudio-content-native- advertising-2017-vcorta.pdf

González, L., & Jouanjean, J. (2017). Digital Trade: Developing a Framework for Analysis. In *Trade Policy Papers, No 205.* OECD Publishing. https://www.oecd-ilibrary.org/trade/digital-trade_524c8c83-en

Gorla, N., & Lau, M. B. (2010). Will negative experiences impact future outsourcing? *Journal of Computer Information Systems, 50*(3), 91–101.

Gràffica. (2019). *Así es el nuevo logo Zara.* Recuperado de https://graffica.info/asi-es-el-nuevo-logo-de-zara/

Grant, R. M. (2010). *Contemporary Strategy Analysis* (7th ed.). John Wiley & Sons.

Greaver, M. F. II. (1999). *Strategic outsourcing: A structured approach to outsourcing decisions and initiatives.* Amacom.

Greefhorst, D. (2009). *Using TOGAF as a pragmatic approach to architecture.* Informatica.

Greenberg, D. (2015). *Truth in Politics Now.* https://prospect.org/article/truthpolitics-now

Grieves, M. (2006). *Product Lifecycle Management. Driving the Next Generation of Lean Thinking.* Mc Graw Hill.

Grisold, Wurm, Mendling, & vom Brocke. (2020). *Using Process Mining to Support Theorizing About Change in Organizations.* Academic Press.

Gritzalis, D. (2003). *Secure Electronic Voting systems.* Dept. of Informatics Athens University of Economics & Business &Data Protection Commission of Greece. doi:10.1007/978-1-4615-0239-5

Grönlund, Å., & Horan, T. (2004). Introducing e-Gov: History, Definitions, and Issues. *Communications of the Association for Information Systems, 15.*

Grossman, R. (2016). The Industries That Are Being Disrupted the Most by Digital. *Harvard Business Review.*

Groza, I. V., & Balint, R. (2022). *Realizarea și planificarea modelului științific pentru produsele noi. Proiect: INOvări și optimizări economice și funcționale în producția industrială de MATeriale pentru energie termică, „INO-MAT", cod SMIS 119412. Realizarea și planificarea modelului științific pentru produsele noi.* Titus Industries SRL.

Grudin, J. (2006, January). Enterprise knowledge management and emerging technologies. In *Proceedings of the 39th Annual Hawaii International Conference on System Sciences (HICSS'06)* (Vol. 3, pp. 57a-57a). IEEE. 10.1109/HICSS.2006.156

Gruia, L. A., Bibu, N., Roja, A., Danaiață, D., & Năstase, M. (2020). Digital Transformation of Businesses in Times of Global Crisis. In *Griffiths School of Management and IT Annual Conference on Business, Entrepreneurship and Ethics* (pp. 43-62). Springer.

Guinan, P. J., Parise, S., & Langowitz, N. (2019). Creating an innovative digital project team: Levers to enable digital transformation. *Business Horizons, 62*(6), 717–727. doi:10.1016/j.bushor.2019.07.005

Gupta, S. K. S., Mukherjee, T., Varsamopoulos, G., & Banerjee, A. (2011), Research directions in energy-sustainable cyber–physical systems. *Sustainable Computing: Informatics and Systems, 1*(1), 57-74.

Gupta, B. B., Chaudhary, P., & Gupta, S. (2020). Designing a XSS defensive framework for web servers deployed in the existing smart city infrastructure. *Journal of Organizational and End User Computing*, *32*(4), 85–111. doi:10.4018/JOEUC.2020100105

Gurbaxani, V., & Dunkle, D. (2019). Gearing Up for Successful Digital Transformation. *MIS Quarterly Executive*, *18*(3), 209–220. doi:10.17705/2msqe.00017

Habibu, T., Sharif, K., & Nichola, S. (2017). Design and Implementation of Electronic Voting System. *International Journal of Computer & Organization Trends*, *45*(1), 7.

Haller, M., Ludig, S., & Bauer, N. (2012). Bridging the scales: A conceptual model for coordinated expansion of renewable power generation, transmission and storage. *Renewable & Sustainable Energy Reviews*, *16*(5), 2687–2695. doi:10.1016/j.rser.2012.01.080

Hamel, G., & Prahalad, C. K. (2000, July). Competing for the Future. *Harvard Business Review*.

Harris, M. D. S., Herron, D., & Iwanicki, S. (2008). *The business value of IT: Managing risks, optimizing performance and measuring results*. CRC Press. doi:10.1201/9781420064759

Harvard Business Review. (2020a). *Reevaluating Digital Transformation During Covid-19* [Research Report]. Harvard Business Review Analytic Services. https://hbr.org/sponsored/2020/11/reevaluating-digital-transformation-during-covid-19

Harvard Business Review. (2020b). *Reconciling Cultural and Digital Transformation to Design the Future of Work* [White Paper]. Harvard Business Review Analytic Services. https://hbr.org/sponsored/2020/10/reconciling-cultural-and-digital-transformation-to-design-the-future-of-work

Harvey, D. (2020). The condition of postmodernity. In Knowledge and Postmodernism in Historical Perspective (pp. 494–507). Taylor and Francis. https://doi.org/10.4324/9781003060963-38/CONDITION-POSTMODERNITY-DAVID-HARVEY.

Harvey, D. (2010). Between Space and Time: Reflections on the Geographical Imagination1. *Annals of the Association of American Geographers*, *80*(3), 418–434. https://doi.org/10.1111/J.1467-8306.1990.TB00305.X

He, Z., Huang, H., Choi, H., & Bilgihan, A. (2022). Building organizational resilience with digital transformation. *Journal of Service Management*. https://fowmedia.com/helping-your-employees-adapt-to-digital-transformation/ https://www.capgemini.com/wp-content/uploads/2019/10/Digital-Report-%E2%80%93-Emotional-Intelligence.pdf

Hee, Sidorova, & Van der Werf. (2013). *Business Process Modeling Using Petri Nets*. Academic Press.

Hepworth, M. E. (1990). Geography of the information economy. *NETCOM : Réseaux, Communication et Territoires. NETCOM*, *4*(1), 266–267. https://doi.org/10.3406/NETCO.1990.1055

He, Z., Wu, Q., Wen, L., & Fu, G. (2018). A process mining approach to improve emergency rescue processes of fatal gas explosion accidents in Chinese coal mines. *Safety Science*, 111.

Hide, L., & Pemberton, D. (2021). Mobilising Collections Storage to Deliver Wide-Ranging Strategic Objectives at the Sedgwick Museum. *Museum International*, *73*(1-2), 110–119. doi:10.1080/13500775.2021.1956753

Hilpert, U. (2021). Regional selectivity of innovative progress: Industry 4.0 and digitization ahead. *European Planning Studies*, *29*(9), 1589–1605. doi:10.1080/09654313.2021.1963047

Hopkin, J., & Rosamond, B. (2017). Post-truth Politics, Bullshit and Bad Ideas: 'Deficit Fetishism' in the UK. *New Political Economy*. Advance online publication. doi:10.1080/13563467.2017.1373757

Horizon Europe. (2022). https://cordis.europa.eu/

Horizon Europe. (2022). https://eismea.ec.europa.eu/programmes/european-innovation-ecosystems_en

Horizon Europe. (2022a). https://research-and-innovation.ec.europa.eu/funding/funding-opportunities/funding-programmes-and-open-calls/horizon-eur
ope_en

Hosiaisluoma, E. (2021). *ArchiMate Cookbook-Patterns & Examples*. Hosiaisluoma.

Hou, J., & Neely, A. (2018). Investigating risks of outcome-based service contracts from a provider's perspective. *International Journal of Production Research*, *56*(6), 2103–2115. doi:10.1080/00207543.2017.1319089

Hou, S. I., Charlery, S. A. R., & Roberson, K. (2014). Systematic literature review of Internet interventions across health behaviors. *Health Psychology and Behavioral Medicine*, *2*(1), 455–481. doi:10.1080/21642850.2014.895368 PMID:25750795

Huang, P. S., Paulino, Y. C., So, S., Chiu, D. K. W., & Ho, K. K. W. (2021). Editorial. *Library Hi Tech*, *39*(3), 693–695. doi:10.1108/LHT-09-2021-324

Huang, P.-S., Paulino, Y. C., So, S., Chiu, D. K. W., & Ho, K. K. W. (2022). Guest editorial: COVID-19 Pandemic and Health Informatics Part 2. *Library Hi Tech*, *40*(2), 281–285. doi:10.1108/LHT-04-2022-447

Human Rights Watch. (2019). *Nigeria: Widespread Violence Ushers in President's New Term*. Human Rights Watch. https://www.hrw.org/news/2019/06/10/nigeria-widespread-violence-ushers-presidents-new-term

Huysmans, P., De Bruyn, P., Benazeer, S., De Beuckelaer, A., De Haes, S., & Verelst, J. (2014). On the relevance of the modularity concept for understanding outsourcing risk factors. *Proceedings of the 47th Hawaii International Conference on System Sciences*.

Huysmans, P., De Bruyn, P., Benazeer, S., De Beuckelaer, A., De Haes, S., & Verelst, J. (2014). Understanding outsourcing risk factors based on modularity: The BSkyB case. *International Journal of IT/Business Alignment and Governance*, *5*(1), 50–66. doi:10.4018/ijitbag.2014010104

Iansiti, M., & Levien, R. (2004). The Keystone Advantage. Harvard Business School Press.

IBM. (2014). *Big Data & Analytics Maturity Model*. Retrieved April 13th, 2020, from https://www.ibmbigdatahub.com/blog/big-data-analytics-maturity-model

IBM. (2014). *Smart Service Oriented Architecture: Helping businesses restructure. Where are you on the Service Oriented Architecture adoption path?* IBM.

IBM. (2019). *Industry 4.0 and Cognitive Manufacturing. Architecture Patterns, Use Cases and IBM Solutions*. Retrieved April 13th, 2020, from https://www.ibm.com/downloads/cas/M8J5BA6R

ICOM-CC. (1984). *Definition of the profession*. Retrieved from https://www.icom-cc.org/en/definition-of-the-profession-1984

ICOM-CC. (2008). *Terminology for conservation.* Retrieved from https://www.icom-cc.org/en/terminology-for-conservation

IDC. (2015). *A Digital Transformation Maturity Model and Your Digital Roadmap.* Retrieved April 13th, 2020, from https://pdf4pro.com/amp/download?data_id=3c6d16&slug=a-digital-transformation-maturity-model-and-your-digital

IDC. (2020). *Worldwide Digital Transformation Strategies.* Retrieved April 13th, 2020, from https://www.idc.com/getdoc.jsp?containerId=IDC_P32570

IEEE. (2019). *Platforms, Present and Future XIV.* Retrieved April 13th, 2020, from https://cmte.ieee.org/futuredirections/2019/11/13/platforms-present-and-future-xiv/

Imbens, G. W., & Kolesár, M. (2016). Robust Standard Errors in Small Samples: Some Practical Advice. *The Review of Economics and Statistics, 98*(4), 701–712. https://doi.org/10.1162/REST_A_00552

IMD. (2020). *The digital business agility perspective.* Retrieved April 13th, 2020, from https://www.imd.org/contentassets/929fc49598cc47ba888277438bd963b6/tc044-16---digital-business-agility-pdf.pdf

Inderwildi, O., Zhang, X., Wang, X., & Kraft, M. (2020). The Impact of Intelligent Cyber-Physical Systems on the Decarbonization of Energy. *Energy and Environmental Science, 3.*

Inditex. (n.d.). *Nuestra historia.* https://www.inditex.com/es/quienes-somos/nuestra-historia

Inman, R. A., Moreira, P. A., Faria, S., Araújo, M., Cunha, D., Pedras, S., & Correia Lopes, J. (2022). An application of the transtheoretical model to climate change prevention: Validation of the climate change stages of change questionnaire in middle school students and their schoolteachers. *Environmental Education Research, 28*(7), 1003–1022. doi:10.1080/13504622.2021.1998382

Interbrand. (2021). *Best Global Brands 2021. The Decade of Possibility.* Interbrand Best Global Brands 2021Report.

International Institute for Democracy and Electoral Assistance. (2014). *A Brief Assessment Report on Electronic Voting and the 2014 Namibian General Elections.* Author.

International Telecommunication Union. (2018). *Measuring the Information Society Report 2018* (vol. 1). ITU. https://www.itu.int/en/ITU- Retrieved from D/Statistics/Documents/publications/misr2018/MISR-2018-Vol-1-E

Ishizaka, A., & Blakiston, R. (2012). The 18C's model for a successful long-term outsourcing arrangement. *Industrial Marketing Management, 41*(7), 1071–1080. doi:10.1016/j.indmarman.2012.02.006

Issa, A., Hatiboglu, B., Bilstein, A., & Buernhansl, T. (2018). Industrie 4.0 roadmap: Framework for digital transformation based on the concepts of capability maturity and alignment. *Procedia CIRP, 72.* Doi:10.1016/j.procir.2018.03.151

Jabangwe, R., Smite, D., & Hesbo, E. (2016). Distributed software development in an offshore outsourcing project: A case study of source code evolution and quality. *Information and Software Technology, 72,* 125–136. doi:10.1016/j.infsof.2015.12.005

Jacobides, M. G., Knudsen, T. R., & Augier, M. (2006). Benefiting from innovation: Value creation, value appropriation, and the role of industry architectures. *Research Policy, 35*(8), 1200–1221. doi:10.1016/j.respol.2006.09.005

Jacques, V. (2006). *International outsourcing strategy and competitiveness: Study on current outsourcing trends: IT, business processes, contact centers.* Publibook.

Janow, M. E., & Mavroidis, P. C. (2019). Digital trade, e-commerce, the WTO and regional frameworks. *World Trade Review, 18*(S1), S1–S7.

Jans, M., Van Der Werf, J. M., Lybaert, N., & Vanhoof, K. (2011). A business process mining application for internal transaction fraud mitigation. *Expert Systems with Applications, 38*(10), 13351–13359. doi:10.1016/j.eswa.2011.04.159

Janssenswillen, G., Depaire, B^., Swennen, M., Jans, M., & Vanhoof, K. (2018). bupaR: Enabling reproducible business process analysis. *Knowledge-Based Systems, 163.*

Janzen, D., & Saiedian, H. (2005). *Test-driven development concepts, taxonomy, and future direction.* IEEE.

Javed, A., & Charles, A. (2018). The importance of social cognition in improving functional outcomes in schizophrenia. *Frontiers in Psychiatry, 9*, 157. doi:10.3389/fpsyt.2018.00157 PMID:29740360

jayi, K. (2014). The ICT culture and transformation of Electoral Governance and politics in Africa: the challenges and prospects. *5th European Conference on African Studies.*

Jay, M. (2010). *The Virtues of Mendacity.* University of Virginia Press.

Jetten, J., Bentley, S. V., Crimston, C. R., Selvanathan, H. P., & Haslam, S. A. (2021). COVID-19 and social psychological research: A silver lining. *Asian Journal of Social Psychology, 24*(1), 34–36. doi:10.1111/ajsp.12465 PMID:33821140

Jiang, X., Chiu, D. K. W., & Chan, C. T. (2022). Application of the AIDA model in social media promotion and community engagement for small cultural organizations: A case study of the Choi Chang Sau Qin Society. In M. Dennis & J. Halbert (Eds.), *Community Engagement in the Online Space.* IGI Global.

Jiménez-Zazo, F., Romero-Blanco, C., Castro-Lemus, N., Dorado-Suárez, A., & Aznar, S. (2020). Transtheoretical model for physical activity in older adults: Systematic review. *International Journal of Environmental Research and Public Health, 17*(24), 9262. doi:10.3390/ijerph17249262 PMID:33322327

Jnr, B. A. (2020). Use of telemedicine and virtual care for remote treatment in response to COVID-19 pandemic. *Journal of Medical Systems, 44*(7), 1–9. PMID:32542571

Johansson, C. (2009). *Knowledge Maturity as Decision Support in Stage-Gate Product Development: A Case From the Aerospace Industry* [Doctoral Thesis]. Department of Applied Physics and Mechanical Engineering, Luleå University of Technology.

Johansson, P., Christian, J., Ola, I., Ola, T., & Larsson, T. (2020). *Take the knowledge path to support knowledge management in product/service systems.* Academic Press.

Johnson, R. B., & Christensen, L. B. (2004). *Educational research: Quantitative, qualitative, and mixed approaches.* Allyn and Bacon.

Jonkers, H., Band, I., & Quartel, D. (2012a). *ArchiSurance Case Study.* The Open Group.

Joppe, M. (2000). *The Research Process.* http://www.ryerson.ca/~mjoppe/rp.htm

Jorgensen, M., Mohagheghi, P., & Grimstad, S. (2017). Direct and indirect type of connection between type of contract and software project outcome. *International Journal of Project Management, 35*, 1573–1586.

Kalhori, S. R. N., Bahaadinbeigy, K., Deldar, K., Gholamzadeh, M., Hajesmaeel-Gohari, S., & Ayyoubzadeh, S. M. (2021). Digital health solutions to control the COVID-19 pandemic in countries with high disease prevalence: Literature review. *Journal of Medical Internet Research*, *23*(3), e19473. doi:10.2196/19473 PMID:33600344

Kalpokas, I. (2019). *A Political Theory of Post-Truth*. Springer Nature Publishers. doi:10.1007/978-3-319-97713-3

Kamal, M. M., Sivarajah, U., Bigdeli, A. Z., Missi, F., & Koliousis, Y. (2020). Servitization implementation in the manufacturing organisations: Classification of strategies, definitions, benefits and challenges. *International Journal of Information Management*, *50*. Advance online publication. doi:10.1016/j.ijinfomgt.2020.102206

Kamkari, K. (2022). Communication model of self-concept and Exercise self-efficacy on mental health of physical education staff in Iranian universities of medical sciences of Tehran. *Majallah-i Ulum-i Pizishki-i Razi*, *28*(11), 14–23.

Kane, G. C. (2017). 'Digital Transformation' Is a Misnomer. *MIT Sloan Management Review*. https://sloanreview.mit.edu/article/digital-transformation-is-a-misnomer/?og=Digital+Leadership+Tiled

Kane, G. C., Palmer, D., Phillips, A. N., Kiron, D., & Buckley, N. (2017). *Achieving Digital Maturity*. MIT Sloan Management Review and Deloitte University Press. https://sloanreview.mit.edu/projects/achieving-digital-maturity/

Kane, G. C., Phillips, A. N., Copulsky, J., & Andrus, G. (2019). How Digital Leadership Is(n't) Different. *MIT Sloan Management Review*, *60*(3), 34–39. https://www.proquest.com/scholarly-journals/how-digital-leadership-is-nt-different/docview/2207927776/se-2

Kappagantula, S. (2018). *Microservices vs SOA — Battle Between The Top Architectures*. Edureka. https://medium.com/edureka/microservices-vs-soa-4d71c5590fc6

Karimi-Alaghehband, F., & Rivard, S. (2012, December). Information technology outsourcing success: A model of dynamic, operational, and learning capabilities. *33rd International Conference on Information Systems (ICIS)*.

Kaur, Singh, Kumar, Gupta, & El-Latif. (2021). Secure and energy efficient-based E-health care framework for green internet of things. *IEEE Transactions on Green Communications and Networking*, *5*(3), 1223–1231.

Kavadias, S., Ladas, K., & Loch, C. (2016). The Transformative Business Model. *Harvard Business Review*.

Keane, J. (2018). *Post-Truth Politics and why the Antidote isn't simply 'Fact-Checking' and Truth*. https://theconversation.com/post-truth-politics-and-why-the-antidoteisnt-simply-fact-checking-and-truth-87364

Kebede, M., & Dumas, M. (2015). *Comparative evaluation of process mining tools*. University of Tartu.

Keith, B., Vitasek, K., Manrodt, K., & Kling, J. (2016). *Strategic Sourcing in the New Economy: Harnessing the Potential of Sourcing Business Models for Modern Procurement*. Springer.

Keyes, R. (2004). *The Post-Truth Era*. St Martin's Press.

Kiseleva, I., Karmanov, M., Korotkov, A., Kuznetsov, V., & Gasparian, M. (2018). Risk management in business: Concept, types, evaluation criteria. *Revista ESPACIOS*.

Kitsios, F., Kyriakopoulou, M., & Kamariotou, M. (2022). Exploring Business Strategy Modelling with ArchiMate: A Case Study Approach. MPDI. *Information (Basel)*, *2022*(13), 31. doi:10.3390/info13010031

Kıyak, A., & Bozkurt, G. (2020). A general overview to digital leadership concept. *Uluslararası Sosyal ve Ekonomik Çalışmalar Dergisi*, *1*(1), 84–95.

Kleppe, A., Warmer, J., & Bast, W. (2003). *MDA Explained: The Model Driven Architecture-Practice and Promise.* Addison-Wesley.

Knowledge Exchange and Fraunhofer. (2017). *Industry 4.0 maturity assessment.* Retrieved April 15th, 2020, from https://www.researchgate.net/profile/Alexander_Kermer-Meyer/publication/317720108_Industry_40_Maturity_Assessment/links/594a68b8aca2723195de5ed1/Industry-40-Maturity-Assessment.pdf

Koenig, J., Rustan, K., & Leino, M. (2016). *Programming Language Features for Refinement.* Stanford University. doi:10.4204/EPTCS.209.7

Kornilova, I. (2017). *DevOps4AI is a culture, not a role!* Medium. Retrieved January 2, 2018, from https://medium.com/@neonrocket/devops-is-a-culture-not-a-role-be1bed149b0

Koskela, L. (2007). *Test driven: practical tdd and acceptance tdd for java developers.* Manning Publications Co.

Kostoff, R. N., & Schaller, R. R. (2001). Science and technology roadmaps. *IEEE Transactions on Engineering Management, 48*(2), 132–143. doi:10.1109/17.922473

Kothandaraman, P., & Wilson, D. T. (2001). The Future of Competition: Value-Creating Networks. *Industrial Marketing Management, 30*(4), 379–389. doi:10.1016/S0019-8501(00)00152-8

Koudelia, N. (2011). *Acceptance test-driven development* [Master Thesis]. University of Jyväskylä, Department of Mathematical Information Technology, Jyväskylä, Finland.

Kowalkowski, C., & Kindström, D. (2014). Service innovation in product-centric firms: A multidimensional business model perspective. *Journal of Business and Industrial Marketing, 29*(2), 96–111. doi:10.1108/JBIM-08-2013-0165

Kozanoglu, D. C., & Abedin, B. (2020). Understanding the role of employees in digital transformation: Conceptualization of digital literacy of employees as a multi-dimensional organizational affordance. *Journal of Enterprise Information Management.*

KPMG. (2016). *Digital Readiness Assessment.* Retrieved April 13th, 2020, from https://www.future.consulting/en/foresight/trend-analyses/analyses/article/new-digital-readiness-assessment-by-2b-ahead-and-kpmg/

KPMG. (2019). *Strategic roadmap.* Retrieved April 13th, 2020, from: https://home.kpmg/xx/en/home/insights/2019/04/strategic-road-map.html

Kramer, T., Heinzl, A., & Neben, T. (2017, January). Cross-organizational software development: Design and evaluation of a decision support system for software component outsourcing. *Proceedings of the 50th Hawaii International Conference on System Sciences.* 10.24251/HICSS.2017.041

Kraus, S., Schiavone, F., Pluzhnikova, A., & Invernizzi, A. C. (2021). Digital transformation in healthcare: Analyzing the current state-of-research. *Journal of Business Research, 123,* 557–567. doi:10.1016/j.jbusres.2020.10.030

Kreiss, S. B. (2014). Digitalisation and Digitization. Encyclopedia of Communication Theory and Philosophy, 3-15.

Krimmer, R., & Schuster, R. (2008). *The E-Voting Readiness Index.* Competence Centre for Electronic Voting and Participation.

Kshetri, N. (2014). Developing Successful Entrepreneurial Ecosystems: Lessons from a Comparison of an Asian Tiger and a Baltic Tiger. *Baltic Journal of Management, 9*(3), 330–356. doi:10.1108/BJM-09-2013-0146

Kumar, S., & Walia, E. (2011). Analysis of Electronic Voting System in Various Countries. *International Journal on Computer Science and Engineering*, *3*(5).

Kupiainen, J. (2006). Translocalisation over the Net: Digitalisation, information technology and local cultures in Melanesia. *E-Learning and Digital Media*, *3*(3), 279–290. doi:10.2304/elea.2006.3.3.279

Kursan Milaković, I. (2021). Purchase experience during the COVID-19 pandemic and social cognitive theory: The relevance of consumer vulnerability, resilience, and adaptability for purchase satisfaction and repurchase. *International Journal of Consumer Studies*, *45*(6), 1425–1442. doi:10.1111/ijcs.12672 PMID:33821146

Kutin, A., Dolgov, V., & Sedykh, M. (2016). Information links between product life cycles and production system management in designing of digital manufacturing. *Procedia CIRP*, *41*, 423–426. doi:10.1016/j.procir.2015.12.126

Kwiotkowska, A. (2022). The interplay of resources, dynamic capabilities and technological uncertainty on digital maturity. *Dynamic Capabilities*, 1-17. doi:10.29119/1641-3466.2022.155.15

Lackey, R. T. (1998). Seven pillars of ecosystem management. *Landscape and Urban Planning*, *40*(1-3), Page21–30. doi:10.1016/S0169-2046(97)00095-9

Lahoti, S. (2019). *Defining REST and its various architectural styles*. packtpub. https://hub.packtpub.com/defining-rest-and-its-various-archi tectural-styles/

Lam, A. H. C., Ho, K. K. W., & Chiu, D. K. W. (2022). (in press). Instagram for student learning and library promotions? A quantitative study using the 5E Instructional Model. *Aslib Journal of Information Management*. Advance online publication. doi:10.1108/AJIM-12-2021-0389

Lambert, D. M., Emmelhainz, M. A., & Gardner, J. T. (1999). Building successful logistics partnerships. *Journal of Business Logistics*, *20*(1), 165–181.

Lambrinoudakis, C., & Kokolakis, S. (2002). Functional Requirements for a Secure Electronic Voting System. *Conference Paper*.

Landscheidt, S., & Kans, M. (2016). Automation practices in wood product industries: Lessons learned, current practices and future perspectives. In *The 7th Swedish Production Symposium SPS*. Lund University.

Langlois, R. N. (2002). Modularity in technology and organization. *Journal of Economic Behavior & Organization*, *49*(1), 19–37. doi:10.1016/S0167-2681(02)00056-2

Lara, L., & Mas, J. (2018). *Por qué unas tiendas venden y otras no en la era digital: Claves del éxito del New Retail*. Libros de Cabecera.

Launchmetrics Content Team. (2019). *La evolución en marketing digital de las marcas de moda y lujo*. Launchmetrics. https://www.launchmetrics.com/es/recursos/blog/la-evolucion-digital-de-las-marcas-de-lujo

Law, Ip, Gupta, & Geng. (2021). *Managing IoT and Mobile Technologies with Innovation, Trust, and Sustainable Computing*. CRC Press.

Lazar, I., Motogna, S., & Parv, B. (2010). Behaviour-Driven Development of Foundational UML Components. Department of Computer Science, Babes-Bolyai University, Cluj-Napoca, Romania. doi:10.1016/j.entcs.2010.07.007

Leandro, G., Miura, D., Safanelli, J., Borges, R., & Cl'audia, M. (2022). *Analysis of Stroke Assistance in Covid-19 Pandemic by Process Mining Techniques* (Vol. 294). doi:10.3233/SHTI220394

Lee, J. Y. (2021). A study on metaverse hype for sustainable growth. *International Journal of Advanced Smart Convergence, 10*(3), 72-80.

Lee, M. Y. H. (2015, July 8). Donald Trump's False Comments connecting Mexican immigrant and Crime. *The Washington Post.* https://www.washingtonpost.com/news/fact-checker/wp/2015/07/08/donald-trumps-false-comments-connecting-mexican-immigrant
s-and-crime/

Lee, J.-N. (2001). The impact of knowledge sharing, organizational capability and partnership quality on IS outsourcing success. *Information & Management, 38*(5), 323–335. doi:10.1016/S0378-7206(00)00074-4

Legislative Council Panel on Home Affairs. (n.d.). *Progress Report on Enhancement of Programming, Audience Building and Collection Management of Public Museums.* Retrieved from https://www.legco.gov.hk/yr17-18/english/panels/ha/papers/ha20171221cb2-553-3-e.pdf

Leisure and Cultural Services Department. (2017). *LCSD Museums Collection Management System.* Retrieved from https://mcms.lcsd.gov.hk/Search/search/enquire?timestamp=1639643297962

Leisure and Cultural Services Department. (2020a). *LCSD Annual Report 2019-2020.* Retrieved from https://www.lcsd.gov.hk/dept/annualrpt/2019-20/en/cultural-services/museums

Leisure and Cultural Services Department. (2020b). *LCSD Customer Appreciation Card.* Retrieved from https://www.lcsd.gov.hk/en/aboutlcsd/forms/appr_card.html

Leisure and Cultural Services Department. (2021). *About Us.* Retrieved from https://www.museums.gov.hk/en_US/web/portal/about-us.html#

Lemos, Sabino, Lima, & Oliveira. (2011). Using process mining in software development process management: A case study. In *2011 IEEE International Conference on Systems, Man, and Cybernetics* (pp. 1181–1186). IEEE. 10.1109/ICSMC.2011.6083858

Leonardi, P. M., Huysman, M., & Steinfield, C. (2013). Enterprise social media: Definition, history, and prospects for the study of social technologies in organizations. *Journal of Computer-Mediated Communication, 19*(1), 1–19. doi:10.1111/jcc4.12029

Leung, H. F., Chiu, D. K. W., & Hung, P. C. (Eds.). (2010). *Service Intelligence and Service Science: Evolutionary Technologies and Challenges: Evolutionary Technologies and Challenges.* IGI Global.

Levin-Zamir, D., & Bertschi, I. (2019). Media health literacy, eHealth literacy and health behaviour across the lifespan: Current progress and future challenges. International handbook of health literacy, 275.

Leyh, C., Bley, K., Schäffer, T., & Forstenhäusler, S. (2016). SIMMI 4.0 - a maturity model for classifying the enterprise-wide it and software landscape focusing on Industry 4.0. In *2016 Federated Conference on Computer Science and Information Systems (FedCSIS).* IEEE.

Li & de Carvalho. (2019). Process Mining in Social Media: Applying Object-Centric Behavioral Constraint Models. *IEEE Access*, 1–1.

Li, L. (2014). Supply Chain Management and Strategy. *Managing Supply Chain and Logistics*, 3-36.

Li, A. S. W., Figg, G., & Schüz, B. (2019). Socioeconomic status and the prediction of health promoting dietary behaviours: A systematic review and meta-analysis based on the theory of planned behaviour. *Applied Psychology. Health and Well-Being*, *11*(3), 382–406. doi:10.1111/aphw.12154 PMID:30884154

Liberatore, M. J., & Miller, T. (2016). Outbound logistics performance and profitability: Taxonomy of manufacturing and service organisation. *Business and Economics Journal*, *7*(2), 1000221.

Li, D., & Mishra, N. (2020). Engaging Suppliers for Reliability Improvement under Outcome based Compensations. *Omega*. Advance online publication. doi:10.1016/j.omega.2020.102343

Liebrecht, C. (2020). *Entscheidungsunterstutzung fur den Industrie 4.0-Methodeneinsatz: Strukturierung, Bewertung and Ablietung von Implementierungsreihenfolgen. Zugl: Karlsruhe, Diss, 2020.* Shaker Verlag G.

Liebrecht, C., Kandler, M., Lang, M., Schaumann, S., Stricker, N., Wuest, T., & Lanz, G. (2021). Decision support for the implementation of Industry 4.0 methods: Toolbox, Assessment and Implementation Sequences for Industry 4.0. *Journal of Manufacturing Systems*, *58*(3), 412–430. doi:10.1016/j.jmsy.2020.12.008

Liebrecht, C., Schaumann, S., Zeranski, D., Antonszkiewicz, A., & Lanza, G. (2018). Analysis of Interactions and Support of Decision Making for the Implementation of Manufacturing Systems 4.0 Methods. *10th CIRP Conference on Industrial Product-Service Systems, Procedia CIRP*, 161-166. DOI: 10.1016/j.procir.2018.04.005

Liinamaa, J., Viljanen, M., Hurmerinta, A., Ivanova-Gonge, M., Luotola, H., & Gustafsson, M. (2016). Performance-Based and Functional Contracting in Value Based Solution Selling. *Industrial Marketing Management*, *59*, 37–49. doi:10.1016/j.indmarman.2016.05.032

Li, K. K., & Chiu, D. K. W. (2021). A Worldwide Quantitative Review of the iSchools' Archival Education. *Library Hi Tech*. Advance online publication. doi:10.1108/LHT-09-2021-0311

Lin, C. Y., Imani, V., Majd, N. R., Ghasemi, Z., Griffiths, M. D., Hamilton, K., Hagger, M. S., & Pakpour, A. H. (2020). Using an integrated social cognition model to predict COVID-19 preventive behaviours. *British Journal of Health Psychology*, *25*(4), 981–1005. doi:10.1111/bjhp.12465 PMID:32780891

Lindgren, E., & Münch, J. (2015). Software Development as an Experiment System: A Qualitative Survey on the State of the Practice. XP 2015: Agile Processes in Software Engineering and Extreme Programming. *International Conference on Agile Software Development*.

Ling, R., & Typhina, E. (2016). Mobile communication. In A. de S. e Silva (Ed.), *Dialouges on Mobile Communication* (p. 222). Routledge. https://www.routledge.com/Dialogues-on-Mobile-Communication/Silva/p/book/9781138691582

Li, S. M., Lam, A. H. C., & Chiu, D. K. W. (2023). Digital transformation of ticketing services: A value chain analysis of POPTICKET in Hong Kong. In R. Pettinger, B. B. Gupta, A. Roja, & D. Cozmiuc (Eds.), *Handbook of Research on the Digital Transformation Digitalization Solutions for Social and Economic Needs*. IGI Global.

Liu, A. (2022). *Rumbaugh, Booch and Jacobson Methodologies*. Opengenus. https://iq.opengenus.org/rumbaugh-booch-and-jacobson-methodologies/

Liu, K., & Chen, M. H. (2017). Research on Innovation Ecosystem Based on Green Management. *Proceedings of International Symposium on Green Management and Local Government's Responsibility*, 174-178.

Liu, S., Chiang, Y. T., Tseng, C. C., Ng, E., Yeh, G. L., & Fang, W. T. (2018). The theory of planned behavior to predict protective behavioral intentions against PM2. 5 in parents of young children from urban and rural Beijing, China. *International Journal of Environmental Research and Public Health*, *15*(10), 2215. doi:10.3390/ijerph15102215 PMID:30309043

Llanos, P. (2019). Zara transforma tienda online y cambia la manera de vender sus colecciones. *Revista ELLE*. https://www.elle.com/es/moda/noticias/a28962635/zara-tienda-online-nueva/

Lockie, S. (2017). Post-truth Politics and the Social Sciences. *Environmental Sociology*, *3*(1), 1–5. doi:10.1080/23251 042.2016.1273444

Logan, M. S. (2000). Using agency theory to design successful outsourcing relationships. *International Journal of Logistics Management*, *11*(2), 21–32. doi:10.1108/09574090010806137

Lo, P., Chan, H. H. Y., Tang, A. W. M., Chiu, D. K. W., Cho, A., Ho, K. K. W., See-To, E., He, J., Kenderdine, S., & Shaw, J. (2019). Visualising and Revitalising Traditional Chinese Martial Arts – Visitors' Engagement and Learning Experience at the 300 Years of Hakka KungFu. *Library Hi Tech*, *37*(2), 273–292. doi:10.1108/LHT-05-2018-0071

Lubis, M., Kartiwi, M., & Durachman, Y. (2017). *Assessing Privacy and Readiness of Electronic Voting System in Indonesia*. International Islamic University Malaysia and Syarif Hidayatullah State Islamic University. doi:10.1109/CITSM.2017.8089242

Lund, S., & Manyika, J. (2016). *How digital trade is transforming globalisation*. International Centre for Trade and Sustainable Development (ICTSD). http://e15initiative.org/wp-content/uploads/2015/09/E15-Digital-Lund-and-Manyika.pdf

Lusch & Harvey. (1994). The case for an off -balance- sheet controller. *Sloan Management Review*, (2), 101–105.

Machlis, G., Force, J. E., & Burch, W. R. Jr. (1997). The human ecosystem. 1. The human ecosystem as an organizing concept in ecosystem management. *Society & Natural Resources*, *10*(4), 347–367. doi:10.1080/08941929709381034

Magretta. (2002). Why Business Models Matter. *Harvard Business Review*.

Maguire, J. (2020). *Microservices vs SOA vs API Comparison*. Devteam. https://www.devteam.space/blog/microservices-vs-soa-and-api-comparison/

Mahendrawathi, E. R., Zayin, S. O., & Pamungkas, F. J. (2017). ERP post implementation review with process mining: A case of procurement process. *Procedia Computer Science*, *124*, 216–223. doi:10.1016/j.procs.2017.12.149

Mahmud, K., Sahoo, A., K., Fernandez, E., Sanjeevikumar, P., & Holm-Nielsen, J. (2020). Computational Tools for Modeling and Analysis of Power Generation and Transmission Systems of the Smart Grid. *IEEE Systems Journal*, *14*(3). doi:10.1109/JSYST.2020.2964436

Mair, P. (2013). *Ruling the Void. The Hollowing of Western Democracy*. Verso.

Maiti, M., & Kayal, P. (2017). Digitization: Its Impact on Economic Development & Trade. *Asian Economic and Financial Review*, *7*(6), 541–549. doi:10.18488/JOURNAL.AEFR.2017.76.541.549

Makarava, Y. (2011). *Critical Assessment of the Relationship between E-governance and Democracy*. Mid Sweden University.

Malagón, P. (2020). Así es la nueva App de Zara. *Libertad Digital*. https://www.libremercado.com/2020-09-16/app-nueva-zara-inditex-6660510/

Mankins, J. C. (1995). *Technology Readiness Levels*. NASA Advanced Concepts Office.

Mannaert, H., Verelst, J., & De Bruyn, P. (2016). *Normalized systems theory: From foundations for evolvable software toward a general theory for evolvable design*. Koppa.

Manning-Franklin, A. (2022). *Functional Domain Driven Design: Simplified*. Antman Writers Series. https://antman-does-software.com/functional-domain-driven-design-simplified

Mansuroğlu, S., & Kutlu, F. Y. (2022). The Transtheoretical Model based psychoeducation's effect on healthy lifestyle behaviours in schizophrenia: A randomized controlled trial. *Archives of Psychiatric Nursing, 41*, 51–61. doi:10.1016/j.apnu.2022.07.018 PMID:36428075

Manyika, J., Bughin, J., Lund, S., Nottebohm, O., Poulter, D., Jauch, S., & Ramaswamy, S. (2014). *Global flows in a digital age: How trade, finance, people, and data connect the world economy*. McKinsey Global Institute. https://scholar.google.com/scholar?cluster=3193246601035535385&hl=en&oi=scholarr

Mapelsden, D., Hosking, J., & Grundy, J. (2002). *Design Pattern Modelling and Instantiation using DPML*. Department of Computer Science, University of Auckland.

Marashi, M. Y., Nicholson, E., Ogrodnik, M., Fenesi, B., & Heisz, J. J. (2021). A mental health paradox: Mental health was both a motivator and barrier to physical activity during the COVID-19 pandemic. *PLoS One, 16*(4), e0239244. doi:10.1371/journal.pone.0239244 PMID:33793550

Marcella, R., & Davies, S. (2004). The use of customer language in international marketing communication in the Scottish food and drink industry. *European Journal of Marketing, 38*(11/12), 1382–1395. https://doi.org/10.1108/03090560410560155

Maroufkhani, P., & Ralf Wagner, R. (2018). Entrepreneurial ecosystems: A systematic review. *Journal of Enterprising Communities: People and Places in the Global Economy*.

Martin, N. (2020). *Data Quality in Process Mining*. Academic Press.

Martínez Barreiro, A. (2008). Hacia un nuevo sistema de la moda. El modelo Zara. *Revista Internacional de Sociologia, 66*(51), 105–122. doi:10.3989/ris.2008.i51.111

Martorell, C. (2009). Y ahora pasamos a publicidad...si usted quiere. El advertainment como alternativa al modelo de comunicación basado en la interrupción. In *Actas del I Congreso Internacional Brand Trends*. Dpto. de Comunicación Audiovisual, Publicidad y Tecnología de la Información. CEU Universidad Cardenal Herrera, Alfara del Patriarca. Recuperado de https://goo.gl/xibOeQ

Marty, E. M., & Moore, J. (2000). *Politics Religion and the Common Good*. Jesse-Bass Publisher.

Marushchak, L., Pavlykivska, O., Khrapunova, Y., Kostiuk, V., & Berezovska, L. (2021). The Economy of Digitalization and Digital Transformation: Necessity and Payback. *2021 11th International Conference on Advanced Computer Information Technologies (ACIT)*, 305–308. 10.1109/ACIT52158.2021.9548529

Maryniarczyk, A. (2016). Rationality and Finality of the World of Persons and Things. *Polskie Towarzyztwo Tomasza z Akwinu*.

Mason, C., & Brown, R. (2014). Entrepreneurial Ecosystems and Growth Oriented Entrepreneurship, Conference: Entrepreneurial Ecosystems and Growth Oriented Entrepreneurship. OECD LEED Programme and the Dutch Ministry of Economic Affairs on Entrepreneurial Ecosystems and Growth Oriented Entrepreneurship, The Hague, Netherlands.

Maurya, A. (2012). *Running Lean: Iterate from Plan A to a Plan That Works*. Retrieved on August 20th, 2022, from https://books.google.ro/books?id=j4hXPn233UYC&redir_esc=y

Maza-Ortega, J., M., Acha, E., Garcia, S., & Gómez-Expósito, A. (2017). Overview of power electronics technology and applications in power generation transmission and distribution. *Journal of Modern Power Systems and Clean Energy, 5*(4), 499 – 514. doi:10.1007/s40565-017-0308-x

McAfee. (2006). *Enterprise 2.0, Version 2.0*. Retrieved on August 20th, 2022, https://andrewmcafee.org/2006/05/enterprise_20_version_20/

McGrath, R. G., & MacMillan, I. C. (1995). Discovery Driven Planning. *Harvard Business Review, 73*(4), 44–54.

McGrath, R. G., & MacMillan, I. C. (2009). *Discovery Driven Growth: a Breakthrough Process to Reduce Risk and Seize Opportunity*. Harvard Business Publishing.

McInery, R. (Ed.). (1998). *Aquinas, T. Selected Writings*. Penguin Books.

McIntyre, A. (2006). *Ethics and Politics: Selected Essays* (Vol. 2). Cambridge University Press. doi:10.1017/CBO9780511606670

McIntyre, A. (2007). *After Virtue* (3rd ed.). University of Notre Dame Press.

McIntyre, L. (2019). *Post-Truth*. The MIT Press.

McKinsey. (2016). Industry 4.0 at McKinsey's model factories. Retrieved April 14th, 2020, from-http://sf-eu.net/wp-content/uploads/2016/08/mckinsey-2016-industry-4.0-at-mckinseys-model-factories-en.pdf

McKinsey. (2017). *A roadmap for a digital transformation*. Retrieved on June 30th, 2021, from: https://www.mckinsey.com/industries/financial-services/our-insights/a-roadmap-for-a-digital-transformation

McKinsey. (2018). *Unlocking success in digital transformations*. Retrieved April 13th, 2020, from https://www.mckinsey.com/business-functions/organization/our-insights/unlocking-success-in-digital-transformations

McKinsey. (2019). *Hannover Messe 2019: The 3 P's of Industry 4.0*. Retrieved April 13th, 2020, from https://www.mckinsey.com/business-functions/operations/our-insights/operations-blog/hannover-messe-2019-the-3-ps-of-industry-40

McKinsey. (2020). *Digital 20/20*. Retrieved on June 30th, 2021, from: https://www.mckinsey.com/business-functions/mckinsey-digital/how-we-help-clients/digital-2020/overview

McKinsey. (2021). *Five Fifty: The skillful corporation*. Available online at https://www.mckinsey.com/business-functions/mckinsey-accelerate/our-insights/five-fifty-the-skillful-corporation

McLee, L., Luke, J., Ong, A., & University of Washington. Museology. (2018). *The Collections Connection: Understanding the Attitudes of Participants in Behind-the-Scenes Museum Tours*. Museology Master of Arts Theses, University of Washington.

Mearsheimer, J. J. (2011). *Why Leaders Lie: The Truth About Lying in International Politics*. Oxford Press.

Medlock, S., & Wyatt, J. C. (2019). Health behaviour theory in health informatics: Support for positive change. *Studies in Health Technology and Informatics*, *263*, 146–158. PMID:31411160

Meffert, J., Mohr, N., & Richter, G. (2020). *How German "Mittelstand" copes with CO-VID-19 challenges.* https://www.mckinsey.com/business-functions/mckinsey-digital/our-insights/how-the-german-mittelstand-is-mastering-the-co vid-19-crisis

Mehta, N., & Mehta, A. (2010). It takes two to tango: How relational investments improve IT outsourcing partnerships. *Communications of the ACM*, *53*(2), 160–164. doi:10.1145/1646353.1646393

Meijers, H. (2013). Does the internet generate economic growth, international trade, or both? *International Economics and Economic Policy*, *11*(1), 137–163. https://doi.org/10.1007/S10368-013-0251-X

Meltzer, J. P. (2016). Maximizing the Opportunities of the Internet for International Trade. *ICTSD and World Economic Forum*, 35. doi:10.1109/APEC.2015.7104315

Memoria anual inditex 2020 y Estado de Información no financiera del Grupo Inditex. (2020). https://www.inditex.com

Méndiz Noguero, A., Regadera González, E., & Pasillas Salas, G. (2018). Valores y storytelling en los fashion film. El caso de Tender Stories (2014-2017), de Tous. *Revista de Comunicación*, *17*(2), 316–334. doi:10.26441/RC17.2-2018-A14

Meng, Y., Chu, M. Y., & Chiu, D. K. W. (2022). (in press). The impact of COVID-19 on museums in the digital era: Practices and Challenges in Hong Kong. *Library Hi Tech*. Advance online publication. doi:10.1108/LHT-05-2022-0273

Methodology for Classification and Market Readiness. (2018). *Methodology for the Classification of Projects/ Services and Market Readiness.* https://cyberwatching.eu/sites/default/files/D2.3%20Methodol ogy%20for%20the%20classification%20of%20projects%20and%20mar ket%20readiness_0.pdf

Mettler, T. (2009). *A Design Science Research Perspective on Maturity Models in Information Systems.* Working Paper. Institute of Information Management, University of St. Gallen.

Mettler, T. (2011). Maturity Assessment Models: A Design Science Research Approach. *International Journal of Society Systems Science*, *3*(1-2).

Meyers, E. M., Erickson, I., & Small, R. V. (2013). Digital literacy and informal learning environments: An introduction. *Learning, Media and Technology*, *38*(4), 355–367. doi:10.1080/17439884.2013.783597

Microsoft. (2016a). *Code Snippets for Design Patterns.* CodePlex.

Microsoft. (2016b). *Model-View-Controller.* https://msdn.microsoft.com/en-us/library/ff649643.aspx

Mielli, F., & Bulanda, N. (2019). Digital Transformation: Why Projects Fail, Potential Best Practices and Successful Initiatives. *2019 IEEE-IAS/PCA Cement Industry Conference (IAS/PCA)*, 1-6. doi:10.1109/CITCON.2019.8729105

Miguel, P. A. C. (2005). Modularity in product development: A literature review towards a research agenda. *Product: Management & Development*, *3*(2), 165–174.

Mikkola, J. H. (2003). Modularity, component outsourcing, and inter-firm learning. *R & D Management*, *33*(4), 439–454. doi:10.1111/1467-9310.00309

Ministerul Educatiei si Cercetarii. (2021). *Registrul Entidin Infrastructurtatilor acreditate si Autorizate Provizoriu din Infrastructura de Inovare si Transfer Tehnologic.* Retrieved on October 26th, 2022, from https://www.research.gov.ro/uploads/sistemul-de-cercetare/in frastructuri-de-cercetare/infrastructura-de-inovare-si-trans fer-tehnologic/2021/registru-entitati-de-inovare-si-transfer -tehnologic-ianuarie-2021.pdf

Mintzberg, H. (1979). *The structuring of organizations.* Prentice-Hall.

Mirosa de Villalobos, R. (2021). Bienvenidos a la era de la venta omnicanal. *Harvard Business Review, 308.* Recuperado de https://www.harvard-deusto.com/bienvenidos-a-la-era-de-la-ve nta-omnicanal

Mitrou, L., Gritzalis, D., Katsikas, S., & Quirchmayr, G. (2009). *E-Voting: Constitutional and Legal Requirements and Their Technical Implications.* Academic Press.

Mittal, S., Romero, D., & Wuest, T. (2018). Towards a Smart Manufacturing Model for SMEs. AICT, 536(2), 155-163.

Mokodir, P. E. (2011). *E-voting readiness in Kenya: A case study of Nairobi County.* University of Nairobi School of Computing and Informatics.

Moore, J. (2014). *Java programming with lambda expressions-A mathematical example demonstrates the power of lambdas in Java 8.* https://www.javaworld.com/chapter/2092260/java-se/java-progr amming-with-lambda-expressions.html

Moreland, C. (2006). *An Introduction to the OMG Systems Modeling Language (OMG SysML).* Object Management Group. ARTiSAN Software Tools.

Morris, R. (2009). *The fundamentals of product design.* AVA Publishing.

Mourtzis, D. (2020). Simulation in the design and operation of manufacturing systems: State of the art and new trends. *Internaional Journal of Production Resources, 58,* 1927–1949. doi:10.1080/00207543.2019.1636321

MPSINC. (2021). *Progamming Languages Conversion.* MPSINC. http://www.mpsinc.com/index.html

Murawski, M., & Bick, M. (2017). Digital competences of the workforce–a research topic? *Business Process Management Journal, 23*(3), 721–734. doi:10.1108/BPMJ-06-2016-0126

Mutibwa, D., Hess, A., & Jackson, T. (2020). Strokes of serendipity: Community co-curation and engagement with digital heritage. *Convergence (London, England), 26*(1), 157–177. doi:10.1177/1354856518772030

Mutz, M., Müller, J., & Reimers, A. K. (2021). Use of digital media for home-based sports activities during the COVID-19 pandemic: Results from the German SPOVID survey. *International Journal of Environmental Research and Public Health, 18*(9), 4409. doi:10.3390/ijerph18094409 PMID:33919180

Myers, M. D., & Newman, M. (2007). The qualitative interview in IS research: Examining the craft. *Information and Organization, 17*(1), 2–26. doi:10.1016/j.infoandorg.2006.11.001

Naderifar, V., Sahran, S., & Shukur, Z. (2019). A review on conformance checking technique for the evaluation of process mining algorithms. *TEM Journal, 8*(4), 1232.

Nadler, D. A., & Tushman, M. L. (1999). The organization of the future: Strategic imperatives and core competencies for the 21st century. *Organizational Dynamics, 28*(1), 45–60. doi:10.1016/S0090-2616(00)80006-6

Nagpal, P., & Lyytinen, K. (2010, December). Modularity, information technology outsourcing success, and business performance. *International Conference on Information Systems*.

Nambisan, P., & Watt, J. H. (2011). Managing customer experiences in online product communities. *Journal of Business Research*, *64*(8), 889–895. doi:10.1016/j.jbusres.2010.09.006

Nanda, A., Xu, Y., & Zhang, F. (2021). How would the COVID-19 pandemic reshape retail real estate and high streets through acceleration of E-commerce and digitalization? *Journal of Urban Management*, *10*(2), 110–124. doi:10.1016/j.jum.2021.04.001

NASA. (2022). *Appendix G: Technology Assessment/Insertion*. Accesat în iunie 2022: https://www.nasa.gov/seh/appendix-g-technology-assessmentinsertion

Naslund, J. A., Aschbrenner, K. A., Kim, S. J., McHugo, G. J., Unützer, J., Bartels, S. J., & Marsch, L. A. (2017). Health behavior models for informing digital technology interventions for individuals with mental illness. *Psychiatric Rehabilitation Journal*, *40*(3), 325–335. doi:10.1037/prj0000246 PMID:28182469

National Election Board of Ethiopia. (n.d.). In *Wikipedia, the Free Encyclopaedia*. https://en.wikipedia.org/w/index.php?title=National_Election_Board_of_Ethiopia&oldid=791863032

Nauman, A. B., Aziz, R., & Ishaq, A. F. M. (2009). Information system development failure and complexity: A case study. In M. G. Hinter (Ed.), *Selected reading on strategic information systems* (pp. 251–275). IGI Publications. doi:10.4018/978-1-60566-090-5.ch017

Nayab, N. (2010). *The Difference Between CMMI vs CMM. Bright Hub PM*. Retrieved April 13th, 2020, from https://www.brighthubpm.com/certification/69744-cmmi-vs-cmm-which-is-better/

Netapsys. (2016). *Les ESB, exemple particulier de Mule ESB*. http://blog.netapsys.fr/les-esb-exemple-particulier-de-mule-esb

Neugebauer, R. (2019). *Digital Transformation* (1st ed.). Springer-Verlag GmbH Germany. doi:10.1007/978-3-662-58134-6

Neumann, G. (2002). Programming Languages in Artificial Intelligence. In Encyclopedia of Information Systems. Academic Press.

Ng, T. C. W., Chiu, D. K. W., & Li, K. K. (2021). Motivations of choosing archival studies as major in the i-School: Viewpoint between two universities across the Pacific Ocean. *Library Hi Tech*. Advance online publication. doi:10.1108/LHT-07-2021-0230

Nickkhou, S., Taghizadeh, K., & Hajiyakhali, S. (2016). Designing a portfolio management maturity model. *Procedia: Social and Behavioral Sciences*, *226*, 318–325.

Nimit, J. (2014). *Model First Approach in ASP.Net MVC 5. C Sharpcorner*. https://www.c-sharpcorner.com/UploadFile/4b0136/model-first-approach-in-Asp-Net-mvc-5/

Ntulo, G., & Otike, J. (2013). *E-Government: Its Role, Importance, and Challenges*. School of Information Sciences Moi University Eldoret.

Nuseibeh, B., & Easterbrook, S. (2000). *Requirements Engineering: A Roadmap. Department of Computing Department of Computer Science*. Imperial College.

O'connor, P. J., Martin, B., Weeks, C. S., & Ong, L. (2014). Factors that influence young people's mental health help-seeking behaviour: A study based on the Health Belief Model. *Journal of Advanced Nursing*, *70*(11), 2577–2587. doi:10.1111/jan.12423 PMID:24720449

O'Meara, M. (2013). *Survey and Analysis of E-Voting Solutions*. Master in Computer Science Trinity College Dublin.

O'Shea, C. (2008). *Así es Amancio Ortega, el hombre que creó Zara*. La Esfera de los Libros.

OECD (2022), Gross domestic spending on R&D (indicator). doi:10.1787/d8b068b4-en

OECD. (2018a). *Oslo Manual 2018. Guidelines for Collecting, Reporting and Using Data on Innovation*. Accesat în iunie 2022, de la https://www.oecd-ilibrary.org/docserver/9789264304604-en.pdf ?expires=1656233036&id=id&accname=guest&checksum=4CCC30AA4DC CE4B7811A0F98517C9292

OECD. (2018b). *R&D intensity by industry: Business enterprise expenditure on R&D as a share of gross valued added, 2018 (or nearest year)*. https://www.oecd-ilibrary.org/social-issues-migration-health /r-d-intensity-by-industry-business-enterprise-expenditure-o n-r-d-as-a-share-of-gross-valued-added-2018-or-nearest-year_ f9722f58-en

OECD. (2020). *Coronavirus (COVID-19): SME policy responses*. OECD Policy Responses to Coronavirus (COVID-19). https://www.oecd.org/coronavirus/policy-responses/coronaviru s-covid-19-sme-policy-responses-04440101/

OECD. (2020). *Research and development statistics*. Retrieved April 13th, 2020, from: https://www.oecd.org/sti/inno/researchanddevelopmentstatisti csrds.html

Oguejiofor, O. O. (2018). Advancing Electronic Voting Systems in Nigeria's Electoral Process. *Afe Babalola University: J. of Sust. Dev. Law & Policy, 9*(2).

Onwe, S., Nwogbaga, D., & Ogbu, M. (2015). *Effects of Electoral Fraud and Violence on Nigeria Democracy: Lessons from 2011 Presidential Election*. Academic Press.

Op't Land, M. (2008). *Applying architecture and ontology to the splitting and allying of enterprises* [Doctoral dissertation, Delft University of Technology]. Gildeprint.

Open Innovation Community. (2017). *Open Innovation Community*. Retrieved on August 20th, 2022, http://openin-novation.net/

Oracle. (2002). *Core J2EE Patterns - Data Access Object*. Oracle. https://www.oracle.com/technetwork/java/dataaccessobject-138 824.html

Oracle. (2016a). *API Management in 2026*. Oracle.

Oracle. (2016b). *Evolution and Generations of API Management*. Oracle.

Ortega, A., & Mera, R. (2019). Plan de comunicación y marketing de Zara - Case Study de The Apartment. *The Apartment*. Retrieved https://www.theapartment.es/blog/plan-de-comunicacion-y-mark e ting-de-zara/

Ortegón Cortázar, L. (2014). *Gestión de Marca. Conceptualización, diseño, registro, construcción y evaluación*. Editorial Politécnico Grancolombiano. Recuperado de https://alejandria.poligran.edu.co/handle/10823/798

Orton, J. D., & Weick, K. E. (1990). Loosely coupled systems: A reconceptualization. *Academy of Management Review*, *15*(2), 203–223. doi:10.2307/258154

Oshri, I., Kotlarsky, J., & Willcocks, L. (2015). *The handbook of global outsourcing and offshoring*. Palgrave Macmillan. doi:10.1057/9781137437440

Osibanjo, O., & Nnorom, I. C. (2008). Material flows of mobile phones and accessories in Nigeria: Environmental implications and sound end-of-life management options. *Environmental Impact Assessment Review*, *28*(2-3), 198–213. https://doi.org/10.1016/j.eiar.2007.06.002

Osterwalder, A. (2010). *Business Model Canvas*. Retrieved on August 20th, 2022, https://strategyzer.com/canvas/business-model-canvas

Osterwalder, A. (2011). *Burn Your Business Plan*. Retrieved on August 20th, 2022, https://www.slideshare.net/Alex.Osterwalder/creativity-world-forum-belgium/undefined

Osterwalder, A., Pigneur, Y., & Smith, A. (2010). *Business Model Generation*. Independently Published.

Osterwalder, A., & Pigneur, Y. (2013). *Business Model Generation: A Handbook for Visionaries, Game Changers, and Challengers*. Wiley.

Ostmeier, E., & Strobel, M. (2022). Building skills in the context of digital transformation: How industry digital maturity drives proactive skill development. *Journal of Business Research*, *139*, 718–730. doi:10.1016/j.jbusres.2021.09.020

Pan, R., Tang, Z., & Da, W. (2019). Digital stone rubbing from 3D models. *Journal of Cultural Heritage*, *37*, 192–198. doi:10.1016/j.culher.2018.11.013

Papavassiliou, G., Ntioudis, S., Mentzas, G., & Abecker, A. (2001). The DECOR approach to Business Process Oriented Knowledge Management (BPOKM). In *DEXA '01 Proceedings of the 12th International Workshop on Database and Expert Systems Applications*. IEEE Computer Society. https://astimen.wordpress.com/2009/10/28/the-decor-approach-to-business-process-oriented-knowledge-management-bpokm

Parnas, D. L. (1972). On the criteria to be used in decomposing systems into modules. *Communications of the ACM*, *15*(12), 1053–1058. doi:10.1145/361598.361623

Parowicz, I. (2018). *Cultural Heritage Marketing: A Relationship Marketing Approach to Conservation Services*. Palgrave.

Parra, R., Nieves, G. D., & Sánchez, H. (2017). 8 Tipos de contenido que funcionan en Instagram. *Postedin*. https://www.postedin.com/blog/8-tipos-de-contenido-que-funcion an-en-instagram/

Parviainen, P., Kääriäinen, J., Tihinen, M., & Teppola, S. (2017). Tackling the digitalization challenge: How to benefit from digitalization in practice. *International Journal of Information Systems and Project Management*, *5*(1), 63–77. doi:10.12821/ijispm050104

Pashkova, V. (2016). *Arendt's political thought: The relationship between truth and politics*. http://hdl.handle.net/1959.7/uws:37675

Patni, S. (2017). *Pro RESTful APIs Design, Build and Integrate with REST, JSON, XML and JAX-RS*. Apress.

Paulk, M. C. (1993). Capability Maturity Model SM for Software, Version 1.1. Software Engineering Institute, Carnegie Mellon University.

Pavel, F. (2011). Grid Database—Management, OGSA and Integration. Academy of Economic Studies Romania.

PayPal. (2013). *Modern Spice Routes The Cultural Impact and Economic Opportunity of Cross-Border Shopping*. Scientific Research - An Academic Publisher. https://www.scirp.org/(S(i43dyn45teexjx455qlt3d2q))/reference/ReferencesPapers.aspx?ReferenceID=1379307

Pecchioli, R. M. (1983). *The internationalisation of banking: The policy issues*. Organisation for Economic Co-operation and Development.

Peckham, J. (2021). *Best smartwatch 2021: The top wearables you should buy today*. TechRadar. https://www.techradar.com/news/wearables/best-smart-watches-what-s-the-best-wearable-tech-for-you-1154074

Peranzo, P. (2021). What is digital transformation and why it is important for business. *Imaginovation*. https://imaginovation.net/blog/what-is-digital-transformation-importance-for-businesses/

Perera, U. T., Heeney, C., & Sheikh, A. (2022). Policy parameters for optimising hospital ePrescribing: An exploratory literature review of selected countries of the Organisation for Economic Co-operation and Development. *Digital Health*, 8. doi:10.1177/20552076221085074 PMID:35340903

Pérez, M. J. (2021). *Zara y la película navideña que sí apetece ver*. Recuperado de https://www.elmundo.es/yodona/moda/2021/12/14/61b89826e4d4d82b338b45af.html

Pérez, R. (2020). *LaLiga, DIA y Desigual entran en el ranking de las 30 marcas españolas más valiosas de 2020*. Recuperado de https://www.kantar.com/es/inspiracion/marcas/laliga-dia-y-desigual-entran-en-el-ranking-de-las-30-marcas-espanolas-mas-valiosas-de-2020

Perroud, T., & Inversini, R. (2013). *Enterprise Architecture Patterns*. Practical Solutions for Recurring IT-Architecture Problems. doi:10.1007/978-3-642-37561-3

Peschl, A., & Schüth, N. J. (2022). Facing digital transformation with resilience–operational measures to strengthen the openness towards change. *Procedia Computer Science*, 200, 1237–1243. doi:10.1016/j.procs.2022.01.324

Peter, M. K., Kraft, C., & Lindeque, J. (2020). Strategic action fields of digital transformation: An exploration of the strategic action fields of Swiss SMEs and large enterprises. *Journal of Strategy and Management*, 13(1), 160–180. doi:10.1108/JSMA-05-2019-0070

Peterson, S. (2011). *Why it Worked: Critical Success Factors of a Financial Reform Project in Africa*. Faculty Research Working Paper Series. Harvard Kennedy School.

Peterson, B. L., & Carco, D. M. (1998). *The smart way to buy information technology: How to maximize value and avoid costly pitfalls*. Amacom.

Pezzini, M. (2018). *The Digital Integration Hub Turbocharges Your API Strategy*. LinkedIn. https://www.linkedin.com/pulse/digital-integration-hub-turbocharges-your-api-strategy-pezzini/

Pezzotta, G., Sassanelli, C., Pirola, F., Sala, R., Rossi, M., Fotia, S., Koutoupes, A., Terzi, S., & Mourtzis, D. (2018). The Product Service System Lean Design Methodology (PSSLDM): Integrating product and service components along the whole PSS lifecycle. *Journal of Manufacturing Technology Management, 29*(8), 1270–1295. doi:10.1108/JMTM-06-2017-0132

Philipp, R. (2020). Digital readiness index assessment towards smart port development. *Sustainability Management Forum | Nachhaltigkeits Management Forum, 28*(1–2), 49–60. doi:10.1007/s00550-020-00501-5

Pinter, H. (2012). Art, Truth and Politics. In J. Elkins & A. Norris (Eds.), *Truth and Democracy*. University of Pennsylvania Press. doi:10.9783/9780812206227.9

Porges, A., & Enders, A. (2016). *Data Moving Across Borders: The Future of Digital Trade Policy*. International Centre for Trade and Sustainable Development (ICTSD). Available at e15initiative.org/wp-content/uploads/2015/09/E15-Digital-Economy-Porges-and-Enders-Final.pdf

Porter Choi, H. S., & Kim, S. H. (2017). A content service deployment plan for metaverse museum exhibitions—Centering on the combination of beacons and HMDs. *International Journal of Information Management, 37*(1), 1519–1527. doi:10.1016/j.ijinfomgt.2016.04.017

Porter, M. (1985). Competitive advantage: Creating and sustaining superior performance. Collier Macmillan.

Porter, M. E., & Heppelmann, J. E. (2014). Wie smarte Produkte den Wettbewerb verändern. *Harvard Business Manager*, 34-60.

Poster, M. (1990). Words without Things: The Mode of Information. *October, 53*, 63–67. doi:10.2307/778915

Prathama, F., Yahya, B. N., & Lee, S.-L. (2021). A Multi-case Perspective Analytical Framework for Discovering Human Daily Behavior from Sensors using Process Mining. *2021 IEEE 45th Annual Computers, Software, and Applications Conference (COMPSAC)*, 638–644. 10.1109/COMPSAC51774.2021.00093

Proença, D., & Borbinha, J. (2016). Maturity models for information systems - A state of the art. *Procedia Computer Science, 100*, 1042–1049. doi:10.1016/j.procs.2016.09.279

Project Management Institute. (2017). PMBOK® Guide – Sixth Edition. Author.

PWC. (2015). *2015 Global Digital IQ® Survey. Lessons from digital leaders 10 attributes driving stronger performance*. Retrieved April 13th, 2020, from https://www.pwc.com/gx/en/advisory-services/digital-iq-survey-2015/campaign-site/digital-iq-survey-2015.pdf

PWC. (2016). *Industry 4.0: Building the digital enterprise*. Retrieved April 13th, 2020, from https://www.pwc.com/gx/en/industries/industries-4.0/landing-page/industry-4.0-building-your-digital-enterprise-april-2016.pdf

PWC. (2018). *Delivering digital change*. Retrieved April 13th, 2020, from: https://pwc.blogs.com/fsrr/2018/11/delivering-digital-change.html

PWC. (2020). *Accelerate digital transformation*. Retrieved April 13th, 2020, from https://www.pwc.co.uk/services/consulting/accelerate-digital.html

Quartiroli, I. (2011). *The Digitally Divided Self: Relinquishing Our Awareness to the Internet*. Silens Publishers.

Quéré, B. P., Nouyrigat, G., & Baker, C. R. (2015). A Bi-Directional Examination of the Relationship Between Corporate Social Responsibility Ratings and Company Financial Performance in the European Context. *Journal of Business Ethics*, *148*(3), 527–544. https://doi.org/10.1007/S10551-015-2998-1

R.P., Bose, Mans, & Aalst. (2013). *Wanna improve process mining results?* Academic Press.

Raddats, C., Kowalkowski, C., Benedettini, O., Burton, J., & Gebauer, H. (2019). Servitization: A Contemporary Thematic Review of Major Research Streams. *Industrial Marketing Management*, *2019*, 207–223. doi:10.1016/j.indmarman.2019.03.015

Rader, D. (2019). Digital maturity – the new competitive goal. *Strategy and Leadership*, *47*(5), 28–35. doi:10.1108/SL-06-2019-0084

Rai, B., & Arokiasamy, P. (2021). Identifying stages of smoke and smokeless tobacco cessation among adults in India: An application of transtheoretical model. *Journal of Substance Use*, *26*(4), 343–350. doi:10.1080/14659891.2020.1807634

Raman, J., Smith, E., & Hay, P. (2013). The clinical obesity maintenance model: An integration of psychological constructs including mood, emotional regulation, disordered overeating, habitual cluster behaviours, health literacy and cognitive function. *Journal of Obesity*, *2013*, 2013. doi:10.1155/2013/240128 PMID:23710346

Ramesh Radhakrishnan, R., & Radhakrishnan, R. (2004). *IT Infrastructure Architecture-Building Blocks*. Sun Microsystem. https://www.opengroup.org/architecture/0404brus/papers/rakesh/abb-1.pdf

Ramírez Barredo, B. (2016). *Los títulos de crédito, marca de las películas* [Tesis doctoral]. Universidad Complutense, Madrid.

Reeve, J. (2021). The Museum as Changemaker. *A Pathmaking Arts Quarterly*, *72*(3), 64-71.

Regazzi, J. J. (2018). The Shifting Sands of the Information Industry. In *Information Retrieval and Management: Concepts, Methodologies, Tools, and Applications* (Vol. 4, pp. 1–23). IGI Global. doi:10.4018/978-1-5225-5191-1.CH001

Regueira, J. (2019). *El nuevo contenido de Zara: ¿apuesta por el Branded Content? No Content No Brand*. Recuperado de https://www.javierregueira.com/contenido-de-zara/

Reis, J., Amorim, M., Melão, N., Cohen, Y., & Rodrigues, M. (2020). Digitalization: A Literature Review and Research Agenda. In Z. Anisic, B. Lalic, & D. Gracanin (Eds.), *Proceedings on 25th International Joint Conference on Industrial Engineering and Operations Management – IJCIEOM* (pp. 443–456). Springer International Publishing. 10.1007/978-3-030-43616-2_47

Remawati, D. (2016). Analisis SWOT Implementasi Green Computing Di Sekolah Kejuruan (Studi Kasus Pada SMK XYZ). *J. Ilm. SINUS*, 23–36.

Rhodes, R. E., McEwan, D., & Rebar, A. L. (2019). Theories of physical activity behaviour change: A history and synthesis of approaches. *Psychology of Sport and Exercise*, *42*, 100–109. doi:10.1016/j.psychsport.2018.11.010

Richardson, C. (2014). *Pattern: Microservices architecture*. https://microservices.io/patterns/microservices.html

Ries, E. (2010). *Introduction to Customer Development at the Lean Startup Intensive at Web 2.0 Expo by Steve Blank*. Retrieved on August 20th, 2022, https://www.slideshare.net/startuplessonslearned/

Ries, E. (2011a). *The Lean Startup: How Today's Entrepreneurs Use Continuous Innovation to Create Radically Successful Businesses*. Retrieved on August 20th, 2022, from https://books.google.ro/books?id=tvfyz-4JILwC&redir_esc=y

Ries, E. (2011b). *The Lean Startup - Google Tech Talk*. Retrieved on August 20th, 2022, from https://www.slideshare.net/startuplessonslearned/eric-ries-the-lean-startup-google-tech-talk

Rodrigues, E. A., Carnevalli, J. A., & Miguel, P. A. C. (2014, January). Modular design and production: An investigation on practices in an assembler and two first-tier suppliers. *Proceedings of the International Conference on Industrial Engineering and Operations Management.*

Rodríguez, R. (2021). Zara tiene más seguidores en Instagram que en Facebook, Twitter y Tik Tok juntos. *Galicia.* Economía. https://www.economiadigital.es/galicia/empresas/zara-tiene-mas-seguidores-en-instagram-que-en-facebook-twitter-y-tik-tok-juntos .html

Rogers, M. E. (2009). *Diffusion of Innovations.* The Free Press.

Rohrbeck, R., & Schwarz, J. O. (2013). *The value contribution of strategic foresight: insights from an empirical study of large European companies.* Retrieved on June 30th, 2021, from: https://www.researchgate.net/publication/236977761_The_Value_Contribution_of_Strategic_Foresight_Insights_From_an_Empirical_Study_of_Large_European_Companies

Roland Berger. (2015). *The digital transformation of industry.* Retrieved April 13th, 2020, from www.rolandberger.com › publications › publication_pdf

Roland Berger. (2018). *Industrie 4.0? Step this way!* Retrieved April 13th, 2020, from www.rolandberger_coo_insights_e

Romero Coves, A., Carratalá Martínez, D., & Segarra Saavedra, J. (2020). Influencers y moda en redes sociales. Análisis de las principales modelos españolas en Instagram. *Revista de Marketing Aplicado.* https://revistas.udc.es/index.php/REDMARKA/article/view/redma.2020.24.2.7053

Rose, J. (2017). Brexit, Trump, and Post-Truth Politics. *Public Integrity, 19*(6), 555–558. doi:10.1080/10999922.2017.1285540

Ross, J. (2017). Don't confuse digital with digitization. *MIT Sloan Management Review.* https://sloanreview.mit.edu/article/dont-confuse-digital-with-digitization/

Rottman, J. W. (2008). Successful knowledge transfer within offshore supplier networks: A case study exploring social capital in strategic alliances. *Journal of Information Technology, 23*(1), 31–43. doi:10.1057/palgrave.jit.2000127

Rubin, D. A. (2003). *Analysis of an Electronic Voting System.* Information Security Institute.

Ruchkin, A., V., & Trofimova, O., M. (2017), Project Management: Basic Definitions and Approaches. *Management Issues / Voprosy Upravleniâ, 46,* 1-10.

Ruiz Vega, A. V., Riano Gil, C., & Aguado Gonzalez, A. (2019). ¿Cómo se forma el estado de flow entre los internautas compradores de moda? un estudio comparativo de los sitios web de Zara y H&M. *International Journal of Information Systems and Software Engineering for Big Companies, 6*(1), 79–95. www.ijisebc.com

Sako, M. (2005). Modularity and outsourcing. In A. Principe, A. Davies, & M. Hobday (Eds.), *The business of system integration* (pp. 229–253). Oxford University Press. doi:10.1093/acprof:oso/9780199263233.003.0012

Sako, M. (2014). Outsourcing and offshoring of professional services. In L. Empson, D. Muzio, J. Broschak, & B. Hinings (Eds.), *The oxford handbook of professional service firms* (pp. 327–347). Oxford University Press.

Samuel, A. (2013). *Investigating the feasibility of implementing the E-Voting system in Ghana*. Department of Computer Science, Kwame Nkrumah University of Science and Technology.

Sanchez, R., & Mahoney, J. T. (1996). Modularity, flexibility, and knowledge management in product and organization design. *Strategic Management Journal*, *17*(S2), 63–76. doi:10.1002mj.4250171107

Sanchez, R., & Mahoney, J. T. (2013). Modularity and economic organization: Concepts, theory, observations, and predictions. In A. Grandori (Ed.), *Handbook of economic organization* (pp. 383–399). Edward Elgar Publishing. doi:10.4337/9781782548225.00031

Sanchez, R., & Shibata, T. (2021). Modularity design rules for architecture development: Theory, implementation, and evidence from the development of the Renault–Nissan alliance "common module family" architecture. *Journal of Open Innovation*, *7*(4), 1–22. doi:10.3390/joitmc7040242

Santos, C., Mehrsai, A., Barrosaa, C., Araújob, M., & Ares, E. (2017). Towards Industry 4.0: An overview of European strategic roadmaps. *Procedia Manufacturing*, *13*, 972–979. doi:10.1016/j.promfg.2017.09.093

Santos, R. C., & Martinho, J. L. (2019). An Industry 4.0 maturity model proposal. *Journal of Manufacturing Technology Management*, *31*(5), 1023–1043. doi:10.1108/JMTM-09-2018-0284

SAP. (2013). GBTM: Global Business Transformation Manager Master Certification (SAP Internal). Business Transformation Academy. SAP.

Sarcar, V. (2016). *Java Design Patterns: A tour of 23 gang of four design patterns in Java*. Apress. doi:10.1007/978-1-4842-1802-0

Sarvari, P. A., Unstundag, A., Cevikcan, E., & Kaya, I. (2018). Technology Roadmap for Industry 4.0. In Industry 4.0: Managing The Digital Transformation (pp. 95-103). Springer.

Sato, D. M. V., De Freitas, S. C., Barddal, J. P., & Scalabrin, E. E. (2021). A survey on concept drift in process mining. *ACM Computing Surveys, 54*(9), 1–38.

Saunders, J. (2014). Conservation in Museums and Inclusion of the Non-Professional. *Journal of Conservation & Museum Studies*, *12*(1), 6. doi:10.5334/jcms.1021215

Savić, D. (2019). From Digitization, Through Digitalization, to Digital Transformation. *Online Searher, 43*(1). https://www.infotoday.com/OnlineSearcher/Articles/Features/From-Digitization-Through-Digitalization-to-Digital-Transformation-129664.shtml?PageNum=2

D. Schallmo, & C. A. Williams (Eds.). (2021). Integrated Approach for Digital Maturity: Levels, Procedure, and In-Depth Analysis. In *The ISPIM Innovation Conference – Innovating Our Common Future*. LUT Scientific and Expertise Publications.

Schallmo, D., Williams, C. A., & Boardman, L. (2017). Digital transformation of business models—Best practice, enablers, and roadmap. *International Journal of Innovation Management*, *21*(08), 1740014. doi:10.1142/S136391961740014X

Schilling, M. A. (2000). Towards a general modular systems theory and its application to inter-firm product modularity. *Academy of Management Review*, *25*(2), 312–334. doi:10.2307/259016

Schilling, M. A., & Steensma, K. (2001). The use of modular organizational forms: An industry-level analysis. *Academy of Management Journal*, *44*(6), 1149–1168. doi:10.2307/3069394

Schmidtab, M., & Åhlundb, C. (2018). Smart buildings as Cyber-Physical Systems: Data-driven predictive control strategies for energy efficiency. *Renewable & Sustainable Energy Reviews, 90*, 742–756. https://doi.org/10.1016/j.rser.2018.04.013

Schmidt, N., Zoller, B., & Rosenkranz, C. (2016). The clash of cultures in information technology outsourcing relationships: An institutional logics perspective. In J. Kotlarsky, I. Oshri, & L. P. Willcocks (Eds.), *Shared services and outsourcing: A contemporary outlook* (pp. 97–117). Springer. doi:10.1007/978-3-319-47009-2_6

Schuh, G., Anderl, R., Gausemeier, J., ten Hompel, M., & Wahlster, W. (Eds.). (2017). Industrie 4.0 Maturity Index. Managing the Digital Transformation of Companies (acatech STUDY). Herbert Utz Verlag.

Schumacher, A., Erol, S., & Sihn, W. (2016). A maturity model for assessing Industry 4.0 readiness and maturity of manufacturing enterprises. *Procedia CIRP - Changeable, Agile. Reconfigurable & Virtual Production, 52*(1), 161–166.

Schumacher, A., Nemeth, T., & Sihn, W. (2019). Roadmapping towards industrial digitalization based on an Industry 4.0 maturity model for manufacturing enterprises. *Procedia CIRP, 79*, 409–414. doi:10.1016/j.procir.2019.02.110

Schunk, D. H., & DiBenedetto, M. K. (2020). Motivation and social cognitive theory. *Contemporary Educational Psychology, 60*, 101832. doi:10.1016/j.cedpsych.2019.101832

Schwertner, K. (2017). Digital transformation of business. *Trakia Journal of Sciences, 15*(Suppl.1), 388–393. doi:10.15547/tjs.2017.s.01.065

Science Museum. (2019). *Unlocking the Secrets: The Science of Conservation at the Palace Museum.* Retrieved from https://hk.science.museum/ms/con2019/activities-EN.html

Science Museum. (2020). *Conservation Laboratory – Unlocking the Secrets of Artefact Conservation.* Retrieved from https://hk.science.museum/web/scm/se/cl.html

Sedik, A., Hammad, M., El-Samie, A., Fathi, E., Gupta, B. B., El-Latif, A., & Ahmed, A. (2022). Efficient deep learning approach for augmented detection of Coronavirus disease. *Neural Computing & Applications, 34*(14), 11423–11440. doi:10.100700521-020-05410-8 PMID:33487885

Seghelmeble, M. (2020). *Análisis de las estrategias de marketing empleadas por Zara que contribuyeron a su desarrollo y renombre internacional en el mercado de la moda del siglo XXI* [Trabajo de investigación]. Monografía, Centro Educativo Particular San Agustín.

Şener, U., Gökalp, E., & Eren, P. E. (2018). Towards a maturity model for industry 4.0: A systematic literature review and a model proposal. In Industry 4.0 From the Management Information Systems Perspectives (pp. 291–303). Academic Press.

Sener, U., Gokalp, E., & Eren, P., E. (2018). Towards a Maturity Model for Industry 4.0: a Systematic Literature Review and a Model Proposal. *Industry 4.0 from the MIS*, 221

Shaikh, A. A., & Karjaluoto, H. (2015). Mobile banking adoption: A literature review. *Telematics and Informatics, 32*(1), 129–142. https://doi.org/10.1016/J.TELE.2014.05.003

Shapin, S. (2019). Is there a Crisis of Truth. *Los Angeles Review of Books.* https://lareviewofbooks.org/article/is-there-a-crisis-of-truth/

Sheikh, O. (2020). *Securing Your Digital Channels With API Gateways.* IBM.

Shiau, W. L., Yuan, Y., Pu, X., Ray, S., & Chen, C. C. (2020). Understanding fintech continuance: Perspectives from self-efficacy and ECT-IS theories. *Industrial Management & Data Systems, 120*(9), 1659–1689. doi:10.1108/IMDS-02-2020-0069

Shmueli, L. (2021). Predicting intention to receive COVID-19 vaccine among the general population using the health belief model and the theory of planned behavior model. *BMC Public Health*, *21*(1), 1–13. doi:10.118612889-021-10816-7 PMID:33902501

Shuaiyin, M., Yingfeng, Z., Jingxiang, L., Haidong, Y., & Jianzhong, W. (2019). Energy-cyber-physical system enabled management for energy-intensive manufacturing industries. *Journal of Cleaner Production*, *226*, 892–903. doi:10.1016/j.jclepro.2019.04.134

Sidell, E. (2020). *Choosing the Right API Gateway Pattern for Effective API Delivery*. NGINX. https://www.nginx.com/blog/choosing-the-right-api-gateway-pattern/

Siedler, C., Sadaune, S., Zavareh, M. T., Eigner, M., Zink, K. J., & Aurich, J. C. (2019). Categorizing and selecting digitization technologies for their implementation within different product lifecycle phases. *12th CIRP Conference on Intelligent Computation in Manufacturing Engineering*, 274–279. DOI: 10.1016/j.procir.2019.02.066

Silva, F. G., Reis da Silva, T., Alan de Oliveira, S., & Aranha, E. (2020). Behavior analysis of students in video classes. In *2020 IEEE Frontiers in Education Conference (FIE)* (pp. 1–8). IEEE. doi:10.1109/FIE44824.2020.9274274

Simone, C, Cerquetti, M., & La Sala, A. (2021). Museums in the Infosphere: Reshaping value creation. *Museum Management and Curatorship, 36*(4), 322-341.

Simon, H. A. (1962). The architecture of complexity. *Proceedings of the American Philosophical Society*, *106*(6), 467–482.

Sîrbu, C., Groza, I. V., Mnerie, D., & Mnerie, G. V. (2022). *Proiectarea şi optimizarea arhitecturii instalaţiei pilot de producere peleţi-bricheţi. Proiect: Inovări şi optimizări economice şi funcţionale în producţia industrială de Materiale pentru energie termică, „INO-MAT", cod SMIS 119412*. Titus Industries SRL.

Sklyar, A., Kowalkowski, C., Tronvoll, B., & Sörhammar, D. (2019). Organizing for digital servitization: A service ecosystem perspective. *Journal of Business Research*, *104*, 450–460. doi:10.1016/j.jbusres.2019.02.012

Soeken, M., Wille, R., & Drechsler, R. (2012). Assisted Behavior Driven Development Using Natural Language Processing. Institute of Computer Science, University of Bremen Group of Computer Architecture. doi:10.1007/978-3-642-30561-0_19

Soobramanien, T. Y., & Worrall, L. (Eds.). (2017). *Emerging Trade Issues for Small Developing Countries: Scrutinising the Horizon*. Commonwealth Secretariat. https://books.google.co.in/books?hl=en&lr=&id=M2VnDwAAQBAJ&oi=fnd&pg=PA111&dq=the+export+of+conventionally+imported+goods+and+services+and+the+growth+of+e-commerce+are+positively+correlated+in+the+Caribbean+islands&ots=zeeBEYxdBU&sig=MeegIUQEdDA8m7OWfzxyRmvR11k&redir_esc=y#v=onepage&q&f=false

SouleD. L.PuramA.WestermanG. F.BonnetD. (2016). Becoming a digital organization: The journey to digital dexterity. Available at SSRN 2697688.

Stam, E. (2015). Entrepreneurial Ecosystems and Regional Policy: A Sympathetic Critique September 2015. *European Planning Studies*, *23*(9). Advance online publication. doi:10.1080/09654313.2015.1061484

Stana, R. A., Fischer, L. H., & Nicolajsen, H. W. (2018). Review for future research in digital leadership. *Context, 5, 6*.

Stark, J. (2015). *Product Lifecycle Management: The Devil Is in the Details* (Vol. 2). Springer International Publishing.

Statista. (2021). *Product lifecycle management (PLM) & engineering software market revenues worldwide from 2019 to 2025(in million U.S. dollars)*. Accesat în iunie 2022, de la https://www.statista.com/statistics/796151/worldwide-product-lifecycle-management-engineering-software-market/

Statista. (2021). *Size of computer-aided engineering (CAE) market worldwide, from 2015 to 2021*. Accesat în iunie 2022, de la https://www.statista.com/statistics/732384/worldwide-computer-aided-engineering-market-revenues/

Stergiou, C., Psannis, K. E., Gupta, B. B., & Ishibashi, Y. (2018). Security, privacy & efficiency of sustainable cloud computing for big data & IoT. *Sustainable Computing: Informatics and Systems*, *19*, 174–184. doi:10.1016/j.suscom.2018.06.003

Stewart, G. (1997). Supply-chain operations reference model (SCOR): The first cross-industry framework for integrated supply-chain management. *Emerald*, *10*(2), 62–67.

Stojanović, Z. (2005). *A Method for Component-Based and Service-Oriented Software Systems Engineering*. Technische Universiteit Delft. doi:10.4018/978-1-59140-426-2

Stolterman, E., & Fors, A. C. (2004). Information technology and the good life. *Information Systems Research*, 687–692.

Storbacka, K. (2011). A solution business model: Capabilities and management practices for integrated solutions. *Industrial Marketing Management*, *40*(5), 707–711. doi:10.1016/j.indmarman.2011.05.003

Störrle, H., & Knapp, A. (2006). *UML 2.0 – Tutorial/Unified Modeling Language 2.0*. University of Innsbruck.

Strauss, L. (2020). 10 estrategias que utiliza Zara para vender más y que puedes llevar a cabo en tu empresa. *Muy Negocios & Economía*. https://www.muynegociosyeconomia.es/negocios/articulo/10-estrategias-que-utiliza-zara-para-vender-mas-y-que-puedes-llevar-a-cabo-en-tu-empresa-481584704915

Strömberg, J., Sundberg, L., & Hasselblad, A. (2020). Digital Maturity in Theory and Practice: A Case Study of a Swedish Smart-Built Environment Firm. *2020 IEEE International Conference on Industrial Engineering and Engineering Management (IEEM)*, 1344–1348. 10.1109/IEEM45057.2020.9309760

Subramaniam, M. (2020). Digital ecosystems and their implications for competitive strategy. *Journal of Organization Design*, *9*(1), 1–10. doi:10.118641469-020-00073-0

Suen, R. L. T., Tang, J., & Chiu, D. K. W. (2020). Virtual reality services in academic libraries: Deployment experience in Hong Kong. *The Electronic Library*, *38*(4), 843–858. doi:10.1108/EL-05-2020-0116

Suiter, J. (2016). Post-Truth Politics. Political Insight, 7(3), 25-27. doi:10.1177/2041905816680417

Sundberg, L., Gidlund, K. L., & Olsson, L. (2019). Towards Industry 4.0? Digital Maturity of the Manufacturing Industry in a Swedish Region. *2019 IEEE International Conference on Industrial Engineering and Engineering Management (IEEM)*, 731–735. 10.1109/IEEM44572.2019.8978681

Sun, X., Chiu, D. K. W., & Chan, C. T. (2022). Recent Digitalization Development of Buddhist Libraries: A Comparative Case Study. In S. Papadakis & A. Kapaniaris (Eds.), *The Digital Folklore of Cyberculture and Digital Humanities* (pp. 251–266). IGI Global. doi:10.4018/978-1-6684-4461-0.ch014

Suominen, K. (2017). *Fuelling Trade in the Digital Era: Policy Roadmap for Developing Countries*. Geneva: International Centre for Trade and Sustainable Development (ICTSD). Available at https,//www.ictsd.org/sites/default/files/research/suominen_fuelling_trade_in_the_digital_era_0.pdf

Șuta, A., I., Mariș, Ș., A., Gomoi, V., S., Molnar, M., & C., F. S. (2022). *Conceperea și depunerea documentației pentru obținerea unui brevet. Proiect: Inovări și optimizări economice și funcționale în producția industrială de Materiale pentru energie termică, „INO-MAT", cod SMIS 119412*. Titus Industries SRL.

Şuta, A. I., & Cazan, A. C. (2022). *Realizarea fizică a eşantioanelor de peleţi/bricheţi. Proiect: INOvări şi optimizări economice şi funcţionale în producţia industrială de MATeriale pentru energie termică, „INO-MAT", cod SMIS 119412.* Titus Industries SRL.

Şută, A., Mnerie, A. V., & Holotescu, C. (2022). *Evaluarea industrială şi testarea indicatorilor tehnici a produselor obţinute. Proiect: Inovări şi optimizări economice şi funcţionale în producţia industrială de Materiale pentru energie termică, „INO-MAT", cod SMIS 119412.* Titus Industries SRL.

Tadelis, S. (2007). The innovative organization: Creating value through outsourcing. *California Management Review, 50*(1), 261–277. doi:10.2307/41166427

Taleb, M., & Cherkaoui, O. (2012, January). Pattern-Oriented Approach for Enterprise Architecture: TOGAF Framework. *Journal of Software Engineering & Applications, 5*(1), 45–50. doi:10.4236/jsea.2012.51008

Tanniru, M., Khuntia, J., & Weiner, J. (2018). Hospital leadership in support of digital transformation. *Pacific Asia Journal of the Association for Information Systems, 10*(1).

Tarhan, A., Turetken, O., & Reijers, H. A. (2016). Business process maturity models: A systematic literature review. *Information and Software Technology, 75*, 122–134.

Tauber, E. M. (1988). Brand Leverage-Strategy for Growth in a Cost-Control World. *Journal of Advertising Research, 28*(4), 26–30.

Tee, R. (2009, January). Product, organization, and industry modularity exploring interactions across levels. *DRUID-DIME Academy Winter Ph.D. Conference on Economics and Management of Innovation, Technology, and Organizational Change.*

Teichert, R. (2019). Digital Transformation Maturity: A Systematic Review of Literature. *Acta Universitatis Agriculturae et Silviculturae Mendelianae Brunensis, 67*(6), 1673–1687. doi:10.11118/actaun201967061673

Terhune, L. (2006). *Smithsonian Museums Make Art Conservation Part of the Show: Lunder Conservation Center allows visitors to see conservators at work.* Federal Information & News Dispatch, LLC. Retrieved from Research Library http://eproxy.lib.hku.hk/login?url=https://www.proquest.com/reports/smithsonian-museums-make-art-conservation-part/docview/190012923/se-2?accountid=14548

Terlouw, L. I. (2011). *Modularization and specification of service-oriented systems* [Doctoral dissertation, Delft University of Technology]. Gildeprint.

Thao, H., & Hwang, J. (2010). *Factors affecting employee performance. Evidence from Petro Vietnam engineering consultancy J.S.C.* Academic Press.

The Open Group. (2006a). *Building Blocks.* The Open Group. https://pubs.opengroup.org/architecture/togaf8-doc/arch/chap32.html

The Open Group. (2006b). *The Enterprise Continuum in Detail.* https://pubs.opengroup.org/architecture/togaf8-doc/arch/chap18.html

The Open Group. (2011a). *Architecture Development Method.* The Open Group. https://pubs.opengroup.org/architecture/togaf9-doc/arch/chap05.html

The Open Group. (2011a). *Phase P: The Preliminary Phase.* The Open Group. https://pubs.opengroup.org/architecture/togaf9-doc/arch/chap06.html

The Open Group. (2011b). *TOGAF 9.1.* The Open Group. https://www.opengroup.org/subjectareas/enterprise/togaf

The Open Group. (2011b). *TOGAF's The Open Group Architecture Framework.* www.open-group.com/togaf

The Open Group. (2011b). *TOGAF's. The Open Group Architecture Framework.* www.open-group.com/togaf

The Open Group. (2011c). *Introduction to Building Blocks.* http://www.opengroup.org/public/arch/p4/bbs/bbs_intro.htm

The Open Group. (2011d). *Foundation Architecture: Technical Reference Model.* http://www.opengroup.org/public/arch/p3/trm/trm_dtail.htm

The Open Group. (2018). *Architecture Patterns.* The Open Group. https://pubs.opengroup.org/architecture/togaf9-doc/arch/chap 22.html

Thomson, L., Kamalaldin, A., Sjödin, D., & Parida, V. (2021). A maturity framework for autonomous solutions in manufacturing firms: The interplay of technology, ecosystem, and business model. *The International Entrepreneurship and Management Journal.* Advance online publication. doi:10.100711365-020-00717-3

Tiong, T. K., Tianyi, G., Sopra, R., & Sharma, R. (2012). The Issue of Fragmentation on Mobile Games Platforms. In R. S. Sharma, M. Tan, & F. Pereira (Eds.), *Understanding the Interactive Digital Media Marketplace* (pp. 89–96). IGI Global. doi:10.4018/978-1-61350-147-4.CH009

Tkachenko, E. (2015). *5 key attributes of requirements testing: Know before you code.* EPAM Systems. https://techbeacon.com/app-dev-testing/5-key-attributes-requirements-testing-know-you-code

TOGAF Catalogs. (2011). *Sample catalogs, matrices and diagrams.* http://www.opengroup.org/bookstore/catalog/i093.htm

TOGAF Directory. (2011). *Using TOGAF to Define & Govern SOA.* https://pubs.opengroup.org/architecture/togaf9-doc/arch/chap 22.html

TOGAF Skills. (2011). *Architecture skills framework.* http://pubs.opengroup.org/architecture /togaf9-doc/arch/chap 52.html

Togaf-Modeling. (2020). *Application communication diagrams.* Togaf-Modeling.org. https://www.togaf-modeling.org/models/application-architecture/application-communication-diagrams.html

Tom, B., & Stalker, G. M. (1998). *The Management Of İnnovation.* Tavistock.

Trad, A., & Kalpić, D. (2013). *The Selection, and Training Framework (STF) for Managers in Business Innovation Transformation Projects - The Background.* Conference on Information Technology Interfaces, Cavtat, Croatia.

Trad, A., & Kalpić, D. (2022). Business Transformation Projects based on a Holistic Enterprise Architecture Pattern (HEAP)-The Basic Construction. IGI.

Trad, A., & Kalpić, D. (2022). Business Transformation Projects based on a Holistic Enterprise Architecture Pattern (HEAP)-The Implementation. IGI.

Trad, A. (2015a). *A Transformation Framework Proposal for Managers in Business Innovation and Business Transformation Projects-Intelligent aBB architecture.* Centeris.

Trad, A. (2015b). *A Transformation Framework Proposal for Managers in Business Innovation and Business Transformation Projects-An ICS's atomic architecture vision.* Centeris.

Trad, A. (2020). *Applied Mathematical Model for Business Transformation Projects: The Intelligent Strategic Decision-Making System (iSDMS). In Handbook of Research on IT Applications for Strategic Competitive Advantage and Decision Making.* IGI Global.

Trad, A. (2021). An Applied Mathematical Model for Business Transformation and Enterprise Architecture: The Holistic Organizational Intelligence and Knowledge Management Pattern's Integration. *International Journal of Organizational and Collective Intelligence, 11*(1), 1–25. doi:10.4018/IJOCI.2021010101

Trad, A., & Kalpić, D. (2016a). *A Transformation Framework Proposal for Managers in Business Innovation and Business Transformation Projects-Basics of a patterns based architecture.* Centeris.

Trad, A., & Kalpić, D. (2016b). *A Transformation Framework Proposal for Managers in Business Innovation and Business Transformation Projects-Design patterns based architecture.* Centeris.

Trad, A., & Kalpić, D. (2016c). *A Transformation Framework Proposal for Managers in Business Innovation and Business Transformation Projects-Enterprise patterns based architecture.* Centeris. doi:10.1016/j.procs.2016.09.158

Trad, A., & Kalpić, D. (2017a). *An Intelligent Neural Networks Micro Artefact Patterns' Based Enterprise Architecture Model.* IGI-Global.

Trad, A., & Kalpić, D. (2018a). *The Business Transformation Framework and Enterprise Architecture Framework for Managers in Business Innovation-Knowledge and Intelligence Driven Development (KIDD). In Encyclopedia of E-Commerce Development, Implementation, and Management.* IGI-Global.

Trad, A., & Kalpić, D. (2018b). *The Business Transformation Framework and Enterprise Architecture Framework for Managers in Business Innovation-Knowledge Management in Global Software Engineering (KMGSE). In Encyclopedia of E-Commerce Development, Implementation, and Management.* IGI-Global.

Trad, A., & Kalpić, D. (2020a). *Using Applied Mathematical Models for Business Transformation.* IGI Global. doi:10.4018/978-1-7998-1009-4

Trad, A., & Kalpić, D. (2020b). *The Business Transformation Framework and Enterprise Architecture Framework for Managers in Business Innovation-Intelligence Driven Development and Operations (IDDevOps).* IGI Global.

Trad, A., & Kalpić, D. (2022). Business Transformation Project's Holistic Agile Management (BTPHAM). *The Business & Management Review, 13*(1), 103–120. doi:10.24052/BMR/V13NU01/ART-12

Trad, A., Nakitende, M., & Oke, T. (2021). *Tech-Based Enterprise Control and Audit for Financial Crimes: The Case of State-Owned Global Financial Predators (SOGFP). In Handbook of Research on Theory and Practice of Financial Crimes.* IGI Global.

Trenerry, B., Chng, S., Wang, Y., Suhaila, Z. S., Lim, S. S., Lu, H. Y., & Oh, P. H. (2021). Preparing workplaces for digital transformation: An integrative review and framework of multi-level factors. *Frontiers in Psychology, 12,* 822. doi:10.3389/fpsyg.2021.620766 PMID:33833714

Tse, H. L., Chiu, D. K., & Lam, A. H. (2022). From Reading Promotion to Digital Literacy: An Analysis of Digitalizing Mobile Library Services With the 5E Instructional Model. In A. Almeida & S. Esteves (Eds.), *Modern Reading Practices and Collaboration Between Schools, Family, and Community* (pp. 239–256). IGI Global. doi:10.4018/978-1-7998-9750-7.ch011

Tseng, M. F., Huang, C. C., Tsai, S. C. S., Tsay, M. D., Chang, Y. K., Juan, C. L., ... Wong, R. H. (2022). Promotion of Smoking Cessation Using the Transtheoretical Model: Short-Term and Long-Term Effectiveness for Workers in Coastal Central Taiwan. *Tobacco Use Insights, 15.*

Tutak, M., & Brodny, J. (2022). Business Digital Maturity in Europe and Its Implication for Open Innovation. *Journal of Open Innovation, 8*(1), 27. doi:10.3390/joitmc8010027

Tutorialspoint. (2022). *Requirement Based Testing.* Tutorialspoint. https://www.tutorialspoint.coM/Roftware_testing_dictionary/requirements_based_testing.htm

Ubiparipović, B., Matković, P., & Pavlićević, V. (2022). Key activities of digital business transformation process. *Strategic Management, 00*(00), 1–8. doi:10.5937/StraMan2200016U

Uhl, L., & Gollenia, L. A. (2012). *A Handbook of Business Transformation Management Methodology, Gower.* SAP.

UN Report. (2019). *The impact of digital technologies.* Retrieved August 25, 2022, from https://www.un.org/sites/un2.un.org/files/2019/10/un75_new_technologies.pdf

UN Sustainable Development Goals. (2022). *Do you know all 17 SDGs?* https://sdgs.un.org/goals

UNCTAD. (2015). *Information economy report 2015-unlocking the potential of e-commerce for developing countries.* United Nations. unctad.org/en/PublicationsLibrary/ier2015_en.pdf

Vaishya, R., Bahl, S., & Singh, R. P. (2020). Letter to the editor in response to: telemedicine for diabetes care in India during COVID19 pandemic and national lockdown period: guidelines for physicians. *Diabetes & Metabolic Syndrome, 14*(4), 687–688. doi:10.1016/j.dsx.2020.05.027 PMID:32442918

Valderrama Santomé, M., Mallo Méndez, S., & Crespo Pereira, V. (2020). Nuevas narrativas en el marketing de moda: estudio de caso Fashion Dramas de Vogue, Redmarka. Revista de Marketing Aplicado, 24(2), 238-250. doi:10.17979/redma.2020.24.2.7161

Valencia, A., Mugge, R., Schoormans, P. L., & Schifferstein, H. N. J. (2015). The Design of Smart Product-Service Systems (PSSs). An Exploration of Design Characteristics. *International Journal of Design, 9,* 13–28.

Van Alstyne, M. W., Parker, G. G., & Choudary, S. P. (2016). Pipelines, Platforms, and the New Rules of Strategy. *Harvard Business Review.*

Van der Aalst, van Dongen, Gu¨nther, Rozinat, Verbeek, & Weijters. (2009). ProM: The process mining toolkit. *BPM, 489*(31), 2.

Van der Borgh, M., Cloodt, M., & Romme, A. G. L. (2012). Value Creation by Knowledge-Based Ecosystems: Evidence from a Field Study. *R & D Management, 42,* 150–169.

Van der Linden, D., Mannaert, H., & De Bruyn, P. (2012, February). Towards the explicitation of hidden dependencies in the module interface. *Proceedings of the Seventh International Conference on Systems (ICONS).*

van Saaze, V. (2011) *Going Public: Conservation of Contemporary Artworks. Between Backstage and Frontstage in Contemporary Art Museums.* Retreived from http://hdl.handle.net/10362/16705

VanBoskirk, S., Gill, M., Green, D., Berman, A., Swire, J., & Birrel, R. (2017). The Digital Maturity Model 5.0. *Forrester.* https://www.forrester.com/report/The+Digital+Maturity+Model+50/-/E-RES137561

Vázquez, S. (2019). ¿Qué es Zara Scenes? *Revista Woman*. https://woman.elperiodico.com/moda/shopping/zara-scenes-nueva-coleccion-cambios-web-compras-tienda-online

VDMA. (2017). *Industrie 4.0 Readiness study*. Retrieved April 11th, 2020, from https://industrie40.vdma.org/en/viewer/-/v2article/render/15525817

Velilla, J. (n.d.). *Estrategias corporativas: el caso de Zara*. Recuperado de https://comuniza.com/blog/estrategias-corporativas-zara

Velu, C., Barrett, M., Kohli, R., & Salge, T. L. (2013). *Thriving in Open Innovation Ecosystems: Towards a Collaborative Market Orientation*. Cambridge Service Alliance Working Paper.

Verner, J. M., & Abdullah, L. M. (2012). Exploratory case study research: Outsourced project failure. *Information and Software Technology*, *54*(8), 866–886. doi:10.1016/j.infsof.2011.11.001

Vishnevskiy, K., Karasev, O., & Meissner, D. (2016). Integrated roadmaps for strategic management and planning. *Technological Forecasting and Social Change*. Advance online publication. doi:10.1016/j.techfore.2015.10.020

Visnjic, I., Jovanovic, M., Neely, A., & Engwall, M. (2017). What brings the value to outcome-based contract providers? Value drivers in outcome business models. *International Journal of Production Economics*, *192*, 169–181. doi:10.1016/j.ijpe.2016.12.008

Visnjic, I., Neely, A., & Jovanovic, M. (2018). The Path to Outcome Delivery: Interplay of Service Market Strategy and Open Business Models. *Technovation*, *72-73*, 46–59. doi:10.1016/j.technovation.2018.02.003

Visual Paradigm. (2019). *TOGAF ADM4AI Tutorial*. *Visual Paradigm*. https://cdn.visual-paradigm.com/guide/togaf/togaf-adm-tutorial/02-togaf-adm.png

Vogel, E. (1985). Pax Nipponica. *Foreign Affairs*, *64*, 752. https://heinonline.org/hol-cgi-bin/get_pdf.cgi?handle=hein.journals/fora64&section=52

Vogler, C. (2016). Anscombe on Two Jesuits and Lying. In L. Gormally, D. A. Jones, & R. Teichmann (Eds.), *The Moral Philosophy of Elizabeth Anscombe*. Imprint Academic.

Voss, C. A., & Hsuan, J. (2009). Service architecture and modularity. *Decision Sciences*, *40*(3), 541–569. doi:10.1111/j.1540-5915.2009.00241.x

Wackerbeck, P., Helmuth, U., Skritek, B., & Putz, A. (2017). *Building the modular bank: Sourcing strategies in the age of digitization*. Strategy& PWC Consulting. Retrieved August 24, 2022, from https://www.strategyand.pwc.com/reports/building-modular-bank

Wald, D., de Laubier, R., & Charanya, T. (2019). *The Five Rules of Digital Strategy*. Boston Consulting Group. https://web-assets.bcg.com/img-src/BCG-The-Five-Rules-for-Digital-Strategy-May-2019_tcm9-220981.pdf

Walker, J. (2019). *Generics & Design Patterns*. Oberlin College Computer Science. https://www.cs.oberlin.edu/~jwalker/langDesign/GDesignPat/

Walter, I. (1988). *Global Competition in Financial Services: Market Structure, Protection and Trade Liberazation*. Ballinger Publishing Company. https://scholar.google.com/scholar?hl=en&as_sdt=0%2C5&q=Global+Competition+in+Financial+Services%2C+Market+Structure%2C+Protection%2C+and+Trade+Liberalization+by+walter&btnG=

Wang, J., Deng, S., Chiu, D. K. W., & Chan, C. T. (2022). Social Network Customer Relationship Management for Orchestras: A Case Study on Hong Kong Philharmonic Orchestra. In N. B. Ammari (Ed.), *Social Customer Relationship Management (Social-CRM) in the Era of Web 4.0*. IGI Global. doi:10.4018/978-1-7998-9553-4.ch012

Warf, B. (2016). Telecommunications and the changing geographies of knowledge transmission in the late 20th century. *Urban Studies (Edinburgh, Scotland), 32*(2), 361–378. https://doi.org/10.1080/00420989550013130

Warner, K., & Wäger, S. R. (2019). Building dynamic capabilities for digital transformation: An ongoing process of strategic renewal. *Long Range Planning, 52*(3), 329–349. doi:10.1016/j.lrp.2018.12.001

Wati, A., Ranggadara, I., Kurnianda, N., Irmawan, D., & Frizki, D. (2019). Enterprise Architecture for Designing Human Resources Application Standard Reference. *International Journal of Innovative Technology and Exploring Engineering, 8*(12).

Weijters, A. (2006). Process Mining with the Heuristics Mineralgorithm. Academic Press.

Wendler, R. (2012). The maturity of maturity model research: A systematic mapping study. *Information and Software Technology, 54*(12), 1317–1339. doi:10.1016/j.infsof.2012.07.007

Westerman, G., Bonnet, D., & McAfee, A. (2011). *Digital Transformation: A Roadmap for Billion-Dollar Organizations*. Harvard Business School Press.

Westerman, G., Bonnet, D., & McAfee, A. (2014). *Leading Digital: Turning Technology Into Business Transformation*. Harvard Business Review Press.

WHO. (2021). *WHO Coronavirus (COVID-19) Dashboard | WHO Coronavirus (COVID-19) Dashboard With Vaccination Data*. https://covid19.who.int/

Wikipedia. (2021a). *Code refactoring*. https://en.wikipedia.org/wiki/Code_refactoring

Williams, C. A., Krumay, B., Schallmo, D., & Scornavacca, E. (2022). An interaction-based Digital Maturity Model for SMEs. *PACIS 2022 Proceedings*, 254. https://aisel.aisnet.org/pacis2022

Williams, B. (2002). *Truth and Truthfulness: An Essay in Genealogy*. Princeton University Press.

Wojewoda, S., & Hastie, S. (2015). *Standish group 2015 chaos report - Q&A with Jennifer Lynch*. Infoq. Retrieved August 20, 2022, from https://www.infoq.com/articles/standish-chaos-2015

World Bank. (2013). *Crowdfunding's Potential for the Developing World*. Info Dev, Finance and Private Sector Development Department, World Bank. Available at https,//www.infodev.org/infodev-files/wb_crowdfundingreport-v12.pdf

World Economic Forum. (2016). *Digital Disruption Has Only Just Begun*. Retrieved on August 20th, 2022, from https://www.weforum.org/agenda/2016/01/digital-disruption-has-only-just-begun/

World Economic Forum. (2020). *The Future of Jobs Report*. Available online at: https://www3.weforum.org/docs/WEF_Future_of_Jobs_2020.pdf

Wu, L., & Park, D. (2009). Dynamic outsourcing through process modularization. *Business Process Management Journal, 15*(2), 244–255. doi:10.1108/14637150910949461

Wynn, M. T., & Lebherz, J. (2022). Rethinking the input for process mining: insights from the XES survey and workshop. In *International Conference on Process Mining* (pp. 3–16). Springer. 10.1007/978-3-030-98581-3_1

Xie, X., Zhang, H., & Blanco, C. (2022). How organizational readiness for digital innovation shapes digital business model innovation in family businesses. *International Journal of Entrepreneurial Behaviour & Research*. Advance online publication. doi:10.1108/IJEBR-03-2022-0243

Yasmin, F. A., Bukhsh, F. A., & Patricio, D. A. S. (2018). Process enhancement in process mining: A literature review. *CEUR Workshop Proceedings, 2270*, 65–72. http://simpda2018.di.unimi.it/

Yasseri, S., & Bahai, H. (2018). System Readiness Level Estimation of Oil and Gas Production Systems. *International Journal of Coastal and Offshore Engineering, 2*(2). Doi:10.29252/ijcoe.2.2.31

Yildiz, M. (2007). Decision making in e-government projects. The case of Turkey. In Handbook of Decision-making. Marcel Dekker Publication.

Yilmaz, G. (2019). Post-truth politics in the 2017 Euro-Turkish crisis. *Journal of Contemporary European Studies, 27*(2), 237–246. doi:10.1080/14782804.2019.1587390

Yılmaz, K. Ö. (2021). Mind the Gap: It's About Digital Maturity, Not Technology. In T. Esakki (Ed.), *Managerial Issues in Digital Transformation of Global Modern Corporations* (pp. 222–243). IGI Global. doi:10.4018/978-1-7998-2402-2.ch015

Yin, R. K. (2009). *Case study research: Design and methods*. Sage Publications.

Yin, R. K. A. (2003). *Case study research; design and methods (3ʳᵈ ed.)*. Sage Publications Inc.

Ylimäki, T. (2006). Potential critical success factors for enterprise architecture. *Journal of Enterprise Architecture, 2*(4), 29–40.

Yu, H. H. K., Chiu, D. K. W., & Chan, C. T. (2022). Resilience of symphony orchestras to challenges in the COVID-19 era: Analyzing the Hong Kong Philharmonic Orchestra with Porter's five force model. In W. Aloulou (Ed.), *Handbook of Research on Entrepreneurship and Organizational Resilience During Unprecedented Times* (pp. 586–601). IGI Global. doi:10.4018/978-1-6684-4605-8.ch026

Yu, P. Y., Lam, E. T. H., & Chiu, D. K. W. (2022). Operation management of academic libraries in Hong Kong under COVID-19. *Library Hi Tech*. Advance online publication. doi:10.1108/LHT-10-2021-0342

Zahra, S. A., & Nambisan, S. (2010). Entrepreneurship in Global Innovation Ecosystems, March 2011. *Academy of Marketing Science Review, 1*(1), 4–17. doi:10.100713162-011-0004-3

Zahra, S. A., & Nambisan, S. (2012). Entrepreneurship and Strategic Thinking in Business Ecosystems. *Business Horizons, 55*(3), 219–229. doi:10.1016/j.bushor.2011.12.004

Zaoui, F., & Souissi, N. (2020). *Roadmap for digital transformation: A literature review*. The 7th International Conference on Emerging Inter-networks, Communication and Mobility (EICM), Leuven, Belgium. DOI: 10.1016/j.procs.2020.07.090

Zara. (2011). *Lucy Chadwick: A Selby film for ZARA*. YouTube. https://www.youtube.com/watch?v=s3gDnnTfE2E&t=136s

Zara. (2019). *Zara scenes | the wake up call*. YouTube. https://www.youtube.com/watch?v=2UqbRq_JMMY

Zara. (n.d.). https://www.inditex.com/es/quienes-somos/nuestras-marcas/zara

Zerbino, P., Stefanini, A., & Aloini, D. (2021). Process Science in Action: A Literature Review on Process Mining in Business Management. *Technological Forecasting and Social Change, 172*, 121021. doi:10.1016/j.techfore.2021.121021

Zerilli, L. M. G. (2020). Fact-Checking and Truth-Telling in an Age of Alternative Facts. *Le foucaldien, 6*(1), 1–22. . doi:10.16995/lefou.68

Zhang, H., Lu, Y., & Guta, S., & Zhao, L. (2014). What motivates customers to participate in social commerce? The impact of techological environments and virtual customer experiences. *Information & Management, 52*(4), 496–505.

Zheng, Y., & Abbott, P. (2013, June) Moving up the value chain or reconfiguring the value network? An organizational learning perspective on born global outsourcing vendors. *Proceedings of the 21st European Conference on Information Systems.*

Zhipeng, C., & Zheng, X. (2018). *A Private and Efficient Mechanism for Data Uploading in Smart Cyber-Physical Systems.* IEEE. doi:10.1109/TNSE.2018.2830307

Zhou, C., Yue, X. D., Zhang, X., Shangguan, F., & Zhang, X. Y. (2021). Self-efficacy and mental health problems during COVID-19 pandemic: A multiple mediation model based on the Health Belief Model. *Personality and Individual Differences, 179*, 110893. doi:10.1016/j.paid.2021.110893 PMID:36540084

Zhou, Z., Wang, M., Ni, Z., Xia, Z., & Gupta, B. B. (2021). Reliable and Sustainable Product Evaluation Management System Based on Blockchain. *IEEE Transactions on Engineering Management.*

About the Contributors

Richard Pettinger is Professor of Management Education at University College London (UCL). He has worked at UCL for over 30 years. He continues to teach foundations and practices of management, project management, leadership and organisational behaviour. He is presently developing research, publication and syllabus based teaching work on the relationships between society, technology, behaviour and business; and on business as a social infrastructure. This is in turn reflecting the seismic shifts in society, business practice and organisational activities and effectiveness as the result of the present global situation, the COVID19 pandemic and its effects on society, activities and business development. To date Richard has published over 50 books and numerous articles on all aspects of leadership, management and business practice. He delivers at least one conference keynote speech a year. He continues to work in many different parts of the world, including Cyprus, China, South Africa, Singapore, Hong Kong, Spain, Italy, Poland, France, Germany, and Romania.

Brij B. Gupta received the PhD degree from Indian Institute of Technology (IIT) Roorkee, India. In more than 17 years of his professional experience, he published over 450 papers in journals/conferences including 40 books and 10 Patents with over 19000 citations. He has received numerous national and international awards including Canadian Commonwealth Scholarship (2009), Best Faculty Award (2018 & 2019), etc. He is also selected in the 2022, 2021, and 2020 Stanford University's ranking of the world's top 2% scientists. He is also a visiting/adjunct professor with several universities worldwide. He is also an IEEE Senior Member (2017) and also selected as 2021 Distinguished Lecturer in IEEE CTSoc. Dr Gupta is also serving as Member-in-Large, Board of Governors, IEEE Consumer Technology Society (2022-2024). At present, Prof. Gupta is working as Director, International Center for AI and Cyber Security Research and Innovations, and Professor with the Department of Computer Science and Information Engineering (CSIE), Asia University, Taiwan. His research interests include cyber security, cloud computing, artificial intelligence, blockchain technologies, social media and networking.

Alexandru Roja is an PhD Lecturer at Faculty of Political, Administrative and Communication Sciences, and researcher at faculty of Mathematics and Computer Science, Babeş – Bolyai University in Cluj – Napoca, Romania. He also currently serves as the Head of Innovation and Digital Transformation at the Transilvania IT Cluster. He is PhD in Strategic Management, based on research scholarship at the Université de Lille in 2010 and Université Pierre Mendès-France (Grenoble II) in 2012. His PhD thesis is focused on the emergence of virtual organizations and the roles of thechnologies in organizational digital transformation. Alexandru has acted as Postdoctoral Researcher at the Université de Lille years 2014-2015. He has completed his Postdoctorate Research in Management at the Bucharest Academy of

Management Studies, focusing on technology start-ups strategies and innovation ecosystems development. He teaches entrepreneurship courses for doctoral students and leads several entrepreneurship related initiatives. Since 2018 he serves as Expert at the European Research Council, European Commission for the Horizon 2020 program.

Diana Cozmiuc has blended her academic career with corporate career. She was an intern at Alcatel, and then university assistant. Her next jobs were as Financial Analyst at Siemens VDO Automotive and then at Continental Automotive. Main achievements have been conducting the planning and analysis process fully for a 250 million Euros per annum, and then 600 million Euros per annum. Whereas daily activities have involved analyzing variances to planning, a more challenging task has been setting the planning targets for marketing and purchasing activities on a worldwide basis. Other challenging tasks include the Budget Booklet, a 200 pages report from the Siemens VDO board to the business CEO covering strategic analyses, market analyses, geographical analyses, variance analyses. Diana Cozmiuc has been involved in reporting for 4 billion euros per annum joint venture since early years. Diana Cozmiuc has been involved in all levels of an international business bottom up managing responsibility centers and their interface with upper management, an international business line and its interface with the Siemens Board, or top bottom, setting targets and questioning variance analyses and scorecard reports. She has also worked closely with all functional areas and their processes, in setting targets, analyzing variances, analyzing key performance indicators, or conducting performance improvement programs. She has also interfaced closely with quotations, business controlling and investment requests in projects or operations. She returned to the university for her PHD, and worked as university assistant for 3 years of the program. One of the articles in the PhD thesis has received 33 citations and is rated pioneer. Other four chapters in books have been republished by the Information Resources Management Association in USA. She was awarded part of the best management authors in Romania 2017-2018 and 2019-2020 by the Management Academy SAMRO. She has taken part in an effort research by SAMRO to help organizations manage the Covid effort published in a book; contribution: digitalization and investing in cloud and automation enterprise architecture. She has worked as ad-hoc reviewer for Emerald Competitiveness Review journal; ad-hoc reviewer for two Information Resources Management Association in USA journals, International Journal of E-Business Research (IJEBR) and International Journal of Digital Strategy, Governance, and Business Transformation (one is indexed Web of Science). She is member of the Harvard Business Review Board. Diana Cozmiuc is currently a Researcher at the Ioan Slavici University.

* * *

Georgette Andraz is a lecturer at the University of the Algarve (School of Management, Hospitality and Tourism- ESGHT) where she teaches Strategic Management and Entrepreneurship. Purposes (Tourism, Hospitality, Business), Georgette Andraz holds a Master's in r in Economic and Business Administration Sciences from the Faculty of Economics, Universidade do Algarve and a PhD in Management Sciences from the Universidade de Évora. Her main research interests are Strategy and Performance and Entrepreneurship. She has participated in several national and international conferences and has coauthored several national and international scientific publications in these areas: Variables Explicativas del Grado de Éxito de las Nuevas Empresas de Base Tecnológica en su Orientación al Mercado (Revista Espacios, 40, 2019); Medición del Éxito A través de los Factores Críticos de las Nuevas Empresas de

Base Tecnológica (Revista Espacios, 39, 2018); Portuguese Manufacturing Industry: Determinants of New Firm Entry (Tourism & Management Studies, 11, 2015), among others.

Pradip Kumar Bala is a professor in the area of Information Systems & Business Analytics at Indian Institute of Management (IIM) Ranchi. He received his B.Tech., M.Tech. and Ph.D. from Indian Institute of Technology (IIT) Kharagpur, India in 1993, 1999 and 2009 respectively. He worked in Tata Steel before joining academics. He also worked as associate professor at Xavier Institute of Management Bhubaneswar and as assistant professor at IIT Roorkee before joining IIM Ranchi in 2012. His teaching and research areas include text mining & NLP, recommender systems, data mining applications, data mining and NLP algorithms, social media analytics and marketing analytics. He has conducted many training programmes in business analytics & business intelligence. He has published more than 100 research papers in reputed international journals, conference proceedings and book chapters. He is also a member of the International Association of Engineers (IAENG). He served as Director In-Charge, Dean (Academics), Chairperson, Post-Graduate Programmes, Chairperson, Doctoral Programme & Research, and Member of Board of Governors of IIM Ranchi.

Subhanil Banerjee is an Economist currently working as Associate Professor and Head of the Economics Department, School of Humanities, K.R. Mangalam University, Gurgaon. Doctor Banerjee has 19 years of experience in Core Social Science Research. His Research interests are International Trade, Foreign Direct Investment, Health Economics, Digital Economy, Sustainable Development, Environmental Economics, Econometrics, History of economics, Gender Studies and others. So far he has 17 publications 9 of them are SCOPUS indexed Journal Articles, 4 of them are International book chapters, 1 Conference proceeding and 3 other peer reviewed publications.

Arakhita Behera is an Economist currently working as Assistant Professor in Economics Department, in the School of Humanities, K.R. Mangalam University, Gurgaon. Doctor Behera has 4 years of both teaching and research experience in Core Social Science Research. His Research interests are Hospitality and Tourism, Environmental Tourism, Consumer Behaviour, Sustainable Development, Environmental Economics, Gender Studies and others. So far he has 8 publications 1 of them are SCOPUS indexed Journal Articles, 3 of them are International book chapters, 1 Conference proceeding and 3 other peer reviewed publications.

Dickson K. W. Chiu received the B.Sc. (Hons.) degree in Computer Studies from the University of Hong Kong in 1987. He received the M.Sc. (1994) and the Ph.D. (2000) degrees in Computer Science from the Hong Kong University of Science and Technology (HKUST). He started his own computer consultant company while studying part-time. He has also taught at several universities in Hong Kong. His teaching and research interest is in Library & Information Management, Service Computing, and E-learning with a cross-disciplinary approach, involving library and information management, e-learning, e-business, service sciences, and databases. The results have been widely published in around 300 international publications (most of them have been indexed by SCI/-E, SSCI, and EI, such as top journals MIS Quarterly, Computer & Education, Government Information Quarterly, Decision Support Systems, Information Sciences, Knowledge-Based Systems, Expert Systems with Application, Information Systems Frontiers, IEEE Transactions, including many taught master and undergraduate project results and around 20 edited books. He received a best paper award in the 37th Hawaii International Conference on

System Sciences in 2004. He is an Editor (-in-chief) of Library Hi Tech, a prestigious journal indexed by SSCI (impact factor 2.357). He is the Editor-in-chief Emeritus of the International Journal on Systems and Service-Oriented Engineering (founding) and International Journal of Organizational and Collective Intelligence, and serves in the editorial boards of several international journals. He co-founded several international workshops and co-edited several journal special issues. He also served as a program committee member for around 300 international conferences and workshops. Dr. Chiu is a Senior Member of both the ACM and the IEEE, and a life member of the Hong Kong Computer Society. According to Google Scholar, he has over 5,500 citations, h-index 39, i-10 index 123, ranked worldwide 1st in "LIS," "m-learning," and "e-services."

Sibanjan Debeeprasad Das is a Ph.D. scholar in the area of Information Systems & Business Analytics at the Indian Institute of Management (IIM) Ranchi. He received his Master of IT with a specialization in Business Analytics from Singapore Management University. In his current role, he works with a team of highly skilled Data Scientists, ML Data Engineers and MLOps engineers. His work experience includes building AI products such as Adaptive Intelligent Apps for Manufacturing and working as a Data Science/Advanced Analytics consultant, developing advanced analytics products, and implementing predictive analytics and business insights solutions. He has the right mix of knowledge on Business processes, Machine Learning and Software Engineering and is passionate about exploring new techniques to provide effective and efficient solutions for business improvements. He believes in sharing knowledge with the community, and has been actively writing books, articles and provides mentoring to the professionals/students interested in transitioning to AI/ML and IT in general.

Suhanee Gupta is pursuing her graduation in Economics from the Department of Economics, School of Humanities, K.R. Mangalam University Gurugram.

Mekuria Hailu received a diploma in mathematics from Debre Berhan Teachers and Vocational Training College in 2010. He also received his Bachelor of Science Degree in Mathematics from Dire Dawa University, Ethiopia, in 2015. Then he received his Master of Science Degree in Information Science & Systems from Addis Ababa University, Ethiopia, in 2020. He has served for 11 years in different private and government organizations in Ethiopia. Currently, Mekuria Hailu is an ICT teacher at Lebu secondary school, Addis Ababa, Ethiopia. His research interest is related to e-services in general and e-Government in particular.

Jong-Seok Kim is a lecturer at Chungnam National University and holds a part-time position as chief executive officer of KJS & Group, South Korea. He is an entrepreneur and founder of KJS Group (now KJS & Group). He obtained his PhD in Business Administration from the University of Manchester Business School, UK. Since then, he has studied innovation management, foresight, and strategic management. He worked as a research associate for a European Union project and as a seminar leader at Manchester Business School, UK. His works have been internationally recognised. He has published many interesting papers in reputable international journals. Recently, he was the recipient of the 2020 Emerald Literati Award: Highly Commended Paper. His ORCid is 000-0002-4560-2553.

Souren Koner (B. Sc., MBA, Ph.D.) has been acting as an Assistant Professor at the Amity University, Raipur, Chhattisgarh since Aug 2019. He has more than 15 years of experience teaching in different management institutions all over India. His area of specialization is Marketing. He has been teaching in the areas of Marketing Management, Advertising, Sales Management, Service Marketing, etc. So far, he has 15 publications. 2 of them are SCOPUS indexed Journal Articles, 1 of them is in UGC care, 3 of them are international book chapters, 2 Conference proceedings, and 7 other peer-reviewed publications.

Apple Hiu Ching Lam obtained her degree of Bachelor of Business Administration (Honours) in International Business from City University of Hong Kong (2016) and degree of Master of Science in Library and Information Management with distinction from the University of Hong Kong (2020). She is a doctoral student in Education at the University of Hong Kong. Her current research interests are social media in library, user education, and the 5E Instructional Model.

Lemma Lessa is an assistant professor of Information Systems at School of Information Science, Addis Ababa University, Ethiopia. He received his doctorate in Information Technology from Addis Ababa University in 2016. He has served over twenty years in the Ethiopian Higher education. His teaching interest is on the socio-technical aspects of Information Systems. His research interest is on e-Services, Information Systems Security Management, Enterprise Systems and IT Governance. He is member of Association for Information Systems (AIS) and current president of the Ethiopian chapter of AIS. He has authored or coauthored over twenty-five articles - four journal articles, and about twenty-one peer-reviewed conference papers. His articles have been presented and published in the International Journal of Leadership and Management, African Journal of Information Systems, proceedings of Americas Conference on Information Systems, International Conference on Theory and Practice of Electronic Governance, and European Conference of Information Systems among others. He also co-authored two peer-reviewed book chapters. He is serving as associate editor and/or PC member for African Journal of Information Systems, the International Conference on e-Democracy & e-Government, International Conference on Digital Transformations and Global Society, Hawaii International Conference on System Sciences, and International Conference on Electronic Governance and Open Society.

Sin Man Li is currently a student of the MSc in Library and Information Management at the University of Hong Kong. She received her B.A. (Hons.) in Digital Media Broadcasting from the City University of Hong Kong. She worked as a Project Coordinator at the Chinese University of Hong Kong to coordinate Multimedia projects and design editorial materials. She is now working as an Executive Manager in the University Grants Committee Secretariat of Hong Kong, assisting in coordinating policy inputs into discussion papers and policy briefs to facilitate discussion on Hong Kong's higher education policy and development. Her research interests are E-learning, Art and culture, and Knowledge Management.

Pratiksha Mishra obtained her Master's degree in commerce in 2019 from Jain deemed to be university, Bengaluru, India. She is currently pursuing her doctoral studies at Amity Business School, Amity University Chhattisgarh, India. Her research interests include Financial Inclusion, Financial Technology, Digitalization, Rural Studies.

Ece Özer Çizer is a doctoral student and Research Assistant at the Department of Business Administration, Faculty of Economics and Administrative Sciences, Yildiz Technical University. Her research interests are digital marketing and sharing economy.

Shweta Patel is a strong believer that people are the biggest asset of any organization which inspired her to take up Master's in Human Resource Development, after completing her graduation in Bachelor of Science from Surat, Gujarat. She has been actively involved in various social activities. Following her interest and passion to enter academics she has gained 7 years of enriching experience in both teaching and professional career. Currently she is pursuing her PhD in Management from Amity University Chhattisgarh.

Kingsley Mbamara Sabastine received his Master's degree in Philosophy from the Catholic University of Eastern Africa, Nairobi-Kenya in 2012. In the same year (2012) he was hired as a lecturer at the Good Shepherd Major Seminary Kaduna, Nigeria and is currently a doctoral student at the Catholic University of Lublin, Poland. His work focuses on ethics, theories of truth and epistemology.

Athena Kin-kam Wong has worked in public museums and archives in Hong Kong for more than ten years. She is responsible for the collection care of paper and photographic materials and extension activities in museum conservation for public and secondary students. She received a bachelor's degree in chemistry from the University of Hong Kong and a master of arts in museology from the University of Sydney. Athena is currently pursuing a master's degree in information management at the University of Hong Kong. Her research interests are digital transition in museums and archives, and the preservation of digital holdings and contemporary artworks.

Kemal Özkan Yılmaz received his BE in Industrial Engineering at Yıldız Technical University. Has started his professional career by joining Beko Elektronik A.Ş -served in organizational development, engineering, and international sales & marketing areas. Simultaneously completed his MA in Management and Organizational Studies at Marmara University. Also occupied executive positions at leading Turkish companies, had systems design and strategy formulation roles, where he also had the chance of being an instructor in deploying TQM philosophies and formulated business development strategies. The related companies' efforts were also crowned with local and international prizes such as the TPM Award. In the meantime, had also started his Ph.D. degree in Contemporary Business Studies Program at Işık University and received the title in 2017. Has been lecturing at the Graduate School of Social Sciences at Işık University since September '18, and he is a full-time faculty member of İstanbul Kültür University – BA Department since September '19. Interests are marketing, management, corporate sustainability, innovation management, digital transformation, and new product development.

Index

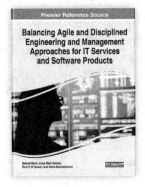

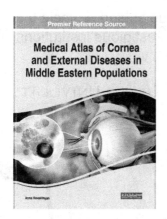

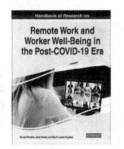
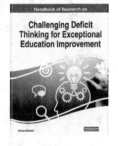

Have Your Work Published and Freely Accessible
Open Access Publishing

With the industry shifting from the more traditional publication models to an open access (OA) publication model, publishers are finding that OA publishing has many benefits that are awarded to authors and editors of published work.

Freely Share Your Research

Higher Discoverability & Citation Impact

Rigorous & Expedited Publishing Process

Increased Advancement & Collaboration

Acquire & Open

When your library acquires an IGI Global e-Book and/or e-Journal Collection, your faculty's published work will be considered for immediate conversion to Open Access *(CC BY License)*, at no additional cost to the library or its faculty *(cost only applies to the e-Collection content being acquired)*, through our popular **Transformative Open Access (Read & Publish) Initiative**.

Provide Up To
100%
OA APC or
CPC Funding

Funding to
Convert or
Start a Journal to
**Platinum
OA**

Support for
Funding an
**OA
Reference
Book**

IGI Global publications are found in a number of prestigious indices, including Web of Science™, Scopus®, Compendex, and PsycINFO®. The selection criteria is very strict and to ensure that journals and books are accepted into the major indexes, IGI Global closely monitors publications against the criteria that the indexes provide to publishers.

Learn More Here:
For Questions, Contact IGI Global's Open Access Team at openaccessadmin@igi-global.com

IGI Global
PUBLISHER of TIMELY KNOWLEDGE
www.igi-global.com

Are You Ready to
Publish Your Research ?

IGI Global
PUBLISHER of TIMELY KNOWLEDGE

IGI Global offers book authorship and editorship opportunities across 11 subject areas, including business, computer science, education, science and engineering, social sciences, and more!

Benefits of Publishing with IGI Global:

- Free one-on-one editorial and promotional support.

- Expedited publishing timelines that can take your book from start to finish in less than one (1) year.

- Choose from a variety of formats, including Edited and Authored References, Handbooks of Research, Encyclopedias, and Research Insights.

- Utilize IGI Global's eEditorial Discovery® submission system in support of conducting the submission and double-blind peer review process.

- IGI Global maintains a strict adherence to ethical practices due in part to our full membership with the Committee on Publication Ethics (COPE).

- Indexing potential in prestigious indices such as Scopus®, Web of Science™, PsycINFO®, and ERIC – Education Resources Information Center.

- Ability to connect your ORCID iD to your IGI Global publications.

- Earn honorariums and royalties on your full book publications as well as complimentary copies and exclusive discounts.

Join Your Colleagues from Prestigious Institutions, Including:

Australian National University

MIT Massachusetts Institute of Technology

JOHNS HOPKINS UNIVERSITY

HARVARD UNIVERSITY

COLUMBIA UNIVERSITY IN THE CITY OF NEW YORK

Learn More at: www.igi-global.com/publish

or Contact IGI Global's Aquisitions Team at: acquisition@igi-global.com

Printed in the United States
by Baker & Taylor Publisher Services